EVOLUTIONARY BIOLOGY

Douglas J. Futuyma

DEPARTMENT OF ECOLOGY AND EVOLUTION
STATE UNIVERSITY OF NEW YORK AT STONY BROOK

EVOLUTIONARY

SINAUER ASSOCIATES, INC., SUNDERLAND, MASSACHUSETTS

BIOLOGY

THE COVER

Adaptive radiation. Lizards have diversified to occupy a variety of habitats. Clockwise from upper right: *Chamaeleo,* the arboreal African chamelion; the mosasaur *Tylosaurus,* a large marine Cretaceous lizard; *Bipes,* a Mexican burrowing lizard; *Basiliscus,* a Neotropical riparian species; *Chlamydosaurus,* the frilled lizard of Australian deserts.

EVOLUTIONARY BIOLOGY

Copyright © 1979 by Sinauer Associates, Inc.

For information address Sinauer Associates, Inc., Sunderland, Massachusetts 01375.

Printed in U. S. A.

Library of Congress Cataloging in Publication Data

Futuyma, Douglas J   1942–
    Evolutionary biology.

    Bibliography: p.
    Includes index.
    1. Evolution. I. Title.
QH366.2.F87   575   78–27902
ISBN 0-87893-199-6

9 8 7 6 5 4 3 2

For Bruce

Contents

Preface

Poor fool, with all this sweated lore,
    I stand no wiser than I was before.
Master and Doctor are my titles;
    For ten years now, without repose,
I've held my erudite recitals
    And led my pupils by the nose.
And round we go, on crooked ways or straight,
    And well I know that ignorance is our fate,
And this I hate.[1]

I would not presume to compare my knowledge to the legendary Faust's — nor yet dismiss what we do know with his cynical despair. But I have been troubled as I wrote this book by the reflection that it is uncommonly presumptuous for any one person to attempt a comprehensive textbook on evolutionary theory. The subject is vast, embracing all of biology and drawing heavily on mathematics, geology, and chemistry; it extends its influence into philosophy, the arts, anthropology, psychology, and the social sciences; and it itself evolves so rapidly that new facts and theoretical formulations make *caveats,* maybe's, and possibly's inevitable on every page. I can excuse my presumption only by the reflection that if I and others in similar positions can dare to teach a course in evolution singlehandedly, I should dare to expose my recitals to a wider audience. The experience of writing has revealed to me my ignorance at every turn, but it has been immensely enlightening; I hope I will not have been the only one to profit by it.

Evolution is, to me, the most exciting subject in biology, partly because it is so all-encompassing, partly because it is the study of dynamic processes rather than static facts, and partly because it's a lot of fun to argue about. Almost any question in evolution — Is geographic isolation a prerequisite for species formation? — is approached more by logic or circumstantial evidence than by direct observation, and so often is never quite resolved. Very often no single "right" answer applies universally. The literature of evolutionary biology reflects the dynamic changes in ideas that the complexity of the subject compels. Today's theoretical formulations prove that what the last

---

[1] Goethe: *Faust,* Part One. Translated by Philip Wayne (Penguin Classics, 1949, p. 43). Copyright the Estate of Philip Wayne 1959; reprinted by permission of Penguin Books Ltd.

generation of evolutionists said could not happen, can; new evidence or, just as often, new ways of viewing old evidence force us to reconsider entrenched ideas.

I have tried to convey the dynamic state of evolutionary biology by focusing on controversial issues, by raising problems without resolving them, by presenting plausible arguments only to finish them with a querulous "but," and by presenting questions after each chapter that admit no easy answers, but will, I hope, stimulate thought, argument, and perhaps even research. These discussion and thought sections are an integral, and perhaps the most important, part of the book.

The result will not please everyone, I suspect. One of the reviewers said "Students want principles and facts, not speculative essays," and some students do indeed feel more comfortable with nice clean facts or nice clean principles (which amount to the same thing) that can be filed and withdrawn without question. But more readers, I suspect, will appreciate that we deal not so much with irrefutable facts as with hypotheses that may be refuted by tomorrow's experiment, not with unquestionable principles engraved on stone tablets, but with concepts formulated by the fallible human mind. If science teaches us anything, it should teach us to doubt, to question every statement, however authoritative the lips from which it spills.

I have written this book at a level set by students in my evolution course, who come prepared with the calculus, with a full course in genetics, with some elementary exposure to ecology and the essential principles of evolution, and with sundry courses in physiology, anatomy, or natural history. In places, therefore, I have assumed some knowledge of plant and animal anatomy, taxonomy, and biochemistry, but I have tried to define terms that may be unfamiliar. Chapters 3 and 4 summarize the elements of genetics, development, and ecology that I consider essential background to the subsequent chapters; but I suggest that a course in genetics (exclusive of population genetics, which I develop from first principles) would be advisable before using this book. I have used mathematics sparingly and have stated all conclusions verbally; familiarity with the most elementary principles of the calculus and statistics should suffice. Still, parts of the text (especially Chapters 14 and 15) may prove somewhat difficult, but there is no reason to expect biology books to be easier than textbooks in physics or chemistry, disciplines whose subject matter is so much simpler.

After the preliminary chapters (1–4), I attempt a summary of some of the salient generalities of historical evolution — the phenomena that the theoretical structure of evolutionary thought is designed to explain. Because of space limitations, I have had to exclude treatment of the origin of life, historical human evolution, and the evolutionary history of specific groups of organisms, subjects which are treated in other books (e.g., Dobzhansky et al. 1977, Stebbins 1974). From historical evolution, I pass to variation within and among contemporary species and then to the mechanisms that govern genetic change within populations. These also engender the origin of new species, consideration of which brings us by a commodious vicus of recirculation to the explanation of many of the major features of historical evolution. As a postlude, I devote two chapters to two of the several special topics I should have liked, given more space, to discuss: coevolution and the social implications of evolutionary thought.

The merits of this book are very largely the contributions of my friends, colleagues, and students, who have enriched me by their hospitality, knowledge, questions, and criticisms; its faults lie in my ignorance and inability to recognize good advice. Parts of the manuscript have been read, invariably to their improvement, by Robert Colwell, Fred Gould, Steven Jay Gould, Conrad Istock, Bruce Levin, Richard Lewontin, Gregory Mayer, David McCauley, Charles Mitter, Sally Shankman, Christine Simon, Michael Soulé, John Vandermeer, David Wake, Sherwood Washburn, B. J. Williams, Edward O. Wilson, and David Woodruff, to whom I am indebted. I should not have begun it without the encouragement of Rob Colwell, nor finished it without the advice and faith of Andy Sinauer. I am thankful to Jodi Kepke and George Wright for their indispensable help in compiling the illustrative material, and to my teachers, especially W. L. Brown, Jr., R. C. Lewontin, and L. B. Slobodkin, and to my students, for teaching me about evolution.

I owe a special debt of gratitude to the many friends who extended their hospitality during this long project: to the late Marston Bates and Nancy Bell Bates in Ann Arbor; to Sally and Paul Shankman, and to the University of Colorado, in Boulder; to Rob Colwell and Mary-Claire King, and to the Department of Zoology of the University of California, Berkeley; to Barbara Bentley in Stony Brook; and to George Wright in San Francisco. And I am inexpressibly grateful to Bruce Smith, whose intangible contributions have made all the difference.

DOUGLAS J. FUTUYMA

PART I

INTRODUCTION

CHAPTER 1

The Origin and Impact of Evolutionary Thought

*O ew'ge Nacht! Wann wirst du
    schwinden?
Wann wird das Licht mein Auge finden?*

—**Mozart,** *Die Zauberflöte*

Old ideas give way slowly, for they are more than abstract logical forms and categories. They are habits, predispositions, deeply engrained attitudes of aversion and preference. Moreover, the conviction persists — though history shows it to be a hallucination — that all the questions that the human mind has asked are questions that can be answered in terms of the alternatives that the questions themselves present. But in fact intellectual progress usually occurs through sheer abandonment of questions together with both of the alternatives they assume — an abandonment that results from their decreasing vitality and a change of urgent interest. We do not solve them: we get over them. Old questions are solved by disappearing, evaporating, while new questions corresponding to the changed attitudes of endeavor and preference take their place. Doubtless the greatest dissolvent in contemporary thought of old questions, the greatest precipitant of new methods, new intentions, new problems, is the one effected by the scientific revolution that found its climax in the "Origin of Species."

So concluded the philosopher John Dewey in his essay "The Influence of Darwin on Philosophy" (1910). A century after the publication of Darwin's book, philosophers could still affirm that "there are no living sciences, human attitudes, or institutional powers that remain unaffected by the ideas that were catalytically released by Darwin's work" (Collins 1959).

The theory of biological evolution is the mature expression of two revolutionary streams of thought antithetical to a world view that had long prevailed. The concept of a changing universe had been replacing the long-unquestioned view of a static world, identical in all essentials to the Creator's perfect creation. Darwin more than anyone else extended to living things, and to the human species itself, the conclusion that mutability, not stasis, is the natural order. A society that had sought the causes of phenomena in the will of God, or in Aristotelian final causes (the purposes for which events occur) rather than effi-

3

cient causes (the mechanisms that cause events to occur), had to face a profound and disturbing realization. Material causes are a sufficient explanation not only for physical phenomena, as Descartes and Newton had shown, but also for biological phenomena with all their seeming evidence of design and purpose. By coupling undirected, purposeless variation to the blind, uncaring process of natural selection, Darwin made theological or spiritual explanations of the life processes superfluous. Together with Marx's materialistic theory of history and society and Freud's attribution of human behavior to influences over which we have little control, Darwin hewed the final planks of the platform of mechanism and materialism — of much of science, in short — that has since been the stage of most Western thought.

ORIGINS OF EVOLUTIONARY THOUGHT

I shall treat the history of the idea of evolution only briefly, for it is described in detail by Greene (1959), Eiseley (1958), and others. It is useful, though, to have some idea of preevolutionary patterns of thought and how they changed, for our modern view may reveal vestiges of ancient thought that bear reexamination.

Although the notion of a dynamic world was not foreign to the Greeks, the nonstatic explanation of the origins of living things offered by Empedocles and Anaximander gave way to the philosophy of Plato, which had a permanent, dominant effect on subsequent Western thought. Foremost in Platonic philosophy was the concept of the εἶδος, the "form" or "idea," a transcendent ideal form imperfectly mimicked by its earthly representatives. The triangles or horses we see in the material world are but imperfect copies of the true, perfect Triangle and Horse that exist in the transcendent world of ideas. Variation is thus a manifestation of the imperfections of the material world. Later Christian alterations of Platonic thought viewed the world of material things not only as imperfect, but corrupted and inferior to the rarefied world of the spiritual and nonmaterial, the fixed world of ideas in the mind of God. Indeed, since God is perfect, the universe He created must in the beginning have included everything on which He saw fit to bestow existence; and He could not have been perfect had the original creation been incomplete. Thus nothing worthy of existence could have come into being after the Creation; and nothing He has seen fit to create could have disappeared or become extinct, for there would then be imperfection in His universe.

In a fascinating book, Lovejoy (1936) traces in detail the development of these ideas, their impact on Western philosophy, and the contradictions to which they give rise; he focuses especially on an important derivative of Platonic thought: the *Scala Naturae,* or Great Chain of Being. Since order is clearly superior to disorder, God's creations must fit a pattern; and since there seems to be a gradation from the nonliving world through lower forms of life and humans to spiritual forms, this pattern must be a rigid vertical series of links from the lowest, most material, to the highest, most spiritual of beings.

This ladder of life must be perfect and have no gaps; it must be permanent and unchanging; and every being must have its fixed place in the series, according to God's plan. The hierarchy must extend to social relations as well: it is right and proper that there be inferior and superior human races, higher and lower classes. And since this order was created by a perfect God, that which is natural is good, and this must be the best of all possible worlds (despite its manifest imperfections). To aspire to change the social order must be immoral, and biological evolution is unthinkable.

The role of natural science, then, was to catalogue the links of the Great Chain of Being and to discover their order, so that the wisdom of God could be revealed and appreciated. One such effort was John Ray's (1701) "The Wisdom of God Manifested in the Works of the Creation," in which the adaptations of creatures to their conditions of life were held up as evidence of the Creator's wise beneficence. The pioneering work of Linnaeus (1735) in classification, on which all our taxonomy rests, was likewise undertaken *ad majorem Dei gloriam,* "for the greater glory of God."

This view of the world began to fall, as Lovejoy demonstrates, partly from inherent philosophical contradictions and partly from the development of empirical science. Hallowed concepts, such as the central position of the earth in the universe, were challenged. Newton demonstrated that physical phenomena needed no more metaphysical explanation than the laws of matter in motion, and Kant and Laplace entertained notions of stellar evolution. A historical view of human relations began to develop, in which the idea that human society had changed from savagery to civilization became more norm than heresy. Rousseau and Hobbes speculated on what the characteristics of humans in their original "state of nature" might have been and came to opposite conclusions. Geology provided unsettling evidence that the earth's crust had indeed changed and that species had indeed become extinct.

That there had developed gaps in the chain of being was most disturbing. Some denied that fossils represented once-living animals or that these species had indeed become extinct. But the evidence was so incontrovertible that new explanations had to be found, chief among which was the supposition of successive creations. In the growing scientific spirit of the eighteenth century, however, others such as Cuvier and Buffon recognized that explanations might better be sought in natural causes. Thus arose the theory that extinctions had been caused by successive geological catastrophes and that the earth was somewhat older (perhaps as much as 70,000 years old, Buffon suggested in 1780) than had been thought. By 1788 a radical new theory, attributed largely to Hutton and Playfair, arose. Geological events could best be explained by the long-continued action of minute, slowly acting forces such as deposition and erosion that we see about us every day. This principle of uniformitarianism, which held that the same processes are responsible for both past and present events, had two major consequences. It implied that the earth was very old, with "no vestige of a beginning — no prospect of an end," as Hutton put it. And it had a major impact on the theory of biological evolution, for uniformitarianism was a chief principle of the geological theories of Charles Lyell, from whom Darwin borrowed the concept to apply to the biological realm. Biological evolution, Darwin said, must proceed by insensibly slight changes rather than drastic, immediate alterations. This view has permeated most of modern evolutionary thought and is only now beginning once again to be questioned.

By the beginning of the nineteenth century, the great age of the earth was widely accepted. Paleontology provided evidence not only of extinctions but of evolutionary change; comparative anatomy strongly suggested genealogical affinities among organisms; and the idea that evolution had indeed occurred was gaining acceptance. All that was lacking was a statement of the mechanisms by which it might have occurred, and this was provided by the first of the truly great evolutionary theorists, Jean-Baptiste de Lamarck. Lamarck's theory of the development of new characteristics by the use and disuse of parts, coupled with inheritance of such acquired characteristics, was (to the detriment of his reputation since) wrong. But his was the first scientific theory of evolution and was quite plausible in view of the long-held, widespread belief that acquired characteristics are inherited.

By the middle of the century evolution was a common topic of discussion (Lovejoy 1959). Chambers' anonymously published "Vestiges of the Natural History of Creation" (1844) was a widely read evolutionary tract; Herbert Spencer's voluminous writings combined philosophy, ethics, sociology, and psychology into a comprehensive evolutionary world view. But Lamarck had been discredited, so no satisfactory mechanism was recognized. Curiously, the concept of the survival of the fittest, used especially to account for extinction, was well entrenched; but no one except William Wells (1818) and Patrick Matthew (1831), in publications that were little read, seems to have realized that it could explain the modification of species. It was not until 1858 that coherent statements, presented jointly by Alfred Russel Wallace and Charles Darwin, were made of the concept of natural selection and the way it could cause evolution. The pieces finally fell into place, and a public now ready for the idea made Darwin's book an instant best-seller in 1859.

CONCEPTIONS AND MISCONCEPTIONS OF EVOLUTION

Like all important concepts, evolution generates controversy. Some find it threatening and oppose it vigorously, some react to the opposition with evangelical fervor, and some use it as a philosophical ground, or often just a rationale, for the most disparate philosophical, ethical, or social views — and in doing so often twist evolutionary concepts into unrecognizable form. For this reason it is important to state precisely what evolution entails and what it does not, both to recognize abuses of evolutionary thought and to prepare for the detailed discussion of evolutionary mechanisms that constitutes the bulk of this book.

Evolution in the broadest sense is the observable or inferable fact of change. Scoon (in Persons 1950) takes evolution to mean "a continuous process of change in a temporal perspective long enough to produce a series of transformations." Thus evolution is all-pervasive. Galaxies, chemical elements, religions, languages, and political systems all evolve. Biological (organic) evolution transcends the lifetime of a single organism; the ontogeny of an individual is not considered evolution. Evolutionary changes are those that are inheritable from one generation to the next. The hereditary factors are the genes; thus biological evolution can be defined more precisely as any change from one generation to the next in the proportion of different genes, of the carriers of different hereditary information. Evolutionary changes can be almost immeasurably slight, or they can be dramatic if genes with large effects replace one another over long periods of time. I consider biological evolution, then, to embrace everything from slight changes in the proportions of different blood types (such as the A, B, O system) in a single population to the successive alterations that led from the earliest proto-organisms to snails, bees, giraffes, and dandelions.

Biological evolution is an empirical fact. We can observe it, at least on a small scale, and can infer it from geological, geographical, anatomical, embryological, and biochemical evidence. The possibility of evolution was entertained before the publication of *The Origin of Species* (Lovejoy 1959). But whether evolution occurs is a different question from how it occurs, what mechanisms are responsible. The answer to this second question was the contribution of Darwin and Wallace, who jointly proposed that inheritable variations arise (although they did not know how) and that forms differ in their susceptibilities to the exigencies of the environment. The difference in their rates of survival and reproduction constitutes natural selection and must lead, by mathematical logic, to alteration in the relative abundance of various forms — evolution. Later additions to evolutionary thought have recognized that the proportions of forms may fluctuate just by chance. This too is evolution, although it is seldom supposed that the most interesting features of organisms, those that best fit them to their environment, have evolved by chance alone. But chance is another mechanism of evolutionary change, in addition to natural selection, which Darwin (and most recent evolutionists) held to be the primary agent of evolution.

These essential tenets of evolutionary theory entail no value judgments, nor any sense of purpose or direction. Variations arise by accident and are no more likely to be useful to the organism than harmful (in fact, they are most often harmful). Which of the variants survives depends solely on which has the greater reproductive rate under the conditions it encounters. Future conditions cannot affect present survival. The enduring variations may increase the organism's complexity or behavioral repertoire, or they may decrease it. They may increase the likelihood of survival through subsequent environmental changes, or they may increase the likelihood of subsequent extinction. They may carry the species further along a path of evolutionary change that it has followed for some time, or they may alter the direction of evolution entirely. Evolutionary change is not universally headed in any one direction (certainly not toward human characteristics); it has followed a bewildering diversity of paths.

Despite the simplicity of these concepts, both opponents and proponents of evolution have held remarkable misconceptions of evolution and of its mechanisms. Many people, especially fundamentalists, seem to think that "evolution" is identical to "the evolution of the human species." Then, since we do not have an incontrovertible series of fossils from hominoid apes to modern humans, they conclude that evolution must be a mere hypothesis rather than a fact. But the reality of a general phenomenon does not depend on the evidence for its application in any specific instance, any more than the laws of organic chemistry are challenged by our ignorance of the precise reaction by which a particular complex compound is formed.

Opponents of evolution often maintain, moreover, that it is merely a "theory" and as such should share the stage equally with other theories of the origin of biological diversity, such as creation. This stand is based on a misunderstanding of what biologists mean by *evolutionary theory,* or what scientists generally mean by *theory.* The *Oxford English Dictionary* lists six major definitions of theory; the popular notion is "in a loose or general sense, a hypothesis proposed as an explanation; hence a mere hypothesis, speculation, conjecture; an idea or set of ideas about something; an individual view or notion." But in science theory means "a statement of what are held to be the general laws, principles, or causes of something known or observed."

By evolutionary theory, then, we mean not mere hypotheses that life has evolved or that natural selection is a major mechanism, but the body of interconnected statements (often expressed in mathematical form) that describe the general processes by which variations arise and are altered in frequency to cause changes of the kind documented by paleontology and systematics. These principles can equally well be termed evolutionary laws, and there is no need to use the word *theory* if it runs the risk of misinterpretation.

The word *law,* however, also has many meanings, including "a rule of conduct imposed by authority" or a custom taken to be right and proper. Thus the laws of evolution are taken by some to be morally binding, as by John D. Rockefeller when he described economic competition as an extension of natural selection and thus a morally proper "law of nature." In science, however, a law is simply, as the *Oxford English Dictionary* puts it, "a theoretical principle deduced from particular facts, applicable to a defined group or class of phenomena, and expressible by the statement that a particular phenomenon always occurs if certain conditions be present." Neither the law of gravity nor the law of natural selection has moral force.

Evolution has nonetheless been used as a framework for the development of ethical systems. An example is the evolutionary ethics of the eminent evolutionist Julian Huxley, who envisions human cultural evolution as an extension of the grand historical pageant of ever more complex and conscious behavior, which will lead inexorably to higher consciousness and humanitarianism. Appealing as this vision may be, it is a transition from scientific explanation to scientifically oriented myth (Toulmin 1957), in which a selected biological order is treated as the "Sovereign Order of Nature" and is proposed as a universal ethical norm. In such transitions a concept is expanded beyond its purely scientific meaning. Because cultural evolution does not entail genes, it cannot be discussed in terms of the data of biology. The extension is made for nonscientific motives; it is an attempt, as Collins (1959) says, to offer "a cosmic backing for the transition from *is* to *ought*." Evolution and natural selection *are,* but whether they *ought* to be is a question of value that falls outside the realm of science.

Both the fact and the mechanism of evolution have been invested with ethical interpretations based on misconceptions. The history of evolutionary change has been seen as directional progress, despite the evidence that species cannot be said to have progressed universally in any direction. But the concept of progress having been imposed on the phylogenetic tree of life, it was easy to see progress as natural, and therefore (by a transition from scientific explanation to scientifically oriented myth) morally incumbent on society. Thus anything could be justified in the name of progress as "natural law": class struggle on the one hand (Marx wanted to dedicate the second volume of *Das Kapital* to Darwin); imperialism, *laissez faire* capitalism, and the hegemony of the "higher" over the "lower" races, on the other. The supposition that the law of nature is natural selection "red in tooth and claw" was taken as moral license for untrammeled economic competition and imperialism, while with more charity but as little logic Kropotkin (1902) and others pointed to the natural evolution of cooperative social behavior in animals as justification for more cooperative economic institutions. Throughout all such thinking runs the supposition that what is natural is good, what is unnatural, bad — a misguided posture that persists today when people justify their opposition to birth control or homosexual behavior by terming them unnatural. This supposition long predates evolutionary theory, however, which has merely served as a scientific rationalization for persistent biases.

THE IMPACT OF DARWINISM

The Darwinian view that evolution not only had occurred, but was caused by the impersonal process of natural selection met strong opposition in both scientific and nonscientific circles, because of its perceived threat to theological doctrine and to the unique position in nature that humans would like to arrogate to themselves. The view offered by Darwin — of a purposeless universe in which life changes, to no ultimate purpose, by the survival of the fittest of random variations; a material world from which we have arisen and with which we are one; a universe that does not care about us and is not going to save us from our follies — such a vision is far less reassuring and less flattering to the ego than the notion of a world created to serve us, the apples of God's eye. It is a view distasteful not only to the theologically inclined, but to the literary tradition that opposes materialism with more transcendent values (see, e.g., Barzun 1958). Seldom are the positive implications of Darwinism acknowledged: that it forces us to view ourselves not as prisoners of a static world order, but as the masters of our fate; that our salvation lies not in Providence, but in ourselves.

I have already noted that evolutionary theory, together with the rest of science, had important consequences in philosophy, chiefly a severe weakening of its theological underpinnings. The Great Chain of Being was demolished — or at least temporalized, for the concept persists in the form of evolutionary progress from "lower" to "higher" forms of life. Innumerable authors, starting perhaps with Spencer, developed evolutionary philosophies in which human progress was seen as one with a grand cosmic, progressive evolution. Some, such as Dewey, saw in the random element of biological evolution an escape from the rigid determinism implied by the physical sciences of the day. But no coherent philosophical system based on evolution has arisen and persisted; as Collins (1959) notes, the scientific data and theories of evolution do not themselves compel any single philosophical conclusion nor settle any major philosophical issues.

For a while the social sciences were strongly influenced by evolutionary thought, especially the concept of natural selection. Spencer held that the ethical principles of society should mimic nature, that social progress would follow from the survival of the fittest. He opposed all forms of welfare, government-supported housing, tariffs, and even governmental postal systems in the name of free competition. His views, which came to be known as social Darwinism, were extremely popular in the United States early in this century and were invoked to support capitalism, imperialism, "eugenic" sterilization of the "inferior" elements of society, and immigration laws restricting the entry of "inferior" races and ethnic groups (Hofstadter 1955). American politics and economics would no doubt have followed these courses anyway, but a theoretical justification from the lips of scientists and philosophers is always helpful.

Anthropology had had an evolutionary viewpoint before Darwin, but this was certainly reinforced by the emergence of the biological theory. It was at first naively interpreted, so that an evolutionary view of culture fell into disrepute when it was discovered that cultures did not necessarily advance along a single line from hunting and gathering toward modern industrialism. Recently cultural anthropology has become more evolutionary, viewing many cultural institutions as nongenetic adaptations to local ecological, economic, and social conditions, much as the properties of different species fit them to different environments.

Evolution has had both direct and indirect influences on psychology. Most directly, perhaps, it has drawn attention to similarities between human and animal behavior (beginning with Darwin's "The Descent of Man" and "The Expression of the Emotions in Man and Animals"), leading to the field of comparative psychology, to the use of animals as experimental models for human behavior, and to speculations about which human behaviors are "innate" derivations from prehuman ancestors. Evolution indirectly influenced the development of Freud's theory, in which neuroses stem from the battle of the ego and the superego against the "natural" urges (the id) inherited from our brutish ancestors. Freud, who believed in the inheritance of acquired characteristics and in the principle that ontogeny recapitulates phylogeny, suggested that each of us repeats the phylogenetic states of sexual development, and even the events of the Oedipus drama as they might once have actually occurred. Carl Jung's psychology is highly evolutionary, being founded on the notion of the collective unconscious, a collection of archetypal images inherited from our evolutionary past.

With the exception of physics and in large part chemistry, the natural sciences have become steeped in evolutionary thought. That the universe has developed over eons is generally accepted; geology has a symbiotic relationship with evolutionary biology; and each of the subdisciplines of biology, from biochemistry to ecology, assumes evolution and interprets data in its light. Most of the practical applications of biology in medicine and agriculture depend on evolutionary biology; dogs and monkeys are used in medical research only because their physiological systems are homologous to those of humans. Evolution is, indeed, the one coherent system of principles that unifies all of biology.

In light of the manifold reactions to evolution, how shall we judge Darwin? His ideas have been bent to good and evil, used for scientific and medical advances and for discriminatory social policies, interpreted and misinterpreted in the most diverse of ethical and unethical contexts. Many people would agree with the conclusion of Greene (1961), a historian of science.

> Darwin's influence on social thought has been a mixed one. Insofar as he opened a vast new perspective by showing man's organic relation to the animal kingdom, he took his place among the few greatest contributors to human knowledge. Insofar, however, as he reinforced Herbert Spencer's emphasis on individual, racial, and national competition as the source of social progress; insofar as he minimized the difference between man and other animals; insofar as he encouraged the idea that the methods of natural science are fully adequate to the study of human nature and society; insofar as he ignored the moral ambiguity of human progress and allowed us to think that science could support itself without philosophy and religion; to this extent he contributed to the growth of misconceptions whose evil effects we still combat.

This judgment of Darwin, I think, is too harsh. Darwin generally avoided discussion of ethical and social issues to which he could not apply his scientific methods. It is largely the legions that followed (and some, such as Spencer, who preceded) him who are guilty of these errors. Evolution, like any concept, can be used for good or ill; the devil can cite Scripture to his purpose.

EVOLUTION SINCE DARWIN

Before embarking on a detailed treatment of modern evolutionary biology, I think it appropriate to sketch the roles that the biological disciplines have played in its development. In support of his views, Darwin drew on systematics, comparative anatomy, embryology, biogeography, and paleontology to contend that evolution had occurred; on the rudimentary genetics of his day to document the existence and origin of variation; and on studies of ecology and behavior to elucidate the agents of natural selection.

All these fields and others have continued to contribute to evolutionary thought. Biological systematics is now practiced in an entirely evolutionary framework; using the data of comparative anatomy, embryology, physiology, behavior, and biochemistry, it has provided an immense amount of information on evolutionary trends, patterns of adaptation, the kinds of evolutionary transformation that organisms' features undergo, the intermediate stages in evolutionary sequences, the patterns of variation within species. The concept of species itself, partly because of systematic studies, has been transformed. Together with the far richer data of paleontology than were available to Darwin, systematics has documented the temporal changes in the diversity and form of organisms and in their patterns of geographic distribution.

Descriptive embryology has continued to provide data on phylogenetic relationships; and some naive ideas, like the notion that ontogenetic change repeats phylogenetic transformation, have been revised. Experimental and genetic approaches to developmental biology have had a very different role, for evolution consists largely of transformations in physical structures and hence of the developmental pathways by which these structures arise. Slight changes in the timing or rate of development of a structure, or of its influence on other embryonic tissues, can have drastic phenotypic effects; conversely, rather extensive changes in an organism's genetic information may not be expressed in the phenotype. From the 1930s into the 1950s some authors (especially Goldschmidt and Waddington) attempted to integrate developmental biology into evolutionary thought, but a satisfying synthesis has yet to be achieved.

The most serious gap in the structure of Darwin's theory was his ignorance of how hereditary variations arise and persist. The development of genetics has been the single most important contribution to evolutionary thought since Darwin, for it has become the basis of the entire theoretical structure of the mechanisms of evolutionary change. The details of heredity are understood in material, molecular terms; from physiological and developmental genetics we have some understanding, albeit far from perfect, of how alterations in the genes are translated into phenotypic form. The statistically predictable segregation of genes in meiosis has enabled the development of a highly sophisticated theory of population genetics, first developed by R. A. Fisher, J. B. S. Haldane, and Sewall Wright, that specifies mathematically which evolutionary events are possible, how fast genetic change can occur, what forces govern genetic variation, and what evolutionary factors can account for various observations. There was a time, early in this century, when genetics seemed at odds with evolution, when the mutation process itself rather than natural selection was invoked as the driving force of evolution. The *modern synthesis* of the late 1930s and early 1940s, however, drew together the thoughts of geneticists, systematists, and paleontologists into a *neo-Darwinian* view, in which the joint action of mutation, whereby variation arises, and selection, whereby it is shaped into coherent adaptive form, is considered sufficient explanation of the evolutionary process.

In the last few decades the mathematical theory of population genetics has become extremely sophisticated; numerous laboratory and some field studies have shown how genetic variation is maintained and altered by natural selection; and there has been an increasing attempt to wed the theory to observations on real populations. Ecological factors as agents of natural selection have been scrutinized, and conversely there has been a recognition that ecological phenomena such as the diversity of species in a community have partly evolutionary, hence genetic, explanations. The synthesis of population genetics and ecology has led to the field of population biology, which treats the influences of population dynamics and other ecological factors on the genetic characteristics of populations, as well as the genetic processes whereby the demographic characteristics of species and their interactions with other species evolve.

Modern population biology has taken uniformitarianism to heart; it restricts itself largely to the processes of microevolutionary change, the alteration of gene frequencies within populations and species. Its units are abstracted symbols — gene frequency, selection coefficient, population size — rather than the material objects with which the systematist is concerned — bones, hemocoeles, tracheids. Excluded from much of the modern evolutionary literature are the problems of developmental biology, which asks how the mutation denoted simply as A in the genetic equations can actually transform a fly's wing. Almost disregarded are the great questions posed by the history of life, on the assumption that study of microevolutionary changes in gene frequencies can, by uniformitarian extrapolation, explain the evolution of angiosperms and mammals from their simpler ancestors. In their enthusiasm for the beauty of genetics, many evolutionists seem to have forgotten that the phenomena crying out for explanation are the historical transformations of the phenotype. I have thus tried to emphasize in this book not only the modern genetic theory of the mechanisms of evolutionary change, but also the phenotypic and historical phenomena that the mechanistic theory is meant to explain, but which fit the theory only awkwardly. Perhaps we may look forward to an "ultra-modern synthesis" in which developmental and molecular biology will come to play their necessary roles.

SUMMARY

Evolution, a fact rather than mere hypothesis, is the central unifying concept of biology. By extension it affects almost all other fields of knowledge and thought and must be considered one of the most influential concepts in Western thought. Its tenets have frequently been misinterpreted — it is often equated with progress — and the objective science of evolutionary biology has often been extended into the subjective realm of ethics and used illegitimately as justification for both pernicious and humanitarian economic, social, and scientific policies.

The recognition that evolution has happened was widespread before Darwin and, together with other advances in science, was a major change in the Western world view, which took a static, orderly Great Chain of Being to be natural and therefore good. Darwin and Wallace, by providing the idea of natural selection, transformed speculation into scientific theory. Most of their ideas have been validated in more than a century of subsequent research, in which evolutionary biology, especially through the growth of genetics, has become an ever more intricate, all-embracing, and sophisticated body of explanatory principles.

FOR DISCUSSION AND THOUGHT

1 Debate the conclusions to which Greene, in the quotation on page 11, comes.
2 Read "Did Man Get Here by Evolution or Creation?" published by the Watchtower Bible and Tract Society, and refute its arguments.
3 There has been quite a bit of debate about just how original and creative Darwin was. Read the books by Himmelfarb (1962) and by Ghiselin (1969) for examples of antithetic views, and draw your own conclusions.
4 One unsettling conclusion that some draw from evolution is that humans are nothing more than animals. This has been termed the *reductionist fallacy*. In what sense is it a fallacy? Does the fact that an organism consists of an evolved set of biochemical systems mean that life is nothing more than molecules in motion, and that all of biology can be explained in biochemical terms?
5 Is it possible to defend ethical principles ("Thou shalt not kill") on scientific grounds? Is it desirable? Are there nonscientific grounds on which such principles can be defended? See the articles by Adler and others in Deely and Nogar (1973).
6 Contrast Spencer's point of view with T. H. Huxley's statement (1893), "Let us understand, once and for all, that the ethical progress of society depends not on imitating the cosmic process, still less in running away from it, but in combating it."
7 Analyze and evaluate Emerson's couplet.
> Striving to be man, the worm
> Mounts through all the spires of form.
8 Among the graffiti I have seen recently is this. "Do you really believe that birds can navigate by the stars because of some random change in the DNA?" The content of the specific question merits debate; but so do the

larger questions it raises. In what sense is evolution random, due to chance? What does *chance* mean? Aren't all events determined by antecedent events, through the laws of matter in motion? As Lovejoy (1936) discusses, part of the allure of the Great Chain of Being was its affirmation of perfect order in a perfectly deterministic universe that precisely obeyed fixed laws. But both determinism and chance are appealing world views for different people. Why?

9 The idealization of progress as good is so much a part of our thinking that it may come as a shock to realize that this is a very recent view, stemming from no earlier than the sixteenth century. Why did this ideal arise, and why is it so appealing? Need we invoke evolution to justify it? How can we objectively determine whether the history of evolution, or of society for that matter, has been one of progress?

MAJOR REFERENCES

The references at the end of each chapter include some that provide basic background on elementary aspects of the subject, but most are major works that provide a comprehensive treatment of the subject and an entry into the technical literature.

Greene, J. C. 1959. *The death of Adam: Evolution and its impact on Western thought*. Iowa State University Press, Ames, Iowa. 388 pages. A scholarly but easily readable description of the history of evolutionary thought.

Eiseley, L. 1958. *Darwin's century*. Doubleday, New York. 378 pages. A beautifully written essay on the same subject.

Lovejoy, A. O. 1936. *The great chain of being: A study of the history of an idea*. Harvard University Press, Cambridge, Mass. 382 pages. An erudite history of the subject for readers with some background in philosophy and history.

Hofstadter, R. 1955. *Social Darwinism in American thought*. Beacon Press, Boston. 248 pages. A study of some social applications of evolution.

PART

PRELUDE TO
THE STUDY OF
EVOLUTION

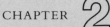

A Synopsis of Evolutionary Theory

Lest the reader become overwhelmed by the welter of detail and speculation that comprises the body of this book, it is best to begin, I think, with a *précis* of evolutionary thought. Most of the following points are already accepted by anyone who has had a modicum of exposure to biology, although they may be couched in slightly unfamiliar terms. They constitute the evolutionary principles on which biologists almost universally agree; I will document their validity in subsequent chapters.

Although it will become apparent that there are innumerable arguments about the relative importance of the factors that compose evolution, about the precise details of the mechanisms by which it occurs, and about the detailed evolutionary history of particular species and groups of organisms, there is no uncertainty about whether evolution occurs, nor what its essential mechanisms are. I state this only because anti-evolutionists have the indefensible habit of pointing to controversies within the evolutionary literature as evidence that evolution *as a process* is not proven, but is merely hypothetical; in fact the controversies treat only specific details of the general mechanisms or particular historical evolutionary events. Evolution is not only a fact; it is the central unifying principle of all biology. Its major, undisputed features may be set out as follows.

1. *Every characteristic* of organisms *has a frequency distribution (Figure 1a)*, for example, body size of a number of individual organisms. This distribution has an average (mean) value and, unless the individuals are identical, a certain degree of variation (the variance or standard deviation; see Appendix I). If two characteristics on each individual are measured, a bivariate distribution can be described (*Figure 1b*); and if *n* characteristics are measured, the position of each individual can be described in an *n*-dimensional space (which can be imagined but not drawn).

For any one of these characteristics the

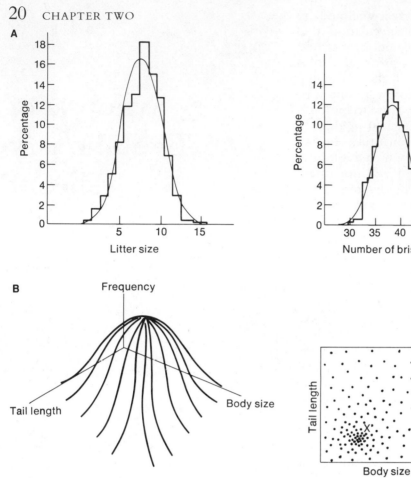

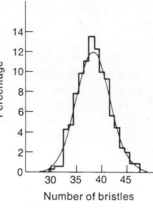

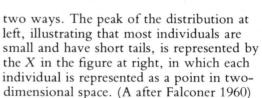

1 Frequency distributions of characteristics. (A) Litter size in mice and number of bristles on the ventral surface of two abdominal segments of *Drosophila*, showing approximate conformity to a normal (bell-shaped) curve. (B) The distribution of variation of two characteristics simultaneously, illustrated in two ways. The peak of the distribution at left, illustrating that most individuals are small and have short tails, is represented by the X in the figure at right, in which each individual is represented as a point in two-dimensional space. (A after Falconer 1960)

variation can be *partitioned*. For example, the variation in body size of several cats, dogs, sheep, and cows, each from several localities, can be partitioned (1) between the orders Carnivora (cats and dogs) and Artiodactyla (sheep and cows); (2) among species; (3) among localities within each species; and (4) among individuals within each locality within each species. Both the mean body size and the variance would be changed by altering the proportions of carnivores vs. artiodactyls, of dogs vs. cats, of cats (or any other species) from a locality where they tend to be large vs. one where they are smaller, or of larger vs. smaller cats from any one locality.

Evolution in a broad sense may be considered any change in the form of the frequency distribution of one or more characteristics that persists for one or more generations. It entails, then, a difference in the frequency distribution of a set of parents vs. their offspring. Clearly evolution cannot exist without preexisting variation. Evolution can consist of changes at each of the levels at which variation exists. So the evolution of body size can be described as changes in the proportions of large vs. small species (e.g., dinosaurs vs. lizards), of large-bodied vs. small-bodied populations within a species (e.g., changes in relative population size of Masai vs. Ituri pygmies), or of large vs. small individuals within a single population of a species.

2. *The form of the frequency distribution remains constant over time if there is perfect heredity and if individuals of all phenotypes have identical numbers of offspring; otherwise it can change.* Change can occur if the phenotype (the magnitude of the particular feature we are studying) of offspring is not identical to their parents (*Figure 2*). If individuals of some phenotypes leave fewer copies of themselves than others, the relative proportions of the different phenotypes will differ between parent and offspring generations. This can occur most obviously if some phenotypes reproduce less because of an inability to survive to reproductive age, because of sterility, or because of a lower fecundity (number of offspring) or ability to obtain mates. Such disparities in reproduction may change the form of the frequency distribution without altering its mean (*Figure 2a*), or they may shift the mean in one direction (*Figure 2b*). Such a directional change in average properties is the aspect of evolutionary change in which we tend to be most interested. Directional evolution can be viewed, then, as a shift in the position of the mean phenotype in the phenotype space.

3. Constancy of the frequency distribution from generation to generation depends on the degree of perfection of heredity, but the *variation exists* in the first place only *because heredity is not perfect.* If it were, there would be only one form of life, and all living things would be precisely identical to the first self-replicating molecule.

By imperfect heredity I mean only that offspring are not identical to their parents. They can differ in two ways: by possessing different genetic information (a different GENOTYPE) that is translated into phenotypic difference or by exposure to different environmental conditions that together with the genotype produce the phenotype. Every genotype has a NORM OF REACTION, a variety of phenotypes that it produces in conjunction with different environments (*Figure 3*). Thus hereditary similarity can derive not only from a genetic similarity between parents and their offspring, but also from the similarity of their environments. This can happen, for example, if parents teach their offspring behaviors that they themselves learned. In certain birds there is "heredity" of song, for instance, based on the learning of song from parents and other nearby birds; local dialects arise not from genetic differences

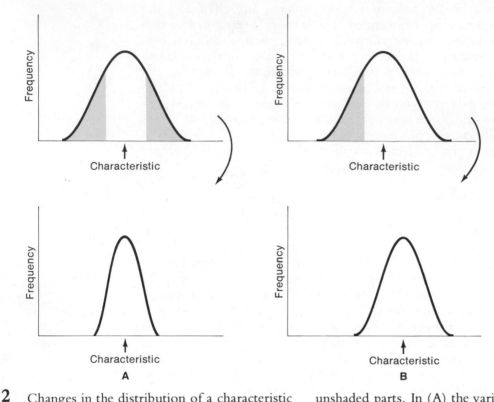

2 Changes in the distribution of a characteristic from one generation (above) to the next (below), that might result if individuals in the shaded part of the distributions (above) do not reproduce as much as those in the unshaded parts. In (A) the variation in the population is altered, but the mean (denoted by the arrow) is not. In (B) both the mean and the variance are altered. Both cases constitute evolutionary change.

among birds, but from different cultural traditions of song. Such a cultural form of "heredity" is most elaborately developed, of course, in humans.

The bearers of heredity with which biologists are mostly concerned are the genes. Genetic heredity is perfect only if there is asexual reproduction and no mutation; otherwise the GENETIC CORRELATION between parent and offspring is reduced by recombination (Mendelian segregation, chromosomal recombination, and sex) and by MUTATION, alterations of the information carried by the genetic macromolecules. Mutations precede recombination as sources of genetic variation, since recombination can only re-assort preexisting variants into new combinations.

4. *The more mutable a unit of heredity is,* the less faithful heredity is, and *the more evanescent any newly arisen variant will be.* Most environmental effects on the phenotype have little long-term evolutionary consequence; they either are not hereditary or they are nullified by any

alteration in the environmental factor that influenced their development. Plants grown from the seed of plants cultivated in rich soil are stunted if grown in poor soil; and the mutability of cultural "heredity" is attested by the almost immediate cultural changes that are the record of human history.

Genetic heredity varies in degree. Individual genes, cistrons, are hardly mutable at all; they are inherited as particles and they only rarely undergo alterations in the copying process. Because they retain their identities over long numbers of generations, genic mutations have a long-term effect on evolution. Aggregations of genes (genotypes), on the other hand, are usually quite short-lived in sexual species since they are altered into new combinations during meiosis and syngamy; thus most of the new genotypes generated by recombination have only a short-lived effect on evolution. Different populations of species

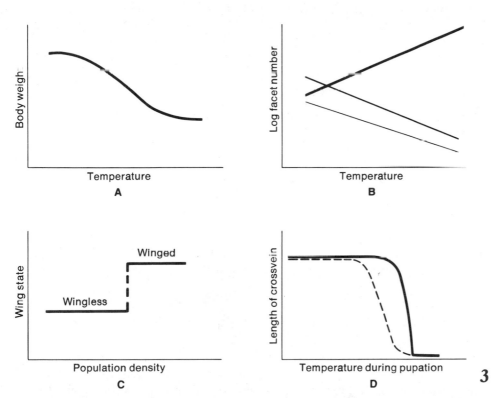

3

Four schematic illustrations of real reaction norms. In each case the phenotype developed depends on environmental conditions; the norm of reaction of a genotype is the variety of different phenotypic expressions it can manifest. (A) The response of body weight to temperature in *Drosophila* and many other insects. (B) Number of eye facets, in relation to temperature, of different mutant genotypes at the *Bar* locus in *Drosophila*. (C) A developmental "switch," as found in some aphids, which develop wings if sufficiently crowded at a critical period in development. (D) Differential sensitivity of two *Drosophila* genotypes to a heat shock that affects the development of the crossvein in the wing. (B after Hersh 1930)

("races") differ genetically from each other, but the consistency of differences among populations from generation to generation can be very low if the genes within them replace one another rapidly (*Figure 4*). Moreover such populations lose their identity if they interbreed, just as individual genotypes within a population do. So we might suspect that little long-term change in the characteristics of a species as a whole stems from the replacement of one kind of population by another kind.

The differences between species, however, have more nearly perfect heredity than those between populations; bluebirds and robins transmit the characteristics of bluebirds and robins, respectively, to their offspring. Since a species has long-term identity as such, a change in the relative abundance of several species (e.g., extinction of some of them) does constitute a long-term change in the characteristics that the species in aggregate possess. For example, average body size in reptiles has declined since the Cretaceous, as the group has come to consist more of lizards and less of dinosaurs.

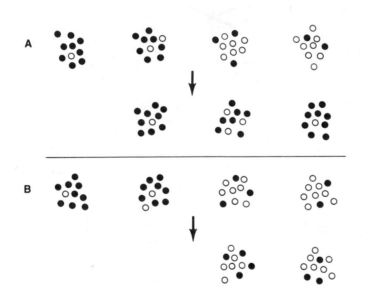

4   Selection within and among populations. The solid and clear circles represent two asexually reproducing genotypes in density-limited populations. Within each population, the "solid" genotype reproduces more than the "clear" genotype; this constitutes *individual* selection within populations. However, populations consisting mostly of the "solid" genotype have a higher extinction rate than those consisting mostly of the "clear" genotype; this constitutes *group* selection, or interdemic selection, in favor of the "clear" genotype. In (A), group selection is weaker than individual selection, so "solid" increases in the species as a whole. In (B), group selection is stronger than individual selection because the extinction rate of "solid"-dominated populations is greater and the reproductive advantage of the "solid" genotype is not as great; thus the "clear" genotype increases in the species as a whole.

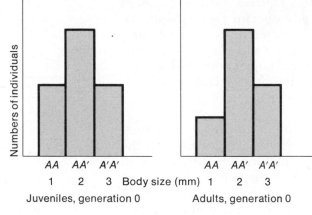 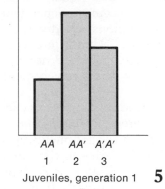

AA AA' A'A'   AA AA' A'A'   AA AA' A'A'
 1   2   3 Body size (mm) 1   2   3   1   2   3

Juveniles, generation 0   Adults, generation 0   Juveniles, generation 1 **5**

The unitary event in evolution, schematically represented. Only half as many *AA* individuals survive to adulthood as do *AA'* or *A'A'* individuals, perhaps because they are smaller. As a result, the frequency of the *A* allele changes from 0.5 to 0.43. Thus both the frequencies of the genes and of the genotypes they compose are altered. During mating the genes are recombined at random to form the next generation, in which *AA* and *AA'* are relatively less abundant and *A'A'* is more abundant than among the juveniles of the previous generation. Mean body size has changed from 2.0 to 2.14. Phenotypic change has followed from the genotypic changes engendered by the interaction of the phenotype (in the previous generation) with the environment.

5. Variation among populations or species requires that the populations differ with respect to the relative proportions of different phenotypes within them. Therefore all evolutionary changes depend on changes *within populations. These consist of alterations of the proportions of the genotypes, hence of the genes, borne by the individual organisms that make up the population (Figure 5)*. These proportions, or GENE FREQUENCIES, remain constant unless forces change them. Some forces destabilize gene frequencies, while others stabilize them, returning them to their original equilibrium state if they should be changed. The forces that stabilize gene frequencies at intermediate values (between 0 and 100 per cent) cause genetic variation to persist within populations. The question of which factors are most important in maintaining this variation is a major controversy.

6. Among these factors is the RATE OF MUTATION; but because mutation rates are so low, their effect on the genetic composition of a population is likely to be overwhelmed by stronger factors.

7. One such factor is chance, or GENETIC DRIFT, whereby the relative abundance of different alleles can increase and decrease at random from one generation to the next, purely by accident. This effect is greatest in small populations, just as a deviation from a 1:1 ratio of heads and tails is likely to be greater if a coin is tossed 10 times than if it is tossed

1000 times. Eventually one allele completely replaces another (becomes fixed in the population) if chance is the only operant factor (*Figure 6*). Since this process occurs independently in different populations of a species, the populations can diverge in genetic and phenotypic composition, by chance. The differences among populations and species that arise by genetic drift are not necessarily either adaptive or maladaptive; they may be effectively neutral.

8. Very commonly the phenotypes in a population *differ consistently in their ability to survive and in the number of offspring they produce*. To the extent that differences in phenotype are founded on differences in genotype, the alleles vary in the rate at which they are replicated. As long as one allele replicates faster than others, it increases in frequency at the expense of the others. It may completely replace other alleles, so that the genetic composition of the population is completely altered and becomes MONOMORPHIC, consisting of one genotype (at this locus) only. This is the traditional view of evolution by NATURAL SELECTION, which is often defined as *the differential survival and reproduction of geno-*

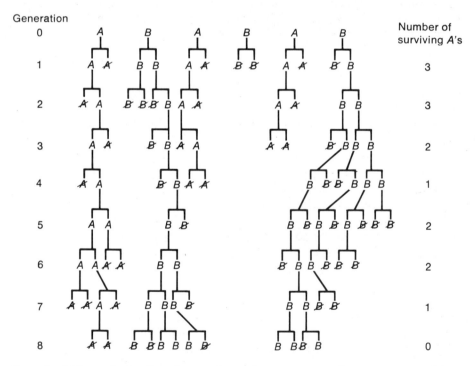

6   Genetic drift: evolution by chance. In each generation each individual of genotypes *A* and *B* produces two identical offspring. Half the juveniles die (population size remains constant), but which ones die is entirely random, not dependent on genotype. (This example was generated by tossing a coin to specify which individuals were to die.) The frequency of *A* fluctuates at random until by chance *A* becomes extinct and *B* is fixed in the population.

types. Often, however, the superiority of an allele holds only in certain environments, or at certain gene frequencies, or in conjunction with certain other alleles. In such cases the allele may not completely replace others but remain at a stable intermediate frequency, so that the population remains POLYMORPHIC, consisting of two or more genotypes; genetic variation persists. Natural populations are highly polymorphic.

When the gene frequencies do change, the rate of change depends on the gene frequencies and on the disparity among genotypes with respect to their fecundity and their capacity for survival. Since all species produce far more offspring than the environment supports, mortality is high in each generation; thus much of natural selection consists of differential mortality. If the individuals that bear different alleles do not differ consistently in their capacity to survive and reproduce, the alleles are not subject to natural selection; they are neutral with respect to each other.

9. In principle these statements apply to any system in which entities that differ from each other have heredity. Not only genes, but genotypes, populations, species, and different culturally transmitted patterns of behavior can vary in their ability to survive and reproduce. Hence *selection can act at each of several levels*. If the direction of evolution caused by selection at one level is opposite that at another level, the outcome depends on the relative strength of the two selective forces (*Figure 4*).

10. The alleles that arise by mutation vary in their effect on the phenotype. Those that have a drastic effect seem unlikely to increase survival and reproduction (the composite of which is termed FITNESS); for since most organisms are intricately constructed systems that are already functioning very well, a drastic change in structure is more likely to disrupt their "machinery" than to improve it. Hence it is commonly believed that advantageous mutations are likely to have slight phenotype effects and that *evolution proceeds by successive slight changes in phenotype and is a gradual process*. This proposition, however, is presently coming into question.

11. Although some mutations (e.g., those that cause death before hatching) are unconditionally harmful, *many mutations are advantageous in some contexts and disadvantageous in others*. Even mutations that might be thought unconditionally deleterious may not be; sterility has evolved in the worker castes of social insects, for example. But there is probably no such thing as an unconditionally advantageous mutation. What is advantageous to a mammal may not be to an insect and *vice versa*.

12. *The advantage* of an allele or genotype *depends on the environment*, which includes not only obvious physical factors (e.g., temperature) and biotic factors (e.g., predators), but other members of the species.

Moreover, each gene must function in an internal environment, the product of the rest of the genotype in conjunction with the external environment. This internal environment has in part been fixed by prior evolution, so that a new allele is advantageous only if it contributes to the function of an established system, and what is advantageous to one species is disadvantageous to others with different structures and ways of life. Moreover the adaptive value of an allele can depend on which of the many genotypes in the population it is associated with. It may lose its potential advantage if a similar mutation has already preempted its functions; or it may be advantageous if associated with a certain allele at another locus, but not others. That is, constellations of alleles at different loci may be advantageous together but not separately. This is most apparent in the case of complex organs or biochemical pathways, in which many components must function harmoniously. In extreme cases the evolution of a complex organ may require the simultaneous conjunction of many mutations, each individually nonadaptive; but in most instances a complex system evolves successively from a simple to an elaborate state, with each new adaptive mutation adding to the effectiveness of a preexisting structure.

13. The necessity for parts of a complex structure to be "co-adapted" to one another may impose *limits on the rate of evolution,* since some crucial gene combinations may be most unlikely to arise. The extent of evolutionary change may also be limited; further change of one part of an organism can be incompatible with the function of other parts. Very often, then, the features of an organism are adaptive compromises between conflicting selective pressures.

14. *The chance that a mutation will occur is not affected by the advantage or disadvantage it confers.* A specific mutation does not arise in response to a need just because it would be advantageous at that time. Still, evolution is not a random process. Natural selection sifts adaptive from nonadaptive genotypes, which have arisen at random, in a highly nonrandom way — more or less as we form meaningful words in a Scrabble game from randomly assorted letters. Similarly the present species composition of the world's biota is a selection of the innumerable species that have arisen in the past; the ones that are with us are those that happened to have the characteristics necessary to survive the vicissitudes of time.

15. *Natural selection is simply differential survival and reproduction, not Providence.* Characteristics selected here and now are not necessarily the best of all possible characteristics; they are simply the most suitable of those currently available. The characteristics evolved under one set of circumstances may prove inappropriate for subsequent environments; yet because prior evolution determines the evolutionary paths that a species can subsequently take, the species may be stuck with inappropriate features. *Organisms cannot evolve adaptations for future events* that are different from past events to which they have already adapted.

Hence selection of currently advantageous genotypes is random with respect to their future advantage. Consequently the direction of evolution can vary as the environment fluctuates; and extinction is a major feature of evolution.

16. *Not all characteristics evolved because they were adaptive.* Genetic drift can fix neutral or even maladaptive traits. Moreover the feature we are looking at may be irrelevant to the organism. We may be looking at the wrong aspect: redness of leaf in the autumn is probably not in itself advantageous to maple trees; what is advantageous is the presence of anthocyanins for other reasons and the breakdown of chlorophyll so that some of the breakdown products can be stored. Every gene has a variety of effects on the phenotype (pleiotropy); some may be advantageous and others neutral or even maladaptive. A neutral or maladaptive allele may have been very closely linked to a highly advantageous allele at a nearby locus and thereby brought to preponderance. For these and other reasons *organisms are not necessarily as well adapted as they might be;* they do not live in the best of all possible worlds.

17. If allele replacement occurs within a single group of interbreeding individuals, ancestral characteristics will disappear as they are replaced (*Figure* 7). The persistence of both ancestral and derived states thus requires allele replacement to occur in some populations and not in others. Thus *the development of diversity requires barriers to gene exchange* (gene flow). If there is limited interbreeding among populations experiencing different genetic changes, the result is GEOGRAPHIC VARIATION among the populations of a species. The amount of genetic (hence phenotypic) difference among these populations depends on a balance between the homogenizing effect of gene flow and the diversifying effects of different selection pressures and genetic drift.

18. *If gene flow* among local populations of a species *is sufficiently restricted, the populations may so diverge that they will not or cannot interbreed;* they have then become different species. This process, called SPECIATION, usually requires that interbreeding be initially restricted by a geographic or topographic barrier; with sufficient divergence the biological characteristics of the populations then impose reproductive isolation. Evolutionary changes in small isolated populations can be exceptionally rapid, and most major evolutionary change may take place in such populations during the process of speciation.

19. *If two populations have become different species, the genetic changes in one are not transmitted to the other,* so they pursue independent evolutionary paths. Ancestral characteristics can persist in one but be replaced by derived characteristics in the other. In speciation lies the genesis of biological diversity.

20. Whether related species diverge in their characteristics depends

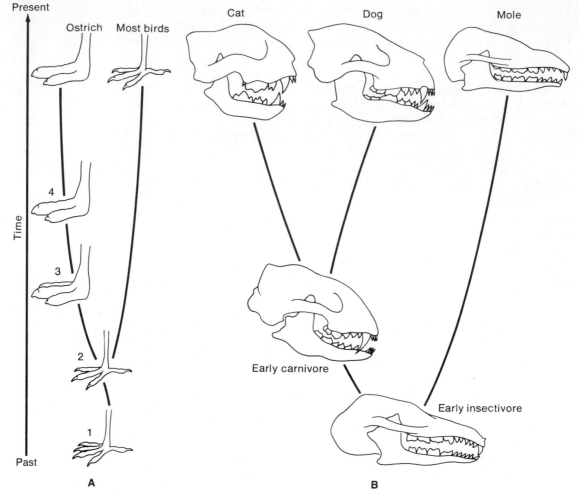

7 Past stages in evolution are evident among contemporary organisms only if speciation occurred in the past and if the species with divergent characters have survived to the present time. (A) Because reptiles have five toes, we might presume that early birds had five toes (stage 1) but that they either did not speciate and evolved directly into four-toed forms (stage 2), or that they did speciate and the five-toed forms have not persisted. Similarly the modern ostrich has only two toes; if it passed through a three-toed stage (3), this form must have evolved directly into a two-toed form rather than branching into two-toed and three-toed species. Thus there are no three-toed modern ostriches. (B) In mammals, however, the reduction in tooth number that is well documented in the fossil record is also evident among Recent forms, for the primitive conditions persist in modern species (mole, dog) along with the derived conditions in other species (cat) to which their ancestors gave rise.

on both their biological properties and the differences in their environments. *Related species,* because of the similarity of their genetic and developmental organization, *may evolve along nearly parallel paths.* Similarity of environment may favor the acquisition of similar characteristics even in unrelated species which become *convergently* similar. But *environmental factors may favor divergence as well.* Among these factors are interspecific competition, which presses species to use different resources; predation, which can favor the acquisition of different mechanisms of protection; and the threat of hybridization with related species, which can favor the development of differences in breeding behavior.

21. Species vary in how long they persist before they become extinct, and some species undergo speciation more rapidly than others. Hence just as alleles differ in their rates of survival and reproduction, different groups of species vary in their "death" and "birth" rates, so that *the distribution of phenotypic characteristics among species taken together changes by a process of species selection.* The replacement of some species by others can cause long-term evolutionary trends.

22. This process of *species selection is also important in determining the species composition of ecological communities,* in which the evolved properties of species determine whether they can stably coexist. Unstable assemblages are transformed to stable combinations by the local extinction of species that have the wrong characteristics. The interactions among species in a community presumably change as the individual species evolve; and evolution tends to stabilize the interactions in some instances and destabilize them in others.

23. *The principles of intraspecific evolution apply to all characteristics,* not only adult morphology, but developmental patterns, physiology, genetic features such as sexual vs. asexual reproduction, features of the life cycle such as fecundity, and behavior. Moreover the phenotypic norm of reaction can itself evolve, so species vary in their flexibility of development, physiology, and behavior. None of the modes of adaptation is universally superior to another, including the flexibility vs. the inflexibility of behavior. The capacity for flexibility of behavior is most elaborately developed in the human species, but it cannot be objectively cited as evidence of any greater evolutionary progress than the features of other species. Evolution has no goal, so progress is not a valid evolutionary concept.

The study of evolution is a study of interplay between heredity and environment. Evolution follows from a cycle of events repeated in each generation: (1) the genetic process, in which programs of genetic information (genotypes) are produced; (2) the epigenetic process, the ontogenetic transformation between genotype and phenotype, in which both genotype and environment play roles; (3) the environmental process, in which the interaction between phenotype and environment determines which phenotypes shall once again transmit genetic information and to what extent.

I shall try to set the stage for a detailed analysis of evolutionary mechanisms by exploring those aspects of genetics, development, and ecology that are prerequisite to understanding the evolutionary process. This analysis is followed by a summary of the major events in evolutionary history that evolutionary theory is supposed to explain: how organic diversity has varied in time and space, how evolution has and has not followed trends, and how innovations of structure and function have arisen. To explain these events, I describe the amount and nature of variation and then consider in detail the factors that maintain variation and change its form. These factors are the core of modern evolutionary theory. After treating them, I speculate on how adequately this theoretical structure explains the major events of evolution. Finally, I consider a few specific evolutionary problems of considerable current interest: the role of evolution in the formation of ecological communities and the evolution of some of the more interesting features of the human species.

MAJOR REFERENCES

Maynard Smith, J. 1975. *The theory of evolution*. 3rd ed. Penguin Books, Baltimore, Md. 344 pages. A nontechnical summary, especially for the nonbiologist.

Stebbins, G. L. 1971. *Processes of organic evolution*. 2nd ed. Prentice-Hall, Englewood Cliffs, N. J. 193 pages. A short but comprehensive treatment for the beginning student.

Sheppard, P. M. 1959. *Natural selection and heredity*. Harper and Brothers, New York. 209 pages. An old but still excellent introduction to the genetic aspects of evolution.

Heredity and Development

If under changing conditions of life organic beings present individual differences in almost every part of their structure, and this cannot be disputed; . . . if variations useful to any organic being ever do occur, assuredly individuals thus characterized will have the best chance of being preserved in the struggle for life; and from the strong principle of inheritance, these will tend to produce offspring similarly characterized.

Much of evolutionary biology has served to confirm and elaborate on Darwin's statement. While the "struggle for life" is a chief theme of ecology, it is in the fields of genetics and development that explanations of the "strong principle of inheritance" and the origins of individual differences are found. We cannot understand evolution without understanding some principles of heredity.

The hereditary factors, the genes, are translated into the material form, the phenotype, through interaction with the environment. The distinction between GENOTYPE and PHENOTYPE is critical. The genotype is but a blueprint for an organism, the set of instructions for development received from its parents. The phenotype is the manifestation, in a series of developmental stages, of the interaction of this information with the physical and chemical factors — the environment — that allow the blueprint to be realized. Because this process of casting a set of instructions into physical form is complex, the array of phenotypes among a group of organisms may be quite different from the array of genotypes. This fact has most important consequences for evolution.

TWO PRINCIPLES OF GENETICS

Among the most important principles of heredity are that the hereditary factors are particles and that the flow of information from genotype to phenotype is unidirectional. Darwin's ignorance of the particulate nature of inheritance led him to abandon his initial (and largely correct) hypothesis that the action of natural selection on hereditary variation sufficiently explains the major events of adaptive evolution. The prevailing view of heredity in his day was one of blending inheritance, for crossing large and small animals tends to produce offspring of intermediate size, just as mixing strong and weak solutions of dye produces an intermediate color. But if hereditary factors are blended, losing their identity and becoming intermediate, the variation (variance) among organisms will be halved in every generation and so will quickly vanish. Mutations would then have to arise at a phenomenal rate to account for variation; but the "principle of inheritance" is strong enough that mutation rates cannot be that great. Darwin was so distressed by this problem that he ultimately conceded that the environment itself might induce hereditary variation — and so came to rank the inheritance of environmentally acquired characteristics as a potent evolutionary force.

But with Mendel's discovery that genes are particles, this problem disappeared. The pink-flowered plant produced by red- and white-flowered parents does not pass on pink-determining factors. Instead red- and white-determining alleles combine anew to yield red- and white-flowered plants. The same principle holds for the many particulate factors that contribute to such continuously varying characteristics as height (*Figure 1*). Thus variation, once it arises, persists.

This persistence of variation renders unnecessary the theory of the

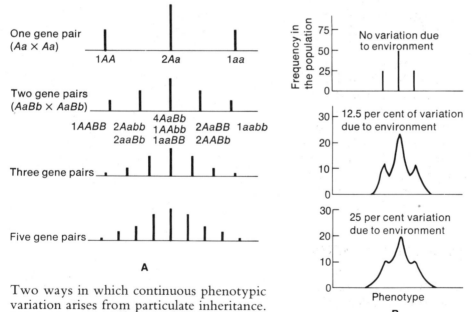

1 Two ways in which continuous phenotypic variation arises from particulate inheritance. (A) As the number of segregating loci, each with a small effect on the phenotype, increases, the number of phenotypic classes increases. (B) Superimposed on this may be a greater or lesser degree of phenotypic variation among the individuals of any given genotype, due to their responses to different environmental conditions. (A after Strickberger 1968; B after Allard 1960)

inheritance of acquired characteristics, now generally recognized to be contradicted by fact. That the inheritance of acquired characteristics should play an important role in evolution is implausible, for if the environmental conditions of the parents in one generation affect the characteristics of their offspring, the same should hold in the next generation. Thus, as the environment fluctuates, there would be no net change. Moreover a system of heredity so subject to environmental influences would not persist very long because it would be as likely to transmit deleterious influences as beneficial ones.

I must, however, insert a caveat. The experience of the parent can influence some characteristics of the offspring in some cases, and a few of these cases are quite puzzling. Because an individual inherits not only chromosomal DNA but also cytoplasm, especially in the egg, alterations of the mother's cytoplasmic characteristics can be transmitted to her offspring. Such CYTOPLASMIC INHERITANCE (reviewed by Jinks 1964, Grun 1976) is commonly based on the transmission of self-replicating cytoplasmic bodies such as mitochondria, chloroplasts, and intracellular virus particles. In species like mammals that nourish their young, variation in the mother's nutritional state, hormones, and antibodies can affect the offspring, causing "maternal effects" that usually persist for only one generation.

A few experimental cases do not seem to entail maternal effects or cytoplasmic inheritance. Durrant (1962) and Hill (1967), working on flax and tobacco respectively, grew plants under varying fertilizer treatments and found that the level of fertilizer affected plant weight (in flax) and height and flowering time (in tobacco). For several subsequent generations they grew offspring from equal numbers of seeds from every parent plant in homogeneous fertilizer conditions. The phenotypic differences persisted strongly in the plants' descendants for at least three generations! Durrant eliminated maternal effects and cytoplasmic inheritance as possible causes and could only suggest that the activity of certain genes had been altered, as it is in different cell types in a developing organism. The unusual twist was that the alteration of gene activity seems to have persisted through meiosis. Such transmission of acquired characteristics appears rare, however.

REPLICATION AND ALTERATION OF THE GENETIC MATERIAL

Heredity exists because the hereditary units are copied from one generation to the next; genetic variation exists because the copying process is not perfect. Copying, or self-replication, occurs on a template. There are several kinds of template in biological systems; for example, cell membranes act as templates to some extent. The template of major interest is DNA, which consists of a double strand made up of four kinds of nucleotide; each nucleotide in one strand is paired with a specific complementary nucleotide in the other: adenine (A) with thymine (T) and cytosine (C) with guanine (G). During DNA replication, the two strands separate and nucleotides are assembled into two complementary strands, forming two double strands identical to the orig-

inal molecule. This precision of self-replication is the essence of heredity.

But mistakes can occur; for example, one kind of nucleotide may
become substituted for another. Two DNA sequences that differ by
one or more substitutions are termed ALLELES. Nucleotide substitution,
one of the events that constitute mutation, is a rare event, probably
occurring at each nucleotide site only about once per 10^7 replications
(Watson 1976). The probability of such mistakes is lessened by the
repairing action of several enzymes, which is one reason for viewing
mutations as unavoidable mistakes rather than adaptive events. Once
a mutation has occurred, a second mutation at the same site is so
unlikely that the mistake will probably be propagated as long as the
DNA molecule continues to be replicated. Thus variation persists. The
range of possible variations is enormous. A section of DNA 1500
nucleotide pairs long has 4^{1500} possible sequences of base pairs, many
more than the number of sequences that have existed in the history of
life.

Much of the DNA in the genome is organized into functionally
independent sequences, each called a GENE (or CISTRON). Many of these
sequences are programs for the production of POLYPEPTIDE molecules,
which singly or in aggregate constitute enzymes and other proteins,
made of linear sequences of AMINO ACIDS. Each amino acid is specified
by a triplet of DNA nucleotides, so a polypeptide 500 amino acids
long is programmed by a gene 1500 nucleotide pairs long.

PROTEIN SYNTHESIS AND THE GENETIC CODE

The DNA is not directly translated into polypeptides, and much of it
does not produce structural proteins. It is first transcribed into RNA,
which consists of a single strand composed of A, C, G, and U (uracil)
rather than T. The RNA formed on the DNA template is thus a kind
of mirror image of one of the DNA strands. There are three kinds of
RNA. Ribosomal RNA (rRNA) combines with protein to form ri-
bosomes, the structures on which protein synthesis is organized.
Transfer RNA (tRNA) is coupled by a specific activating enzyme to
one of the 20 kinds of amino acid that are commonly joined together
into proteins; thus there are about 20 specific kinds of tRNA. Messenger RNA (mRNA) contains in its nucleotide sequence the information,
copied from the DNA cistron from which it was transcribed, for a
specific sequence of amino acids, the polypeptide for which that cistron
codes. The polypeptide may or may not be assembled with other such
polypeptides, with the same or different amino acid sequences, to form
a polymer of two or more polypeptide chains. Some of the proteins
thus formed, like collagen and myosin, are constituents of cells and
tissues; some, like insulin, are hormones; many are enzymes that are
almost entirely responsible for the chemical composition and biochemical activities of the organism. Probably at least 600 to 800 different
enzymes are present at any time in a single bacterial cell, and the DNA
content of such a cell probably codes for about 2000 to 4000 different
polypeptides (Watson 1976).

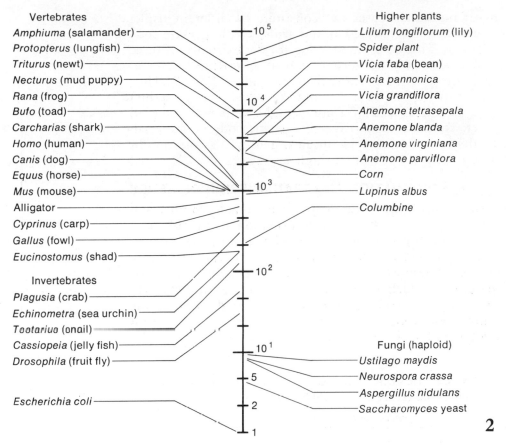

Vertebrates

- Amphiuma (salamander)
- Protopterus (lungfish)
- Triturus (newt)
- Necturus (mud puppy)
- Rana (frog)
- Bufo (toad)
- Carcharias (shark)
- Homo (human)
- Canis (dog)
- Equus (horse)
- Mus (mouse)
- Alligator
- Cyprinus (carp)
- Gallus (fowl)
- Eucinostomus (shad)

Invertebrates

- Plagusia (crab)
- Echinometra (sea urchin)
- Teetariue (snail)
- Cassiopeia (jelly fish)
- Drosophila (fruit fly)

- Escherichia coli

Higher plants

- Lilium longiflorum (lily)
- Spider plant
- Vicia faba (bean)
- Vicia pannonica
- Vicia grandiflora
- Anemone tetrasepala
- Anemone blanda
- Anemone virginiana
- Anemone parviflora
- Corn
- Lupinus albus
- Columbine

Fungi (haploid)

- Ustilago maydis
- Neurospora crassa
- Aspergillus nidulans
- Saccharomyces yeast

2

Approximate haploid DNA content per cell in various organisms, expressed as multiples of the amount in *Escherichia coli* (4×10^{-12} mg). (After Holliday 1970)

The amount of genetic information in even a "simple" organism is thus considerable, and it is even greater in the more "complex" organisms. The DNA of mammals, about 800 times as much as in *Escherichia coli,* could code for more than 2 million different proteins. But some amphibians have more than 25 times as much DNA as mammals (*Figure 2*). It is hard to imagine that they are 25 times as biochemically complex, so it seems likely that much of the DNA in most organisms is not translated into protein and may be doing little except replicating itself. Some of the DNA is repetitious; for example, there are as many as 1000 identical copies of the *r*RNA genes, and the genes for each kind of histone protein in the chromosomes of eukaryotes are reiterated 250–500 times. But most structural genes, like those coding for the polypeptides that make up hemoglobin, are present only as single copies per gamete. Each of the bands visible in the giant salivary chromosomes of *Drosophila* appears to correspond to a single gene locus; since there are about 5000 bands, there may be about that many genes or, at most, twice that many. This leaves a lot of the DNA unexplained, and it may well be nonfunctional.

Four kinds of nucleotide bases can constitute 64 different triplet combinations, yet there are only 20 amino acids, so the genetic code must be DEGENERATE. That is, even allowing for the existence of three triplets (codons) that specify the termination of an *m*RNA message, most amino acids are specified by more than one codon (*Table I*). Thus a mutation (e.g., from UCU to UCC, both of which specify serine) may not change the amino acid of a polypeptide, so it may have no phenotypic effect. As far as is known, the genetic code is universal; triplets specify the same amino acids in all organisms, as one might expect if all living things stem from a single ancestor.

RECOMBINATION AND TRANSMISSION OF GENES

If every organism transmitted to its offspring its intact genome, and if every individual received its genetic information from only one parent, the array of genotypes would be identical from generation to generation, changing only by mutation. This is the case in exclusively asexual organisms. But most organisms engage in some degree of genetic exchange: sex, which forms new combinations of alleles. In haploid organisms like bacteria which have only a single copy of the genome, recombination tends to be imprecise; in bacterial conjugation one cell extrudes a variable length of its single piece of DNA into the

TABLE I. The genetic code. Amino acids specified by the 64 nucleotide triplets in *m*RNA

SECOND NUCLEOTIDE

FIRST NUCLEOTIDE	U	C	A	G	THIRD NUCLEOTIDE
U	UUU UUC } Phe / UUA UUG } Leu	UCU UCC UCA UCG } Ser	UAU UAC } Tyr / UAA Chain End / UAG Chain End	UGU UGC } Cys / UGA Chain End / UGG Trp	U C A G
C	CUU CUC CUA CUG } Leu	CCU CCC CCA CCG } Pro	CAU CAC } His / CAA CAG } Gln	CGU CGC CGA CGG } Arg	U C A G
A	AUU AUC } Ile / AUA / AUG Met	ACU ACC ACA ACG } Thr	AAU AAC } Asn / AAA AAG } Lys	AGU AGC } Ser / AGA AGG } Arg	U C A G
G	GUU GUC GUA GUG } Val	GCU GCC GCA GCG } Ala	GAU GAC } Asp / GAA GAG } Glu	GGU GGC GGA GGG } Gly	U C A G

Note: The names of the amino acids abbreviated in the table are: Ala, alanine; Arg, arginine; Asn, asparagine; Asp, aspartic acid; Cys, cysteine; Gly, glycine; Glu, glutamic acid; Gln, glutamine; His, histidine; Ile, isoleucine; Leu, leucine; Lys, lysine; Met, methionine; Phe, phenylalanine; Pro, proline; Ser, serine; Thr, threonine; Tyr, tyrosine; Trp, tryptophan; Val, valine. (From Ayala 1976)

other. But in the eukaryotes — organisms in which most of the DNA, along with histones and other proteins, is organized into discrete chromosomes within a nuclear membrane — sex is a highly organized process, so exquisitely arranged that its genetic consequences can be described with a mathematical precision that lends to genetics, including evolutionary genetics, a most elaborate, sophisticated mathematical theory.

Most eukaryotes, at least when they are producing gametes, are DIPLOID; they possess two copies of the entire genome (or at least most of it). Quite a few species, especially of plants, have higher ploidy (being tetraploid with four copies, octaploid with eight, and so on), but I will ignore them for the present. Sex in diploids consists of the recombination of two haploid nuclei into a single diploid zygote. Haploid sex cells are formed from diploid antecedents during meiosis; each pair of homologous chromosomes is paired in synapsis and then separated into the two haploid daughter cells, usually independently of the segregation of other such homologous pairs. For any pair of genes the probability that a given gamete will receive one gene rather than the other is usually one-half. (In a few cases one allele has a greater probability than another of being transmitted; such cases are often called MEIOTIC DRIVE.) Thus half the gametes of an AA' parent carry A and the other half carry A'. If this is also true of the other parent and if the gametes combine at random, the proportion of each genotype among the offspring is readily predicted. Thus in the mating $AA' \times AA'$ the probability that two A or two A' gametes will unite is $(\frac{1}{2})^2$; that of two A' gametes is the same, and that of a union between A and A' is $2 \times \frac{1}{2} \times \frac{1}{2}$, since the union of A egg with A' sperm and of A' egg with A sperm are independent events. Thus the genotypes AA, AA', and $A'A'$ appear among the progeny in the ratio 1:2:1.

This is just a special case of a union of gametes among which a proportion p carry A alleles and a proportion q $(= 1 - p)$ carry A' alleles (*Figure 3*). Suppose that among a large number of randomly mating individuals, half of each sex has genotype AA, and the other half AA'. Of all the eggs produced, 75 per cent would bear the A allele (p = proportion of A = 0.75), and 25 per cent A' (q = 0.25), and the same is true of the sperm. If eggs and sperm unite at random, the chance that a zygote will be formed from an A egg is p = 0.75, and the chance that it will incorporate an A sperm is also p = 0.75. So the probability that a homozygous AA zygote is formed is $p^2 = 0.75^2 = 0.5625$. Similarly, $A'A'$ zygotes are formed with probability $q^2 = 0.25^2 = 0.0625$, and heterozygotes (AA') with probability $2\,pq = 2 \times 0.75 \times 0.25 = 0.3750$. Thus 56.25 per cent of the offspring will be AA, 37.50 per cent will be AA', and 6.25 per cent will be $A'A'$. This is an instance of the HARDY-WEINBERG LAW (*Chapter 10*), which underlies all of evolutionary genetics.

As long as two gene loci (the sites that particular genes occupy) lie on nonhomologous chromosomes and so segregate independently in meiosis, the same principles can be brought to bear very simply (*Box*

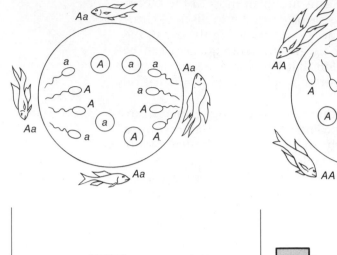

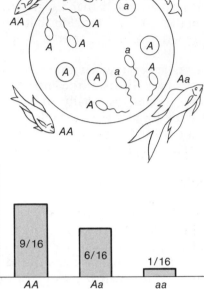

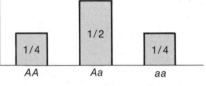

3

If gametes unite at random in the gene pool, the proportion of each genotype formed is readily predicted, on average, whether the alleles are equally frequent as in a Mendelian cross of F_1's (left) or are different in frequency (right).

A). The calculations are only slightly more complicated when loci lie on the same chromosome. They would segregate together as a unit, except that during meiotic synapsis each homologous chromosome is replicated into two daughter chromosomes (chromatids) still joined at the centromere and the strands of the two homologues commonly break and rejoin in the process of CROSSING-OVER. It is a marvel that the homologous chromosomes manage to line up precisely, gene for gene, so that the exchange is almost always exactly reciprocal. When it is not, when there is unequal crossing-over, one chromosome loses a bit of material (a deficiency) and part of the other becomes duplicated (*Figure 4*). This rare process is important in evolution for augmenting the length of chromosomes and the amount of DNA and for forming repeated copies of certain genes. The frequency R of crossing-over between two gene loci is often roughly proportional to the distance between them on the chromosome, but often it is not. Crossing-over is regulated by a lot of factors that we know little about. In males of most species of flies, for example, homologous chromosomes line up in synapsis, but there is no crossing-over.

Recombination forms an array of genotypes among progeny that is different from that of their parents but nonetheless statistically predictable. The progeny of parents with identical but highly heterozy-

Genotype Frequencies for Two Independently Segregating Loci

Let half of each sex have the genotype $AABB'$ and half $AA'BB'$. Four kinds of gamete, AB, AB', $A'B$, and $A'B'$, are formed, with frequencies g_{AB}, $g_{AB'}$, $g_{A'B}$, and $g_{A'B'}$ respectively. In this instance $g_{AB} = (\frac{1}{2})(\frac{1}{2}) + (\frac{1}{2})(\frac{1}{4}) = \frac{3}{8}$, since the probability is $\frac{1}{2}$ that the gamete will come from an $AABB'$ individual, and $\frac{1}{2}$ that it will come from an $AA'BB'$ individual; and $\frac{1}{2}$ of $AABB'$'s gametes are AB, while only $\frac{1}{4}$ of $AA'BB'$'s gametes are AB. Similarly $g_{AB'} = (\frac{1}{2})(\frac{1}{2}) + (\frac{1}{2})(\frac{1}{4}) = \frac{3}{8}$, $g_{A'B} = (\frac{1}{2})(0) + (\frac{1}{2})(\frac{1}{4}) = \frac{1}{8}$, and $g_{A'B'} = (\frac{1}{2})(0) + (\frac{1}{2})(\frac{1}{4}) = \frac{1}{8}$.

The proportion of each genotype among the progeny is then easily found if the gametes unite at random. For example, the frequency of $AABB$ is $(g_{AB})^2 = (\frac{3}{8})^2 = \frac{9}{64}$. That of $AABB'$ is $2(g_{AB})(g_{AB'}) = 2(\frac{3}{8})(\frac{3}{8}) = \frac{18}{64}$ since either egg or sperm can carry either haploid genotype. The frequency of $AA'BB' = 2(g_{AB})(g_{A'B'}) + 2(g_{AB'})(g_{A'B}) = \frac{12}{64}$ since a double heterozygote can be formed by the union of either AB with $A'B'$ or AB' with $A'B$.

In general, if the frequencies of A and A' are p and q ($= 1 - p$), respectively, and those of B and B' are r and s ($= 1 - r$), respectively, the gamete frequencies are $g_{AB} = pr$, $g_{AB'} = ps$, $g_{A'B} = qr$, $g_{A'B'} = qs$.

gous genotypes are immensely more variable in genotype and often in phenotype. The cross $AA'BB'CC'DD' \times AA'BB'CC'DD'$ yields $3^4 = 81$ different genotypes. Conversely the progeny of a cross between greatly disparate homozygous genotypes are often more homogeneous ($AABBCCDD \times A'A'B'B'C'C'D'D'$ yields $AA'BB'CC'DD'$ exclusively), and they are often phenotypically intermediate. Recombination is both a diversifying force that gives rise to immense numbers of new genotypes and a homogenizing force that prevents an extreme genotype from seeing its own unique identity mirrored by its offspring.

Equal crossing-over   Unequal crossing-over   **4**

Ordinarily, as at left, crossing-over is reciprocal. But in unequal crossing-over, at right, a segment of one chromosome, marked by locus B, is transferred to the other chromosome. Thus one chromosome suffers a deficiency, which if large causes inviability. The other chromosome bears a tandem duplication for one or more loci such as B. Such a duplication may persist and is sometimes advantageous.

THE EXPRESSION OF THE GENOTYPE: THE PHENOTYPE

The genome, the aggregate of genes, of every cell is replicated before cell division; mitosis, like meiosis, is so orderly that the full genome is inherited intact by each daughter cell. Yet enzymes and proteins differ from cell to cell in a multicellular organism and from time to time in every organism. Clearly, then, much of the genetic information in a nucleus is not expressed at any given time; it is in a REPRESSED state and is not being transcribed into *m*RNA. Similarly the organism at each stage of ontogeny carries in repressed state the genetic program for other developmental stages, and each sex carries much of the genetic program that is expressed only in the other sex. The phenotypic differences between the sexes are not necessarily due to genes on the sex chromosomes.

If we are to understand the formation of the phenotype, we need to know at the very least how genes are brought out of their repressed states at specific times in specific cells. This mechanism is largely a mystery and is partly understood only in bacteria. The synthesis of certain enzymes, such as β-galactosidase which hydrolyzes lactose into galactose and glucose, is greatly increased, or induced, when lactose enters the cell. Other enzymes are repressible rather than inducible. The synthesis of the enzymes that catalyze the production of certain amino acids proceeds until the amino acid concentration rises to a high level, and then the production of enzyme is repressed.

The β-galactosidase system illustrates one mechanism of GENE REGULATION. This enzyme is produced in concert with two other enzymes. At least one of these, a permease that enhances the flow of lactose into the cell, is functionally related to the activity of β-galactosidase. The three enzymes are produced together by translation of a single, polycistronic messenger RNA transcribed from adjacent loci on the bacterial chromosome. Most of the time these loci are not active because a repressor protein is bound to a short segment of the β-galactosidase DNA, termed the OPERATOR. This protein is produced by a repressor locus that lies elsewhere on the chromosome, disassociated from the β-galactosidase complex, which is termed an OPERON. Lactose induces β-galactosidase synthesis by binding with the repressor protein, thereby dissociating it from the operator and leaving the operon free to begin *m*RNA transcription (*Figure 5*). In repressible systems the opposite happens. A repressor molecule, such as the amino acid that is the end product of the biosynthetic pathway in which the enzyme is involved, binds with an inactive repressor protein to make it capable of binding to the operator, thus turning off *m*RNA transcription.

Thus regulation of gene activity depends on specific substances in the cell's environment, substances derived either from the outside world or from the activity of other genes. Little is known about the regulation of gene activity in eukaryotic cells. In some cases the inducing substances that derepress certain genes are controlled by hormones or are themselves hormones. In principle, then, it is possible to see how differentiation could proceed by the derepression of different

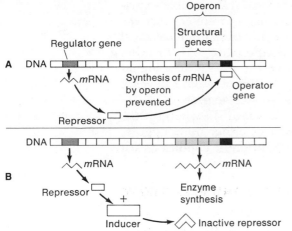

5

Model of gene regulation in bacteria. (A) The structural genes in the operon are inactive, since the operator gene is bound by a repressor. (B) An inducer substance from outside the cell binds to the repressor, derepressing the operator. Transcription of *m*RNA at the structural genes associated with the operator then occurs. (After Strickberger 1968)

genes in various cells early in development, alteration of the tissues' chemical environment by the enzymes and hormones then produced, further derepression of other genes in reaction to the difference in chemical milieu that has been built up, and so on in hierarchical fashion.

Ontogeny entails not only cell differentiation, but also growth. Growth involves an increase in mass, which depends on the enzymes (therefore on specific gene activity) that metabolize nutrients into cell constituents, and usually on cell division as well. Hormones are among the factors that regulate cell growth and differentiation, but why different cells vary in their response to these factors and differ in their rates of proliferation is little understood. Probably regulation of the activity of genes that control cell division is involved.

DOMINANCE AND OVERDOMINANCE

An understanding of development must lie largely in an unraveling of the processes that control gene activity. The effect of a structural gene (one that produces a protein) on the phenotype depends on when it is turned on, for how long, how much of its protein product is made, and what the activity of this product is, once synthesized. We might imagine, for example, that if an enzyme converts substance x into pigment y, then the amount of enzyme produced could affect the amount of y and so the degree of pigmentation. If excess x is present, pigmentation could be proportional to the amount of enzyme present, and an individual with two copies of the enzyme locus could be more highly pigmented than one with only a single copy. Then the pigmentation of the genotype Aa would be intermediate between aa and AA; neither allele is dominant. But if even a single A allele produced enough enzyme to convert all of a limited supply of x into y, genotypes Aa and AA would have the same phenotype, and allele A is said to be

DOMINANT over the recessive *a*. (This is one of several possible explanations of dominance; see Chapter 15.) Imagine, moreover, that the
enzymes provided by alleles *A* and *a* are both functional but are most
active under slightly different conditions of temperature, pH, or other
variables. The heterozygote could then convert *x* into *y* at a higher
overall rate than either homozygote would, and so it would develop
a more extreme phenotype. This is a possible explanation of OVER-
DOMINANCE (similar to HETEROSIS, or "hybrid vigor"), whereby the
phenotype of the heterozygote transcends that of either homozygote.
That variant forms of the same enzyme have different conditions for
optimal activity is well established.

POLYGENES AND PLEIOTROPY

If the phenotype of the heterozygote is precisely intermediate between
that of two homozygotes, there is no dominance and the alleles have
ADDITIVE effects. More than one biochemical pathway can produce a
precursor substance *x,* and many enzymes can act in sequence or in
parallel to convert *x* to *y*; thus many gene loci can contribute to the
production of a single characteristic. It is common for each gene to
contribute to the phenotype independently of other genes, so that the
phenotype is the sum of the contributions of the individual loci. Such
additivity among loci is common for body size, bristle number in flies,
and many other POLYGENIC characteristics.

But pathways are so complex that it is just as common for the
effect of one locus on the phenotype to depend on the genotype at one
or more other loci, so that the whole is not equal to the sum of its
parts. This effect, known as EPISTASIS, can have many bases (*Figure 6*).
For example, the *white* locus in *Drosophila* causes an incapacity to form
the precursor to the eye pigments formed subsequently by enzymes
produced at other loci. "Suppressor alleles" at one locus can nullify
the expression of mutations at other loci. For example, a *td* mutant in
Neurospora produces an inactive enzyme incapable of synthesizing tryptophan. The suppressor mutation *su-td* at another locus restores tryp-
tophan-synthesizing ability, perhaps by lowering the intracellular concentration of Zn^{++} or other ions to which the mutant enzyme seems
sensitive (Wagner and Mitchell 1964). Moreover different loci can
produce the same end product, so activity at one locus can compensate
for inactivity at another if both enzymes are simultaneously regulated
by the same end product. Many biochemical bases of epistasis are
possible, and I have not even mentioned the possibilities that lie above
the strictly biochemical level, in developmental patterns.

Almost every characteristic of an organism is polygenic; it depends
on the action of many genes. Even the activity of an enzyme molecule
whose structure has not been altered by mutation can be changed by
selection, which changes the genetic and hence chemical milieu in
which the enzyme acts. Thus the gene's effect on the phenotype and
consequently its adaptive value depend greatly on other elements in
the genome. The genotype as a whole constitutes a GENETIC ENVIRON-

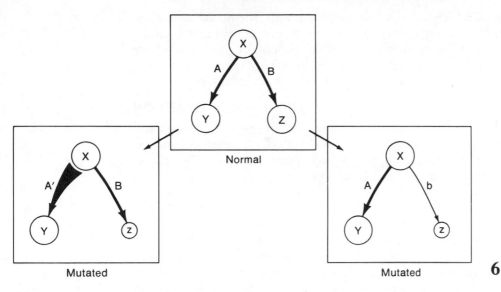

Normal

Mutated   Mutated   **6**

If the enzymes A and B compete for a limited amount of precursor X, a mutation at one locus that produces hyperactive A enzyme may increase product Y at the expense of product Z; a mutation at the other locus that produces an inactive enzyme B may have the same effect. Either an excess of Y, or a deficiency of Z, might be deleterious — or advantageous. This is one of several ways in which loci may interact epistatically.

MENT in which each allele functions; as such it can determine the prospects for survival of the organism that carries it.

Conversely most genes are PLEIOTROPIC; they have more than one measurable effect on the phenotype. The simplest level of pleiotropy is illustrated by a mutation of a gene that codes for transfer RNA. It alters the amino acid composition of every protein and is likely to be lethal. At the next level a polypeptide which in monomeric form acts as an alanine dehydrogenase performs the activity of glutamic dehydrogenase when aggregated into a polymer; a mutation in one polypeptide affects two reactions. Any enzyme that produces a precursor for two or more biochemical pathways can, if altered, affect all those pathways. A given substance x can react with different materials in different cells to yield different products. Thus alteration of the enzyme that produces x has multiple effects. Conversely the failure to produce a material that is a necessary ingredient of many tissues alters all those tissues. The *achondroplasia* mutant in the rat causes an inability to suckle, faulty pulmonary circulation, occlusion of the incisors, and arrested development, all stemming from an organism-wide abnormal development of cartilage early in ontogeny.

The developmental interactions by which pleiotropy is effected are often most obscure (thereby proving how little we understand development). It is hard to see how the mutation *su-pr,* which suppresses the *purple* eye mutation in *Drosophila,* could also enhance the expression of *Hairy wing,* which causes excess bristles to form. In any case, pleiotropy like epistasis reflects the staggeringly complicated, interconnected nature of biochemistry and development. A mutation whose most obvious phenotypic effect may seem advantageous may have disruptive effects on development that render it maladaptive.

THE NORM OF REACTION

Because genes exert their effects on the phenotype via biochemical reactions, they do not operate *in vacuo.* Their effects must depend on the chemical and physical milieu in which these reactions take place. Because this milieu varies, the phenotypic expression of a gene also varies. This milieu can be altered by variation in the genetic environment or sometimes in precisely the same ways by variation in the external environment: temperature, nutrition, and the like. For example, altered phenotypes called phenocopies that are exactly similar to the altered wing veins caused by such mutants as *crossveinless* and *Curly* can be produced in "normal" flies by a temperature shock during critical periods of development. Of course, an organism is not just its genes; it is constructed of materials derived from the environment, under environmental conditions that strongly influence the rate of biochemical reactions. If we could produce many individuals with exactly the same genotype and raise them under an enormous range of environmental conditions, we would find that every gene varies to a greater or lesser degree in its phenotypic expression. Some characteristics, like body weight in animals or growth form in plants, would vary more than others, such as the number of vertebrae in a mammal or the structure of a cell membrane. The more invariant characteristics are said to be more highly CANALIZED, or DEVELOPMENTALLY BUFFERED, into a more restricted set of developmental channels.

Different genotypes are canalized to different extents, so that certain alleles are more variable in penetrance (the fraction of individuals in which their phenotypic effect is manifest) and in expressivity (the magnitude of the phenotypic effect) than others. Each genotype has its own NORM OF REACTION, a variety of expressions under different environmental conditions (*Figure 3, Chapter 2*). In some cases a wide reaction norm, the production of greatly different phenotypes under different environmental conditions, is disadvantageous or pathological. Our tendency to develop scurvy when deprived of vitamin C is not advantageous. But very often, especially if the environment fluctuates greatly, the most advantageous genotype is the one whose phenotypic expression varies with prevailing conditions. Obvious examples are the ability of a weasel to develop pigmented fur in summer and white fur in winter and the capacity of many plants to produce tough, heavily waxed sun leaves and more delicate shade leaves. Similarly, some behavioral traits are more readily altered by interaction with the environment than others, and more readily in some species than in others. This capacity for alteration of behavior, which we call learning, is immensely greater in human genotypes than elsewhere — so great that we do not know its limits.

Because each aspect of the phenotype is a product not of the genes alone or the environment alone, but of the interaction between the two, it is fallacious to say that the origin of a characteristic is "genetic" or "environmental." We can only ask whether the *differences* among individuals are attributable more to genetic differences or to environmental differences. The answer is likely to depend on the particular group of individuals we examine. If they are genetically homogeneous, much of the variation is environmental; if they have all been exposed to similar, homogeneous environmental conditions, more of the variation will be genetically based. The amount of variation stemming from each of these factors will differ from one character to another. These principles may seem obvious, but they are often ignored. For example, innumerable arguments have dealt with the question of whether aggressive behavior or sex role stereotypy is "innate" in humans — whether, as it is often phrased, these behaviors are "genetically determined." But this question is meaningless. These behaviors, like all characteristics, are both genetic and environmental; both genes and environment are prerequisite to their existence. We can meaningfully ask only whether the behavior varies more as a function of genetic or environmental variation, and to what degree.

DEVELOPMENTAL AND FUNCTIONAL INTEGRATION

In the developing embryo of *Drosophila* the ectoderm differentiates into many tissues, including the central nervous system, the lining of the foregut and hindgut, the tracheae and Malpighian tubules, and the hypodermis that underlies the cuticle. Probably by producing a chemical inducer, the hypodermis evokes the differentiation of the mesoderm into heart, muscles, fat bodies, and other tissues. In the mutant *Notch-8* the central nervous sytem develops to thrice its normal size; the ventral hypodermis, the lining of the foregut, the heart, the pharyngeal muscles, and the fat bodies fail to differentiate. Even derivatives of the endoderm are affected. The anterior and posterior rudiments of the midgut do not join.

The most probable interpretation of this syndrome is that the normal allele at the *Notch* locus produces a substance that inhibits the differentiation of much of the ectoderm into nervous tissue. Because the inhibitor is absent in the *Notch-8* mutant, tissue that normally develops into hindgut, tracheae, and so forth becomes part of the nervous system, which becomes so large that it prevents the midgut rudiments from joining. And because part of the ectoderm has differentiated into nervous tissue rather than hypodermis, the induction of mesodermal tissues fails to take place (T. Wright 1970).

Such stories, the stuff of developmental biology, illustrate that the development of an organism is an integrated process in which the formation of each structure or chemical constituent depends not only on a linear sequence of steps, but on cross-influences among many such sequences. The first and most critical test of a mutation's likelihood of survival is whether it disrupts the coordinated, integrated pattern of development. In Figure 7 the pathways leading to products

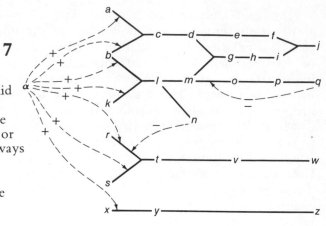

7

A hypothetical scheme of some pathways leading to end products *j, q, w,* and *z.* Solid lines indicate steps catalyzed by enzymes; broken arrows indicate positive or negative influences of a material on the production or rate of reaction of an enzyme. Some pathways (e.g., those leading to *j* and *q*) are tightly integrated; others (e.g., that to *z*) are relatively independent, uncoupled from the rest of the system.

j and *q* are tightly integrated. A mutation of the enzyme governing step *l-m* affects both pathways, and the advantage of an increase of *q* may be nullified by the disadvantageous hypertrophy of *j.* Thus the earlier a gene (such as the one governing step *k-l*) acts in a developmental pathway, the more likely a mutation is to disrupt the integrated pattern, and so to be deleterious. But some genes can act early in development, yet govern several pathways in a balanced fashion. A gene producing a hormone (e.g. *α*) can alter many pathways in concert, so that the ratios of the products (*j, q, w*) remain balanced and development proceeds harmoniously. For example, the balance between the number of X chromosomes and autosomes switches the development of an insect from one balanced mode, leading to one sex, into another. The titer of juvenile hormone determines whether a young termite develops into the worker or the soldier caste (Lebrun 1961). The many morphological and behavioral differences between these develop in concert, illustrating that multiple harmonious changes in many characteristics can follow from single slight alterations of controlling factors.

In general, mutation of a gene that acts late in a developmental pathway may be less likely to disrupt development, because it affects only a small subsystem of the larger developmental whole. The developmental system as a whole consists of subsystems; mutations in subsystems (pathway *x–z*) that are only loosely linked to one another are less likely to be harmful than mutations in tightly integrated subsystems.

Moreover, within a developmental subsystem the effect of a mutation depends on how the locus fits into the homeostatic regulation of the entire pathway. Mutations that govern rate-limiting steps (*m–o*) are more likely to have significant phenotypic effects than mutations in biochemical steps (*o–p* or *p–q*) that do not regulate the rate of production of the end product.

Integration of the organism is functional as well as developmental.

In their fully developed form functionally related structures must act harmoniously. This is the second test of the adaptive value or disadvantage of a genetic change. An increase in body size is maladaptive unless compensatory changes in the surface area of gills, lungs, intestines, and other organs maintain a constant ratio between the mass of metabolizing tissue and the area over which gases, nutrients, and wastes are exchanged.

Thus constraints on the adaptedness of new genotypes are imposed by the requirements of developmental and functional coadaptation. Only when a genotype does not violate these constraints does it have a chance of increasing the organism's ability to cope with the physical and biological contingencies of the external environment. The more tightly integrated the subsystems of development are, and the more exacting the functional interdependence is among structures, the more likely it is that a genetic change will be deleterious. Unfortunately our knowledge of development is so abysmally insufficient that we can seldom predict the variety of phenotypic effects of any given mutation, how complex the genetic control of any characteristic is, how developmentally difficult it would be to achieve a desirable change in the phenotype, or what the pleiotropic consequences of such a change would be. We are far from understanding which paths of evolutionary change are available to a species at any point in its evolutionary history and which paths are closed because of the complexity of genetic and developmental change that would be required. The mysteries of development hold the keys to a full understanding of evolution.

SUMMARY

The phenotype of an organism is the consequence, in a succession of developmental events from conception to death, of the interplay between its genotype and its environment. The resemblances among relatives, and indeed among all the members of a species or of related species, stem in large part from their possession of genes in common — particles of DNA, located mostly in the chromosomes. Because the units of heredity are particles that replicate themselves in their passage from cell to cell and from generation to generation, once genetic variation arises, it persists. Variations in the genetic program arise by errors in replication; variations acquired by an organism during its lifetime through interacting with its environment are not inherited. Because each genotype has a norm of reaction, a variety of possible phenotypes formed under different environmental conditions, phenotypic variation has both genetic and environmental causes. The mechanisms of development, whereby genotype and environment interact to form the phenotype, are important to evolutionary studies, but they are not well understood.

FOR DISCUSSION AND THOUGHT

1 What evidence is there, besides the universality of the genetic code, that all life is monophyletic (has a single origin)?

2 Arthur Koestler, in *The Case of the Midwife Toad,* recounts the history of experiments purporting to show that acquired characteristics can be inherited. Read and evaluate these experiments, and discuss the evidence that would be needed to judge whether transmission of acquired traits occurs. (See also Chapter 15.)

3 Discuss the general principles that would enable you to predict which characteristics of an organism should be phenotypically labile (like β-galactosidase production in *E. coli*) and which should be strongly canalized (like vertebral number in mammals).

4 Evidence is developing that some genes in the genome are not discrete, but overlapping, the way *live* and *very* overlap in the word *livery.* If this is common, what are the consequences of mutation?

5 Discuss possible functions for the great amount of DNA that is not translated into polypeptides. Does it necessarily have a function?

6 Discuss the problems posed by meiosis and by the machinery of protein synthesis. How could these have evolved?

7 Is the DNA code imposed on organisms as their necessary mechanism of heredity, or has the genetic code itself evolved? Might there have been other molecular mechanisms of heredity or other meanings for nucleotide triplets? Is DNA the best possible basis of heredity, or is it quite arbitrary that this is the one that most organisms possess?

8 Discuss the developmental and evolutionary implications of dosage compensation, by which a single gene copy (say, of an allele on an X chromosome in a male *Drosophila*) produces the same phenotype as a gene pair (say, two such alleles in the female).

9 Referring to developmental systems such as those hypothesized in Figure 7, discuss how a complex of developmental pathways might best be constructed to maximize the chances of developing the proper phenotype in the face of the mutational and environmental perturbations to which development is inevitably subjected.

MAJOR REFERENCES

Herskowitz, I. H. 1977. *Principles of genetics.* 2nd ed. Macmillan, New York. 836 pages. Any of various genetics textbooks provide sufficient background for the material in this chapter; this is one of the most comprehensive.

Waddington, C. H. 1956. *Principles of embryology.* Allen and Unwin, London. 510 pages. Out of date in many ways, but still a thought-provoking classic that treats the relation between development and evolution.

Many features of organisms are adaptations to their environment. Indeed much of biology, whether it be biochemistry, anatomy, physiology, or ecology, consists of the study of adaptations, the ways in which organisms manage to survive and reproduce in the face of the innumerable contingencies that beset them. But what is the "environment" to which an organism is adapted?

We commonly think of the environment as the potpourri of obvious events external to the organism that affect its physiological state, its survival, or its reproduction: climate, soil, water, and so forth. Clearly it must also include other species: predators, pathogens, and prey, for example. But the other members of an individual's species are also important features of its environment, clearly so in social animals but no less true of less social species that must find mates and cope with competing individuals. Environment, then, is an all-embracing word for the totality of the factors that influence the activities, achievements, and ultimate fate of an animal or plant. My aim in this chapter is to review those aspects of ecology — the study of interactions between organisms and their environment — that most pertain to evolution; to characterize, if possible, the concept of environment.

THE ECOLOGICAL NICHE

Each of a great many factors extrinsic to a species affects its ability to survive and reproduce and thus in part determines whether it can persist in a particular locality. For example, a clam might tolerate a certain span of temperatures and feed on plankton in a specific size range. Hutchinson (1957) popularized the notion of drawing two such variables as the axes of a two-dimensional graph in which a single point represents a locality with a particular temperature and a particular prey size. Part of this two-dimensional space then represents the possible environments in which the species can persist (*Figure 1a*).

 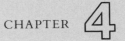

The Ecological Context of Evolutionary Change

We may add a third axis representing salinity, thus defining a three-dimensional space (*Figure 1b*) containing a region representing tolerable combinations of temperature, prey size and salinity. To take into account other environmental factors, we would have to draw many more axes; although we cannot do this, we can more or less conceive of doing it. We thus have an *n*-dimensional space, with one axis for each of *n* environmental factors. Within this space is a region consisting of a cloud of points, each representing a particular combination of temperature, prey size, salinity, copper concentration, starfish abundance, and so forth that together constitute environments conducive to the survival and reproduction of the clam population. Hutchinson calls this region, consisting of the set of possible environments in which the species can persist, the fundamental ECOLOGICAL NICHE of the species.

A species may have a narrow niche with respect to some environmental factors but a much broader tolerance of other factors; it is in some respects SPECIALIZED and in other respects GENERALIZED. The interaction of environmental factors may affect a species' ability to survive; a desert plant can withstand higher temperatures if water is plentiful than if the soil is very dry. And one genotype may have a somewhat different ecological niche from another, having different tolerance limits; and therein lies much of the basis of evolutionary change. In some contexts it is useful to focus only on a single niche axis, as for example when we speak of the niche overlap of two species

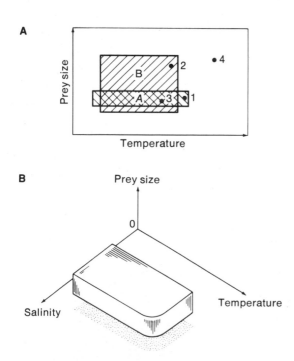

1

(A) Two dimensions of the ecological niche of each of two species, A and B; perhaps they are estuarine bivalves, each restricted to a range of temperatures and of particle sizes they can ingest. Each point in the space represents a possible environment, a combination of particle size and temperature. If a locality presents only environments in the neighborhood of point 3, the species compete intensely. If there are some warm microhabitats (near point 1) to which A is adapted and some large prey items (near point 2) as well as smaller ones (points 1, 3), the species should be able to coexist. If the only microhabitats are near point 4, neither species will occur. (B) Three dimensions of the niche of a species are represented. If a locality contains combinations of the three variables that lie within the solid figure (the species' niche), the species may persist. Notice that the combination of high temperature and high salinity is inimical to the species.

of seed-eating birds, referring only to their common use of a single variable such as the size of the seeds they eat.

ENVIRONMENTAL PATTERNS

Any one environmental factor, any one axis of the niche, has several different features to which an organism may have to respond. For instance, there is its average value. We commonly presume that some average conditions are "harsher," or more difficult to adapt to, than others; we suppose that the tundra and the deep sea are harsh environments, while tropical rain forests and coral reefs are ostensibly more benign. But life in a tropical forest would be as difficult for a caribou as life on the tundra for a spider monkey, and a deep-sea lanternfish would undoubtedly find a coral reef as hostile as a parrotfish would find the abyssal depths. Thus an environment cannot really be considered harsh for the organisms that persist in it. The larvae of the ephydrid fly that inhabit pools of crude oil seem to do quite well; oil is an uncongenial environment only for organisms that are not already adapted to it. Of course, achieving that adaptation is more likely for some species than for others. Indeed the factors that influence this likelihood are among the chief concerns of evolutionary biology.

The average state of the environment, however, is not the only thing that organisms adapt to. Almost all factors vary, and the pattern of variation requires patterns of adaptation different from those that suffice in a constant environment.

There are several important ways of describing the variation in an environmental factor. First, we must distinguish SPATIAL variation from TEMPORAL variation. The factors that affect white-tailed deer in Maine do not appreciably affect the survival of deer in Florida. Local populations, exposed to different states of the environment, may take very different evolutionary paths.

An environmental factor in any one locality may vary over time. The ABSOLUTE RANGE of the variable may be great or small, and it includes extremes that may be quite rare; a really cold winter once a century or an outbreak of a defoliating insect that ordinarily poses little threat to a tree species may have a dramatic impact on a population. Such unusual events must be part of the evolutionary experience of all species, yet we have no way of assessing their impact on evolution. Indeed we have only the slightest theoretical understanding of what impact an unusual, one-time catastrophe has on the genetic composition of a population, or how long such an event may leave its imprint on a species' genes (Lewontin 1966).

Organisms typically evolve adaptations to the usual range of variation, the variance. For example, estuarine species of crabs can acclimate to a wider range of salinities than the more exclusively marine species (*Figure 2*). Such PHYSIOLOGICAL ACCLIMATION is one of several ways in which the phenotype of an individual animal or plant may change during its life to meet the demands of a changing environment (Thoday 1953, Schmalhausen 1949, Levins 1968, Slobodkin 1968).

Turning on repressed gene loci (e.g., inducible enzymes in bacteria) and changes in behavior (including learning) are other ways of coping with environmental changes. Yet another alternative is to escape an environmental challenge by dormancy or by actively moving to a more tolerable environment.

As important as the magnitude of the environmental variation is its predictability (Slobodkin 1968). For an intertidal barnacle the succession of immersion and dessication is a highly predictable problem easily solved by keeping the shell closed when the tide is low. Many birds circumvent the predictable autumnal decline in insect abundance by migrating. But the exact time of winter's arrival is uncertain, so birds have taken advantage of the high correlation between the less predictable event and a more predictable event, the change in photoperiod, which is used as a cue to schedule migration. Some environmental events are still more unpredictable and require a very different adaptive response. Rainfall in many deserts is so undependable from one year to the next that most desert plants germinate and flower not in response to photoperiod, but in direct response to the availability of water.

Some patterns of environmental variation are impossible to adapt to fully. Arctic lemmings are the most famous case of populations that fluctuate dramatically in density (Elton 1942) and are an unpredictable resource for predators that depend on them. The strange life history of the cicadas that emerge once every 13 or 17 years may have evolved to escape predation, since they constitute a most unpredictable food supply for less long lived predators (Lloyd and Dybas 1966).

A most important aspect of environmental variation is its frequency. The effect of the frequency of oscillation depends on what aspect of the organism is affected and the "response time" of the

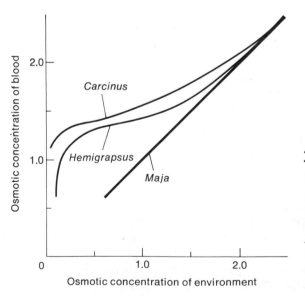

2

Differences in physiological homeostasis among species of crabs. The osmotic concentration of the blood conforms to that of the environment in *Maja*, a marine crab. *Hemigrapsus* and *Carcinus*, shore crabs that experience greater fluctuations in salinity, regulate their blood osmoconcentration. (After Prosser and Brown 1961)

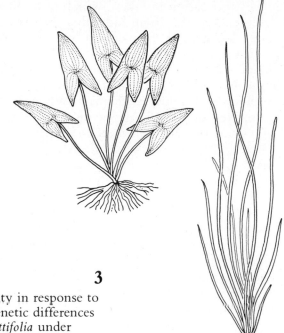

3

Adaptive phenotypic plasticity in response to
ecological conditions: nongenetic differences
in leaf form in *Sagittaria sagittifolia* under
terrestrial (left) and aquatic (right) conditions.

physiological function. For example, a lion, which because of its large
size has a relatively low metabolic rate and can store large quantities
of fat, may require a large meal only every few days; but if food is
available that infrequently to a shrew, it will quickly die because of its
far higher rate of metabolism. Thus an environment that is effectively
constant for a large cat is lethally inconstant for a small homeotherm.

MacArthur and Levins (1964) distinguish between environmental
variation in which the period of fluctuation is less than the life span of
the organism and that in which the period is greater. They term such
variation FINE-GRAINED and COARSE-GRAINED, respectively. Fine-grained
variation may be met by the homeostatic capacities I have discussed;
the organism changes its physiology or behavior but not irreversibly.

For an organism with a shorter generation time these same environmental changes may be coarse-grained. If the environment is likely
to be constant for the individual's lifetime, the organism may develop
one of several irreversible alternative phenotypes. Thus if the rotifer
Brachionus is exposed early in life to the chemicals that exude from a
predatory species of rotifer (*Asplanchna*), it develops long protective
spines (Gilbert and Waage 1967). In many semiaquatic plants the shape
of the mature leaves depends on whether they are submerged during
their development (*Figure 3*). Thus the distinction between fine-grained
and coarse-grained environmental variation holds for both spatial and
temporal heterogeneity. The variety of plant species in a forest is a

fine-grained mosaic to a leaf-eating monkey, which in its lifetime can sample all the available plants. But to an individual gall wasp, which undergoes its entire development in the plant on which it hatches, the forest is a coarse-grained environment.

The variety of adaptations to environmental change is impressive; yet there are environmental variations to which no simple physiological acclimation or developmental response is adequate. When the tolerance limits of individuals are exceeded, they may be unable to reproduce or they may die, and the population may fluctuate in numbers or become extinct.

It is sometimes hard for us to grasp how universal, pervasive, and ineluctable environmental change really is. The seemingly permanent and reliable features of our world are transitory when viewed in long enough perspective; even continents move. An immensely long history of environmental change has shaped the evolution of today's organisms, which continue to evolve in a world that is even now changing at a perhaps unprecedented rate. The environment changes on a physiological time scale, within the lifetime of individual organisms; on an ecological time scale, over which populations fluctuate in density while undergoing only slight genetic changes; and on a geological time scale, over which the genetic properties of species may change greatly.

ECOLOGICAL CHANGES: SUCCESSION

Even without such catastrophes as fire and flood, a species' environment is far from static. The fungi, fly larvae, and beetle larvae that develop in a fallen fruit occupy a most ephemeral resource. To reproduce, they must disperse widely to find other suitable resources that are distributed unpredictably in space or time. Hutchinson (1951) used the term FUGITIVE SPECIES to describe such forms, which persist only by continual dispersal from one place to another.

By extension the same term can describe the pioneer species, such as the dandelion, that colonize newly opened environments. Pioneer species of plants usually persist for one or a few generations before they are replaced by plant species that require for their germination and growth the organic matter, moisture, and shade provided by their predecessors. Analogous patterns of ecological succession occur in virtually all environments. A distinct pattern of microbial succession occurs in decaying vegetation, feces, and carrion; and sessile animals and plants undergo succession in the marine environment. Succession in different communities may have different causes (Connell and Slatyer 1977). Most species, then, have only a temporary tenure in any given locality, so they find the environment more or less inconstant.

The replacement of species by other species can form a complex pattern. A classical view is that the vegetation in a terrestrial community passes through a predictable, orderly sequence and arrives at a stable, self-perpetuating climax community, such as a beech–maple forest. But the ultimate species composition may actually be unpredictable, since the outcome of competition between species may depend on which gets there first (Levin 1974). So sites with similar physical characteristics may develop different associations of plants, and a species of bird that occupies climax forests may encounter an unpredictable variety of plants and insects from one place to another. This seems to be especially true of some tropical forests, where many species have inexplicably patchy distributions (Wilson 1958, Mac-Arthur 1972). Moreover, the climax community is not a static, homogeneous formation of vegetation that stretches unbroken for miles. Because patches are continually opened by winds or defoliating insects, providing gaps of light that are invaded by fugitive species (Horn 1975), a forest is a mosaic of patches in various stages of succession, with the abundances and spatial distributions of the component species changing kaleidoscopically over time.

POPULATION GROWTH

In any locality the density and persistence of the population of a species depends on its capacity to increase in numbers and on the factors that limit its abundance. An understanding of these population dynamics (treated at length by Slobodkin 1961, Wilson and Bossert 1971, Emlen 1973, and others) is essential to an appreciation of evolutionary theory.

The per capita rate of increase r of a population is the difference between the per capita birth rate b and the death rate δ at any time. Thus the rate of change dN/dt of population size N is

$$dN/dt = bN - \delta N$$

or, since $r = b - \delta$,

$$dN/dt = rN.$$

Then N_t, the population size at time t, depends on r and the initial size N_0:

$$N_t = N_0 e^{rt}.*$$

As long as the birth rate exceeds the death rate, the population will increase in an exponential fashion (*Figure 6*); if unchecked, as Darwin noted, the descendants of a single pair of any species would cover the earth in short order. The rate of increase r actually depends on the age structure of the population, for a population in which most members are at the height of their reproductive powers grows more rapidly than one made up largely of either juvenile or senile individuals. However, if the environment remains constant so that each age class has a specific birth rate and death rate that do not change, the population ultimately attains a stable age distribution; the entire population grows at a rate r, but the proportion of the population made up of each age class remains constant. In such a population the rate of increase r is affected by (1) *Survival*. The higher the fraction l_x of newborns that survive to each age x, the greater the growth rate (*Figure 4a*). The l_x curve

---

* Most of the algebraic symbols used in this book are defined in the List of Symbols in Appendix II. In this equation, as elsewhere, e is the base of natural logarithms, 2.718. In the next equation, c is a constant expressing the slope of the relation between dN/dt and N.

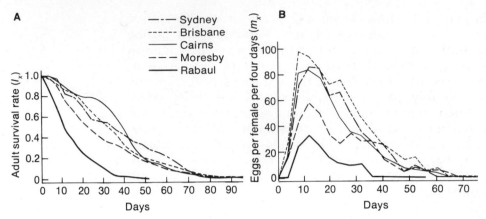

4 Genetic variation in (A) survivorship (l_x) and (B) fecundity (m_x) schedules in *Drosophila serrata*. For strains taken from five Australian localities, the figures show the fraction of adult flies that survive to different ages and the fecundity per surviving female per four-day interval, at 25°C. (From Birch et al. 1963)

determines δ. (2) *Fecundity*. The higher the number of offspring m_x produced by an average female of age x, the higher the birth rate b (*Figure 4b*). (3) *The age at which reproduction begins* (*Figure 5*). An individual who reproduces early in life is more likely to have grandchildren by some time t in the future than one who reproduces at an advanced age. In a growing population a female who reproduces early thus has a greater REPRODUCTIVE VALUE than a female who reproduces later in life; she contributes more to the future population size (Fisher 1930, Slobodkin 1961). This is equivalent to saying that organisms with a short generation time have a rapid rate of increase.

Clearly a population's potential rate of increase may be very different from its actual rate of increase. The death rate δ may actually be great enough to equal the birth rate b, so that $r = 0$. Thus the ACTUAL RATE OF INCREASE r is different from the INTRINSIC RATE OF NATURAL INCREASE r_m, which is the per capita rate of increase that a population

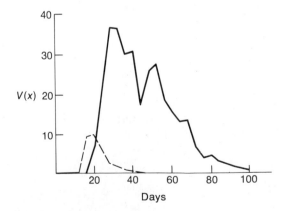

5

The $V(x)$ function for *Drosophila serrata* at 20°C (solid line) and at 25°C (dashed line). $V(x)$ is the contribution of a female of age x to the rate of population growth. Despite the 10-fold difference in total production of offspring at these temperatures, both $V(x)$ functions give the same value of r_m because of the importance of offspring produced early in life at 25°C. (After Lewontin 1965)

with a stable age distribution would have in a given environment if it were utterly free of those factors such as predation and scarcity of food that reduce population growth. This r_m differs from one environment to another; a bacterial culture given nutrients *ad libitum* grows more rapidly at higher temperatures than at lower.

By measuring age-specific values of survivorship l_x and fecundity m_x under optimal conditions, one can estimate the intrinsic rate of natural increase. An important generalization emerges from such data. Species whose fugitive, colonizing mode of existence promotes population growth usually have higher potential growth rates than do species whose more stable environments provide little opportunity to utilize their potential for growth.

THE EFFECT OF DENSITY ON POPULATION GROWTH

As a population grows, the age-specific birth and death rates change, so the actual rate of increase r is not constant. The population by its growth may deplete its resources, poison its environment with metabolic wastes, or engender the buildup of populations of predators or pathogens. Usually, then, the death rate increases as the density increases, and the birth rate drops as the resources needed to make eggs or seeds become scarce. Then the actual growth rate r declines from its intrinsic value r_m and may instead be

$$r = r_m - cN.$$

A little algebra yields the LOGISTIC EQUATION of population growth in a limited environment

$$dN/dt = r_m N (K - N)/K$$

which describes the sigmoid growth of a population in which growth slows down and finally stops when the equilibrium density K is reached (*Figure 6*). In this equation K is the CARRYING CAPACITY of the environ-

Idealized growth of a population. Curves E_1 **6** and E_2 are exponential growth, in which survivorship and fecundity are not reduced by crowding. The growth rate is lower under some environmental conditions (e.g., low temperature, curve E_2) than others (e.g., higher temperature, E_1). A simple model of density-dependent growth yields the logistic curves L_1 and L_2 for higher and lower temperatures respectively. In reality the equilibrium population size K might be different at different temperatures; and few populations would show such smooth curves or such constant equilibrium densities, even under the most constant environmental conditions.

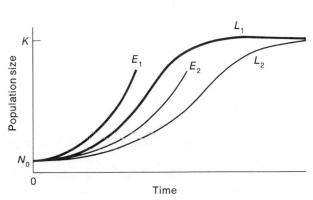

ment, the density attained when population growth is stopped by limiting factors, whatever they may be.

"The causes which check the natural tendency of each species to increase are most obscure," Darwin said, and they still are. The factors that limit population density in nature are the subject of a long-standing controversy (Andrewartha and Birch 1954, Cold Spring Harbor Symposium 1957, Slobodkin 1961, Den Boer and Gradwell 1970) that is summarized in many ecology books (e.g., Krebs 1972, Emlen 1973, Ricklefs 1976). Andrewartha and Birch (1954), among others, argued that many populations are decimated by environmental changes such as inclement weather before they become dense enough to cause a scarcity of resources or a buildup of natural enemies. Such controlling factors, which are not directly influenced by a population's density, are termed DENSITY-INDEPENDENT, and they commonly keep populations far below carrying capacity. Such populations probably include many fugitive species, whose temporary environment may disappear before the population becomes too dense. Many species of temperate-zone insects certainly experience very high mortality from inclement weather.

Another school of thought, influenced by observations of relatively stable vertebrate populations and perhaps by the philosophical conviction that the biotic world must be orderly and stable, has denied the importance of density-independent limiting factors. Nicholson (1958), Lack (1954), and Slobodkin, Smith, and Hairston (1967), among others, have argued that a population's density is usually near its carrying capacity, set by DENSITY-DEPENDENT factors that operate with increasing intensity as density increases (as in the logistic equation). These factors can include a shortage of energy, nutrients, nest sites, or other resources; predation and disease; and, in some social animals, behavioral patterns such as territoriality.

High density, then, is inimical to an individual's prospects for survival and reproduction. Organisms have evolved many mechanisms to mitigate its ill effects, most notably the tendency to disperse when they are crowded. Many species of aphids (plant lice) that are wingless under favorable conditions give birth to winged offspring if they are crowded or if the quality of their food plant deteriorates, as it may if it is heavily infested. As successive generations of the plague locust *Schistocerca gregaria* experience more and more crowding, these grasshoppers undergo a hormonal change that profoundly affects their physiology, morphology, and behavior. They store fat, develop long wings, become more gregarious, and finally depart in enormous swarms, flying downwind to land, with luck, in greener pastures.

Too sparse a population, however, may be as inimical to an individual's reproductive success as too dense a population. The joint activity of a number of animals or plants may ameliorate the environment. Fishes find protection from predators by schooling, birds may find food more effectively by watching other members of a flock than by foraging singly; even *Drosophila* larvae, by burrowing through their food medium, promote the growth of the yeast on which they feed. These and many other advantages of cooperative behavior are discussed by Wilson (1975*a*). At the very least, members of sexually reproducing species need one another for mating and have evolved some exotic mechanisms for finding and recognizing mates.

THE BIOTIC ENVIRONMENT: PREDATORS AND PREY

All species serve as food for some other species — a banal fact with the profound implication that predation, parasitism, and disease are overwhelmingly important aspects of life for all species. Many features of organisms are adaptations to escape predators and pathogens and to capture, as food, species that have evolved elaborate capabilities of escape and defense.

In many cases predator-prey interactions are unstable and result in extinction. A predatory South American cichlid fish (*Figure* 7) introduced into Gatún Lake in Panama has locally extinguished several other species of fishes (Zaret and Paine 1973). But some species manage to coexist with their predators, for several reasons. Predators are often unable to find all the potential prey; some remain inaccessible. A prey species may survive through effective dispersal; populations of prey, and consequently of their predators, may be decimated locally, but build up elsewhere as vacant localities are colonized. In some instances some of the age classes in a prey population are resistant to predators. The moose that are attacked by wolves are not so much the reproductive age classes, which can effectively defend themselves, but the defenseless young and the infirm old. Many predators focus their food-seeking efforts to a disproportionate degree on the most common kinds of prey. Whether by moving to places where a particular food type is concentrated or by forming a SEARCH IMAGE that enables them to find a common prey type more effectively, they lessen their impact on uncommon species of prey (*Figure 8*).

Although no species is entirely free of predation, all have escaped some of their potential predators and parasites by evolving mechanisms of defense. They may become inaccessible by hiding, fleeing, being too big (or small) to eat. They may avoid notice by crypsis — imitating the background in form, color, and pattern. Many plant and animal species possess spines, stinging hairs, protective armor, or noxious chemicals that render them unpalatable to some predators. Distasteful animals frequently advertise their unpalatability by a warning (apose-

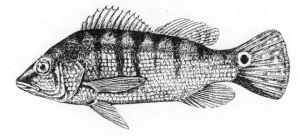

7

Cichla ocellaris, a South American species recently introduced into Gatún Lake in the Panama Canal Zone, where it is extinguishing native species. (From Sterba 1962)

matic) coloration or pattern; after a few encounters with such distasteful organisms, naive predators quickly associate the color pattern with their unpleasant experience and refrain from attacking such prey for some time. Aposematically colored species are often the models in systems of mimicry, convergence to a common color pattern in otherwise dissimilar species. In some cases (Batesian mimicry) a palatable species masquerades as an unpalatable one, while in other cases (Müllerian mimicry) several unpalatable species converge in appearance, each species gaining protection from its similarity to the other ones (*Figure 9*).

A community of organisms displays a bewildering variety of defense mechanisms, which continue to evolve in diversity. If two species share the same defense system, the population of a predator adapted to counter this defense grows as a function of the combined abundance of the two species of prey. Hence each suffers from the presence of the other, and the evolution of different defense systems is thereby favored (Slobodkin 1974). The complex issue of coevolutionary changes in predator-prey systems is one of the topics of Chapter 18.

COMPETITION AND RESOURCE UTILIZATION

A further consequence of the diversity of protective devices is that all consumers have a more or less specialized diet, since the features required to pursue and capture one prey may be antithetical to those required to capture others; a bird cannot be simultaneously an effective woodpecker and an effective flycatcher. The breadth of a species' diet is thus constrained by the differences among potential food items.

The variety of foods taken by a species is further influenced by the variability of the food supply. The diet of species may be quite specialized if their resource is consistently available (MacArthur 1972, Levins 1968), while a fluctuating food supply commonly favors the evolution of a more catholic diet. And if resources are scarce, so that

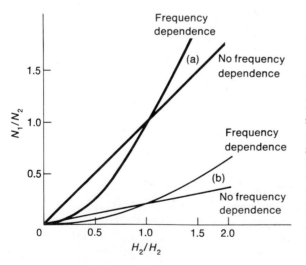

8

Frequency-dependent predation, or "predator switching." The ratio of two prey species (or genotypes) in the diet (N_1/N_2) is plotted against the ratio available (H_1/H_2). The straight lines show cases of no switching (a) when the predator prefers neither prey and (b) when it has a fivefold preference for prey type 2. The curved lines show switching; each prey species is taken disproportionately more as it becomes relatively more abundant. (After Murdoch and Oaten 1975)

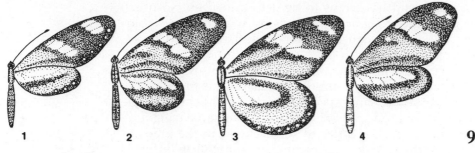

9

Two forms of mimicry in tropical American butterflies. Species 1–4 are all distasteful and are similar to each other (Müllerian mimicry). Species 5 and 6 are palatable, but resemble the unpalatable species (Batesian mimicry). (Redrawn from *Mimicry in plants and animals*, by W. Wickler. Copyright 1968 by W. Wickler. Used with permission of McGraw–Hill Book Company)

the population is food-limited, the ability to use a variety of foods is advantageous, while a population limited by factors other than food can afford to evolve a specialized diet (Emlen 1966).

The variety of resources used by a species is also affected by competition with other species. If the populations of two species are limited by resources and compete for them, one species may win and exclude the other from a region. This has been observed repeatedly in the laboratory and in the field. For example, within a few years after the parasitic wasp *Aphytis lingnanensis* was introduced from Asia to control olive scale in California, it had replaced the species *A. chrysomphali* that had flourished after its introduction some years before (DeBach 1966). The effects of competitive exclusion are often manifest in geographical distributions. The altitudinal distribution of the salamander *Plethodon glutinosus* is greater in mountain ranges where it is the only species than in ranges where it is excluded by *P. yonahlossee* and *P. jordani* from the higher altitudes (Hairston 1951).

When two species compete, which one wins may depend on subtle environmental factors. Brown (1971) found that the chipmunk *Eutamias dorsalis,* a highly aggressive species, excludes *E. umbrinus* from low altitudes by chasing it away from piles of seeds. But at higher altitudes the more arboreal *E. umbrinus* can get more food than *E. dorsalis,* by quickly returning to the seeds from which *E. dorsalis* chases it. It can do this because the branches of trees at these altitudes form an interlocking network and thus provide an effective highway.

Under some conditions either species may win, depending only on which has the initial numerical predominance or the greater rate of increase. For example, each of the small islands of the West Indies has on the average only one species of the diverse lizard genus *Anolis.* But just which species an island has seems to be determined not so much by its habitat or proximity to other islands as by which species colonized it first and excluded subsequent invaders (Williams 1969).

Competing species can coexist if the intensity of intraspecific competition is greater than the intensity of interspecific competition. As Darwin put it, "The struggle will almost invariably be most severe between the individuals of the same species, for they frequent the same districts, require the same food, and are exposed to the same dangers." Thus species may coexist if they differ to some extent in the kinds of resources they use: if their ecological niches are somewhat different (review by Schoener 1974). Reynoldson (1966) found that species of triclad flatworms coexist only where the diversity of prey animals is high enough to provide each species with a food supply that it can largely call its own (*Table I*). On the larger islands of the West Indies, several species of anoline lizards coexist by foraging in slightly different microhabitats (Schoener 1968). Since each species can survive competition only by differing somewhat from all others, a community contains more species if each is specialized for a different resource than if the species are more generalized.

Many other factors permit competing species to coexist. For example, predation on a superior competitor may prevent it from eliminating inferior competitors (Slobodkin 1961, Paine 1966, Janzen 1970, Connell 1970), and competitively inferior fugitive species can persist by colonizing new areas as they are eliminated by competition in others (Horn and MacArthur 1972, Levins and Culver 1971).

Beginning with Darwin, evolutionists have invoked competition

TABLE I. Food organisms in the guts of triclad flatworms (Per Cent)

	Polyscelis nigra	Polyscelis tenuis	Dugesia lugubris	Dendrocoelum lacteum
Oligochaeta				
Lumbriculus	28.6	40.0	14.3	—
Naididae	31.4	2.8	21.4	78.6
Tubificidae	5.7	31.4	10.0	3.6
Lumbricidae	2.8	8.6	5.7	—
Arthropoda				
Insecta	14.3	2.8	1.4	3.6
Asellus	—	2.8	—	—
Others	11.4	5.7	1.4	10.7
Mollusca				
Lymnaea	—	—	15.7	—
Hydrobia	—	—	11.4	—
Ancylus	—	—	1.4	—
Other	5.7	5.7	1.4	3.6

(After Reynoldson 1966)

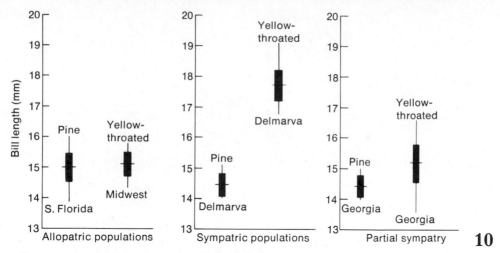

Character displacement in bill lengths of pine warblers (*Dendroica pinus*) and yellow–throated warblers (*D. dominica*). Bill lengths of the species are similar in allopatric populations, somewhat different where the species are broadly sympatric. The long bill of the yellow-throated warbler in the Delmarva area is associated with its habit of probing deep into pine cones for insects, thus reducing competition with the pine warbler. Vertical lines represent range of variation, crossbars the mean, and black rectangles one standard deviation on each side of the mean. (After Ficken, Ficken, and Morse 1968)

between species as a potent force leading to their diversification. Species that have very similar behavior and morphology in isolation are sometimes quite different from each other in geographic regions in which they coexist. In the absence of a competitor a species may display ECOLOGICAL RELEASE and occupy a broader range of habitats. This may be reflected in its morphology. The males of many woodpeckers have larger bills than the females and feed in somewhat different parts of trees. This dimorphism is far more pronounced in *Centurus striatus,* the only species of woodpecker on Hispaniola, than in species that inhabit continents, where several species of woodpeckers usually coexist (Selander 1966).

A greater difference between sympatric (in the same geographic locality) than allopatric (in different localities) populations of two species is termed CHARACTER DISPLACEMENT (Brown and Wilson 1956). Quite a few examples are known (*Figure 10*), although not as many as might be expected (see Grant 1972). Character displacement is an incipient manifestation of the divergence of species into different niches, which reaches its full flowering in the dramatic cases of diversification that we call ADAPTIVE RADIATION. Impressive examples are provided by the Hawaiian honeycreepers (*Figure 11*) and by the cichlid fishes in the Great Lakes of Africa (*Figure 1, Chapter 16*).

OTHER INTERACTIONS AMONG SPECIES

For reasons that may be as attributable to unconscious political biases as to the intrinsic importance of the subject, competition and predation

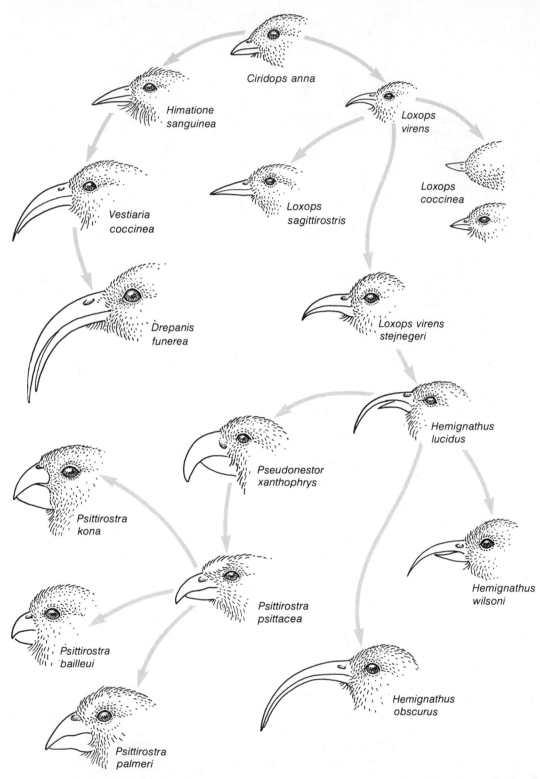

Ciridops anna

Himatione sanguinea

Loxops virens

Vestiaria coccinea

Loxops sagittirostris

Loxops coccinea

Drepanis funerea

Loxops virens stejnegeri

Pseudonestor xanthophrys

Hemignathus lucidus

Psittirostra kona

Hemignathus wilsoni

Psittirostra psittacea

Psittirostra bailleui

Hemignathus obscurus

Psittirostra palmeri

11 Adaptive radiation in the Hawaiian honey
creepers (Drepanididae). Major forms are
shown; some similar species have been
omitted. Arrows indicate presumed
phylogenetic relationships, illustrating how
very different bill shapes may have evolved
through intermediate conditions. This family,
including thin-billed foliage gleaners, long-
billed nectar feeders, woodpeckerlike forms
such as *Hemignathus wilsoni,* and thick-billed
seed eaters, fills many of the ecological roles
that on continents are filled by a variety of
families of birds. (After Bock 1970)

have preoccupied ecologists to an extreme degree. We tend to forget
that other kinds of interactions among species are widespread and
important. Amensalism, wherein one species benefits from another
without affecting it in turn, is so common that we often do not notice
it: owls roosting in abandoned woodpecker holes, seeds being dispersed by adhesion to the fur of mammals. In some cases, as with seed
dispersal, the evolution of the one species' characteristics has been
considerably affected by the properties of the other.

More elaborate, highly coevolved interactions among species frequently take the form of MUTUALISM, in which each of two or more
species benefits from their association. Frequently the species are highly
dissimilar and each provides a service of which the other is incapable.
Chief among such interactions are the relations between microorganisms and multicellular plants or animals. Nitrogen-fixing bacteria
enhance the rate of growth of legumes and other plants, and bacteria
facilitate digestion in vertebrates, as do protozoans in wood-eating
roaches and termites. The roots of most vascular plants develop an
intimate association with fungal mycelia, forming a mycorrhiza; from
this root-fungus combination the fungus derives carbohydrate synthesized by the plant and the plant derives a more rapid uptake of water
and nutrients. Mutualistic interactions between plants and animals,
especially in pollination and seed dispersal, are an important feature of
all terrestrial communities, particularly tropical ones (*Figure 12*; Orians
et al. 1974).

HIGHER-ORDER INTERACTIONS

We are accustomed to thinking in simplistic, linear terms: the more
abundant mice are, the more abundant the weasels that feed on them.
But interactions among species may be far more complex than this.
The effect of one species on another can depend on what other species
are present (Vandermeer 1969, Wilbur 1972, Neill 1974). Smith (1968)
provides a remarkable example of the intricacy of multispecies interactions (*Figure 13*). In Panama giant cowbirds lay their eggs in the
nests of oropendolas (large orioles). Because the young cowbirds de-

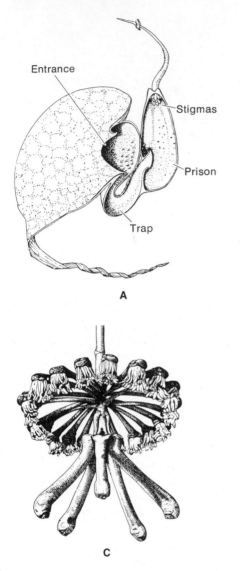

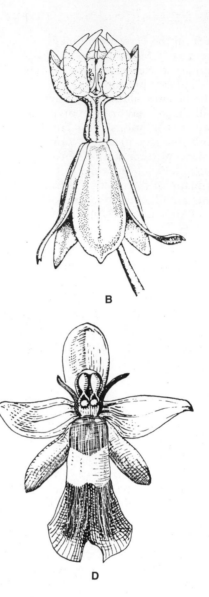

12 Intricate flowers, favoring outcrossing by specialized pollinators. (A) *Aristolochia grandiflora* (Aristolochiaceae), pollinated by flies that enter a trap and are held inside by hairs until pollination is accomplished. (B) *Asclepias curassavica* (Asclepiadaceae), in which the pollen masses (pollinia) attach to the feet of butterflies or other insects and then enter the slit leading to the internal stigmatic chamber when the insect loses its foothold. (C) *Marcgravia* (Marcgraviaceae), in which flowers deposit pollen on a bird's head while it drinks from the nectar containers suspended below. (D) *Ophrys muscifera* (Orchidaceae), which is pollinated by male wasps of a particular species when they attempt to mate with the insectlike flower. (A, B from Proctor and Yeo 1972; C, D from B. J. D. Meeuse, *The story of pollination*. Copyright © 1961 The Ronald Press Company, New York)

prive the oropendola nestlings of food, oropendolas are often aggressive toward cowbirds and remove cowbird eggs from their nests. Consequently the cowbirds, like many other such parasitic birds, have evolved egg mimicry. The color pattern of some cowbirds' eggs resembles that of the hosts'. Smith found that some oropendolas are

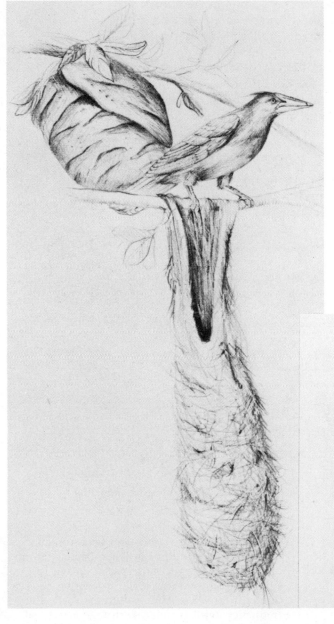

13

Chestnut-headed oropendola (left) and giant cowbird (below) at an oropendola nest. (From Ricklefs 1976; courtesy of Chiron Press, Inc., and Joel Ito)

tolerant of cowbirds and are parasitized by cowbirds that lay nonmimetic eggs. Apparently this is because the survival of young oropendolas can be enhanced by sharing the nest with cowbird nestlings. These, being more precocious than oropendolas, eat the larvae of botflies, which commonly cause high mortality in oropendolas. But oropendolas are intolerant of cowbirds if they nest in trees that harbor nests of bees or wasps, which are aggressive toward botflies and thus reduce the incidence of botfly attacks on the young oropendolas. In the consequent absence of botflies the disadvantage of associating with a young cowbird outweighs the advantage, and the behavior of the adult oropendola changes appropriately.

Thus the effect of one species on another depends on the environment provided by yet another species; the net effect of several species can differ from the simple sum of the independent effects of each. Such synergistic interactions are probably widespread in nature and constitute important variation in the environment to which a species adapts.

COMMUNITY STRUCTURE

The interactions of species with one another and with the physical environment follow rules that we understand only dimly, but whose existence we surmise from a certain regularity of structure in com-

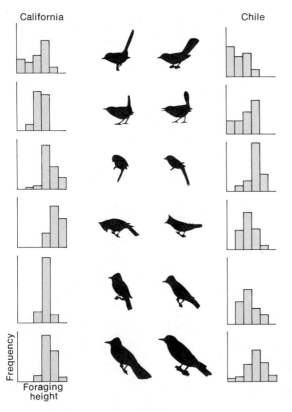

14

Similarity of ecological roles in birds of California chaparral (left) and of a similar vegetation community in Chile (right). Some of these ecological counterparts, drawn in silhouette to the same scale, belong to different families. The histograms show the frequency with which they forage at several intervals (0–6″, 6″–2′, 2′–4′, 4′–10′, 10′–20′, >20′) above ground. (From Cody 1974a)

MacArthur and Wilson's model of species diversity on islands or similar isolated patches of habitat. The number of species S on the island increases as new species immigrate and decreases as species already present become extinct. When the rates of immigration (I) and extinction (E) are equal (where the curves cross), the number of species is at an equilibrium $\hat{S}$. The greater the number of species on the island, the smaller the number of immigrants that are new; hence the immigration curve declines. Even if the probability of extinction of each species is constant, the more species there are, the more extinctions there will be; hence the extinction curve rises. Immigration rates are likely to be higher for near (I_N) than far (I_F) islands; extinction rates are likely to be greater on small (E_S) than large (E_L) islands. Hence $\hat{S}$ should be lowest on distant small islands ($\hat{S}_1$),

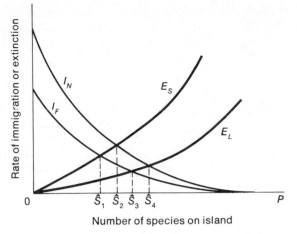

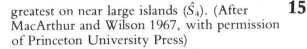

greatest on near large islands ($\hat{S}_4$). (After MacArthur and Wilson 1967, with permission of Princeton University Press)

15

munities. Cody (1974*a*) finds that the shrubby chaparral vegetation in climatically similar regions of California and Chile supports surprisingly convergent bird faunas (*Figure 14*), made up of unrelated species that are similar in both morphology and feeding behavior.

Much of the effort of community ecologists has been devoted to showing that regular patterns of community structure exist (but see Simberloff 1978 for a critique of these efforts); although the identity of the species varies from place to place, the number of species, their relative abundance, and the manners in which they differ from one another in their patterns of resource use are often quite predictable from the major climatic and topographic features of the environment (see papers in Cody and Diamond 1975 for examples).

Among such regularities are the numbers of species on islands. MacArthur and Wilson (1967) point out that the number of species S on an island changes as new species enter at some immigration rate I and as resident species are lost by extinction at rate E. Immigration I decreases as S increases, because the more species on the island, the fewer the immigrants that are new additions to the biota (*Figure 15*). Extinction E increases with S, since with any given probability of extinction, the more species there are, the more extinctions there will be. The immigration and extinction functions cross where $I = E$. At this point the number of species S is at EQUILIBRIUM.

The further an island is from a source of new species, the lower the immigration is; thus the equilibrium number of species should decrease with increasing distance from continents. On islands with small populations the probability of extinction — whether by competition, predation, weather, or chance — is greater than on large islands, so small islands have lower species diversity.

This theory does not explain all variations in species diversity among islands, of course; ecological succession, species interactions, and the evolution and diversification of species after colonization are clearly important (Wilson 1969). Nevertheless the relation between species number and distance from source, and especially between species number and island size, conform very well to this theory. As the model assumes, small isolated populations do indeed have high extinction rates (Diamond 1969). This is true also on continents. For example, a local population of the butterfly *Euphydryas editha* has twice undergone extinction and reestablishment in the course of 15 years (Ehrlich et al. 1975). Thus the principles of island biogeography apply not only to islands, but to any localized area. The number of species in a region approaches an equilibrium determined by the rates at which species enter the community and become extinct.

SUMMARY

Organisms must adapt not only to the average states of the environment, but to its pattern of variation as well. The amplitude, frequency, and predictability of environmental fluctuations affect the pattern of adaptation or even preclude adaptation altogether. Environmental variation is universal, especially because the other organisms with which a species interacts are such an important part of its environment. Among these organisms are other members of its own species, which act as mates, social consorts, or competitors. The rate of population growth commonly declines as the density of a population increases, because of the increasing scarcity of resources or the increasing prevalence of predators or disease. The "struggle for existence" so engendered has important effects on the evolution of reproductive rates, life history characteristics, dispersal mechanisms, feeding habits, defense mechanisms. Indirectly or directly, it affects all the characteristics of organisms. Interactions with other species, including predation, competition and symbiosis, affect a species' abundance and distribution in time and space and have led to many of the adaptations that make up the diversity of the living world.

FOR DISCUSSION AND THOUGHT

1   Marston Bates (1960) points out that it can be difficult to distinguish between an organism and its environment: when does my food cease to be part of my environment and become part of me? In general is it possible to distinguish between organism and environment, or is it desirable? What are the evolutionary, sociological, and philosophical implications of thinking instead of an organism-environment complex as a unit?

2   The deserts of southwestern North America contain many species of bees. Each appears in years of high rainfall and feeds on only a few of the many species of plants that also appear in response to rain. Do these species experience a variable, unpredictable environment? What we perceive as different environments may or may not be different to another organism (Colwell and Futuyma 1971), so how can we measure the heterogeneity of an environment? Is an apparently novel change, such as exposure to an insecticide, always as novel as we might think?

3   Can you find in the literature any cases in which the reasons why a species has not become adapted to an environment are really understood? What factors limit the geographic range of a species? Over a geographic transect the environment usually changes gradually, but the edge of a species' range is often abrupt. Why? Why should it be incapable of adapting enough to extend just a few more miles?

4   Hairston, Smith, and Slobodkin (1960) argued that the density of a population cannot be limited by physical factors (e.g., weather, salinity), even if these appear harsh, since the species is adapted to these factors. But, they say, such factors can limit the geographic and local distribution of a species. Discuss.

5   If organisms are supposed to be adapted just to the environments they usually encounter, why can ostriches be kept outdoors all year in a New York zoo, or magnolias in a yard in Michigan? Why don't magnolias occur naturally in Michigan?

6   In a resource-limited environment do organisms of different ages compete as much with each other as with others their own age? Do different age classes specialize on different resources and so avoid competition? How could this specialization affect the evolution of life cycles? (See Istock 1967.)

7   Haldane (1956) argued that any characteristic that makes an individual less subject to the factor that limits population density is advantageous. Thus a food-limited species should evolve to be less dependent on its limiting resource, perhaps by expanding its diet. Does this always happen? Can organisms simultaneously adapt to factors that do not limit population density? Does evolution necessarily lead to an increase in population size?

8   Every species is subject to at least one kind of predator that can overcome the prey's defenses. Can the possession of defense mechanisms such as cryptic coloration and noxious chemicals then be a sufficient explanation for the coexistence of a prey species with such a predator?

9   Does the continuous coevolution of species, such as predator-prey coevolution, imply that natural communities are now more diverse (in terms of the variety of defense systems, kinds of niches occupied, number of species) than they were in the Eocene? in the Jurassic?

10   Because a "jack of all trades is master of none," a generalized species tends to lose in competition to a specialist, if they are competing for the resource on which the specialist specializes. Why is it, then, that a community contains both generalized and specialized species? How can they coexist?

11   Although a community of competing species is stable only if each occupies a somewhat distinctive niche, I have avoided saying that species evolve in such a way as to insure the stability of the community. Why did I not make such a statement? Do species actually evolve in such a way that community stability is enhanced? If so, do they evolve *in order to* enhance stability?

12   Dobzhansky (1950), Sanders (1969), and others have suggested that in some regions (e.g., hot springs, shallow waters, the Arctic) organisms adapt primarily to the physical features of the environment (e.g., temperature), while in other communities (e.g., tropical rain forests, coral reefs) organisms adapt primarily to biotic features. Discuss the validity and implications of this distinction.

MAJOR REFERENCES

Wilson, E. O., and W. H. Bossert. 1971. *A primer of population biology.* Sinauer, Sunderland, Mass. 192 pages. Contains an exceptionally lucid introduction to the simpler mathematical formulations of population ecology.

Pianka, E. R. 1978. *Evolutionary ecology.* (2nd ed.) Harper & Row, New York. 397 pages. An introduction to modern ecology from an evolutionary perspective.

Whittaker, R. H. 1975. *Communities and ecosystems.* Macmillan, New York. 383 pages. An introductory but comprehensive survey of community ecology.

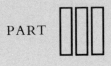

PART III

HISTORICAL
EVOLUTION

The study of biology may be divided into two modes. The approach taken by biochemists, physiologists, and developmental biologists is a functional one that asks, How does it work? What are the mechanisms by which an organism develops and maintains itself? The other approach to biology is to ask, How and why has life come to be this way? How have these mechanisms come into being? This question, of so much greater concern to a biologist than to a physicist or chemist, is a historical one; it recognizes that the phenomena we study are the products of historical development, in which past events may have determined the present state of affairs. Just as a political scientist cannot explain the present patterns of national boundaries or of tensions among peoples without reference to the history of, say, colonialism, a biologist may require a knowledge of the past to understand why the kidney of marine fishes seems designed more for life in fresh than in salt water or why tapirs inhabit the forests of tropical America and southeastern Asia but not Africa. One of the major questions in biology, in fact, is this: to what extent are the properties of organisms — structure, physiology, behavior, geographical distribution, ecological role — explicable solely in terms of present conditions, and to what extent are they the legacy of the past?

Before examining the mechanisms of evolution, we will do well to survey the history of life, for several reasons. The past may be the key to the present, in that the mechanisms of evolution operate within a framework of historical constraints. Patterns of historical events may suggest the existence of evolutionary mechanisms that may not be apparent from the study of contemporaneous organisms. And, most important, we ought to know what phenomena the body of evolutionary theory is designed to explain.

CHAPTER 5

The History of Biological Diversity

THE FOSSIL RECORD

Of the images of the living world that we carry with us from child-hood, few can be as impressive as that of the dinosaurs. However romanticized our images of such creatures may be, their existence reveals a number of profoundly important facts. Many kinds of di-nosaur are known from only one or a few specimens, from rocks deposited more than 100 million years ago. Consider the immensity of time that is spanned. The earliest archaeological evidences of human agriculture are only about 12,000 years old, less than 1/8000 of the time since *Tyrannosaurus* walked the earth. The physical environment has changed drastically since then, for dinosaurs could not have sur-vived the winters that now buffet the regions they occupied. And we have such a fragmentary record of so many species that there must have been many more forms of life of which we have no knowledge.

In this light, statements about evolution that at first seem farfetched become entirely plausible. Evolutionists are forever invoking unknown environmental changes to explain extinctions or the origin of peculiar adaptations; yet we know that enormous changes in both the physical and biological environment have occurred continually for billions of years. The time available for evolution has been great enough for the most improbable of events, such as rare mutations, to have occurred repeatedly. And evolutionists need not apologize if they cannot find fossils of most of the ancestral forms of modern organisms that they postulate existed; the probability of finding such fossils is very low. Organisms without hard skeletons can be fossilized only under excep-tional conditions or not at all. Even organisms with fossilizable parts are unlikely to be preserved unless they occupy habitats like swamps or estuaries where their remains can become buried in sediments. These sediments must persist without metamorphosis or erosion for millions of years if we are to discover their contents, and they must be exposed in localities accessible to investigation. Moreover, there are theoretical reasons to suppose that many ancestral forms, the intermediate stages in evolutionary sequences, existed for such a short time that their preservation was unlikely. The resulting incompleteness of the fossil record, more often than not, forces us to resort to inferences from the present to reconstruct the past. Indeed, in view of the factors that hinder the paleontologist's efforts, it is remarkable how good the fossil record is.

A BRIEF SUMMARY OF MAJOR EVENTS IN THE HISTORY OF LIFE

The Precambrian era

The appearance of invertebrate animal forms marks the beginning of the Phanerozoic time, starting with the Cambrian era (*Table I*). Before this event, dated by isotope analysis as 570 to 600 million years before present (myBP), life had a long history, most of which is quite un-known. The oldest fossils, algal forms in South African chert, are 3.4

TABLE I. The geological time scale

Era	Period	Epoch	Millions of Years from Start to Present	Major Events
Cenozoic	Quaternary	Recent	0.01	Repeated glaciations in northern hemisphere; extinctions of large mammals; evolution of *Homo*; rise of civilizations.
		Pleistocene	2	
	Tertiary	Pliocene	12	Radiation of mammals and birds; flourishing of insects and angiosperms. Continents in approximately modern positions.
		Miocene	25	
		Oligocene	36	
		Eocene	58	
		Paleocene	63	
Mesozoic	Cretaceous		135	Mass extinctions of marine and terrestrial life, including last dinosaurs. Angiosperms become dominant over gymnosperms. Continents well separated.
	Jurassic		181	Dinosaurs abundant; birds, mammals, angiosperms appear. Gymnosperms dominant. Continents drifting.
	Triassic		230	Increase of reptiles, first dinosaurs; gymnosperms become dominant. Continents begin to drift apart.
Paleozoic	Permian		280	Continents aggregated into Pangaea; glaciations. Marine extinctions, including last trilobites. Reptiles radiate, amphibians decline.
	Carboniferous (Pennsylvanian and Mississipian)		345	Extensive forests of early vascular plants, especially lycopsids, sphenopsids, ferns. Amphibians diverse; first reptiles. Radiation of early insect orders.
	Devonian		405	Fishes and trilobites diverse. First amphibians and insects.
	Silurian		425	Invasion of land by primitive tracheophytes, arthropods.
	Ordovician		500	First agnathan vertebrates.
	Cambrian		600	Sudden appearance of most marine invertebrate phyla; primitive algae.
Pre-cambrian				Trace fossils of marine algae, especially Cyanophyta. Origin of life in the dim past.

billion years old; and chemical analyses of carbonate sediments 3.2 billion years old indicate that photosynthetic organisms must have existed at that time (J. Schopf 1974). So the Precambrian development of life spans a lot more of the earth's history than the entire period with which the invertebrate or vertebrate paleontologist usually deals. It seems likely that life arose soon after the earth became cool enough to support it, about 4.5 billion years ago (Gould 1978).

This early history and the origin of life, which are tangential to the purposes of this chapter, are reviewed by Cloud (1976). It is generally agreed that the prebiotic earth had an atmosphere rich in H_2, N_2, CO_2, H_2O, and sulfurous gases, but lacked free oxygen and was therefore a reducing atmosphere. In the absence of O_2 and O_3 ultraviolet light penetrated in far greater strength than it now does and by photochemical reactions could generate organic compounds as it does in laboratory experiments. How these compounds became assembled into self-replicating bodies and what the nature of these forms was are unknown. These first forms must have given rise to the first cells, which were almost undoubtedly heterotrophic prokaryotes, perhaps bacteriumlike in nature, from which chemosynthetic and photosynthetic autotrophs evolved. With the evolution of photosynthesis by organisms not unlike the blue-green algae (Cyanophyta) that are the earliest fossils, the chemical environment was radically changed. Photosynthesis produced oxygen and hence an oxidizing atmosphere, the ozone (O_3) screen against ultraviolet light developed, and aerobic respiration became the norm for most subsequent forms of life.

From the time of the first fossils until about a billion years ago, the Cyanophyta were the dominant forms of life, leading J. W. Schopf (1974) to call the Precambrian the age of blue-green algae. As long as 2 billion years ago, algae very like modern blue-green genera such as *Nostoc* and *Oscillatoria* existed. Their evolution, in both biochemical and morphological properties, has apparently been extremely conservative. They are living fossils, compared to which horseshoe crabs and opossums are newcomers.

The first eukaryotes, green algae, apparently arose about 1.6 billion years ago, and by the end of the Precambrian were more prevalent than blue-greens. Algae continued to diversify in later periods; for example, coccolithophorids appeared first in the Triassic and diatoms in the Cretaceous. Thus ancient groups continued to diversify and evolve actively long after their origin.

The Paleozoic era

The first invertebrates are found in the fragmentary record of the Ediacara assemblage of Australia. Deposited toward the end of the Precambrian about 700–600 myBP, it contains traces of soft-bodied coelenterates and rather questionable annelids, and an arthropod that might be a trilobite (*Figure 1*) (Cloud 1968; see also Cloud 1976). The recorded history of animals really begins at the beginning of the Cambrian, and it begins with a bang. To be sure, the beginning of the

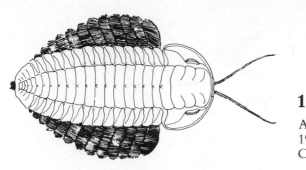

1

A trilobite, *Triarthrus eatoni*. (From Levi–Setti 1975, *Trilobites: A photographic atlas*. Copyright The University of Chicago Press)

Cambrian covers quite a few million years, which is a lot of time for some most elaborate evolution; but the fact remains that the beginning of the metazoan record is marked by the "sudden" appearance of all of the animal phyla that have fossilizable skeletons, except for the Chordata. Coelenterates, arthropods, echinoderms, molluscs — all are present in profusion. Some authors (e.g., Durham 1971) hold that their appearance was preceded by a long evolution in which the phyla came into being and differentiated from one another, but it seems likely that they diversified rapidly after a late Precambrian origin (Cloud 1976, Gould 1976). The origin of invertebrate phyla is one of the great problems of evolution; but rapid diversification, or adaptive radiation as it is usually called, occurs throughout the fossil record and thus may not be too implausible an interpretation of this great event.

The rest of the Cambrian is marked by increasing diversity of the orders and families of marine invertebrates (*Figure 7a*) and by a fairly large number of extinctions in the mid-Cambrian. Whole families of invertebrates became extinct in the late Ordovician as well; and it is in this period that the first chordates, the jawless (agnathan) ostracoderms (*Figure 2a*), are recorded. These are followed in the Silurian by other agnathan "fishes," as well as the first vertebrates with jaws, the placoderms (*Figure 2b*). It seems likely that the two other great groups of fishes, the Chondrichthyes (sharklike fishes) and Osteichthyes (bony fishes) originated at about this time, but their remains are not found until later. In the Silurian the first fossilized terrestrial organisms appear, two groups of vascular plants related to the modern psilopsids and club mosses.

In the Devonian, marine invertebrates, especially trilobites, continued to diversify, and in the middle of the period the first bony fishes (*Figure 2c–e*) appeared in the record, followed some time later by the first of the Chondrichthyes. On land vascular plants became more diverse, and the first tentatively identified insect (Collembola?) is found. At the very end of the period the first amphibian fossils, whose structure resembles the crossopterygian bony fishes, appeared. But quite a few groups of invertebrates, including most of the trilobites, and most of the agnathan and placoderm "fishes" became extinct.

Carboniferous times, usually divided into the Mississippian and

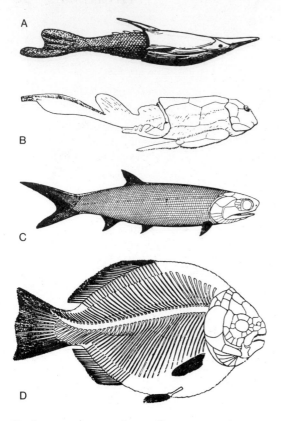

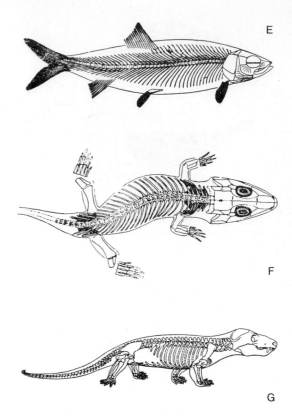

2 Extinct representatives of some vertebrate groups. (A) *Pteraspis,* a Devonian ostracoderm; (B) *Bothriolepis,* a Devonian placoderm; (C) *Palaeoniscus,* a Permian chondrostean; (D) *Dapedius,* a Jurassic holostean; (E) *Leptolepis,* a Jurassic teleost; (F) *Diplovertebron,* a Carboniferous seymouriamorph amphibian with reptilian features; (G) *Cynognathus,* a Triassic therapsid, or mammal-like reptile. (From Romer 1960, *Vertebrate paleontology,* Copyright The University of Chicago Press)

Pennsylvanian periods, are marked by no exceptional marine events, but the terrestrial record is rich. A great diversity of vascular plants, many extinct by the end of the period (*Figure 3*), is evident. In the Pennsylvanian a number of orders of insects proliferated; of these only the roaches have survived to the present. The fossil record of these times contains many groups of amphibians. Almost all subsequently became extinct without issue, but they include the ancestors of the reptiles, the Seymouriamorpha (*Figure 2f*), which are sometimes classified as cotylosaur reptiles.

In the Permian both the reptiles and the insects diversified greatly; among the insects the first groups (Neuroptera, Mecoptera, Coleoptera) with complete metamorphosis (inclusion of the pupal stage) appear. But it is the transition from the Permian to the Triassic that is marked by another of the paleontologist's great mysteries, the mass extinction that reduced the diversity of marine life — plankton, reef-building corals, and benthic invertebrates — to a degree that suggests a worldwide catastrophe of unprecedented magnitude (*Figure 7a*).

It may not be coincidental that the environment of the earth probably changed drastically at this time because of the movement of the continents by continental drift. Before the Permian the continents were apparently aggregated into two great masses, one somewhere toward the south pole and one at higher latitudes. These continents seem to have drifted toward each other in the Paleozoic, coming together in the Permian to form the supercontinent Pangaea (*Figure 4*). From present-day effects of land masses on climate, it seems probable that continental climates, with drastic fluctuations in temperature, prevailed during this period. Moreover the joining of the continents must have brought into contact two separately evolved sets of species, and the unprecedented interactions among them may have caused extinctions. The reduced number of species after the mass extinction seems to fit the number predicted by the general relation between species diversity and area (*Chapter 4*), based on the reduced area of the continental shelf

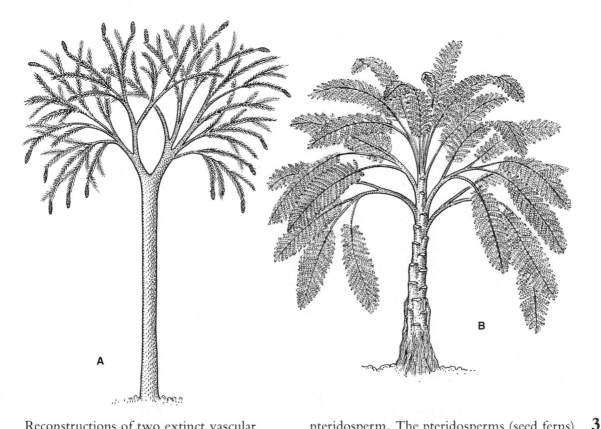

Reconstructions of two extinct vascular plants. (A) *Lepidodendron,* a Carboniferous lycopsid, and (B) *Medullosa,* a Carboniferous pteridosperm. The pteridosperms (seed ferns) are entirely extinct. (From Delevoryas 1962; with permission, *The Botanical Review* and *Palaeontographica*)

3

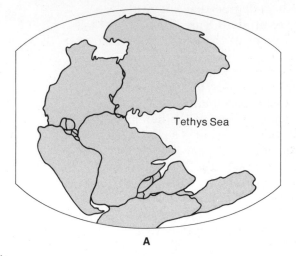
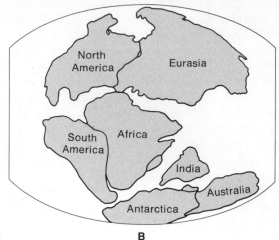

4 Possible positions of the continents. (A) At the end of the Permian, when there was probably a single continent, Pangaea. (B) At the end of the Triassic, when they were beginning to drift toward their present positions. (Modified from Keast et al. 1972, after Dietz and Holden)

occupied by the marine groups in question (T. Schopf 1974, Simberloff 1974).

The Mesozoic era

During the Mesozoic, Pangaea fragmented into the northern continent Laurasia and the southern continent Gondwanaland. Laurasia split into Eurasia and westward-drifting North America, and Gondwanaland fragmented into South America, Africa, Antarctica, Australia, and the subcontinent of India, which drifted northward and collided with Eurasia. The continents probably were approaching their present positions by the end of the Mesozoic (*Figure 4*), but the details of movement and the times at which the pieces were joined and detached are subject to debate among geologists (see discussions in Keast et al. 1972). After the great Permo-Triassic extinction, many marine invertebrate groups began to rediversify, a process that continued through the Triassic into the Jurassic. The Triassic is most notably the beginning of the age of reptiles that persisted through the Mesozoic era. The first fossils that can be termed mammals appeared in the late Triassic, but their definition as mammals is quite arbitrary, for they could equally well be called therapsid reptiles (*Figure 2g*). Several orders of insects (Orthoptera, Thysanura) first appeared as fossils at this time, but they probably had evolved long before.

In the Jurassic other orders of insects, notably the "advanced" orders Diptera (true flies) and Hymenoptera (sawflies and parasitic wasps), first appeared, as did a few fossils that are probably the first angiosperms (flowering plants). A major diversification of reptiles, including all sorts of dinosaur, dominates the terrestrial vertebrate record, and some major groups of amphibians became extinct. The first birds and several groups of mammals appeared at this time. The marine scene featured a reradiation of Chondrichthyes, including modern sharks, and a great diversity of holostean bony fishes, the group that survives at present only as bowfins (*Amia*) and gars (Lepisosteidae). The modern bony fishes, the teleosts, began to appear at this time.

The diversity of teleosts increased throughout the Cretaceous; the fauna was dominated by the soft-rayed fishes, especially clupeoids (represented today by herrings, anchovies, and the like), but some of the more advanced spine-finned fishes (like modern perches and basses) were represented as well. On land diverse reptile groups held sway, only to perish at the end of the Cretaceous. Several groups of mammals existed, including the ancestors of the modern orders; but this period was to see the last of such early mammalian groups as the triconodonts and symmetrodonts. The adaptive radiation of the angiosperms proceeded rapidly during this period (*Figure 7b*). Quite a few modern families such as Salicaceae (poplars) and Platanaceae (sycamores) appeared quite early in the Cretaceous, and by the end of the period a taxonomically rather modern flora had appeared. Insects likewise diversified at this time; indeed many modern genera can be traced to the end of this period. But the end of the Cretaceous was another time of widespread extinction, although not as catastrophic as the Permo-Triassic. The ruling reptiles perished, as did many marine groups, including the ammonites (nautilus-like cephalopods) and a great many planktonic Foraminifera. The end of this period was marked by a regression of the shallow epicontinental seas that had covered much of what is now dry land and by a cooling of the climate.

The Cenozoic era

In the first two epochs of this era, the Paleocene and Eocene, most of the modern orders of birds and mammals became differentiated, although some of them were not to survive to the present. Teleosts and angiosperms continued to diversify, and the diversity of Foraminifera rebounded from its late Cretaceous nadir. The continents arrived at their present positions early in the era, but they were connected and disconnected as the elevation of land and sea changed; for example, from about the Eocene to the Pliocene Central America formed a series of islands between North and South America. During the Pleistocene climates changed significantly. In the northern latitudes there were repeated glaciations during which temperatures dropped, a drier climate prevailed over much of the earth, and sea level fell. Interglacial times were warmer and wetter, and the level of the seas was higher. Despite these climatic changes, the only massive extinctions occurred among the large birds and mammals — mammoths, giant sloths, and so forth. According to one school of thought, they were the first of many victims of the human species (see Martin and Wright 1967).

CHANGES IN DIVERSITY

This survey of major events in the history of life has been most superficial, of course; more detailed expositions are available from Valentine (1973), Romer (1960), Banks (1970a, b), Kummel (1970), and Harland et al. (1967). My concern is not so much to offer a detailed history as to examine the generalizations about evolution that the historical record may help formulate. One approach is to ask, What determines the diversity of organisms? If it is constant through time, why is it constant? If not, why not?

Diversity very often means the number of species in a particular group, such as birds, in a given region. The differences among closely related species, however, are often in nonskeletal characters. For this and other reasons species are difficult to distinguish or even define in fossil material, so the paleontologist counts genera, families, or other higher taxa more often than species.

Whether we count species, genera, or other units, the number N at any time t may change at the rate dN/dt. Like the number of individuals in a population, this number is a function of the rate of input and the rate of loss. The rate of input or origination O is the number of new taxa per preexisting taxon, per unit time, and it is attributable either to the immigration of new taxa from outside the area of concern or to the evolution of new taxa by splitting. Splitting of one phyletic line into two or more occurs by speciation; if the allocation of species into genera is about the same for each geological time period, the rate of origin of new genera is proportional to that of species. The rate of loss E is determined by the probability that a species (or genus) will become extinct during the period dt. Thus the diversity changes over time just as a biological population changes in number:

$$dN/dt = ON - EN = N(O - E).$$

If $O = E$, diversity is constant, but it remains constant over a long term only if a homeostatic, negative feedback process holds it at equilibrium, just as the size of a population remains constant only if density-dependent factors balance birth and death rates (*Chapter 4*). Diversity-dependent controlling factors exist only if, as diversity increases, the rate of origination declines and/or the extinction rate increases (*Figure 5*). I shall first treat the factors that affect rates of origination and extinction and then return to the question of whether they tend to be balanced.

PATTERNS OF ORIGINATION

In the early days of paleontology it was presumed that new forms proliferated most rapidly during times of intense geological activity, times at which great climatic and other environmental change stimulated evolutionary activity. Such simple explanations have now been largely discarded (Henbest et al. 1952, Westoll 1954), for although

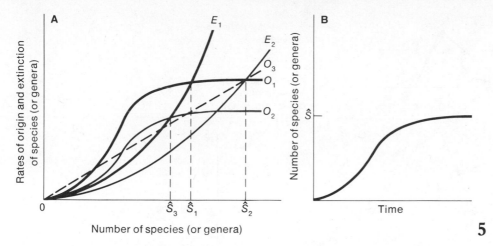

(A) Both the number of new species (or genera, or higher taxa) originating per unit time and the number of extinctions per unit time must increase as the number of species already present rises. If rates of extinction (E) increase more rapidly with species number than do origination rates (O), an equilibrium number of species ($\hat{S}$) will exist when $O = E$. Whether origination rates are constant (O_3) or decline as diversity increases (O_1, O_2), $\hat{S}$ will be lower if extinction rates are greater ($\hat{S}_1 < \hat{S}_2$) or if origination rates are lower ($\hat{S}_3 < \hat{S}_1$). (B) The change in species diversity over time, according to the equilibrium model. It is debatable whether the world biota is at or below $\hat{S}$.

orogeny (the formation of mountains), continental drift, and climatic changes affect evolution, the periods of diversification are scattered throughout evolutionary time and bear no simple relation to times of mountain building or other geological changes.

In a highly influential book Simpson (1953) has stressed that diversification often follows soon after the "invention" of critical new characteristics that permit their possessors to occupy new ADAPTIVE ZONES. By an adaptive zone he means a way of life in which a group of species engages that is very different from other such groups, a set of similar ecological niches. For example, owls as a group can be said to occupy an adaptive zone different from that of hawks; although the species of owls differ from one another in what they eat and where they nest and so have different niches, they have a rather homogeneous set of niches compared to those of other groups.

Once a species develops such an adaptation, it gives rise by speciation to a multitude of forms that differ from one another only in degree and which occupy the niches that make up the adaptive zone. Thus in quite a few cases almost instantaneous (in geological terms!) radiation follows shortly after the first few members of a group appear in the fossil record. The various groups of "fishes" (*Figure 6*), for example, follow this pattern; in each class the rate of proliferation of new orders and families was high soon after the class first appeared in the record but then dropped rather quickly. It is generally presumed that the

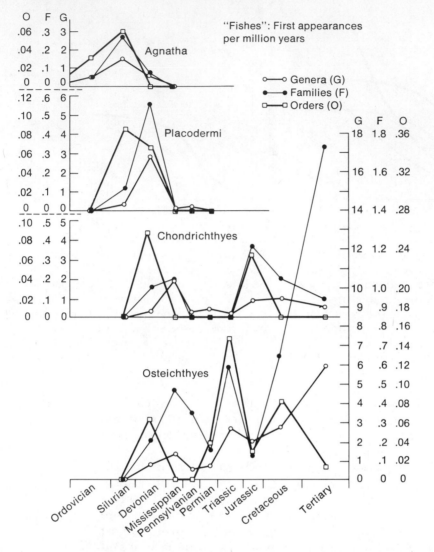

6 Rates of origination of new orders, families, and genera in each of the four classes of fishlike vertebrates, from the Ordovician to the Tertiary. The three major peaks in the Osteichthyes (bony fishes) correspond roughly to the rise of the subclasses Chondrostei, Holostei, and Teleostei. (After Simpson 1953, by permission of the publisher)

distinctive complex of morphological features that define each of these major groups is the critical adaptation, the ticket to entry into a new adaptive zone, that permitted the subsequent adaptive radiation — the acquisition of a bony skeleton by the agnathans or of jaws by placoderms, for example. In some instances, as with the Osteichthyes, there are successive peaks of diversification, attributable to successive evolutionary innovations. In the Osteichthyes (*Figure 2c–e*) these peaks correspond to the superorders Chondrostei (sturgeons are modern-day examples), Holostei (gars, etc.), and Teleostei (modern bony fishes), in which successive improvements in the structure of the tail and the jaw mechanism (see Romer 1960 for details) were associated with new locomotory and feeding activities.

Temporal patterns of diversity (*Figure 7*), however, suggest that immediate but short-lived episodes of proliferation are not the only pattern. In some instances, as with the teleosts, a high rate of proliferation of new families and genera continued for a very long period before it dropped off. In other instances the diversification of a group did not really begin until long after the group first appeared in the fossil record. A variety of mammals, including the ancestors of marsupials and placentals, was present rather early in the Cretaceous, but the great, geologically instantaneous radiation of mammals did not get under way until the Paleocene, more than 75 million years later.

This pattern suggests that adaptive radiation at least sometimes depends not only on the acquisition of new adaptations, but also on the availability of ecological opportunities in which the adaptations can be exercised. Of course, we can tell in retrospect that some ecological opportunities were available long before something that used them evolved; there were empty niches. Flying insects, a potential food source for large volant predators, existed long before pterodactyls, birds, or bats evolved, although one can argue that dragonflies and spiders took a toll. There is every reason to believe that unfilled ecological niches exist at present. For example, sea snakes are diverse in the tropical parts of the Indian and Pacific oceans, but they do not exist in the Atlantic; fish-eating and vampire bats inhabit the New World tropics, but not those of the Old World. It is hard to prove that there is no future for sea snakes in the Atlantic or for blood-drinking bats in Africa, but the recent expansion of some species' ranges, without obvious competitive exclusion of other species, suggests that ecological vacancies may persist for millions of years without any species' evolving to fill them. The cattle egret (*Bubulcus ibis, Figure 8*), which throughout the Old World hunts insects stirred up by grazing ungulates, is thriving in both North and South America after having arrived in the 1930s. It associates mostly with domestic cattle, but there were large ungulates such as bison on the plains of both continents until recently. Thus new species do not evolve just because ecological opportunity awaits them. Nature does not so abhor a vacuum that it creates species just because they fit in.

In many instances a taxon fitted for an adaptive zone seems not to diversify until preexisting occupants of the zone become extinct. Many cases of ECOLOGICAL REPLACEMENT in the fossil record are commonly assumed to exemplify competitive exclusion of a less "efficient" by a more "efficient" group. The rodents are often thought to have been competitively superior to the multituberculates, an early order of mam-

7 Chronological distribution and very approximate changes in diversity of some major groups of organisms. (A) Total counts for marine invertebrates. (B) Very approximate changes in the diversity of groups of plants. (C) Fishlike vertebrates other than Osteichthyes. (D) Reptiles. (E) Placental mammals. (A after Valentine 1973; B after Arnold 1947; C, D, E after Romer 1960, *Vertebrate paleontology,* Copyright The University of Chicago Press)

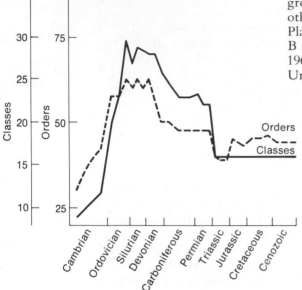

WELL-SKELETONIZED MARINE SHELF INVERTEBRATES **A**

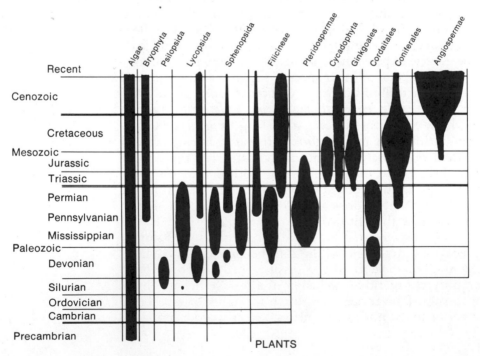

PLANTS **B**

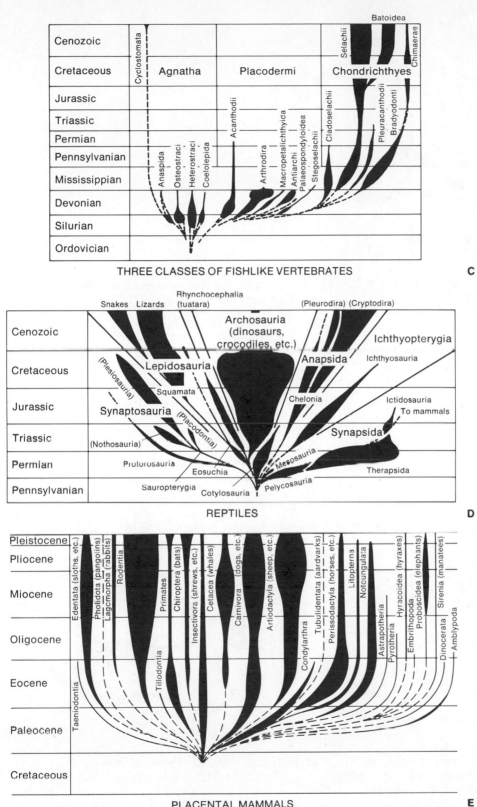

THREE CLASSES OF FISHLIKE VERTEBRATES   **C**

REPTILES   **D**

PLACENTAL MAMMALS   **E**

91

8

The cattle egret, *Bubulcus ibis,* which a few decades ago invaded the New World from the Old World, and has become established without obvious competition from other species. (After Henry 1971)

mals that had rodentlike teeth (*Figure 9*). In the Rocky Mountains, where the early rodent and late multituberculate fossils occur, the diversity of the one group drops as that of the other rises. Similarly the crocodiles appear as fossils in the Upper Triassic at just about the time that the morphologically similar but distantly related phytosaurs (*Figure 10*) became extinct. In many cases the supposedly competing groups are unrelated; in others the extinction of an older group is attributed to the greater efficiency of its descendants. For example, Simpson (1953) argued that competition by carnivores such as the Mustelidae (weasels, etc.) caused the extinction of the Miacidae from which they arose.

Competition is often a plausible explanation for the replacement of

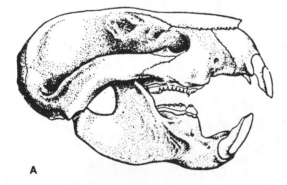
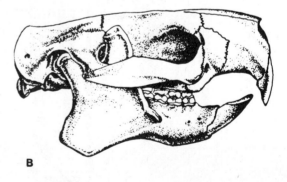

9   Ecological replacements in time. (A) A Paleocene multituberculate, *Taeniolabis.* (B) An Eocene rodent, *Paramys.* (From Romer 1960, *Vertebrate paleontology,* Copyright The University of Chicago Press)

one adaptive radiation by another but is probably never provable. We must beware of falling into the logical fallacy of *post hoc, ergo propter hoc,* the supposition that because event B followed event A, it was caused by event A. This is not to deny the importance of interspecific competition; but the fossil record is seldom precise enough to enable us to say that the extinction of one group coincided exactly with the proliferation of another, much less that the events had any causal relationship. Colbert (1949) argues that the Crocodilia invaded their present adaptive zone shortly *after* the phytosaurs had already become extinct and only then evolved their special adaptations for aquatic life. Similarly the mammals may have replaced the reptiles at the end of the Mesozoic not because they were competitively superior, but because the demise of ruling reptiles, due to entirely different causes, offered mammals ecological space.

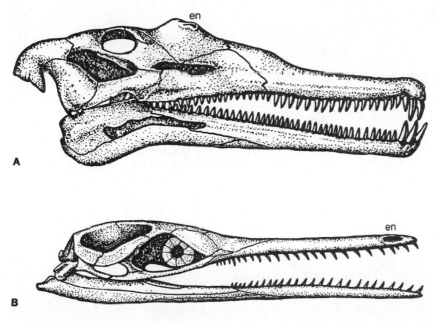

Ecological replacements in time. (A) A phytosaur, *Machaeroprosopus.* (B) A crocodilian, *Geosaurus.* Note the differences in skull structure, especially the position of the external nares (*en*). (From Romer 1960, *Vertebrate paleontology,* Copyright The University of Chicago Press)

Sometimes most of the members of an ecologically diverse group have been extinguished, except for a few species that then give rise to a second adaptive radiation comparable to the first, as in the ammonites (*Figure 11*), a great group of extinct cephalopods. Among the Crustacea the number of families of ostracods peaked in the Carboniferous, dropped during the Permo–Triassic extinction, and has been increasing ever since (Cisne 1974). The members of a second radiation are not

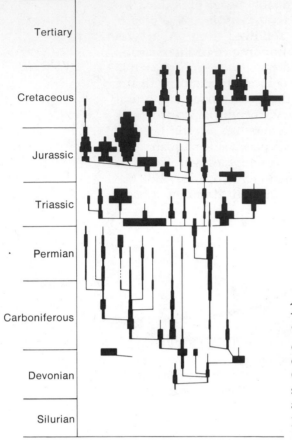

Tertiary

Cretaceous

Jurassic

Triassic

Permian

Carboniferous

Devonian

Silurian

11

The history of 41 superfamilies of ammonite cephalopods, showing mass extinctions at the end of Devonian, Permian, Triassic, and Cretaceous periods. In several instances a single group persisted, diversifying again after a mass extinction. The width of each bar is proportional to the number of genera known for that time period. (After Newell 1967)

identical to the first, but in some instances they come pretty close. The globorotaliid foraminiferans diversified greatly in the Eocene and Miocene after major extinctions in the Paleocene and Oligocene, respectively, and in both periods of diversification keeled species developed from unkeeled ancestors (Lipps 1970). Such repetition of evolutionary trends within a closely defined group has been termed ITERATIVE EVOLUTION (*Figure 12*). Like parallel and convergent evolution (*Chapter 7*), iterative evolution suggests that throughout geological time, and in different parts of the earth, similar ecological opportunities become available and are eventually utilized by phyletic lines that evolve the appropriate adaptive machinery.

The fossil record thus suggests that a group diversifies when ecological opportunity is present, either because empty niches are available or because it has adaptations that enable it to displace other species. But it would be wrong to conclude that the rate of origin of new species depends on available ecological opportunities. Species may fail to arise when niches are empty; conversely speciation proceeds apace when communities are saturated with species, even if the newly formed species are doomed to almost immediate extinction through competitive exclusion. Most such species are not fossilized, since their extinction follows so quickly on the heels of their genesis. Thus it is difficult to tell from the fossil record whether the rate of speciation in a group declines as the number of species increases. All other things equal, though, we can say that the diversity of a group may depend on its speciation rate; so the factors that determine speciation rates (*Chapter 16*) must bear importantly on the history of diversity.

PATTERNS OF EXTINCTION

Those who feel deprived of the opportunity to see a living dodo or who bewail the imminent extinction of the whooping crane are likely to forget that extinction has been the fate of the vast majority of all the species that have existed. Romer (1949) estimated that no more than 1 per cent of Mesozoic tetrapods are represented by living descendants; Simpson (1952) suggested that of about 500 million species that may have existed over all evolutionary time, perhaps 2 million are alive now. There are actually two kinds of extinction. TAXONOMIC EXTINCTION, or pseudoextinction, is the transformation of a single evolutionary line, so that its earlier and later members bear different names; thus *Homo erectus* (*Pithecanthropus*) became extinct only in the sense that it became *Homo sapiens*. TRUE EXTINCTION is extinction without issue; *Tyrannosaurus rex* has left no descendants. This is by far the more common kind of extinction. Indeed many authors (e.g., Eldredge and Gould 1972) believe that taxonomic extinction is extremely rare; they argue that the transformation of a species into a very different-looking form is far less common than the splitting of one species into two, one of which (the daughter species) is greatly changed, while the other (the parent) remains unchanged.

Despite the prevalence of extinction, very little is known about its causes. The greatest speculation surrounds those dramatic periods in history in which mass extinctions occurred. Valentine (1973) summa-

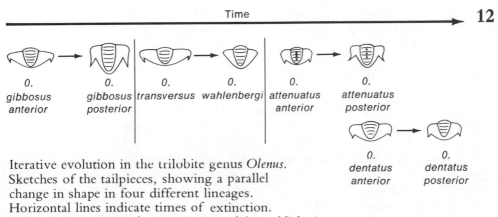

Iterative evolution in the trilobite genus *Olenus*.
Sketches of the tailpieces, showing a parallel change in shape in four different lineages.
Horizontal lines indicate times of extinction.
(After Simpson 1953, by permission of the publisher)

rizes the hypotheses that have been proffered. Some authors have postulated major changes in the physical environment — decreases in oxygen, changes of temperature, and the like. Others have focused on biological factors. Martin (Martin and Wright 1967), for example, believes that the extinction of large mammals in the Pleistocene is attributable to prehistoric overkill by early human populations. As we have seen, the idea that the extinction of major groups followed from competition with "superior" forms that then replaced them is prevalent. Extinction rates have often been high when biotas that evolved in different places came into contact. For example, South America was isolated from North America throughout most of the Tertiary and was the site of an enormous adaptive radiation of endemic mammals that resembled rhinoceroses, elephants, tapirs, horses, rodents, rabbits, and camels. The last of these became extinct in the Pleistocene, after a fauna of North American ungulates and carnivores crossed the Central American land bridge that formed in the Pliocene.

Predation and competition probably played a role in their demise, but disease should not be ruled out. Diseases to which one population has evolved resistance can devastate nonresistant forms, as is borne out by human history. The conquest of Amerind peoples by the Spaniards was facilitated by epidemics of smallpox and measles, and McNeill (1976) argues that disease played a role in most of the major events of human history. Within the last few decades the American chestnut has been almost eliminated by the fungus *Endothia parasitica,* which is a rather benign parasite on the Chinese chestnut; and the Dutch elm disease *Cerastosomella ulmi* is rapidly devastating the American elm.

The mechanisms of extinction can thus be many, but all are generally presumed to follow from a change in the physical or biotic environment. In the past some authors believed that species or higher taxonomic groups underwent a life cycle analogous to that of an individual, whereby after flourishing they degenerated into extinction. For example, the oyster *Gryphaea* supposedly evolved itself to death by developing so coiled a shell that the valves could no longer open and starvation ensued. Gould (1972) has shown that there is no factual basis for this myth. The idea that organisms evolve themselves to extinction by a kind of evolutionary momentum is no longer taken seriously. But evolution within a species can nevertheless lead to its extinction, by ecological feedback (*Chapter 12*).

Simpson (1953) and Williams (1975) have pointed out that extinctions may vary in kind, depending on the form of the environmental change. Some changes, such as volcanic eruptions, are such that a species simply has no possibility of evolving resistance. Other changes can in principle be survived by a resistant genotype, but they occur so rapidly and in such magnitude that there is no time for such a genotype to arise; Simpson cites the case of the extinction of bird species on Caribbean islands by the mongoose. In such instances the species survives only if a resistant genotype is already present when the catastrophe occurs. A few blight-resistant American chestnut trees may already have been present when the fungus swept through; if so, the survival of the species depends on these trees. Other changes, such as the increasing concentration of pollutants that we are now experiencing, may occur slowly enough that resistant genotypes may arise by mutation and recombination (at least in species with large populations and short generation time) before the environment becomes completely intolerable.

But many of the environmental changes causing extinction evoke no adaptive response, simply because they do not impose any new selective pressures. In resource-limited species many individuals fail to survive and reproduce because of insufficient resources. Genotypes that can use other resources would be advantageous, were it not for countervailing pressures, including the preoccupation of these other resources by other species. Thus the species is always under pressure to expand its niche but cannot. Now if the limiting resource diminishes, the population decreases, but individuals are dying at a greater rate *for the same reason as before*: the cause of mortality, to which the species has already adapted as far as possible, remains the same. Thus there is no more basis for genetic change now than there was when the resource was more abundant — even if the resource becomes so rare that the population becomes extinct.

It is customary to say that species meet extinction "because they are overspecialized." I do not think this is very informative, because the only way to demonstrate that a species was overspecialized is to point to its extinction. Every species is specialized in some way to some extent and is thus susceptible to extinction in some way. Nevertheless one can often infer from a species' morphology whether it led a specialized or generalized life-style. Paleontologists point to many cases in which a major episode of extinction in a taxonomic group follows the group's diversification into a host of morphologically bizarre forms, each presumably leading a peculiarly specialized way of life. The ammonite cephalopods and the dinosaurs are such cases. The ax of extinction sometimes falls more heavily on the morphologically unusual species than on their less flamboyant, presumably less specialized relatives. Among plankton only the morphologically simple species, whose morphology and widespread geographic ranges suggest that they were ecologically generalized, survived the late Cretaceous extinction (Lipps 1970). Bretsky (1969) discovered that in the late Devonian era, offshore communities of benthic invertebrates changed more drastically in composition than did inshore communities, which are typically populated by more generalized species.

But the evidence that specialized forms are more prone to extinction is not strong. Simpson (1953) stresses that unspecialized ammonites and dinosaurs persisted along with the more bizarre species but also became extinct at about the same time. Flessa, Powers, and Cisne (1975) assumed that specialization in arthropods is proportional to the degree to which the appendages are differentiated to perform different functions; by this criterion specialized orders have persisted just as long as generalized orders.

In the modern world the major cause of species' extinction is the destruction of their habitats. If this has been a common cause of extinction, the survival or extinction of a species may depend not on intricate adaptations, efficiency, competitive ability, or even genetic flexibility, but on the kinds of habitats it occupied. Habitat generalists could then well survive better than habitat specialists, but extinction might otherwise be almost random with respect to species' overall adaptations (Gould and Eldredge 1977). Extinction rates could then be virtually constant over long periods of time, as Van Valen (1973) claims they are for most taxonomic groups. If the extinction rate is constant, the logarithm of the number of taxa that survive for t years, plotted against t, is a straight line with negative slope. Van Valen finds that this holds approximately for a great many groups (*Figure 13*), so he concludes that although bivalve families have a lower rate of extinction than mammal families, for example, these rates have been effectively constant over long periods of time.

This finding implies that as the evolution of a group proceeds, the adaptedness of its members neither increases nor decreases. It also suggests that major, unique environmental events account for rather little extinction. Van Valen proposes that a constant rate of extinction implies a constant rate of deterioration of the environment, caused by the continual evolution of other species (predators and competitors) with which each species interacts.

Van Valen's hypothesis has many important implications; one is that the species in an ecosystem should be continually evolving new

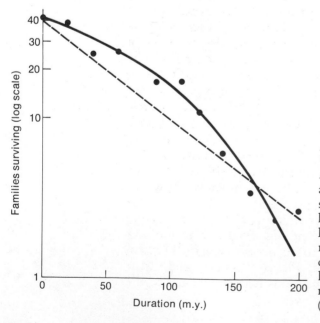

13

A survivorship curve, the paleontological analogue of the l_x curve in demography, showing the number of echinoid families that have survived for x million years. The dashed line shows a theoretical constant extinction rate, as postulated by Van Valen. The data correspond better to the curved line, however, showing that extinction rates have not been entirely constant in this instance. (Modified from Raup 1975)

ways of dealing with each other — a proposition that has been challenged (*Chapter 18*). In any case constant rates of extinction may have other explanations. If each of a great multitude of environmental changes, in the physical as well as the biotic environment, bears with it a small risk of extinction, the law of large numbers establishes an average rate of extinction that, given the imprecision of the fossil record, may well appear constant even if it is fluctuating slightly over time. Moreover extinction rates may not be all that constant. Sepkoski (1975) has suggested that the lognormal distribution of lengths of the Phanerozoic periods is responsible for the linear form of Van Valen's semilogarithmic plots. And Simpson (1953), who performed an analysis similar to Van Valen's, has argued that among the Carnivora recently evolved genera have not survived to the present in as great numbers as would be expected if their extinction rate had been the same as that of earlier genera. On the other hand, more genera of molluscs have survived from the Mesozoic than might be expected. Whether most extinctions in the fossil record have happened at a constant rate thus remains to be determined.

IS DIVERSITY IN EQUILIBRIUM?

As rates of origination and extinction change, so does the diversity of a group. Various patterns of diversity are illustrated in the fossil record (*Figure 7*). Some phyletic groups (CLADES) increased gradually to maximal diversity and then as gradually decreased to extinction. Some rose very rapidly to their peak and then gradually declined, while others showed the opposite pattern. In some groups, such as the ammonites (*Figure 10*), there were two or more peaks of diversity.

The explanations for such patterns fall into two classes. The IDIOGRAPHIC approach looks for a special explanation for each group. It asks what events were responsible for a group's proliferation and what environmental changes caused its decline. Its explanations are cast in such terms as the acquisition of special adaptations or the occurrence of particular environmental catastrophes. Recognizing that the biology of ammonites is different from that of dinosaurs, it seeks explanations for extinction that are different in detail for each.

The more general NOMOTHETIC approach casts explanations in more general terms, holding that broadly applicable ecological and evolutionary laws should apply equally to ammonites and dinosaurs. But the practitioners of this approach have different views about the history of diversity. Valentine (1970), for example, holds that the diversity of higher taxa (orders, classes, phyla) has been quite stable over most of Phanerozoic time, but that families and genera have fluctuated in number more drastically. The origination and extinction rates of families of benthic marine invertebrates have tended to balance one another, so that there has been a high turnover in the composition of the fauna (*Figure 14*). The number of fossilized genera rose steadily since the end of the Cretaceous; so Valentine, extrapolating from the number of species per genus in extant genera, concludes that the number of species has risen dramatically through the Cenozoic, reaching an all-time peak at the present time (*Figure 15a*). If this is true, one is tempted to wonder whether there is any limit to the possible number of species or at least if such a limit has ever been reached. Perhaps most communities are undersaturated with species and are below the theoretical equilibrium diversity.

Many authors believe that diversity-dependent factors rather closely regulate the diversity of species in communities, so that most communities at most times are at an equilibrium diversity. In this view, for example, the lower diversity in temperate than tropical regions is due not to recent Pleistocene extinctions, but to ecological factors that determine a lower species equilibrium. The belief that most communities are at equilibrium stems from such observations as the consistent relation between island area and species number ($S = cA^z$, where A is area and c and z are constants) and the convergent similarity of unrelated bird faunas (*Chapter 4*) — as if biotas have had enough time to conform to a pattern, determined *a priori* by ecological laws, of how many species can coexist in a given environment and what the ecological relations among the species must be.

Some adherents to the equilibrium view hold that most historical

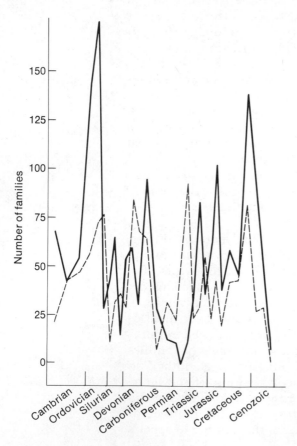

14

Biotic turnover: numbers of originations (solid line) and extinctions (broken line) of families of benthic invertebrates over geological time. (After Valentine 1973)

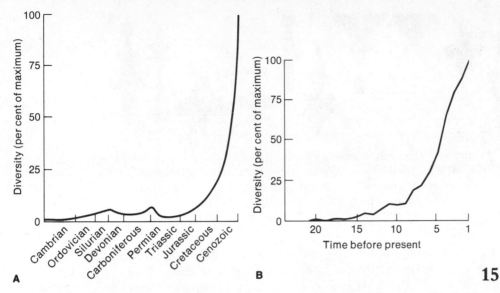

A

B

15

(A) The fossil record documents an apparent increase in the number of marine benthic invertebrate species over geological time.
(B) A similar curve emerges if the accidents of sampling an imperfect fossil record are simulated. The simulation is based on the assumption that there has been no real directional change in species number, but that originations and extinctions occur at random, at effectively constant rates. (After Raup 1972; Copyright 1972, the American Association for the Advancement of Science)

changes in diversity have been merely random fluctuations about an equilibrium value. A rather iconoclastic paper by Raup et al. (1973; see also Gould et al. 1977) exemplifies this position.

Raup et al. simulated the evolution of a major taxonomic group on a computer by specifying that in each time period a lineage could remain unchanged, branch into two daughter species, or become extinct. Each daughter lineage could do the same, so that the diversity of the group could increase or decrease as speciations or extinctions occurred. The critical feature of the simulation was that each species at each time underwent these events at random. The only constraint was that the overall probability of speciation was equal to that of extinction, so that the overall diversity of the group should be at an equilibrium. Note that increases in diversity are not contingent on any special circumstances such as the opening of a new ecological niche, nor is extinction due to any special environmental changes.

Some of their results are illustrated in Figure 16. Their randomly generated clades showed the same kinds of pattern that appear in the fossil record (cf. *Figure 7*); they proliferate rapidly and then become extinct gradually as did the trilobites, increase slowly and become extinct rapidly as did the dinosaurs, have multiple peaks of diversity as did the ammonites, and even replace each other as if there had been competition. Raup et al. conclude that many of the historical changes

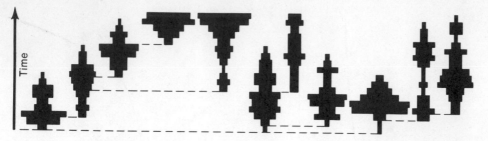

16 Phylogeny and temporal variation in diversity of 11 lineages (clades) generated on a computer by random speciations and extinctions. The width of a bar reflects the number of species. Many of the changes in diversity actually observed in the fossil record are mimicked by these random clades. (After Raup et al. 1973)

in diversity may be attributable to long successions of random events. Not all, however. Their simulations did not include simultaneous mass extinctions like those of Permo-Triassic time, such extremely rapid proliferations as some groups (e.g., therapsid reptiles) experienced, nor what they call the "coelacanth effect," the long-term survival of a single species in a group that otherwise became extinct long ago. These kinds of event may require special biological explanations.

Raup (1972), extending this approach, disputes Valentine's contention that species diversity has been increasing in time. He points out that important biases in the fossil record may suggest an increase in diversity even if it has remained constant. For example, since young rocks are less likely than old rocks to have been destroyed by the ravages of time, the last occurrences of a group of species are more likely to be found than the first occurrences, and the geographic coverage of young rocks will be greater. Moreover, if an extant group is known from just a single fossil, its geological range is from that time until the present. If it were not extant, its geological range would be very brief instead of covering all subsequent time.

By computer Raup randomly generated times of first and last occurrences for 2000 "species." The "true" species diversity increased rapidly, and then dropped to a steady equilibrium. When he sampled these species as if they had been fossilized but simulated all the geological biases that he argues must exist, the apparent species diversity seemed to increase through time (*Figure 15b*), just as Valentine's curve does. There is no reason, Raup argues, to suppose that the apparent changes in diversity through time are due to anything more than the accidents of sampling from an imperfect fossil record.

Is there actually any evidence that origination and extinction rates balance, so that diversity can be equilibrated? Several analyses suggest that there may be, but I believe that the question is still unresolved. There is some evidence that the structure of certain communities has remained remarkably constant through long periods of time, despite turnover in species composition. Walker and Laporte (1970) found a similar diversity of ecologically equivalent species in the Ordovician and Devonian (70 million years apart). even though the Devonian genera were entirely different from the Ordovician. In both faunas the ostracods dominated the supratidal, trilobites the midintertidal, and so on. Levinton and Bambach (1975) found an even more striking congruence of Silurian bivalve communities with living communities in Massachusetts. Apparently no new ways of partitioning resources in this community type have evolved in 440 million years!

T. Schopf (1974) and Simberloff (1974) calculated that the area of continental shelf accessible for benthic invertebrates declined at the end of the Permian and that the species diversity of invertebrates dropped accordingly. Both before and after the mass extinction, diversity followed the species-area relation $S = cA^z$; diversity, they argue, had simply dropped to the new equilibrium determined by the reduced area.

Taking a different approach, Webb (1969) found that in the mammal fauna of North America over the last 12 million years, both extinction and origination have proceeded at a high rate — a turnover rate of one-fifth to one-third of the existing genera per million years, on average. Extinction rates have tended to fluctuate in concert with origination rates, so that the diversity has remained rather constant. Mark and Flessa (1977) have performed a similar analysis on long-term changes in the generic diversity of mammals and of brachiopods. They too find a correlation between rates of origination and extinction, but the correlation is rather weak; they conclude that it will be difficult to determine by this approach whether diversity is in equilibrium.

Thus we probably cannot yet judge whether species diversity has been at equilibrium for long periods of geological time. Although particular community types such as the marine benthos have been relatively unchanged for eons, it seems likely that the greater global diversity of organisms has changed as continents have drifted and as adaptive breakthroughs have opened entirely new adaptive zones. For instance, no one would deny that the evolution of insects, which number more than 800,000 described species, greatly increased the diversity of the world's fauna. The evolution of wings opened entirely new ways of life, but the number of such ways of life achieved by a group after its fundamental body plan has been evolved may be limited. Cisne (1974) claims that the diversity of orders of aquatic arthropods, for example, reached its maximum in the Mesozoic and has since remained unchanged. Such a conclusion may not hold for other groups; the contemporary diversity of rodents, Hymenoptera, or composites may well be comparable to the earlier stages of crustacean evolution (*Chapter* 7). The variety of adaptive forms yet to be evolved must remain unknown to us; but there is certainly no reason to suppose that the present is any more a culmination of evolutionary history than previous times.

SUMMARY

Great geological and climatic changes — continental drift, orogeny, changes in sea level, glaciations — have marked the history of the earth. The composition of the earth's biota has changed continually over the last 3 billion years, as whole groups of organisms have become extinct and others have arisen. Rates of proliferation have varied irregularly and are often greatest soon after the origin of novel adaptations that permit occupation of new adaptive zones. Groups may flourish most after competing groups have become extinct; it is possible, but largely unproven, that such replacements are directly caused by competition. Proliferation depends on speciation rates, which are influenced by many factors besides ecological opportunity. Extinction rates may be fairly constant for each taxonomic group, although there have been several episodes of mass extinction. In general, there is little evidence of overall improvement in adaptation during the history of life. Many changes in diversity may require no special explanation but follow directly from random processes of origination and extinction. Whether the diversity of organisms is presently at equilibrium and has been so throughout much of history is highly debated.

FOR DISCUSSION AND THOUGHT

1   What is meant by *historical accident,* or the proposition that an event is contingent on history? What kinds of event are not contingent on history?

2   How does one account for the observation that higher taxa (phyla, classes) persist longer through evolutionary time than lower taxa (genera, species)?

3   If we survey existing organisms, is there any way we can recognize major innovations that could lead to future diversification? Can we identify existing species that are the progenitors of future higher taxa in new adaptive zones?

4   Do organisms ever vacate a region by actively moving out, rather than by local extinction? Why or why not?

5   Argue for and against the proposition that unfilled niches exist.

6   Is there any reason to suppose that the probability of a species' undergoing speciation depends on the existing level of species diversity?

7   Account for the persistence of relict species (e.g., coelacanth, *Sphenodon,* ginkgo) that are members of otherwise extinct groups.

8   Cite instances in which the evolution of a group of organisms provides a stimulus, or opportunity, for the evolution of other groups. Does diversity beget diversity?

9   Refute the proposition that species evolve to avoid extinction. Are evolutionary rates likely to increase immediately before extinction? If so, why? If not, why not?

10   If the extinction rate of a group is effectively constant over time, it would seem that recently evolved species are no better adapted than older species. Similarly, if species tend to evolve to be more highly specialized, and if specialized species have a high risk of extinction, it would seem that the overall course of evolution is toward higher susceptibility to extinction. Do these statements contradict the supposition that natural selection results in greater adaptation of organisms to their environments?

11   What does *random extinction,* as used by Raup et al., mean? Doesn't every extinction have a cause?

12 In what sense, if any, can groups with more species be considered more successful than groups with fewer species? Is high diversity evidence of better adaptation?

MAJOR REFERENCES

Raup, D. M., and S. M. Stanley. 1971. *Principles of paleontology*. Freeman, San Francisco. 388 pages. An introduction to the principles and methods of modern paleontology, with a discussion of many aspects of evolution as revealed by the fossil record.

Valentine, J. W. 1973. *Evolutionary paleoecology of the marine biosphere*. Prentice-Hall, Englewood Cliffs, N.J. 511 pages. An ecological and evolutionary discussion of marine paleontology, including many of the topics of this chapter.

Simpson, G. G. 1953. *The major features of evolution*. Columbia University Press, New York. 434 pages. A seminally important discussion of evolution from the paleontological perspective.

Spatial Patterns of Diversity

The historical record raises questions such as, Why has diversity varied over time? Has it usually been at an ecological equilibrium? Why have particular groups existed at some times but not others? Precisely comparable questions apply to the spatial distribution of organisms at present. Why, for example, is the diversity of most groups of organisms greater in the tropics than at higher latitudes? Why are lungfishes found in South America, Africa, and Australia, but not elsewhere? These are among the problems that form the subject of biogeography, a topic of great historical importance. As is well known, the peculiarities of geographic distribution were among the most critical observations that led Darwin to his formulation of evolution, and the subject was a major concern of Alfred Russel Wallace.

The relation between the temporal and spatial distributions of organisms is intimate. An understanding of present patterns of distribution depends strongly on a knowledge of historical changes in climates, geography, and species' distributions. Indeed the relationship between the fields of geology and biogeography is a mutualistic one. The geologist's evidence of the past distribution of continents and land bridges provides the routes of migration that biogeographers need to explain the biotic similarities of different regions, and patterns of biotic similarity are often the best evidence for past geological events. For example, the Great Lakes of North America presently drain into the Atlantic via the St. Lawrence River. But the composition of their fish fauna is very similar to that of the Mississippi River, indicating that until recently the lakes drained into the Gulf of Mexico.

This case exemplifies one concern of biogeography: the mapping of distributions of particular taxonomic groups that, in concert with geological and paleontological evidence, will elucidate how they arrived at their present distributions. This historical, taxonomic approach grades into

another major concern: the search for general evolutionary and ecological rules that govern geographic patterns of species diversity and
the conditions that favor the origin of new taxonomic groups. These
are again idiographic and nomothetic approaches.

1

A

B

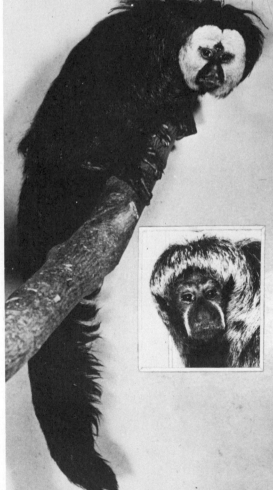

C

D

Some endemic Neotropical forms. (A) An
edentate mammal, the giant anteater
(*Myrmecophaga*), representative of an order
now limited to the Neotropics. (B) The
barred antshrike (*Thamnophilus doliatus*) in the
family Formicariidae, one of several families
of suboscine birds limited to or originating in
the Neotropics. (C) A platyrrhine primate,
the saki monkey *Pithecia*. (D) *Nidularium,* a
member of the Neotropical family
Bromeliaceae. (B from Haverschmidt 1968,
C from Walker 1975, D from Engler 1930)

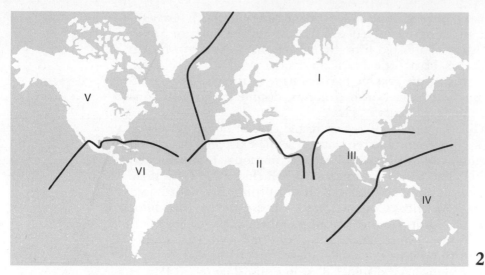

2

The zoogeographic regions recognized by A. R. Wallace: Palearctic (I), Ethiopian (II), Oriental (III), Australian (IV), Nearctic (V), and Neotropical (VI). Note that the borders (which, to be sure, are rather arbitrary) do not necessarily demarcate the continents.

PATTERNS OF GEOGRAPHIC DISTRIBUTION

The earliest biogeographers quickly noticed that the distributions of many species coincide to a greater or lesser degree. Often a simple ecological explanation is ready at hand; the composition of the flora changes rather abruptly, for example, as one goes from the Great Plains to the Rocky Mountains, for the drainage and soil types likewise change. But there is often no ecological basis for the concordant distributions of unrelated higher taxa. The Neotropics bear a unique biota (*Figure 1*) of edentates (anteaters, sloths, armadillos), platyrrhine primates (capuchin monkeys, etc.), cavioid rodents (guinea pigs, chinchillas, and many others), suboscine birds (antbirds, ovenbirds, etc.), dendrobatid frogs, several families of catfishes, bromeliad plants, and many other endemic groups. The Neotropics have thus been termed a BIOGEOGRAPHIC REALM, with a biota that distinguishes it from other such realms. The realms are not all that easy to define, because different workers recognize different numbers of biotic assemblages. The fauna of Central America is distinctive enough from that of South America, by virtue of its many endemic genera and a few endemic families, to be given equal rank by some biogeographers. But many biogeographers would subscribe, with varying reservations, to a classification like that of Darlington (1957), who rather closely follows Wallace in recognizing six zoogeographic realms (*Figure 2*). The realms do not correspond exactly to continental boundaries, and their borders do not necessarily lie where one might expect; the biota of Central America is very different from that of North America.

Such classifications imply that each major region has been the locale of an evolutionary history quite independent of evolution elsewhere, that endemic families have evolved and dominated the biota of South America, for example, affected little by evolutionary events in North America. There has been little biotic exchange with other regions. Presumably climatic or geographic barriers have in the past prevented much interchange with, say, North America, despite the present connection between the continents.

Such a conclusion, although correct in the case of South America, is too hasty. The biogeographic regions mask the existence of a small but significant minority of groups whose distributions are not limited to a single realm. The South American fauna includes the Cricetidae (rice rats, deer mice, and the like) which are prolifically distributed throughout North America. Conversely the opossum, the armadillo, and the tanagers of North America are all members of primarily Neotropical groups. Some groups, such as pigeons and owls, are highly diverse in most of the realms.

Finally some groups have disjunct distributions. The lizard family Iguanidae is diverse in North and South America but has a few species in the Galápagos Islands, Madagascar, and Fiji and Tonga. Many disjunct distributions follow one of several patterns. Eastern North America and eastern Asia share many elements, such as the tulip tree (*Liriodendron tulipifera*) and the skunk cabbage (*Symplocarpus foetidus*). Many plants are distributed in the polar or temperate regions of both North and South America (Raven 1963, Cruden 1966); *Plantago maritima* is found in both northern North America and in Patagonia and Tierra del Fuego.

Many groups inhabit only the southern continents. Africa, Australia, and South America all have lungfishes, ratite birds (ostrich, rhea, emu, and others), and the plant family Proteaceae. Australia and South America share the chelydid turtles, the southern beech (*Nothofagus*), and the marsupials. Africa and South America share pelomedusid turtles, pipid frogs, characid and cichlid fishes, and the Podocarpaceae. The caviomorph rodents (guinea pigs, porcupines) of South America, a highly diverse group from which the North American porcupine is derived, are extremely similar in many skeletal and muscular features to the African porcupines (Hystricidae). Together they form the classical suborder Hystricomorpha, with a South American–African distribution.

THE GENESIS OF BIOGEOGRAPHIC PATTERNS

Such distributions pose problems with which biogeographers (e.g., Cain 1944, Good 1947, Hedgpeth 1957, Darlington 1957, Hubbs 1958, Udvardy 1969) have been concerned. How have they come into being? This question usually entails asking where a group originated and how it managed to get where it is today.

Croizat and his school (1974) argue that this is a misguided approach. They contend that a group with a wide distribution becomes differentiated *in situ* as the geographically separated populations evolve into different species. Many fishes and invertebrates on the Pacific coast of Central America are specifically distinct from their close relatives in the Caribbean, but one would not argue that either fauna is derived from the other by migration and subsequent differentiation. Rather each pair of related species differentiated after the Central American isthmus arose and split a formerly continuous population in two. They feel that this model of geographic speciation should apply to other cases of faunal or floristic differentiation and that the migration of a group from its center of origin is a will-o'-the-wisp. But if a monophyletic group stems from a single ancestral species and if, as there is good reason to believe (*Chapter 16*), most species differentiate from their ancestors in a very restricted geographic locality, a group must have spread from its ancestor's site of origin.

But determining where a group originated can be difficult. It is often presumed that the ancestors of a group originated where the group is now most diverse, that the murid rodents (Norway rat and house mouse are familiar members), for example, arose in southeastern Asia where they now exhibit the greatest adaptive radiation. Willis (1922) expressed this idea explicitly in his age-and-area hypothesis, which held that the center of the region inhabited by a group, from which species would have emanated in all directions, was the site of its origin; and it was here, where evolution had been proceeding longest, that the greatest diversity would exist. It has since been realized that this is an overly simplistic view. Extinction can alter patterns drastically; for example, there is good fossil evidence that the Equidae (horses) arose and diversified in North America, but they are now extinct there and are limited to the Old World. Moreover, the geographic barriers that engender species formation (*Chapter 16*) may be more prevalent in some regions than others, resulting in a local proliferation of the group far from its site of origin. There are as many species of drosophilid flies in the Hawaiian Islands as in the rest of the world combined, but the family surely did not originate there.

A more reliable way of inferring regions of origin might be to compare the distribution of the more "primitive" members of a group with that of more "derived" forms. (This requires that we know how to tell primitive from derived forms, a problem treated in Chapter 7.) Thus, for example, the murid rodents of Australia are mostly endemic genera that are structurally more specialized than those of Asia, which are presumed to resemble more the ancestors of the family. This is one line of evidence that the Australian murids were derived from those of Asia, rather than *vice versa* (Simpson 1961a). But there is no assurance that species most like the ancestor of their group have survived at the group's site of origin rather than elsewhere, as the horses bear witness. Very old groups often have *relictual* distributions: they persist only in small, sometimes scattered, portions of their former range. The tuatara (*Sphenodon*) is a most primitive reptile, the sole survivor of a family that flourished from the Triassic to the mid-Cretaceous (*Figure 3*). It has survived only on New Zealand, but the family did not originate there; fossils are known from Africa, Eurasia, and North America.

THE BIOTA OF SOUTH AMERICA

Despite these complications, evidence from taxonomy and paleontology can enable us to formulate reasonably parsimonious interpretations of past patterns of movement. As an example, I will discuss the peculiar biota of South America, a subject treated by Simpson (1950), Darlington (1965), Cracraft (1974*a*), and Keast et al. (1972), among others.

South America was isolated from North America throughout most of the Cenozoic, as indicated by geological evidence and by the peculiarly endemic biota it supports. A series of Central American islands may have served as stepping stones for the interchange of some species, but there was no continuous land connection. This is proven by the complete absence from South America of any of the very diverse families of North American freshwater fishes — minnows, sunfishes, darters, and the like. The fish fauna of South America consists of a great many endemic groups of gymnotids (e.g., the electric eel) and catfishes and many kinds of characids (e.g., the piranha) and cichlids, two groups that also occur in Africa.

The mammalian fauna of South America consists of ancient, moderately old, and young groups. The edentates are known from the late Cretaceous and Paleocene of both North and South America, which were joined at that time. They diversified in South America in the Cenozoic, where endemic families arose that did not make their way to North America. Even if fossil evidence were lacking, we would deduce that contemporary edentates stem from South American ancestors because anteaters, sloths, and armadillos are so very different from one another that they must be the end products of a very old adaptive radiation whose intermediate forms have long since become extinct. Only recently a single species, the nine-banded armadillo, has invaded North America. Its relationship to a diverse group of South American armadillos indicates that the movement has been to the north from

3

The tuatara, *Sphenodon punctatus,* the sole surviving member of the ancient family Sphenodontidae. It has a relictual distribution, being limited to small islands along the coast of New Zealand. (From *The life of reptiles* by Angus Bellairs, Universe Books, New York, 1970)

South America, for otherwise the nine-banded armadillo must have arisen in North America (from unfindable ancestors), migrated to South America, and given rise there to a diverse group of species, while not diversifying at all in the north. Thus paleontological evidence and taxonomic evidence alike indicate that the contemporary edentates evolved in South America long ago. The South American marsupials are likewise a very old South American group, from which the opossum *Didelphis virginianus,* like the nine-banded armadillo, has recently moved to the north.

The same kinds of argument apply to the young South American groups, such as the cricetid rodents. These are quite similar to the cricetids of North America, where they are very diverse; some of the South American forms (such as *Oryzomys,* the rice rat) are congeneric with North American species. These facts suggest that cricetids only recently invaded South America from North America. Paleontology supports this conclusion; the group existed in North America since the Oligocene but first appears in the South American fossil record in the Pliocene, when the seas were bridged by the Central American isthmus.

The moderately old South American mammals include the platyrrhine primates (capuchin monkeys, marmosets) and the caviomorph rodents. Both groups are clearly related to primates and rodents elsewhere, and both orders clearly originated outside South America; for the most primitive primates, the prosimians (lemurs, galagos, tarsiers), are in the Old World tropics, and the most primitive living rodent, the mountain beaver (*Aplodontia*), is North American. Thus they invaded South America but not as recently as the cricetids, for the New World and Old World primates are very different from one another and the caviomorphs are very different from North American rodents. Nor are they as ancient in South America as the edentates and marsupials; they consist not of a few disparate groups but of adaptive radiations of which various intermediate forms still exist. Thus they are thought to have arrived in South America in the mid-Tertiary, when the continent was an island.

Many authors think these mammals came from North America by hopping from one Central American island to another. There was an early Tertiary prosimian primate fauna in North America, although it did not have particularly platyrrhine characteristics. These could have developed after the first prosimians arrived in South America, where they gave rise to monkeys that are convergently similar to those of the Old World. The problem is even more tantalizing in the case of the caviomorphs. These are so similar in structure to the Old World hystricomorphs (*Figure 4*) that some authorities (e.g., Hershkovitz in Keast et al. 1972, Hoffstetter 1972) believe that the ancestor of the caviomorphs was carried from Africa to South America on a raft of floating vegetation across the Atlantic when the ocean was narrower than it is now. But other authorities (Wood and Patterson 1971, Patterson and Pascual 1972) hold that the similarity of Old World hystri-

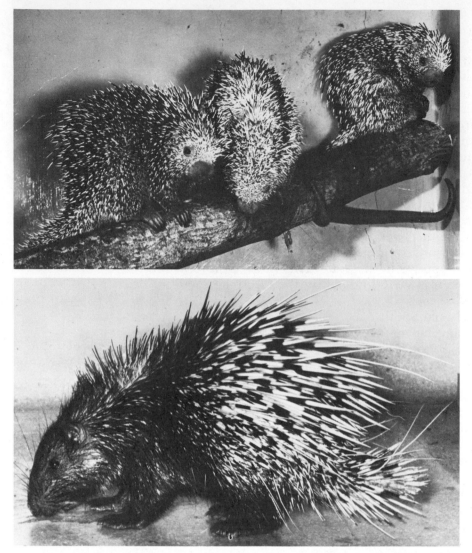

4   A problem in zoogeography and systematics. (A) American porcupines (e.g., the Neotropical prehensile-tailed porcupine *Coendou prehensilis*) and (B) Old World porcupines (e.g., the Malayan porcupine *Hystrix brachyurum*) may be unrelated and convergently similar. If they do form a monophyletic group, it is difficult to account for their distribution. (From Walker 1975)

comorphs and New World caviomorphs is an extreme instance of convergent evolution and that a North American rodent invaded South America and there developed caviomorph features. The problem of whether the hystricomorphs and caviomorphs are related demonstrates forcefully that the entire study of biogeography depends on an understanding of phylogenetic relationships.

Dispersal to South America via the north, as for the Cricetidae, accounts for the distribution of many disjunctly distributed groups, such as the tapirs, which presently occupy tropical America and south-eastern Asia but left fossils in the north. But in the case of old groups such as lungfishes, disjunct distributions may equally well have come about by continental drift. Not only were the southern continents contiguous well into the Mesozoic and much closer to one another in the early Cenozoic than they are now, but they lay at low enough latitudes to support a warm-adapted biota. Antarctica and the other southern continents had a common flora (the *Glossopteris* flora) and fauna. In the Triassic the therapsid reptile *Lystrosaurus,* as well as other large reptiles and amphibians, occupied Antarctica, Africa, and India. Continental drift seems the simplest explanation for the distribution of quite a few groups — the southern beech *Nothofagus,* ratite birds, leptodactylid frogs, certain insects. For example, Brundin (1965) has shown that the midges (Chironomidae) of the southern continents are more closely related to one another than they are to northern forms. The characid fishes exist in South America and Africa almost certainly because of continental drift. If they dispersed through North America, they left neither fossils nor surviving species, which is most unlikely. In such instances it is not useful to ask whether the group arose in South America or in Africa, for it was widespread over both before they became separate land masses.

THE IMPACT OF THE PLEISTOCENE ON GEOGRAPHIC DISTRIBUTIONS

More recent events, the glacial and interglacial events of the Pleistocene, have left their mark on the distributions of plants and animals throughout the world; I emphasize here their effects in eastern North America.

Fossil evidence indicates that in the late Cretaceous and early Tertiary, a broad band of warm-adapted vegetation lay across the northern part of the North American continent, across Alaska and the Bering land bridge into eastern Asia (Graham 1964). The Eocene vegetation of the northern United States included forms such as cycads, figs, manihot, and canna that are now limited to the tropics and subtropics. Many genera, including sassafras and dogwood, stretched from western Europe through Asia to eastern North America. There may have been a North Atlantic connection from North America through Greenland to Europe, facilitating the movement of elms, birches, maples, and other northern forms. A slow cooling trend in the early and mid-Tertiary changed the complexion of the flora. By the Miocene the vegetation had changed to a rather modern temperate flora. Some forms, such as manihot, had been pushed south to the tropics; others persisted in southern Florida. A few members of primarily tropical groups adapted to the cooler conditions; for example, the pawpaw (*Asimina triloba*) of the southeastern United States is a temperate member of an otherwise tropical family (Annonaceae). Across much of the northern part of the continent, a tropical flora had been replaced by the cold-adapted Arcto-Tertiary Geoflora of such plants as hickories, beeches, and maples.

Somewhere between a million and a half million years ago, the cooling trend of the Tertiary took on violent fluctuations. Ice sheets advanced to the south at least four times in North America. The most recent glaciation, the Wisconsin, began about 65,000 years ago and reached its furthest extent in the period 30,000–18,000 B.P. Climates became cooler and drier to the south (Deevey 1949, Kurtén 1971); spruce and fir pollen has been found in Pleistocene deposits in Florida, Louisiana, and Texas, and walruses occurred along the South Carolina coast.

Deevey (1949) and Kurtén (1971) review the effects of these events on biogeography and evolution. Some northern forms have persisted in the south as relics of the ice age; thus there are beeches in the mountains of Guatemala, yew in northern Florida, and crossbills in Hispaniola. Many warm-adapted forms retreated into two or more widely separated refuges and were eliminated from intervening areas. On a broad scale this may account for the disjunct distribution of formerly widespread groups that are now restricted to eastern Asia and eastern North America, such as tulip trees and alligators. Within North America refuges in the southeast and the southwest may account for the present distribution of species like the scrub jay and the burrowing owl, found now in western North America and in peninsular Florida. Thus continuously distributed species were broken into populations that could not exchange genes and so could evolve independently. Some such pairs of populations have diverged enough to be termed different subspecies, while others apparently have become different

Western diamondback
rattlesnake

Eastern diamondback
rattlesnake

5

The western and eastern diamondback rattlesnakes, *Crotalus atrox* and *C. adamanteus,* closely related species that differentiated in southwestern and southeastern refugia during the Pleistocene. (Photographs from Klauber 1972; map after Conant 1958)

species. The eastern and western diamondback rattlesnakes (*Crotalus adamanteus* and *C. atrox*), for example, may have developed in this way from their common ancestor (*Figure 5*).

This discussion of Pleistocene climatic events returns us to a persistent problem. Is biotic diversity at equilibrium, or are species still expanding their ranges from glacial refuges, so that the current composition of communities is still affected by history?

In some cases community composition may still be recovering from the Pleistocene. For example, some species of bog-inhabiting beetles — species that fly little and so disperse very slowly — reach their northern limit along the southernmost frontier of the Wisconsin glaciation, even though apparently suitable habitats exist immediately to the north of this border (Reichle 1966). But there is a strong feeling among ecologists that the Pleistocene has not left the species composition of most communities in disequilibrium. Opossums, cardinals, and other species have rapidly been moving north in historic times because the climate has become warmer and human activities have altered habitats. Their movement indicates that species can expand their ranges very rapidly to occupy suitable habitats almost as soon as they become available. The species equilibrium should therefore be reached rapidly, although the equilibrium itself changes as environments change (*Figure 6*). Still it is not certain whether this equilibrium has actually been attained.

SPATIAL PATTERNS OF SPECIES DIVERSITY

In recent years, especially, biogeographers have been concerned not only with the historical events that contributed to the present distribution of various groups of organisms, but with broad patterns of species diversity (MacArthur 1972). In almost any group of plants or animals, for example, the number of species on an island bears a regular

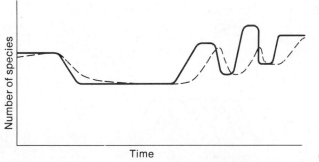

6

As environments change, the theoretical equilibrium number of species ($\hat{S}$) determined by the intersection of origination and extinction rates (*Figure 5, Chapter 5*) fluctuates (solid line). The actual number of species (broken line) approaches the theoretical number, but probably lags behind, especially when increasing. The theoretical equilibrium number $\hat{S}$ is probably approached more closely if environments persist for long periods of time than if they fluctuate every few thousand years or so.

relationship to island size. Another prominent pattern is that of latitudinal gradients in diversity. A square mile of seemingly homogeneous environment is likely to support no more than 30 or 40 species
of woody plants in a temperate broadleaf forest, but in the Amazon
Basin it typically has more than 200 species of trees as well as innumerable species of vines and epiphytes. Almost any group of insects,
birds, amphibians, or reptiles shows a comparable pattern. Within the
tropics diversity is usually far richer in the wet lowlands than at high
elevations or in regions with sparse or highly seasonal rainfall.

The formulation of a coherent theory of species diversity has preoccupied many ecologists for about 15 years. Most of the ecological
theories of species diversity depend at least partly on the competitive
exclusion principle, treating ways in which species at the same trophic
level can coexist in large numbers without eliminating some by competition (*Figure* 7). Theories of the high diversity in tropical rain forests
include the proposition that greater overlap in resource use is possible

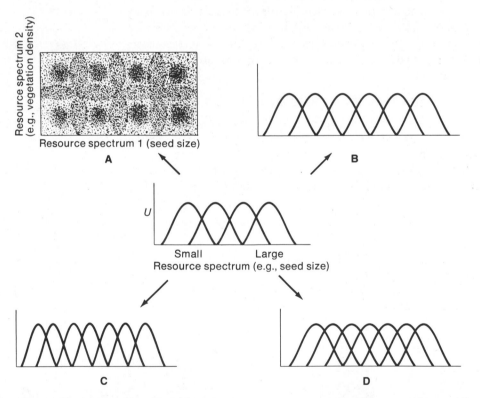

7 If one dimension of a species' niche is represented by its utilization (U) of each resource on a spectrum of resources, the number of competing species that can coexist will increase if there is a wider variety of resources (B), if each species is more specialized but overlaps with others to the same proportional degree (C), if the overlap in resource use is proportionally greater (D), or if the species are segregated by more than one kind of resource spectrum (A).

Possible evidence of differences in specialization associated with differences in species diversity. At each of several sites in the Panama Canal Zone (C) and in Puerto Rico (P), where the diversity of bird species is lower, measurements were taken of bird diversity and of the amount of foliage at several levels above the ground. (A) When only three levels of foliage are recognized, the relation between bird and foliage diversity is weak. (B) The relationship is strong, however, if foliage is assigned to two levels (0–2′, >2′) in Puerto Rico and to four levels (0–2′, 2′–10′, 10′–50′, >50′) in the Canal Zone. This suggests that the bird community "recognizes" more layers in the Canal Zone forest and so consists of more specialized species, each foraging in a more narrowly defined layer of foliage. (Modified from

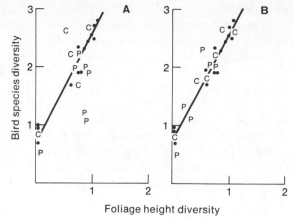

MacArthur, Recher, and Cody 1966, *The American Naturalist,* Copyright The University of Chicago Press)

8

if the environment is relatively constant or if the populations are partly limited by predation. Janzen (1970) and Connell (1970), for example, suggest that host-specific predation by insects and other herbivores on the seeds and seedlings of trees is more prominent in the tropics and prevents competitively superior tree species from monopolizing space. Higher species diversity also follows if species can partition more kinds of resource (niches have more resource axes), if each kind of resource is more diverse (niche axes are longer) or if many species are highly specialized. There is suggestive evidence that each of these possibilities may partly account for high tropical diversity (*Figure 8*), as well as a general impression among biologists that many tropical species, such as the oilbird (*Steatornis*) which feeds almost exclusively on palm fruit, are quite highly specialized. But even though tropical species in some groups such as papilionid butterflies (Scriber 1973) are more specialized than their temperate-zone counterparts, there is no strong evidence that this is generally the case. Certainly many species of the tropical rain forests are no more specialized than species elsewhere.

Ecologists, then, seek to explain geographic variation in the diversity of species by theories based on the population dynamics of interacting species. They are evolutionary theories only in the sense that they postulate the degree of ecological specialization that is likely to evolve under specified environmental conditions or (and this is not quite the same thing) the properties a species must have evolved if it is to persist in stable combination with other species. But to what extent do variations in species diversity depend on the *history* of evolution?

One very plausible proposition is that some environments are more difficult to adapt to than others. Hot springs, to take an extreme example, support only a few species, mostly of prokaryotes, perhaps because cell membranes must have very special properties to function at high temperatures. May not the same principle hold for other nasty environments — those that are excessively cold or dry? Opponents of this view point to the high diversity of deep-sea invertebrates or to the greater number of terrestrial species than of aquatic species, and what could be a harsher environment than dry land for life, which evolved in water? There is no such thing as a harsh environment, they argue; all environments are equally harsh — or benign — to the species adapted to them. If an environment has existed for a reasonably long time, species will have adapted to it and diversified as greatly as the laws of interspecific interaction allow. Communities are saturated with species.

On the other hand, the invasion of an environment that presents special physiological problems does depend on the evolution of key adaptations that are often quite complex and may arise in evolution only infrequently. Each group that happens to acquire such an adaptation may be limited by its body plan to a restricted range of adaptive radiation. Thus only three animal phyla — chordates, arthropods, and molluscs — have fully adapted to terrestrial existence. How much greater would terrestrial diversity be if there were terrestrial echinoderms? Is it purely historical accident that they have not evolved adaptations to terrestrial life? Similarly very cold places do not have ants, possibly because the activities of the colony would go on so slowly in cold ground. Perhaps ants could function efficiently in cold environments — by building nests in sunlit sites on the surface of the ground, for example — but they have not done so. Is their distribution limited by their physiology and behavior?

The implication of this line of thinking — that some environments are so harsh that many creatures simply have not adapted to them yet — is that given enough time, appropriate adaptations will arise, so that species diversity on mountain tops and at high latitudes may someday be as great as that in a tropical wet forest. Unfortunately this is an untestable proposition; and while we may be able to show that the distributions of some groups are limited by their physiology, it is hard to prove that vacant adaptive zones really await ants on mountain tops or monkeys in temperate forests.

Historical biogeography suggests that low species diversity in some places is due to accidents of dispersal; but this, too, amounts to the proposition that there are vacant niches. The diverse freshwater fish fauna of North America includes minnows (Cyprinidae), suckers (Catostomidae), sunfishes (Centrarchidae), darters (Percidae), and other groups that do not extend past southern Mexico. Almost no members of the even more diverse South America fauna of catfishes, gymnotids, characids, and cichlids extend north past Panama. The freshwater fauna of Central America, cut off from both continents for most of the Tertiary, is sparse not for ecological reasons, but because of history.

THE TROPICS: CRADLE OR MUSEUM?

This heading, borrowed from G. L. Stebbins's book *Flowering Plants: Evolution above the species level* (1974), is a *mot juste* for a controversy that bears on a great deal of evolutionary thought and is most germane to an exploration of diversity. The polar views are expressed by Matthew (1915) and Darlington (1957). Matthew argued that most of the major groups of mammals had originated in the temperate zone and spread from there; Darlington maintained that mammalian taxa, and just about every other major group of vertebrates, originated in large areas with warm, equable climates, specifically the Old World tropics. More generally the question is whether the rate of evolutionary diversification, entailing both proliferation of species and the birth of major new adaptations, is greater in the tropics (tropical rain forests in this context) than elsewhere. If so, can this explain the higher species diversity of the tropics?

A priori either hypothesis seems reasonable. If the tropics have experienced less drastic environmental change over evolutionary time than extratropical regions have (which is not at all certain), extinction rates may have been lower, and members of ancient groups, which could well have arisen elsewhere, could persist. Among plants, for example, cycads and the primitive angiosperm family Winteraceae are now limited to the tropics. The supposedly more inconstant environment of the temperate zone may select for generalized adaptations that can be carried over into tropical communities, while the specialized properties that some tropical species evolve might restrict their ability to adapt to new, extratropical climatic and biotic conditions. Thus may Matthew's hypothesis be rationalized.

Conversely, the more specialized species of the tropics might be more sensitive to environmental change, more prone to extinction, than their more generalized temperate counterparts, so that living fossils (such as *Metasequoia* and the horseshoe crab *Limulus*) could be expected in the temperate zone, especially where the variety of competitors or predators is more restricted (e.g., *Sphenodon* in New Zealand). And evolutionary rates could be greater in biotically more complex environments if much of evolution entails adaptation to a diverse, kaleidoscopically changing biotic environment. Speciation rates, if greater in the tropics, could not only contribute to greater species diversity, but might enhance the rate at which species arise with combinations of characteristics that enable them to expand into other environments. Indeed a complex biotic environment could be a testing ground for adaptations that confer generalizable competitive superiority. This argument would uphold Darlington's view.

Much of the evidence brought to bear on this question is indirect. Fossil evidence is seldom adequate to pin down the site of origin of a group, and inferring the site of origin from present patterns of distribution can be difficult. From the current distributions of mammals,

Matthew and Darlington came to opposite conclusions. Axelrod (1966) and Stebbins (1974) have the same information on the distribution of angiosperms; yet Axelrod argues that most angiosperm families arose in the tropics and spread from there, while Stebbins believes that the rain forests are museums for groups that arose elsewhere.

Wilson (1961) described a TAXON CYCLE that he reconstructed from patterns of evolution evident in the ants of Melanesia. Isolated islands are colonized by ecologically generalized species adapted to ecologically marginal habitats. Meeting little competitive resistance from a species-poor native fauna, they invade the rain forest and evolve into ecologically specialized species. These species tend not to colonize other islands; they eventually become extinct and are ultimately replaced by new colonizing species that repeat the cycle (*Figure 9*). Greenslade (1968) described a similar taxon cycle for the birds of the Solomon Islands, as did Ricklefs and Cox (1972) for the West Indian bird fauna. This model could apply more broadly to the tropical-temperate comparison, if species of the tropical rain forest are indeed more specialized than other species and thus less capable of invading other habitats. But there is no clear evidence that tropical species are more specialized.

Axelrod (1967) and Stebbins (1974) argue not only that tropical rain forests accumulate species that last a long time before extinction, but that speciation rates are higher in ecologically marginal environments — especially arid regions — than in more moderate regions. On the other hand, tropical species often seem to have inexplicably patchy distributions and may be fragmented into populations by barriers that less specialized species can readily cross (Janzen 1967, Willis 1974). Thus it is not clear whether speciation rates differ between equable and inequable environments.

The best evidence on the subject is paleontological and comes from

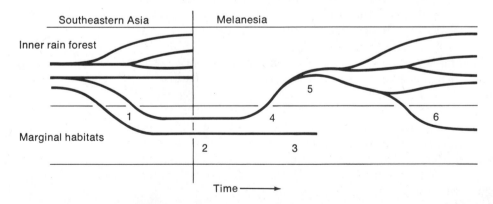

9 The taxon cycle in Melanesian ants, according to Wilson. In southeastern Asia, species adapt to marginal habitats (1), then invade similar habitats on Melanesian islands (2), where they become extinct (3) or give rise to species adapted to the rain forest (4), where they may diversify (5); some of these species may again adapt to marginal habitats (6) and colonize other islands. (From Wilson 1959)

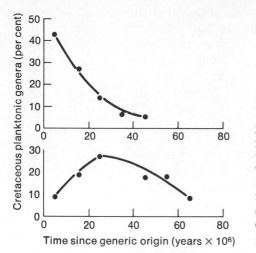

10

Age of genera of Cretaceous Foraminifera in low latitudes (0°–50° N, top) and high latitudes (north of 50° N, bottom). The diverse fauna of the low latitudes contained far more "young" genera, suggesting greater rates of origination in or near the tropics. (From Stehli, Douglas, and Newell 1969; Copyright 1969 by the American Association for the Advancement of Science)

marine invertebrates. The diversity of families and genera of brachiopods and Foraminifera is greater at low than high latitudes (Stehli, Douglas, and Newell 1969). This diversity is made up of cosmopolitan groups that extend throughout cold and warm waters and of endemic warm-water groups. The age of origin of the cosmopolitan genera, as determined from the fossil record, is greater than that of the warm-water endemics (*Figure 10*). The average age of species, on the other hand, does not vary with latitude (Stehli, Douglas, and Kafescioglu 1972). Thus speciation rates seem to be about equivalent, but the rate of origin of new adaptive types, if the morphological features that define genera and families are any indication, is greater in the tropics. But so is their rate of extinction, for the same diversity gradient that exists today is shown by Permian brachiopods and Cretaceous Foraminifera. Thus a latitudinal gradient in taxonomic diversity has existed as an equilibrium condition for at least 270 million years, and it does not require a historical explanation.

Still, the origin of the major new adaptive types that gave rise to the cosmopolitan groups is uncertain. Stehli et al. (1972) suspect that they have expanded from a tropical site, but the paleontological data are inadequate to test the hypothesis. I wonder whether the argument about where new dominant groups arise isn't really a misconceived question. Surely major taxonomic groups, with adaptations appropriate to particular environments, have arisen in each environment. And such groups as rodents, beetles, or composites, with general adaptations that enable them to inhabit a great variety of environmental conditions (Brown 1959), presumably owe their versatility to a characteristic (like the ever-growing incisors of rodents) whose evolution was a most unlikely event. Such events presumably depend more on the vagaries of mutation and recombination than on ecological circumstance.

If so, this argument like many others finds a resolution quite different from either of the polar positions. Darlington and Matthew conceived their answers within a frame of reference that emphasized the environmental factors favoring different kinds of adaptation, and they implicitly assumed that such adaptations would inevitably arise. But if the emergence of general adaptations depends on rare genetic events, the question of where cosmopolitan groups originated must be answered within a different frame of reference, a different paradigm (Kuhn 1962), in which genetic events rather than ecological factors are most important. The latitudinal diversity gradient, on the other hand, seems attributable more to ecological factors than to historical genetic accidents. This distinction between explanations for evolutionary phenomena — the ecological, largely ahistorical vs. the genetic, largely historical — recurs throughout evolutionary biology.

SUMMARY

The geographic distributions of major groups of organisms coincide in greater or lesser degree to form common patterns. Each such pattern gives evidence of past events — changes in the forms of land masses, in the connections among them, and in climates — that have affected the dispersal and distribution of species. From paleontological and taxonomic evidence it is often possible to infer where a group arose and what its avenues of dispersal have been. This study of historical biogeography often conflicts with ecology. Both sciences seek to explain the geographic distribution of species and its corollaries such as geographical variations in species diversity, but the nonequilibrium view of historical biogeography differs from the equilibrium view of ecology, which seeks answers more in current ecological conditions than in history.

FOR DISCUSSION AND THOUGHT

1 By what mechanisms could amphitropical plant distributions have arisen?
2 Evaluate Croizat's argument against traditional biogeography.
3 What evidence could justify the statement in this chapter that the Drosophilidae did not originate in the Hawaiian Islands, the region of their greatest species diversity? (The Drosophilidae do not have a useful fossil record.)
4 Except for volant birds and bats, there are almost no native vertebrates in New Zealand. The fauna includes one frog, a few lizards, *Sphenodon,* and several large flightless birds (kiwis and moas, the latter recently extinct). But there are no snakes, freshwater fishes, or terrestrial mammals. Account for this peculiar situation.
5 In what ways do arguments from the fossil record and from biogeography depend on a taxonomy that is a correct reflection of phylogenetic relationships?
6 Read the papers by Southwood (1961) and Strong (1974b) concerning the proposition that plant species with a longer fossil record and wider geographic distribution are hosts for more insect species than plant species with a more limited temporal and spatial distribution. Evaluate these papers, and discuss their implications for the problem posed in this chapter and Chapter 5 of whether communities are in evolutionary equilibrium.

7 In some cases it can be shown that species are physiologically incapable of surviving temperatures beyond the borders of their range. Do such observations prove that cold regions have low species diversity because of the harsh physical environment they present?

8 Analyze faunal exchange between North and South America (see, e.g., Darlington 1957), and evaluate its relevance to the Darlington–Matthew controversy.

9 Eurasian species of early successional herbaceous plants — weeds — have become successfully established in North America in great profusion. Why should this be, and what bearing does it have on questions posed in this chapter?

10 It is sometimes suggested that species from species-poor biotas such as islands are less capable of invading new regions than are species from diverse communities such as continents. What are the implications of this hypothesis, and what evidence is there in its favor? See Wilson (1965).

11 From Sanders' data on the species diversity of benthic communities, Slobodkin and Sanders (1969) proposed the TIME-STABILITY HYPOTHESIS, which states that diversity is greatest in the most stable environments that persisted longest. What is the relation between this hypothesis and the historical and ahistorical hypotheses considered in this chapter?

12 How distinct are the idiographic and nomothetic approaches? Cite instances in which they may be mutually dependent. How distinct is the difference between ecological and genetic explanations for evolutionary phenomena?

MAJOR REFERENCES

Darlington, P. J., Jr. 1957. *Zoogeography: The geographical distribution of animals.* Wiley, New York. 675 pages. The major reference on the subject; a major compilation of information, invested with a controversial interpretation. Antedates the general acceptance of continental drift.

Good, R. 1947. *The geography of flowering plants.* Longmans, Green and Co., London. 403 pages. A major source of information on phytogeography.

Udvardy, M. D. F. 1969. *Dynamic zoogeography.* Van Nostrand Reinhold Co., New York. 445 pages. Historical zoogeography with an ecological perspective.

Cracraft, J. 1974. Continental drift and vertebrate distribution. *Ann. Rev. Ecol. Syst.* 5:215–261. A review of vertebrate zoogeography in the light of continental drift.

MacArthur, R. H. 1972. *Geographical ecology.* Harper & Row, New York. 269 pages. Patterns of species diversity and distribution from the viewpoint of the theory of species interactions and community structure.

In contemplating the diversity of life, we cannot help asking, What are the major features of the evolution that has produced such variety? How rapidly does evolution happen? Do all the independent paths of evolution describe a common trend? Does evolution as a whole have a direction? Are there limits on the variety of forms that organisms may attain? How do the distinctive characteristics of the major groups evolve?

These questions are the subject of this and the following chapter. The conventional answers to most of them depend, as do the two previous chapters, on inferences about phylogeny made by systematists and paleontologists. This chapter summarizes some of the conclusions of generations of evolutionists about rates and directions of evolution and ends with a discussion of phylogenetic inference and taxonomy, the basis of their conclusions.

ANAGENESIS AND CLADOGENESIS

Most of evolution consists of two processes: ANAGENESIS, change within a specific lineage, or species; and CLADOGENESIS, the splitting of a species into two or more, which diverge in their characteristics by anagenesis within one or both lines (*Figure 1*). The traditional view is that after the speciation process both sister species diverge from their common ancestor at approximately equal rates (*Figure 1a, c*). But for several reasons, including the rarity with which fossil series show prolonged gradual anagenetic change within single species, a different view has now received much acceptance (*Figure 1b, d*): a species changes rapidly as it comes into existence but quite slowly thereafter; and when it subsequently undergoes speciation, it remains largely unchanged while producing a daughter species that diverges (Mayr 1963, Eldredge and Gould 1972, Gould and Eldredge 1977; *Chapter 16*). If this more recent view, evolution by PUNCTUATED EQUILIBRIA, is more often correct than the

CHAPTER 7

Rates and Directions of Evolution

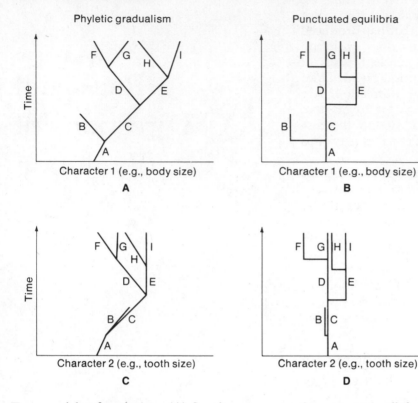

Phyletic gradualism

Time

Character 1 (e.g., body size)

A

Punctuated equilibria

Character 1 (e.g., body size)

B

Time

Character 2 (e.g., tooth size)

C

Character 2 (e.g., tooth size)

D

1 Two models of evolution. (A) Species are presumed to diverge gradually after speciation. (B) Evolutionary change is presumed to happen rapidly during speciation but not thereafter, and the parent species remains unchanged, so that the sequence A, C, D, G would appear to be a long-lived species in the fossil record — and a "living fossil" at that. (C) and (D) For the same two phylogenies, different characters evolve at different rates; for example, B and C hardly diverge at all in character 2, but do so in character 1. In all four diagrams, G and H exhibit convergent evolution from more different ancestors D and E; because of their similarity, they could easily be classified together into a polyphyletic group, if their ancestry was unknown. F and H illustrate parallel evolutionary change. In (A) the sequence C, E, H illustrates reversal of evolution, as does the series C, E, H in (B). In both phylogenies the sets F and G, C through I, and A through I form a nested series of monophyletic groups.

older view of PHYLETIC GRADUALISM, it may alter some of the conclusions about evolutionary rates which are discussed in this chapter.

RATES OF EVOLUTION

Simpson (1953) distinguishes several rates of evolution. Most germane to our purposes are rates within a single lineage (e.g., within species F, in Figure 1a, from its origin to the present) of particular CHARACTERS or complexes of characters; and TAXONOMIC RATES, rates at which species with different characteristics replace one another. The two kinds of rate are probably related to each other quite intimately, especially if the model of punctuated equilibria is more generally correct. In Figure 1b, for example, the rate of change of body size in the group as a whole depends on how frequently speciation takes place. If we had fossils only of A and E, the measurable rate of evolution would be greater than if E had never arisen and we had fossils only of A and D. Because paleontological materials seldom include time series of fossils within single lineages (e.g., a series of points along line segment C in the figure), but rather consists of specimens more widely scattered in time, most evolutionary rates calculated from fossil material are average rates over long periods and often measure differences among different species. They are therefore likely to be very inexact. If we had only a fossil of B and one of E from the phylogeny in Figure 1b, and if we assumed that B was the ancestor of E, we would calculate a large change in body size over a long period rather than the lesser change over a short period that was actually entailed in the divergence of E from C, its real ancestor.

RATES OF EVOLUTION IN SINGLE CHARACTERS

At the most detailed level, we may measure the rates at which genes that control specific characters change in frequency. Such changes can be very rapid. To cite only the most famous example, the allele for industrial melanism in the moth *Biston betularia* went from about 0 to 98 per cent in some populations in about 50 years.

By comparing the amino acid sequences of mammalian hemoglobins and knowing from the fossil record the approximate time at which the mammal groups diverged from one another, Kimura (1968) calculated that there has been on average about one amino acid substitution per 28 million years. If this rate is extrapolated to the entire genome, the average rate of molecular evolution must have been about one DNA nucleotide substitution, somewhere in the genome, every 1.8 years. This is far higher than the maximal rate at which evolution could supposedly happen by natural selection (*Chapter 13*). From this result Kimura argued that much of evolution at the molecular level is random rather than adaptive.

If most amino acid substitutions really are adaptively neutral, the rate of substitution should be independent of the rate of environmental change and thus fairly constant. Langley and Fitch (1974) have calculated that the rate of nucleotide substitution seems to have been more constant over long than over short time intervals, but that the rates of substitution in different mammalian groups have been only about twice as variable as a truly random process with a constant probability of change (such as radioactive decay). If several proteins are considered together, the variations balance out, so that the overall rate at which mammals have diverged in amino acid sequence appears quite constant and quite uniform among groups. There appear to be some exceptions, though. The proteins of the higher primates appear to have evolved more slowly than those of other mammals; for example, the difference in amino acid sequences between humans and anthropoid apes is so slight that it suggests that they diverged only 4 million years ago, whereas paleontological evidence suggests that they separated at least 15 million years ago (Uzzell and Pilbeam 1971, Goodman et al. 1974).

For only a few morphological characteristics — especially those of fossils — is the genetic basis known; thus for most characteristics a description of the rate of phenotypic change must suffice. The proportional rate of change must be calculated if evolutionary rates in organisms of different average sizes are to be compared. Haldane (1949) proposed the DARWIN as a measure of evolutionary rates, defining it as a change by a factor of e (the base of natural logarithms, 2.718) per million years. For example, the average rate of change of the height of the molars in horses throughout the Tertiary was 40 millidarwins, or 4 per cent. The change in body size from the smallest to the largest of the ceratopsian dinosaurs (e.g., *Triceratops*) was on average 60 millidarwins (a factor of 6×10^{-8} per year). The highest rate Haldane found among the limited fossil data available to him was for human skulls, which he calculated to have changed at a rate of 300 millidarwins over the last half million years (an estimate now considered too high).

These are average rates from fossils widely scattered in time, and they can obscure more rapid changes that occur over shorter intervals — changes that could hardly be expected to continue at such high rates for very long. Kurtén (1959), for example, examined sequences of Cenozoic mammals that were probably members of single lineages and found that their dimensions changed quite slowly in some instances (23 millidarwins on average), much more rapidly in others (500 millidarwins), and extremely rapidly in a few cases. For short periods in the postglacial era the skeletal dimensions of bears and other mammals changed at a rate of over 12 darwins.

Studies of contemporaneous populations have revealed instances of rapid changes that are small in absolute magnitude, but enormous on a proportional scale. House sparrows (*Passer domesticus*), introduced into North America from Europe about 100 years ago, have become geographically differentiated into races that are adapted in size and coloration to different North American environments (Johnston and Selander 1964). Some of their skeletal dimensions have diverged from those of the European populations at rates of 50 to 300 darwins. Woodson (1962) compared specimens of butterflyweed (*Asclepias tuberosa*) from each of several populations in 1946 and again in 1960 and found that in some populations the basal and terminal angles of the leaves had changed in form, at rates of 333 and 5662 darwins, respectively (R. R. Sokal, personal communication). If these changes are genetically based rather than environmentally induced, the apparently low rates of evolution evident from fossil material may greatly underestimate the rapidity with which populations can evolve over short periods.

Although evolution can occur only if there is preexisting variation, it is unclear whether the rate of evolution depends on the amount of variation or whether the variation is usually more than sufficient for rapid evolution. Kluge and Kerfoot (1973), comparing closely related races and species of reptiles and birds, found that the characters that differed most greatly among populations were the ones that were most variable within populations, suggesting that less variable characters evolve more slowly (see also Sokal 1976). In time series of Pleistocene voles (*Microtus*) Guthrie (1965) similarly found that the most variable tooth characteristics evolved most rapidly and that the variation did not decrease as evolution proceeded. But in a 4.5-million-year sequence of fossils of the radiolarian *Pseudocubus vema* (*Figure 2*), the rate of evolution was not clearly related at any time to the amount of variation at that time (Kellogg 1973). Mean size increased because of both the elimination of smaller animals early in this history (which reduced variation) and the appearance of larger individuals later in the series (which increased the variation).

MOSAIC EVOLUTION

Figure 3 illustrates the evolution of two tooth characteristics in horses. Each character evolved faster at some times than others and more

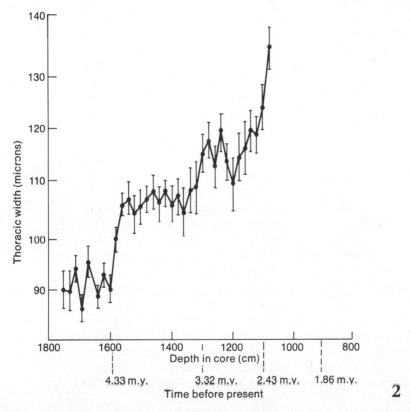

2

Changes in size in the radiolarian *Pseudocubus vema*. There is no simple relationship between the rate of evolution at any time and the amount of variation (represented by vertical bars) at that time. (Adapted from Kellogg 1973)

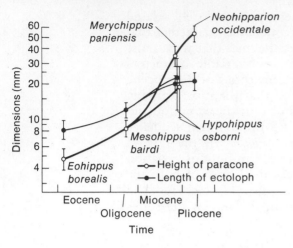

3

Changes in two dental characters in a phylogeny of five genera of horses. The rates of change vary between characters, among phylogenetic sequences, and among time periods. (After Simpson 1953, by permission of the publisher)

rapidly in one lineage than another; and in each lineage the two characters often differed in their rate of change. A species does not evolve through time by a progression of concerted changes in all its attributes; nor does a daughter species diverge from its parent species in all its genes. Rather every species changes rapidly in some features but not at all in others. Thus every species is a mosaic of characteristics that have persisted unchanged from its remote ancestors and of characters that have recently evolved. Schaeffer (1956), analyzing the changes in jaws and fins that marked the transition from chondrostean to holostean fishes, found that the primitive and advanced states of different characteristics were seldom correlated with each other. The progression from chondrostean to holostean features occurred at different rates for different characters.

The most recently evolved forms of life bear primitive features; most of our biochemical pathways are virtually identical to those of protozoans. Groups that arose long ago include species that may have originated in the last million years and have new characteristics. To say that a species such as the horseshoe crab or the ginkgo is primitive does not mean that it has long since stopped evolving; in fact, it may have arisen by speciation quite recently. Rather a primitive form is one whose ancestor branched off early from the rest of a group, and retains one or more specific ancestral features. There is a lot to be said for abandoning the terms *primitive* species and *advanced* species and to refer instead to ANCESTRAL states and DERIVED states for each character separately. Thus compared to Carboniferous amphibians, frogs have an ancestral number of aortic arches (2), but a derived number of fingers (4); humans have a derived number of aortic arches (1), but an ancestral number of digits (5).

Slowly evolving characters, or CONSERVATIVE characters as taxonomists call them, are frequently shared in a similar condition by all or most of the members of a major group — indeed they are often the characteristics that define the group. Thus the form of the incisors in modern rodents as diverse as beavers and kangaroo rats is much like that in the earliest fossil rodents. Having once arisen, such characteristics probably change little because they serve, with little modification, for adaptation to a great variety of environmental conditions. Rodents' incisors are a GENERAL ADAPTATION (Brown 1959) to most of the conditions rodents face. But the form of the legs and the number of toes vary more among species and have evolved more rapidly; the long hind legs and reduced number of toes in kangaroo rats are SPECIAL ADAPTATIONS to desert conditions. Similarly the ovules in angiosperms can have one of several arrangements — on a central column or on the ovarial partitions, for instance — which are quite conservative characters that are independent of the plant's ecology.

Not only would there be little advantage in evolving from one kind of ovule arrangement to another, but such a rearrangement might entail a drastic and unlikely remodeling of the pattern of development. The development of the characteristic is highly canalized (*Figure 4*) and is not easily changed. Canalization, a little-understood phenomenon discussed at greater length in Chapter 15, may also be responsible for the conservative nature of some apparently trivial features that seem to have no adaptive value. All the species in the fly family Sepsidae, for example, have a single bristle, pointed anteriorly, on the rear margin of the posterior thoracic spiracle. Among termites, primitive and advanced species alike have a tiny, probably functionless tooth on the right mandible (*Figure 5*; Emerson 1961).

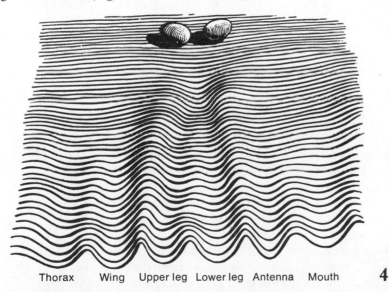

Thorax   Wing   Upper leg   Lower leg   Antenna   Mouth   **4**

Part of an "epigenetic landscape" in which the developmental pathway of a part of an organism is represented by the path taken by a rolling ball. Several paths are possible, and slight alterations of the landscape can shunt development from one path to another. The deep channels leading to leg development illustrate that the development of leg parts is highly canalized. (From Waddington 1956a)

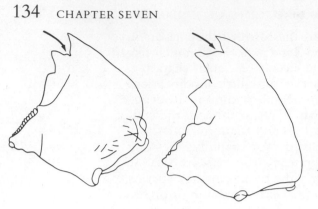

5

A phylogenetically conservative character, the subsidiary tooth (indicated by arrow) on the right mandible of the most primitive known termite, *Archotermopsis wroughtoni* (left) and of *Protohamitermes globiceps,* a member of the most advanced termite family, the Termitidae (right). (After Emerson 1961)

CHARACTER COMPLEXES AND THE ORIGIN OF NEW TAXA

Different characters often evolve mosaically at different rates, but this does not mean that an organism is a hodgepodge of ancestral and derived states combined at random. If a derived state of character B arises in a species whose ancestor already evolved a derived state of character A, all subsequent species will share the derived states of both characters, unless there is a subsequent reversal of evolution. Thus cladogenesis imposes correlations among characteristics (*Chapter 17*).

Very often the transformation of a single feature permits a species and its descendants to invade a new adaptive zone (*Chapter 5*). Transformation of the astragalus, a tarsal bone, to permit efficient running and leaping is the key adaptation of the two major orders of hoofed mammals, the Artiodactyla and Perissodactyla. In the Perissodactyla this transformation took less than 5 million years (Radinsky 1966); but such transitions to new adaptive zones, marking the origin of new families or orders, often happen so fast that they leave little trace in the fossil record. Simpson (1953) has called this QUANTUM EVOLUTION.

Occupancy of a new adaptive zone usually entails changes in other features, a complex of functionally interrelated characteristics. The flight of birds involves not only modification of the forelimbs into wings, but the possession of a sternal keel, hollow bones, air sacs, a rigid vertebral column, and a host of other features. Thus an innovation in one feature may change a species' environment and bring to bear new selective pressures to which the other characteristics quickly respond. These complexes of characters commonly evolve rapidly soon after the new adaptive zone is entered and then more or less reach a plateau. Westoll (1949) and Schaeffer (1952) found that a complex of characters distinguishing lungfishes and coelacanths from their rhipidistian ancestors evolved rapidly soon after their origin in the Devonian but very slowly thereafter (*Figure 6*).

TAXONOMIC RATES OF EVOLUTION

As in most analyses of evolutionary rates in the fossil record, the measurement of the rate of evolution in lungfishes and coelacanths is not based on evolution within single, gradually evolving lineages. The Permian species are not necessarily direct descendants of the Carboniferous species; many speciation events may have intervened. If so, and if the model of evolution by punctuated equilibria is more accurate than that of phyletic gradualism (*Figure 1*), the apparent rate of evolution will be proportional to the rate at which new species replace old species and to the average degree of divergence of a new species from its ancestor.

The rates at which taxa originate and are extinguished are called TAXONOMIC RATES of evolution (Simpson 1953); these rates may explain why some groups appear to evolve more rapidly than others. For example, genera of bivalves have survived on average about 78 million years, but an average carnivore genus has survived for 8 million years. Thus at any time a higher proportion of bivalves than carnivores is more ancient, so that their rate of evolution appears to be about a tenth that of the carnivores. The members of any group vary in survival time, so that some may survive for a very long time and be viewed as living fossils. One suspects that exceptionally ancient forms — the living coelacanth *Latimeria,* the ginkgo, the blue-green algae — may have unusual general adaptations that have enabled them to escape extinction. There is no reason to suppose that they have fallen into evolutionary stagnation and are incapable of further evolution.

The rate of divergence of a group from its common ancestor is greater if speciation rates are higher, if the punctuated equilibrium view is correct. Stanley (1975a) assumed that the number of species in a recently arisen group increases exponentially, as does the size of a population, following the equation $N_t = N_0 e^{Rt}$, where $R = S - E$, the rate of speciation less the rate of extinction. For each of several groups

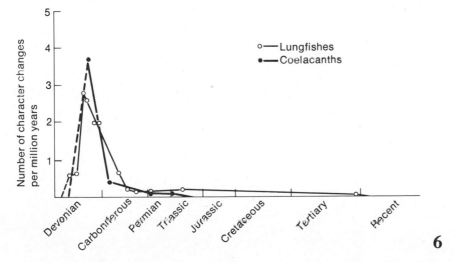

6

Rates of morphological change in lungfishes and coelacanths. The rate of change is greatest soon after the origin of the group and becomes very low thereafter. (After Schaeffer 1952)

of bivalves and mammals, the time t since origin is known from the fossil record; N_0 is assumed to be 1; E is calculated from the average life span of fossil species; and N_t is the number of existing species. Solving for S, Stanley concludes (*Table I*) that speciation rates have been higher in families of mammals than in families of bivalves, perhaps accounting for their higher rate of morphological evolution.

Chapter 5 presented evidence that among aquatic arthropods the diversity of morphological forms reached a plateau, as if the group had radiated into all the available ecological niches. Then if new species can with the passage of time diverge less and less from their ancestors as the variety of empty niches declines, the rate of morphological evolution in a group will diminish, especially if speciation rates also decline as the diversity of a group increases. But on this point there is no direct evidence (*Chapter 5*).

DIRECTIONS OF EVOLUTION

Discussions of the direction of evolutionary change often use the word *trend* in two distinct but related senses. It may mean a consistent, monotonic directional change within a phyletic line in one or more characteristics, as in the popular notion that cranial capacity increased steadily in the hominid line. Or it may mean a prevailing inclination of change in many independent groups, as in the tendency of many mammals to evolve toward larger size.

TABLE I. Estimated rates of speciation S, extinction E, and increase in diversity R

	t (million years)	N (species)	R	$\bar{R}$	E	S
				(per million years)		
Bivalvia				0.07	0.17	0.24
Veneridae	120	2400	0.06			
Tellinidae	120	2700	0.07			
Mammalia				0.20	0.50	0.70
Bovidae (cattle, antelopes)	23	115	0.21			
Cervidae (deer)	23	53	0.17			
Muridae (rats, mice)	23	844	0.29			
Cercopithecidae (OW monkeys)	23	60	0.18			
Cebidae (NW monkeys)	28	37	0.13			
Cricetidae (mice)	35	714	0.19			

(After Stanley 1975*a*)

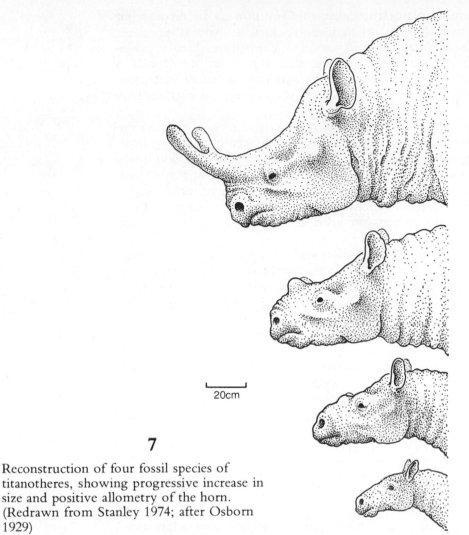

20cm

7

Reconstruction of four fossil species of
titanotheres, showing progressive increase in
size and positive allometry of the horn.
(Redrawn from Stanley 1974; after Osborn
1929)

Trends within phyletic lines

Temporal sequences in the fossil record, whether comprising a single
unbranched lineage or not, sometimes show progressive changes in
one or more features. In the titanotheres, for example, body size and
the size of the nasal horn increased (*Figure* 7), and this trend proceeded
in parallel in three subfamilies.

Many authors have seen in such trends evidence of a purpose or
goal in evolution; it is often argued, for example, that the increasing
complexity of the nervous system from coelenterates to mammals to
Homo sapiens is a reflection of a drive toward higher consciousness. If
higher consciousness were a goal of animal evolution in general, greater
horn size could as well be a goal of titanothere evolution in particular.

But this hypothesis places the cause of evolution in an Aristotelian final cause, rather than an efficient cause such as natural selection, which is not sentient and has no vision of the future. Hence evolutionists deny that evolution has goals, that organisms are driven by vitalistic forces toward predestined ends. And to the extent that progress connotes movement toward a goal, the concept of evolutionary progress must be denied.

Trends can be explained in purely mechanistic terms. An environmental factor may change consistently in one direction for a long time, so that the most advantageous phenotype likewise changes. Evolution within the species or in interacting species may itself cause the optimal phenotype toward which the species evolves to change directionally. If a larger horn conferred superiority in contests between male titanotheres, every slight increase in horn size would make even larger horns advantageous.

Gould and Eldredge (1977), following Wright (1967), have emphasized that although a group as a whole may show a long-term trend in one direction, the various species at any one time may evolve in different directions but survive differentially. Thus some species may at any time be evolving smaller body size and others larger; but if larger species survive longer, the overall trend is toward larger size (*Chapter 17*).

For many adaptations one can envision an optimal condition, given the various constraints posed by a species' way of life and by the features it has inherited. Because trees compete for light, greater height is advantageous as long as it does not expose the tree to excessive wind stress; thus there is an optimal height. As Stanley (1973) points out, a trend (in either sense) exists if species more often than not are on one side of the optimum when they come into existence. Stanley explains COPE'S RULE, the trend toward larger body size, in this manner. The first invaders of an adaptive zone usually stem from generalized rather than highly specialized ancestors. Generalized species are unlikely to be large, for large size entails specializations that preclude adaptations to a very different way of life; bears are unlikely to become fossorial (burrowing), or elephants arboreal. Thus the generalized progenitors of new orders and families are more likely to be below the optimal mean body size for the adaptive zone occupied by their descendants, as long as adaptation to this way of life entails reasonably large size.

This way of looking at evolutionary trends ascribes a certain degree of irreversibility to evolution; it implies that complex adaptations constrain the possible avenues of further evolution, including a return to the ancestral state. Thus organisms are to a degree prisoners of their adaptations; the more elaborate the commitment to one mode of existence, the less freedom they have to adopt others. One common reason for this is that specialization often entails the loss of complex structures; and once lost, they are unlikely to be regained in the original form, a principle known as DOLLO'S LAW. Snakes, descended from lizardlike ancestors, are unlikely to regain legs; they are far more likely to adapt further to snakish ways of life.

Although some trends do exist, the absence of consistent direction is as common in evolution as is direction. The Equidae are classical textbook examples of directional change in size, tooth structure, and toe number, but Simpson (1953) notes that there was not a straight line of evolution from the eohippus to the modern horse. Rather a complex pattern of branching produced many genera throughout the Cenozoic, each with its special adaptations. Their evolution was not orthogenetic in one direction; it entailed reversals in body size, persistence of ancestral features in some genera alongside derived features in others, stepwise rather than gradual transitions, and divergence in different evolutionary directions (*Figure 8*).

Trends among species

The genera of horses, far from showing consistent evolution in any one direction, displayed adaptations to different ways of life, an adaptive radiation within their ecological niches. Cope's law of body size

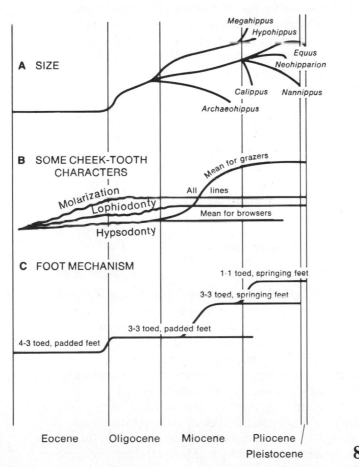

Diagram of the evolution of some characteristics in the Equidae (horses). Sustained directional evolution is the exception rather than the rule. (From Simpson 1953, by permission of the publisher)

may apply widely, but it is nonetheless violated by many species that have become smaller in size. Adaptive radiation, rather than consistent unidirectional change, is the norm for every group. Even such a small group as the Falconiformes, for example, includes not only falcons adapted for rapidly chasing their prey, but eagles and hawks that hunt by soaring, forest hawks that strike quickly from a perch, ospreys and fishing eagles that dive for fish, the long-legged secretary bird that runs after its prey, vultures and caracaras that feed on carrion, and even an African vulture that feeds mostly on the fruit of the oil palm.

If adaptive radiation rather than consistent direction is the norm for each group, it is certainly true of life taken as a whole. Any attempt to find a goal for all of evolution, a universal trend, a most advanced species toward which evolution has been pointed (such as *Homo sapiens*), will fail. The features that different species evolve may increase or decrease their abundance, their reproductive rate, or the efficiency with which they find food, use resources, or do anything else. Species may become more or less ecologically flexible, more or less social in behavior, more or less elaborate in the structure of their brain or behavior. The degree to which behavior is learned rather than instinctual is as much a special adaptation to the unpredictability of the environment and to the social organization in which the individual functions as the ability to acclimate to changes in salinity is a special adaptation to variations in salinity.

It is meaningless to point to *Homo sapiens* or to any other species as most advanced, most highly evolved, most successful, or closest to the goal or end product of evolution. Anthropologists have abandoned the culture-bound notion that Western society is the most advanced of cultures and have instead embraced a cultural relativism that sees each cultural norm as an adaptive, dignified, independent expression of human behavior. Similarly we can replace the antiquated notion of a ladder of life, or great chain of being, by an evolutionary relativism that recognizes adaptive radiation and celebrates diversity instead of trying to make the world into our own image.

HOMOPLASY

Adaptive radiation, divergence into different niches, is the salient aspect of evolution viewed as a whole; but this does not mean that all organisms are at all times in every way becoming more and more different from each other. If for no other reason than that different species are exposed to similar environmental conditions, they often develop similar adaptations. Very often different species achieve similar adaptations but in different ways; different structures become modified to the same adaptive end. The vinelike habit has evolved independently in many groups of plants, in which different structures have been modified as tendrils or holdfasts (*Figure 9*); the mouthparts of the true bugs (Hemiptera) and of biting flies such as mosquitoes (Diptera) have become modified for piercing and sucking in different ways (*Figure 10*).

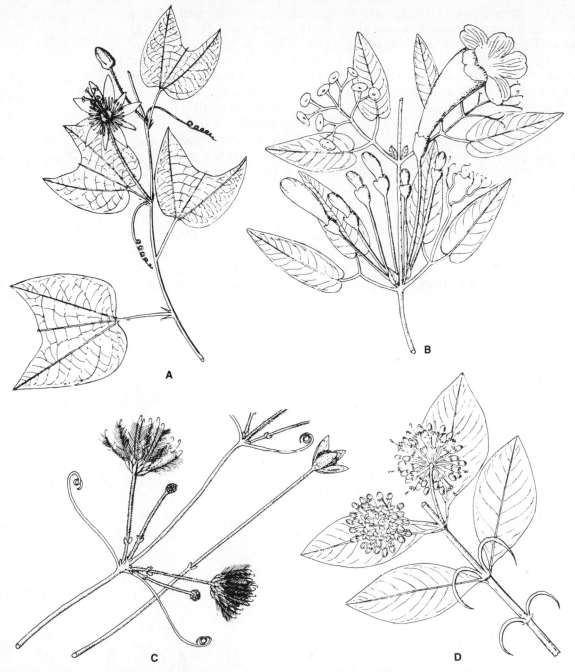

Different evolutionary paths to the same end: modifications for climbing in vines. (A) Stipules modified into tendrils in Passifloraceae. (B) Leaflets modified into tendrils and into suckers in Bignoniaceae. (C) Leaves modified into tendrils in Ranunculaceae. (D) Inflorescences modified into hooks in Rubiaceae. (From Hutchinson 1969)

9

The mouthparts of bugs and biting flies are often both termed beaks; they are similar but not identical structures that evolved independently, probably on different genetic bases, in the two groups. They are an instance of CONVERGENT EVOLUTION. This very common phenomenon can occur in almost any feature of organisms. The eyes of vertebrates and cephalopods, the wings of insects, birds, and bats are easily recognized as convergent, for they are more similar in name than in structure. (Wings of birds and bats are similar insofar as they are modified forelimbs, but completely different insofar as they are wings, since the modifications for flight are entirely different.)

In many instances convergent evolution results in such similar features that we would be hard put to say whether they had the same genetic bases if we did not know that they had evolved independently in unrelated groups. Herbaceous plants have evolved repeatedly from woody ancestors; the leaflessness and growth form of the New World Cactaceae and some Old World Euphorbiaceae are extremely similar adaptations to arid conditions; leglessness has evolved in snakes and in several unrelated groups of lizards; the mosasaurs and the Cretaceous

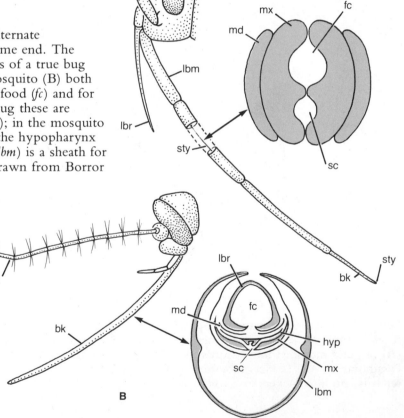

10 Convergent evolution, or alternate evolutionary paths to the same end. The piercing-sucking mouthparts of a true bug (Hemiptera, A) and of a mosquito (B) both have channels for ingesting food (*fc*) and for injecting saliva (*sc*). In the bug these are formed by the maxillae (*mx*); in the mosquito by the labrum (*lbr*) and by the hypopharynx (*hyp*). In both, the labium (*lbm*) is a sheath for the other mouthparts. (Redrawn from Borror and DeLong 1960)

bird *Hesperornis,* alone among the vertebrates, evolved an intramandibular joint between the angular and splenial bones. Hole-nesting birds, whatever their phylogenetic affinities, typically lay white rather than colored or spotted eggs; cave animals of whatever phylum tend to have reduced eyes and pigmentation. Rensch (1960) lists dozens of other patterns of convergent adaptation to similar environments.

Convergent evolution is one form of HOMOPLASY. A characteristic is homoplasious in two species if it was not possessed by all the ancestors intervening between them and their most immediate common ancestor. A characteristic is HOMOLOGOUS in two species, on the other hand, when it has been inherited by both, through both lines of descent, from their common ancestor. Presumably it then has the same genetic basis in both species.

Convergent evolution grades into two other forms of homoplasy that are themselves not easily distinguished from each other: PARALLEL EVOLUTION and EVOLUTIONARY REVERSAL. Parallel evolution occurs when a feature evolves independently in closely related species, but how closely related they need be before it is parallelism rather than convergence is unclear and probably immaterial. Parallel evolutionary tendencies are evident in almost every major group. Female winglessness has evolved repeatedly in the moth family Geometridae, as in other families of moths; colonial, social behavior has arisen quite a few times among bees and in other aculeate Hymenoptera as well; cuckoos seem inclined toward nest parasitism.

It appears from the fossil record that the diagnostic characteristics of many, if not most, of the higher taxa have had multiple origins. Mammals, for example, arose from therapsid reptiles. In a paleontological context a mammal is defined by the structure of the lower jaw, which consists of a single bone, the dentary, that articulates with the squamosal bone of the skull. By this definition mammals evolved independently from at least three and perhaps as many as five different lines of therapsids (*Figure 11*; see Olson 1959, Simpson 1959, Crompton 1963). Similarly, the holostean fishes evolved from at least three different chondrostean stocks (Schaeffer 1956, 1965), and each of the traditional suborders of rodents arose several times from the generalized early rodents, the protrogomorphs (Wood 1959). Many of the higher taxa in the classifications that are presently used are thus not CLADES, sets of species derived from a single remote ancestral species, but GRADES, sets of species that share a level of evolutionary organization attained repeatedly by different related lines (*Figure 12*).

Very complex structures are unlikely to evolve in precisely the same form in unrelated groups — or in the same phyletic line after they have been lost, as Dollo's law states. But less complex characters may display evolutionary reversal, and a complex character may degenerate and return to its original state. Winglessness in insects can be either a primitive condition, as in springtails (Collembola), or a derived condition, as in lice, fleas, and the many wingless species in almost every insect order that have evolved from winged ancestors. In ele-

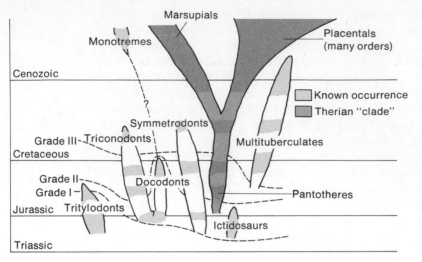

11 Polyphyletic origin of mammalian groups from the therapsid reptiles, according to Simpson. Features of the skull and lower jaw define three levels (grades) of approach to full mammalian organization; each level was attained independently by several groups. One of these groups, the therian clade (see Figure 12), includes most of the modern mammals. (After Simpson 1959)

phants the general trend toward greater size was reversed in several lines that evolved dwarf species, and reversals in the structure of the teeth accompanied the change in body size in each instance (Maglio 1972). Of the several molars possessed by primitive Carnivora, the cats (Felidae) retain only the first; but in the lynx the second molar has reappeared (Kurtén 1963). Thus some lost structures may indeed be regained.

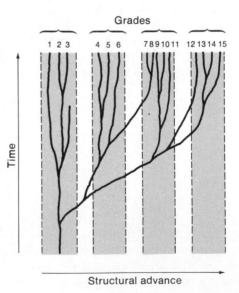

12 Grades and clades. A group of species (e.g., 1, 2, 3) with a recent common ancestor forms a clade; a group with the same level of structural organization (e.g., 7–11) forms a grade. Members of a clade may belong to different grades because of differential evolutionary rates. Note that this diagram assumes gradual evolutionary change in structure, which may not always be the case. (Modified from Simpson 1961*b*, with permission of the publisher)

   That parallel evolution should be common is not surprising. If related species have similar patterns of development, they are likely to be modified in similar ways if subjected to similar selection pressures. It is hard to know the degree to which homoplasious phenotypes in related species have the same genetic basis. Throckmorton (1965) has suggested that peculiarities of the form of the reproductive tract in species of *Drosophila,* whose phylogenetic relationships have been determined by chromosome analysis (*Figure 13*), may have arisen repeatedly when identical genes carried in low frequency in some species are expressed in the homozygous condition in others. It is equally plausible that parallelisms or reversals may be independent alterations of the same developmental pathway in each of several species, that is, they may be switched over into another channel by any of several mutations, with the same ultimate phenotypic effect (*Figure 4*). In the same vein the reappearance of the second molar in the lynx probably did not require a slew of new mutations to form a new tooth *ex nihil.* Whether a tooth develops may depend on slight variations in the concentration of some substance that induces tooth differentiation. A slight mutation could suppress tooth development in the Felidae, and an equally slight compensatory mutation in the lynx could allow the manifestation of the molar-forming potential that may be present but suppressed in the other cats. That such potentials remain immanent in developmental systems for long periods of evolutionary time is well

13

Independent origins of forms of the ejaculatory bulb in species of the *Drosophila repleta* group. The phylogeny of these species is known from chromosomal evidence. Note the identical bulb shape in, for example, species 3, 5, 14, 17; 4, 10; 1, 16, 19. (Redrawn from Throckmorton 1965, in *Systematic Zoology,* vol. 14)

known; for example, frogs have lacked teeth in the lower jaw since the Jurassic, but it is possible to induce their development experimentally (Hecht 1965). The probable directions in which a characteristic will evolve must depend on both the external environment and the potentialities and limitations of the developmental program.

PHYLOGENETIC INFERENCE AND CLASSIFICATION

All the examples of rates and directions of evolutionary change discussed so far have been presented on the assumption that it is possible to infer the phylogenetic history of species correctly. If the phylogeny of species is known, it is possible to identify the directions and the relative rates of evolution of some of their characteristics (*Figure 14*). But one should not accept phylogenetic statements on faith, for inferring phylogenetic histories can be difficult. For this reason I think it appropriate to close this chapter with some consideration of how it might be done.

Phylogenetic inference is intimately linked with the construction of taxonomic classifications; both have been the province of generations of systematists, on whose work almost all of biology rests. Systematists generally have four tasks. The first is to distinguish among kinds of organisms, generally by categorizing them into species. The definition of species and the difficulties in recognizing species are treated in Chapter 9.

The second task is to group species into more inclusive sets, to establish a classification. The classification consists of higher CATEGORIES such as genus and family. Each set of real organisms that hold a particular categorical rank is a TAXON. Thus the Rosaceae and the Felidae are taxa in the category family. Deciding to what categorical rank a taxon belongs (whether it should be a genus or a family) is largely a matter of deciding what name to bestow on it. This entails the third task of the systematist, nomenclature, a procedure governed by purely legalistic sets of rules.

Taxonomists often disagree whether a group should be ranked as a family, subfamily, or genus, or whether the species in it should be combined into a single family (or genus, or whatever) or several. Some workers, for example, would combine all the cats (except the cheetah, *Acinonyx jubatus,* because of its nonretractable claws) into a single genus *Felis,* while others, focusing on differences more than similarities, would split them into the smaller cats (*Felis, sensu stricto*), the larger cats like lions (*Panthera*), and the bobtailed cats (*Lynx*). These variations in nomenclature sometimes affect analyses of evolutionary problems. For example, genera of mammals have arisen at greater rates than those of molluscs, implying that they evolve faster. But Schopf et al. (1975) point out that how finely species are divided into genera depends on their morphological complexity. Because the number of characteristics that can be used to define genera is greater for mammals than for molluscs, the apparent difference in evolutionary rate could be an artifact of nomenclature.

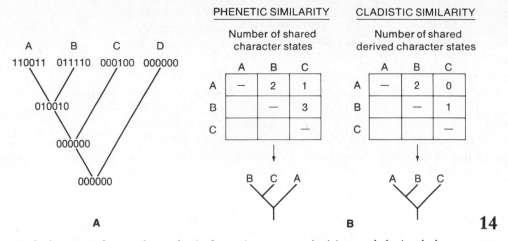

A phylogeny of some hypothetical species. (A) Assume that this phylogeny is correct and is known *a priori,* but that the direction of evolution of each character is not known. Then this phylogeny implies that 0 is the primitive state for most or all of the six characters, since it is typical of C and D; that the fourth character from left exhibits homoplasy; and that characters evolve at different rates, since A and B possess both primitive and derived character states. (B) The phylogeny of A, B, and C is inferred from knowledge of the direction of evolution of each character (instead of *vice versa*). D is an "outside group," so its character states are taken to be primitive. The cladistic similarity, found by counting the number of derived character states held in common, gives the correct phylogeny, whereas the simple phenetic similarity does not.

The fourth occupation of many systematists is to determine the evolutionary relationships among the taxa. But if one is to argue that families A and B had a more recent common ancestor than either had with family C, each family must itself be homogeneous. If the members of a family have had a single common ancestor, the family is a MONOPHYLETIC group, otherwise it is POLYPHYLETIC.

Existing biological classifications do not necessarily reflect evolutionary history in this sense. The members of a taxon are not always descended from a single ancestral species; conversely all the descendants of a single ancestor are not necessarily included in the same taxon. Nor do all systematists believe that they should be. There are three major schools of thought on how classifications should be constructed.

The PHENETIC school, championed especially by Sokal and Sneath (1963; see also Sneath and Sokal 1973), argues that it is impossible to be certain that a phylogenetic reconstruction is correct, so we should not even try to base classification on phylogeny. Organisms should be classified for convenience, as we classify books in a library, on purely phenetic criteria: how similar they are to one another. The classification may often reflect phylogeny, but need not if many characteristics have evolved convergently or if some species have diverged from others at different rates. Thus even if A and B are phylogenetically more related to each other than to C, A may be classified with C if it has converged toward it, or it may be put in a separate taxon from B if it has rapidly evolved many derived characteristics while B has retained primitive features. The phenetic school of thought is not widely embraced by systematists.

Most existing classifications have been erected by adherents to the traditional school of thought, the EVOLUTIONARY SYSTEMATISTS (see, e.g., Simpson 1961, Mayr 1969a, Bock 1973). They hold that classifications should simultaneously reflect two relationships, genealogical and genetic. Suppose that both species A and B have many primitive features, that C has many derived features (because of its more rapid evolution), but that B and C spring from an immediate common ancestor. Species B is genealogically closer to C, but genetically more similar to A, if their primitive features are determined by the same genes. The evolutionary systematists would classify B with A rather than with C, just as the pheneticists would. For example, both birds and crocodilians are derived from the archosaurian reptiles that included the dinosaurs; the common ancestor of turtles, lizards, and archosaurs was a more remote ancestral reptile. Placing birds in one class and crocodilians in another, together with turtles and lizards, does not reflect the genealogical affinity of crocodilians with birds, but their phenetic (and presumed genetic) affinity with turtles and lizards. Similarly (Mayr 1965) evolutionary systematists would place *Homo* in a separate family from the chimpanzee and orangutan, even though *Homo* and chimpanzee share a more recent common ancestor than either has with the orangutan (*Figure 15*).

The third school of thought, the CLADISTIC school of PHYLOGENETIC SYSTEMATICS (e.g., Hennig 1965, 1966; Farris, Kluge, and Eckardt 1970; Nelson 1972, 1973; Cracraft 1974b), which is gaining adherents, argues that classification should strictly reflect genealogical affinities as far as possible. All descendants of an ancestor should be placed in the same taxon; all the members of a taxon should stem from the same ancestor. Thus they might combine all reptiles together with birds into one class, and the birds, crocodilians, and dinosaurs into a subclass, within which the birds would constitute, say, an order. The classification consists of nested series of monophyletic groups (*Figure 1*).

Evolutionary systematists, too, would like to know which species constitute monophyletic groups; but they do not necessarily want this knowledge reflected by taxonomic classifications. But whether or not the classification reflects phylogeny, how can phylogeny be known?

THE LOGIC OF PHYLOGENETIC INFERENCE

If each pair of species that arose from a common ancestor diverged from each other at a constant rate that was the same in both species and the same for all evolving characters, the phenetic similarity between species would directly measure their relative time since divergence, and phylogenetic trees (CLADOGRAMS) would be easy to construct. The closest approximation to such data seems to be in the amino acid sequences of proteins. Proteins diverge at different rates;

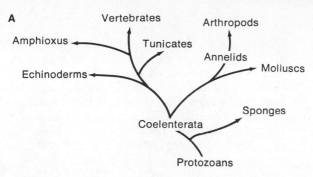

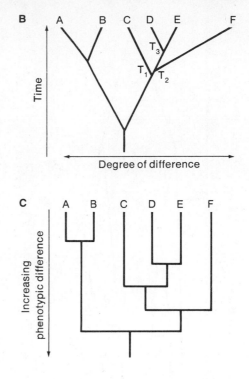

Three diagrammatic ways of expressing relationships among organisms.
(A) A classical phylogenetic tree, expressing only the notion of which groups may have shared common ancestors (e.g., molluscs and annelids from the same stock, tunicates and vertebrates from another). It conveys no information on the time of divergence nor on the amount of phenotypic divergence; these are expressed in the phyletic diagram B, in which the Recent species A–F are placed at the top and are joined at points thought to correspond to their common ancestors (e.g., T_1, T_2, T_3). The angles between the lines reflect the degree of phenotypic difference. This diagram corresponds to the commonly accepted phylogeny of the hominoids (A, B = gibbon species, C = orangutan, D = gorilla, E = chimpanzee, F = human).
(C) A possible phenogram for these species, in which the only information conveyed is degree of phenotypic (phenetic) difference. C joins the cluster D + E at a lower level than D joins E, indicating that C is more different from D and E than they are from each other. (B from Mayr 1965, in *Systematic Zoology*)

15

mammalian fibrinopeptide A, for example, has evolved more rapidly than cytochrome *c*. But for any given kind of protein, divergence seems to have occurred, at least over long time periods, at a fairly constant rate. Although some instances of reversal and convergence are known (see, e.g., Goodman et al. 1974), phylogenetic trees calculated from the number of nucleotide substitutions conform fairly closely to those that have been traditionally accepted (although there are some discrepancies) (*Figure 16*; see Fitch and Margoliash 1970). But for nonmolecular data, phenetic similarity often does not reflect phylogeny because of homoplasy and variation in evolutionary rates.

In recent years the logical procedures by which phylogenetic relationships might be inferred have been clarified and formalized (Wagner 1961; Edwards and Cavalli-Sforza 1964; Camin and Sokal 1965; Farris 1967, 1969, 1970, 1973; for applications of these procedures see Lundberg 1972, Lynch and Wake 1975). This area is one of active research and controversy, but the general ideas are more or less as follows.

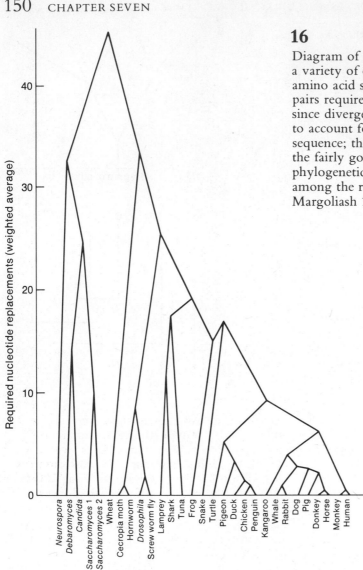

16

Diagram of phylogenetic relationships among a variety of organisms, based on differences in amino acid sequence of cytochrome *c*. Some pairs require fewer nucleotide replacements since divergence from their common ancestor to account for the difference in amino acid sequence; these join near the bottom. Note the fairly good correspondence to classical phylogenetic trees, with some exceptions among the reptiles and birds. (After Fitch and Margoliash 1970)

The problem is to determine which species form monophyletic groups. The first step is to identify a series of independent characters held in common by some or all of the species. Each character can have two or more character states. Thus "toes on the front foot" might be a character, with toe numbers 1, 4, and 5 constituting three states.

It is important for a given character in one species to be homologous to that in another species if the states of these characters are to be compared, because these characters are considered evidence that the species share sets of genes derived from their common ancestor. Indeed characters are defined as homologous if they have been derived with or without modification from a character of the species' common ancestor. Thus the single toe of a horse is taken to be homologous to the middle finger of a human hand; but many frogs have a bony thumblike structure (the prepollex) that is not homologous to any of our digits.

In some instances it is fairly easy to determine whether characters are homologous by their anatomical design; the "beak" of a biting fly is not homologous to that of a true bug, for the relations of the labrum, mandibles, maxillae, and labium are quite different in the two cases (*Figure 10*). But the spatial relationships among the bones of a bird's wing identify them as humerus, carpals, and so forth. Deciding what point of reference to use in judging spatial relationships can be quite a problem (Inglis 1966); for example, one cannot arbitrarily decide that the most anterior tooth in a mammal is an incisor, for cattle and many other mammals lack upper incisors. The development of a feature often provides better evidence on homology than its final form. On this basis the ear ossicles of mammals are believed to be homologous to jaw elements in reptiles.

Ideally one wants to decompose organisms into separate characters, each of which evolves independently of the others; then each character gives independent evidence of phylogenetic relations. Thus a superficially single character, such as number of teeth in a mammal, would more properly be refined into separate characters: number of incisors, canines, and so on. Conversely two seeming characters may be just two aspects of a single feature if they are consistently correlated; differences in shape are often a simple consequence of differences in size, because of allometric growth (*Chapter 8*), and are not independently varying characteristics. But knowing what characters are independent is difficult in practice. Is each bristle on a fly a separate character, or are all the bristles together a single character? And unit characters may be hard to define even conceptually. Developmental biology tells us that organisms consist not of independently formed unit characters, but of interacting developmental pathways, and the interactions among the pathways can themselves change in evolution. The shape of a deer's antlers may be correlated with body size, so that in the evolution of some species shape and size are but two aspects of a single character; yet in the evolution of other species growth rates can change, a new relationship between shape and size can evolve, and a single character has become decomposed into two. The relationship between the formal procedures of systematics and the principles of developmental biology has been tenuous.

Assume, however, that a series of independent homologous characters can be identified. Now each character displays in some species a primitive state (e.g., five toes in tetrapods) and in some species a derived state (e.g., one toe in horses). If there have been no reversals of evolution, the derived state appears in all the descendants of the species in which it first arose. Therefore the members of a monophyletic group share derived character states.

A given character state can arise more than once, however; and such homoplasious character states can lead us to mistakenly combine species into the wrong groups. But such homoplasious characters (including those that undergo evolutionary reversal) do not agree with the true phylogeny, whereas all nonhomoplasious characters must agree with it and therefore with each other. If all the characters are evolutionarily independent, then, different homoplasious characters give different phylogenetic trees, whereas all nonhomoplasious characters give the same phylogenetic tree. The only systematic factor tending to make some groups of species share more derived states than others is thus common ancestry, and the best guess of the true phylogeny is this: species that share the highest number of derived character states constitute monophyletic groups (*Figure 14*).

It can be shown that the phylogenetic tree derived by this criterion is the most parsimonious tree, requiring the smallest number of evolutionary changes, including parallelisms and reversals. It is not necessarily the true phylogeny, just the most likely one based on available information. The fewer homoplasious characters are used, the greater the probability of getting the right phylogeny; so it is best to give little weight *a priori* to those most likely to be homoplasious. Biological judgment may help. Character states thought to be special adaptations to particular environments may be suspected of homoplasy: color patterns, prehensile tails in arboreal animals, succulent leaves in desert plants. Very often such characteristics vary more among species than more generally adaptive character states; thus leaf shape and petal color are far more variable among related plants than the number of floral parts and their spatial arrangement. Nevertheless even general adaptations such as the position of the ovary in angiosperms or body segmentation in animals are subject to homoplasy.

DETERMINING THE POLARITY OF EVOLUTIONARY CHANGE

The crux of this method of phylogenetic inference is the specification of primitive and derived character states. It might be supposed that this is most easily accomplished by recourse to the fossil record, but in fact this is seldom the case (Schaeffer, Hecht, and Eldredge 1972). Only if a character state is ubiquitous among the fossil members of a group known to have included the ancestors of a modern taxon can the fossils' feature be assessed as the ancestral state. For example, fossil therapsids have a single occipital condyle, so the double occipital condyle of the placental mammals may be judged to be a derived state. But our knowledge that mammals descended from therapsids is based not on their temporal distribution in the fossil record, but on the same kinds of inference from their ancestral characteristics that are used to assess the phylogeny of existing forms that have no fossil record.

When the phylogeny of existing species is assessed, it is sometimes assumed that the character state shared by only a few species is derived. Thus among the monkeys and apes only the owl monkey (*Aotus*) is nocturnal, so its nocturnal habit may be a derived state. But very often only a few relict species possess the primitive state; the jawless condition of the early vertebrates, for example, persists only in the few species of modern cyclostomes. More often, perhaps, the ecological function of a characteristic gives a clue to which state is derived. A peculiarly specialized adaptation is more often derived than ancestral; the raptorial forelegs of mantids and the webbed feet of the platypus are specializations for ways of life that the ancestors of insects and mammals surely did not practice.

Embryology provides some important clues to the polarity of evolutionary change. The principle involved is stated most naively in Haeckel's *biogenetic law* that "ontogeny recapitulates phylogeny." As stated, this law is flatly wrong (S. J. Gould 1977); developmental stages are not replicas of ancestral adult forms. Butterflies are not descended from larviform ancestors. Rather, *von Baer's law* is closer to the truth. The early developmental stages of a characteristic tend to be more similar among related species than the later stages; the characteristics that differentiate the taxa are embryologically later accretions on a fundamentally similar developmental plan.

To the extent that this principle holds, species may share similar embryonic patterns that develop further toward the derived state in some species than in others. For example (Rensch 1959), the metacarpals develop first as separate elements in cattle, as they do in other mammals, and only later fuse into the cannon bone, evidence that the fused condition is derived from the unfused. That the rudiments of teeth develop in birds and anteaters, only to be resorbed, gives evidence that toothlessness is a derived state. Persistence of such rudiments in the adult as vestigial structures likewise provides such evidence. The frog *Ascaphus truei* has functionless tail muscles but no real tail, evidence that frogs are derived from tailed ancestors and that *Ascaphus* branched off early in the evolution of frogs.

Nevertheless developmental patterns do not infallibly specify which characteristics are primitive, for early developmental stages have their own special adaptations. The cotyledons of a plant do not resemble ancestral leaves; the horny beaks of tadpoles and the comblike milk teeth of young bats are special adaptations to the juvenile environment, not representatives of ancestral conditions.

The single most useful criterion of which character state is primitive within a group of species is its condition in related forms outside the group. If species A, B, and C are known to share a more recent common ancestor with each other than they do with species D, E, or F, then the character state that is ubiquitous among D, E, and F is most likely to be primitive within A, B, and C (*Figure 17*). This procedure applies the principle of parsimony, since it assumes that the character state has not arisen independently in the two groups of species. In Figure 17, for example, the six-legged condition is assumed to be primitive *within* the Lepidoptera (butterflies and moths), since this is the condition in other insect orders; thus the four-legged condition of the Nymphalidae must be derived.

This criterion of the direction of evolutionary change requires that we know *a priori* which species fall within the group to be analyzed and which outside — which insects are lepidopterans and which not. This judgment, which is based in fact on overall phenetic similarity, assumes then that taxa evolve at roughly equal rates over long time spans — an assumption that is reasonable at least sometimes.

Only recently have systematists begun to state explicitly the criteria by which they erect classifications and to develop explicit algorithms by which primitive and derived characteristics can be identified. Many aspects of this recent phylogenetic methodology are just formal statements of principles that many systematists have long used without stating them explicitly. These methods will probably help to elucidate

17 The lepidopteran families (A, B, C) are known to be more closely related to each other than to the insect orders D, E, and F. If, as in phyletic scheme B, the character state 1 is postulated to be primitive *within* the Lepidoptera (to have been possessed by ancestor H), then reversals must have occurred in characters 1, 3, and 4 (shown by underscoring). The phylogenetic tree A is more parsimonious; there is no homoplasy if it is assumed that state 0 is primitive within the Lepidoptera (ancestor H) and is primitively shared with the outside groups D, E, and F. Phylogeny A is in fact the accepted scheme.

	Mandibles present (0), absent (1)	Scales present (1), absent (0)	Legs 6 (0), 4 (1)	Wing veins many (0), few (1)	Larval crochets present (1), absent (0)	Larval silk glands present (1), absent (0)
(A) Micropterygidae (biting moths)	0	1	0	0	1	1
(B) Sphingidae (sphinx moths)	1	1	0	1	1	1
(C) Nymphalidae (butterflies)	1	1	1	1	1	1
(D) Trichoptera (caddisflies)	0	0	0	0	0	1
(E) Coleoptera (beetles)	0	0	0	0	0	0
(F) Neuroptera (antlions, etc.)	0	0	0	0	0	0

C 111111 B 110111 A 010011 D 000001 E 000000 F 000000
110111
010011 (H)
000001
000000
000000
A

A 010011 B 110111 C 111111 D 000001 E 000000 F 000000
110111
111111 (H)
000001
000000
000000
B

relationships in such groups as the birds, snakes, and higher plants, in which phylogenetic affinities have long been puzzling. In some instances traditional phylogenetic schemes may be modified substantially, but in most cases I suspect that the traditional views will be largely upheld. If so, many of the conclusions about rates, directions, and paths of evolutionary change presented in this chapter and the next are likely to hold.

SUMMARY

To judge how rapidly and in what directions evolution occurs, it is necessary to reconstruct phylogenetic histories, determine when in geological time each major phylogenetic event occurred, and infer the characteristics of ancestral forms. Such phylogenetic inferences are largely reflected by the structure of taxonomic classifications. The reconstruction of phylogenies is fraught with difficulties, especially distinguishing uniquely derived from homoplasious (convergent, parallel, or reversed) characteristics and determining the direction of evolution of a characteristic. These problems can be overcome in many instances, however, thus making it possible to identify characters that change very slowly and those that change rapidly; some that reverse direction and others (especially complex characteristics) that are irreversible; some that change at highly variable rates and others, especially the composition of protein molecules, that change at relatively constant rates. The existence of trends and universal directions in evolution is much less apparent than radiation in innumerable directions. The emergence and diversification of each major group is marked not by adherence to a universal trend, but by the acquisition and improvement of one or more special features that fit it to its special way of life.

FOR DISCUSSION AND THOUGHT

1 Why are reproductive characters such as secondary sexual characteristics useful for distinguishing species but not particularly useful for assessing phylogenetic relationships among species and higher categories?

2 Much of the higher classification of organisms, developed in pre-Darwinian times, has remained essentially unchanged by the introduction of evolutionary thought into systematics. Why?

3 Higher taxa are defined by phylogenetic systematists as sets of species that share uniquely derived character states; yet these taxa often include species that do not have the definitive character states. For example, a distinguishing feature of the insect order Hemiptera is partially sclerotized forewings; yet many hemipterans such as bedbugs possess no wings at all. Explain how their affinities can be determined.

4 Is it possible to draw a phylogenetic tree to describe genealogical relationships among intraspecific groups such as subspecies (e.g., the human "races")? What are the difficulties of such an attempt?

5 This discussion of phylogeny has assumed that diversity arises by irreversible splitting of lineages among which interbreeding ceases. How can a phylogenetic diagram accommodate cases in which there is hybridization among species or genera or in which hybridization produces new, hybrid species?

6 To say that characters of two species are homologous is to imply that they are the products of homologous genes. But if we knew the nucleo-

tide sequences of such genes, how would we tell whether they were homologous, given that they may have come to differ greatly in sequence? See Fitch (1970).

7   Discuss ways in which biogeographic information can help determine phylogenetic relationships among species and in which phylogenetic information can explain biogeographic patterns.

8   Are evolutionary reversals due to back mutations in the same genes in which mutations produced the former character states?

9   If plants with parietal placentation of the ovules can be found in environments as different as deserts and rain forests, why don't all plants have the same kind of placentation?

10   Explain and discuss the taxonomic relevance of the observation that a character which varies among species in one taxon can be invariant among species of another taxon. The number of antennal segments can be sexually dimorphic within species of Hymenoptera, for example, yet is diagnostic of families of Hemiptera.

11   It is commonly asserted (e.g., Wilson 1975a) that since sexual division of labor is very common among the primates, the ancestor of the human species must also have behaved this way. Some authors would go further and assert that such a genetically based pattern of behavior has been inherited by modern humans with little change. Apply the principles of this chapter to evaluate these arguments.

12   Analyze the literature on the systematics of a taxon in which you are interested; in discussing the primates, for example, you might account for (a) the greater diversity of some families (e.g., the monkeys in the Cercopithecidae) than others (e.g., tarsiers, Tarsiidae); (b) the phenetic gaps among groups (that between the tarsiers and everything else); (c) the features that distinguish each family (those of the aye-aye *Daubentonia*); (d) instances of directionality vs. adaptive radiation; do the primates fall onto a single ordered spectrum? (e) the supposed primitiveness of the lemuroids compared to the anthropoids; (f) their biogeographic patterns (why are the lemurs restricted to Madagascar?); (g) the correspondence between the primitiveness of a group and the time of its representation in the fossil record.

13   Review the history of classification of a taxonomically difficult group to determine why authorities differ in their judgments of phylogenetic affinities and how they justify their conclusions. An illuminating case is that of the frogs. Compare treatments by Griffiths (1963), Inger (1967), Kluge and Farris (1969), and Savage (1973).

14   How can one tell whether a reduced character is vestigial (hence a derived state) or incipiently evolving (hence an ancestral state with reference to some others)?

15   What is the evidence for the seemingly plausible assertion that horseshoe crabs and other living fossils have evolved slowly because their environment has not changed much?

16   Discuss the assumptions of Stanley's analysis of the relative rates of evolution (speciation in particular) in mammals and bivalves.

17   Any discussion of whether there is a direction or goal to evolution is likely to involve semantic casuistry, and the discussion in this chapter is probably no exception. Discuss in depth the problem of evolutionary direction and the consequences of your conclusions.

MAJOR REFERENCES

Mayr, E. 1969. *Principles of systematic zoology*. McGraw-Hill, New York. 428 pages. Principles of classification from the viewpoint of evolutionary systematics.

Simpson, G. G. 1953. *The major features of evolution*. Columbia University Press, New York. 434 pages. The classic discussion of rates and directions of evolution, from a paleontological perspective.

Gould, S. J., and N. Eldredge. 1977. Punctuated equilibria: the tempo and mode of evolution reconsidered. *Paleobiology* 3:115–151. A still controversial reinterpretation of the fossil record emphasizing evolution by punctuated equilibria.

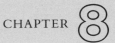

The Origins of Evolutionary Novelties

We are so used to recognizing the relationship of greatly different organisms that we sometimes forget how remarkable it is that forms of life entirely different in shape, size, behavior, and life-style have actually arisen from common ancestors. The differences between ostriches and hummingbirds, springtails and bees, bamboos and orchids are staggering. They are so impressive that evolution is difficult for many people to accept. Evolutionists themselves must question whether the processes of mutation and selection studied in *Drosophila* bottles can account for such transformations, for the origin of orders, classes, and phyla.

Since Darwin, many evolutionists have held that great evolutionary changes are the summation of miniscule alterations that we can study within our lifetimes in the field and laboratory and that evolution has always proceeded by these observable processes. This, the philosophy of uniformitarianism, was taken by Darwin from its counterpart in geology, which held that observable processes could utterly transform a landscape by their persistent action over vast stretches of time; the Grand Canyon is but an eroded roadbank writ large. This gradualistic view of evolution holds that the great differences among organisms are only magnifications of the slight variations evident in all species.

SALTATION OR GRADUAL CHANGE?

If the gradualistic view (e.g., Mayr 1960) is true, such evolutionary "innovations" as the bird's feather or the angiosperm's carpel must have had antecedents in other structures and must have evolved through many intermediate states. Some evolutionists, especially Goldschmidt (1940), have held that such features arose by SALTATIONS, mutations that transform the phenotype drastically in a single step. Goldschmidt pointed to the existence in *Drosophila* of mutations such as *vestigial,* which has minute wings, or HOMEOTIC mutations that transform one organ into

159

another. The mutation *bithorax* (*Figure 1*) transforms the halteres into a second pair of wings, an especially interesting mutation because the halteres of the true flies are homologous to the metathoracic wings of other insects. Goldschmidt recognized, of course, that the true flies differ from other insects in many other respects, so an order does not spring into existence by a single mutation — a *bithorax* fly is still a fly in other respects — but a drastic mutation (a macromutation) might be the evolutionary innovation that places the creature in a new ecological context to which its other features become adapted.

Gradualists have supported their view by arguing that most large mutations such as *bithorax* seem deleterious; they seem to disrupt developmental patterns greatly. Fisher (1930) held that most organisms are so exquisitely constructed that, like delicate machines, they might be improved by very fine alterations, but not by major ones, which would be like adjusting a delicate watch by hitting it with a hammer. But to show that many large mutations are deleterious is not to prove that they all are; Goldschmidt, in fact, proposed that drastically altered forms were "hopeful monsters." A few, finding themselves in unusual environmental circumstances, would have just the right requisites for survival, and would engender new families or orders.

Although Goldschmidt's argument has been very unpopular almost since he proposed it, a *rapprochement* between it and the gradualist view has recently been developing, and it may prove to be one of the most interesting developments in evolutionary thought (*Chapter 17*). Certainly some mutations that have discrete effects on the phenotype are not necessarily deleterious; this is true of dominant and recessive alleles at any locus. Impressive phenotypic differences among species are sometimes based on very few gene substitutions; in the composite genus *Layia,* for example, two major genes produce dramatically different flowers (*Chapter 9*). And slight changes in developmental mech-

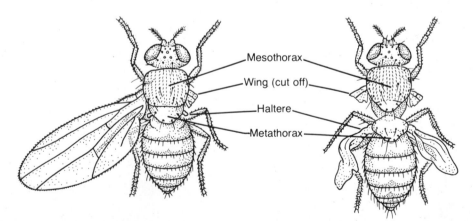

1 A wild-type (left) and a *bithorax* mutant (right) of *Drosophila melanogaster*. In the mutant the metathorax has a mesothoracic form, including the development of halteres into wing-like structures. (After Waddington 1956*b,* in part)

anisms can become magnified during ontogeny into very great differences in adult form. For example, juvenile characters of insects are maintained after a molt if the titer of juvenile hormone is high; if it is low, adult characteristics develop. By simple experimental alterations of the level of juvenile hormone, it is possible to produce miniature adults, giant juveniles, and animals with mixtures of adult and juvenile features (summary in S. J. Gould 1977). Thus important phenotypic consequences can be effected by slight genetic changes in the production or timing of a single hormone. Similarly, in some salamanders differences in the level of thyroxin determine whether a larva metamorphoses or retains its larval form when it is sexually mature. I shall return later in this chapter to a more extended discussion of the proposition, advanced by Løvtrup (1974), Frazzetta (1975), and Gould (1977), that slight changes in developmental pathways can have virtually saltatory phenotypic effects.

PHENOTYPIC GAPS AND SERIES

The difference in opinion between gradualists and their opponents stems partly from differences in their interpretation of the array of phenotypes among organisms. Gradualists emphasize the many cases in which series of intermediate forms connect radically different organisms. A brilliant tropical butterfly differs greatly from other insects; it has greatly expanded wings covered with small colorful scales, no mouthparts other than a coiled proboscis, and in some instances only four rather than six functional legs. But butterflies with semifunctional forelegs exist; the size and color of the wings vary imperceptibly among species, from the dullest moths to the glamorous morphos; and the most primitive moths, the Micropterygidae, have a full set of mouthparts, including chewing mandibles (*Figure 2a*). The maxillae of these moths are slightly elongated, forming a rudimentary proboscis. The wing venation of some micropterygids (*Figure 2c*) is much like that of primitive caddisflies (order Trichoptera), some of which have mothlike wing scales.

At a less pronounced level of divergence, cases like the Hawaiian honeycreepers (*Chapter 4, Figure 11*) illustrate the connecting links among unlike but related species. Within a single species, moreover, individuals sometimes have the diagnostic characteristics of related species or even genera. The form and number of teeth in mammals are important for classification; yet in a single sample of the deer mouse *Peromyscus maniculatus,* Hooper (1957) found variant tooth patterns typical of 17 other species of *Peromyscus.* Among fossils of the extinct rabbit *Nekrolagus,* Hibbard (1963) found one with the premolar pattern characteristic of modern genera of rabbits; and the *Nekrolagus* pattern is occasionally found in living species.

Gradualists have been embarrassed by the fossil record, in which series of intermediate forms leading to major new taxa are most uncommon. Fossils from here and there can be pieced together to form series, as in the succession of horses with decreasing numbers of toes;

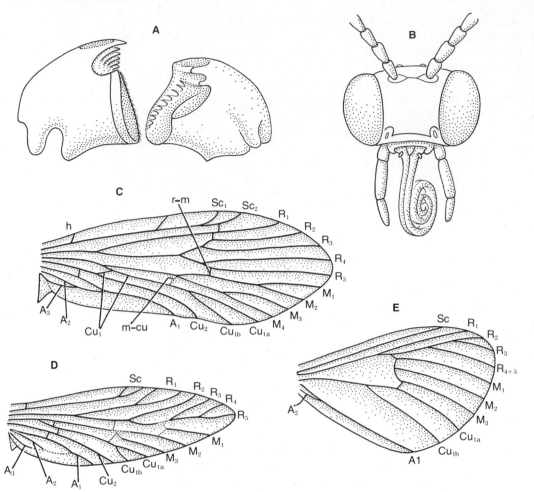

2 Evidence of gradual evolution of the higher categories. Primitive moths such as Micropterygidae have mandibles (A) like those of many other insect orders; most Lepidoptera, however, lack mandibles and have the maxillae elongated into a sucking tube (B). The relationship of the Lepidoptera to the caddisflies (Trichoptera) is evident in the wing venation. C is the forewing of a primitive caddisfly (Rhyacophilidae), D that of a primitive moth (Eriocraniidae), E that of an "advanced" moth (Geometridae). (After Imms 1957)

but seldom if ever can these be shown to be steps on a direct line of descent (Eldredge and Gould 1972, Gould and Eldredge 1977; *Chapter* 7). Intermediate forms, such as the therapsids with both reptilian and mammalian features, or *Archaeopteryx,* with reptilian and avian characteristics, are plentiful and demonstrate conclusively the phylogenetic origins of higher taxa; but they are mosaics of ancestral and derived character states rather than true intermediates. *Archaeopteryx* demonstrates that birds arose from reptiles but does not tell us whether feathers evolved gradually.

Gradualists have traditionally attributed the paucity of intermediate fossils to failures of preservation and to the other imperfections of the fossil record. Simpson (1953), however, advanced another view, championed especially by Eldredge and Gould (1972; Gould and Eldredge 1977). Major evolutionary changes may occur by QUANTUM EVOLUTION, to use Simpson's term, or by punctuated equilibria, to use Eldredge and Gould's (*Chapter 7, Figure 1*). New characters, including those diagnostic of major new groups, may evolve very rapidly when a small population is geographically isolated and becomes a new species (*Chapter 16*). Because of the rapidity of change, intermediate forms, if they existed, are not preserved in the fossil record. Among the great numbers of species, those with characteristics that adapt them to a new, long-lasting adaptive zone are the progenitors of major new groups. The extinction of the others, by a process of species selection, together with the rapidity of phenotypic change during speciation, yields gaps in the fossil record. Thus the theory of punctuated equilibria suggests that major changes can happen very rapidly, but it still does not tell us whether they occur through intermediate morphological steps.

THE ADAPTIVE CONTEXT OF EVOLUTIONARY INNOVATIONS

Proponents of a thoroughgoing gradualistic view of evolution must address a question that at first seems damaging to their position. Given that natural selection depends only on present adaptation, what advantage can there be to the earliest, slightest manifestations of a trait that is adaptive only when fully developed? Can there be adaptive value to slight featherlike structures that are insufficient for flight? And if the function of a complex organ depends on the congruence of interdependent parts, how can slight alterations of each part be adaptive if they do not arise in concert?

These questions have never been fully resolved (Frazzetta 1975); indeed they may not have to be if evolution is less gradual than is generally thought. But they are not the insuperable barriers to a gradualist position that they may seem. Even slight phenotypic differences often have pronounced adaptive effects, and even a slight selective advantage suffices to replace one allele by another, over a long enough time (*Chapters 13–15*). Moreover the interdependence of the components of a complex organ is often not very great in the earlier, simpler stages of its evolution. The function of a vertebrate eye requires precision of form of lens, iris, retina, humors, muscles, and nerves; but an incipient eye can be quite advantageous with only a few of these features, as the less elaborate photoreceptors of platyhelminths and annelids bear witness. Eyes could well have evolved by successive slight alterations (Eakin 1968).

Mayr (1960) has stressed that the major changes in organisms' features follow most often from an intensification, diminution, or change of function. Intensification of function is the adaptive basis for the progressive elaboration of horns as weapons in ungulates, crests and plumes as sexual signals in birds, petals as advertisements to pollinators. The increased use of the molars for grinding tough, silica-laden grasses is associated with the development of a higher tooth crown in horses and in many rodents.

Conversely, many evolutionary alterations stem from the reduction or loss of a structure's function, as in wingless insects, chlorophyll-less parasitic plants, or legless snakes. In some instances the loss of a structure is clearly adaptive; snakes can presumably move and burrow better without legs. Whether there is always a positive advantage to the loss of a structure, such as the eyes in cave fish, has long been debated. Perhaps mutations leading to vestigiality are neither adaptive nor maladaptive, or perhaps the loss is adaptive because useless structures take up space in the body and require energy and materials that can be put to better use.

A very common basis of evolutionary transformations is the change of a characteristic's function, often associated with entry into a new adaptive zone or a new environment. The wings of auks and several other birds are used in the same way in both air and water, and in penguins they have become entirely modified for underwater flight. The leaves, stipules, flower parts, or adventitious roots of vines can serve as holdfasts or tendrils. In animals a change in behavior is often the first step in the development of a new life-style, and changes in morphological and physiological characteristics follow. Many species of mice and ground squirrels burrow but have few of the modifications for subterranean life that gophers and other more exclusively fossorial rodents possess.

Quite often a structure serves both old and new functions, as in many plants (e.g., maples) in which the ovary wall both protects the seeds and disperses them. Thus an evolutionary innovation may have developed because of an original function, to the point where a new function becomes possible. The ability of an electric eel to kill prey by shocking is an elaboration of the much weaker electric fields generated by other gymnotid knife-fishes, which use their electricity for orientation and communication in murky waters. And the capacity to produce a charged field is itself a hypertrophied development of the properties of muscle and nervous tissue in general.

PREADAPTATION

In principle any fish might have developed the ability to generate an electric field, by modification of the muscles that in all fishes are preadapted to produce a charge. Indeed electrical properties have evolved in at least four groups of fishes. The term *preadaptation* must be used circumspectly; it does not mean evolving structures in anticipation of future need (Bock 1959). Fishes did not evolve lungs so that they could someday invade the land; rather the lungs that had already evolved served them well when new environmental problems, such as the drying of ponds, appeared. Thus the function of a characteristic can change more rapidly than its structure. The kea (*Nestor notabilis*), a New Zealand parrot that rips through the skin of sheep with its sharp beak to feed on fat, has a beak much like those of other parrots. Thus most parrots could be said to be preadapted for carnivory. Progenitors of a group usually display preadaptations for that group's distinctive adaptations. Epiphytic plants such as bromeliads, which do not have access to wet forest soil, developed from arid-adapted ancestors. The early hominids, unlike most mammals, may have been preadapted for the development of complex cerebral functions because they had opposable thumbs and erect posture, which freed the hands for delicate manipulations that require elaborate nervous control.

FUNCTIONAL INTERACTIONS

A structure's assumption of a new function may free other structures for different functions. Liem (1973) espouses this principle to account for the enormous adaptive radiation of the cichlid fishes, which vary greatly in the structure of the jaws and teeth (*Chapter 16, Figure 1*) and in their feeding habits. They include species that feed on filamentous algae, plankton, other fishes, and even the scales of other fishes. In most fishes the premaxillary and maxillary bones, which bear the teeth, must perform the dual functions of collecting and manipulating food; deeper within the gullet the pharyngeal bones bear teeth that help hold the prey. In the cichlids the articulations and musculature of the pharyngeal bones make them more versatile, so they can manipulate prey. Thus the premaxillary and maxillary are freed for the specialized food-gathering tasks that vary so greatly among the species.

Because of the functional interactions among the parts of an organism, an alteration of one part is often accompanied or followed by alterations of others. Very often compensatory changes occur. The reduction of leaves to spines in cacti is accompanied by the development of photosynthetic capacity in the stem. The bower birds have duller plumage than the related birds of paradise, but their sexual display is based on an elaborately constructed courtship arena of leaves and other objects. Among the most important compensatory changes are those associated with body size (Thompson 1917, Huxley 1932, Gould 1966). The volume (or weight) of an object increases as the cube of its linear dimensions, but the surface area increases only as the square (assuming identical shapes). Because the weight supported by a structure is proportional to its cross-sectional area, the legs of elephants must be thicker than those of gazelles, and pelicans must have proportionately larger wings than sparrows. And because most physiological processes (digestion, gas and heat exchange, and many others) occur over surfaces, most organs differ in shape between large and small species to maintain adequate surface-volume ratios. This is often accomplished by expansion of a structure to a large size (as in the ears of jack rabbits, which dissipate heat), by dissection into multiple units (as in the kidney), or by folds and convolutions (as on the surfaces of gills or intestines).

Although an adaptive advantage can often be ascribed to a new trait even in its incipient stages, there are some puzzling instances for which a gradualistic explanation is not easily found. The shell of a turtle is constructed in part by the rib cage, which surrounds the pectoral girdle. In all other vertebrates the girdle (scapula, coracoid, and clavicle) is outside the rib cage. The intermediate steps and their advantage are not obvious. And although natural selection is undoubtedly responsible for the origin and elaboration of most of the characteristics that distinguish higher taxa, some novelties may not be advantageous in themselves but developmental consequences of other changes. Evolutionary changes in body size, for example, can have nonadaptive consequences for the size and shape of other structures (*Chapter 12*).

REGULARITIES OF PHENOTYPIC TRANSFORMATION

The previous sections address the *why* of major morphological changes, the adaptive reasons for their occurrence; but it is important to understand the *how* as well. Heritable phenotypic changes must be based on alterations of DNA, which through the processes of development are translated into phenotypes. Thus to understand how mutations affect the phenotype, what degree of genetic change is required to effect a given phenotypic change, or which conceivable evolutionary paths are open or closed to a species, it is necessary to understand the mechanisms of enzyme action, the regulation of gene activity, and growth and differentiation. Our ignorance of these topics is so profound that it is not yet possible to specify the biochemical and developmental bases of most evolutionary changes; but it is possible to develop speculative black-box models that describe in a very formal, nonmolecular way the major patterns of evolutionary change. Such models suggest that although development is a complex process, conceptually simple changes in developmental events can have dramatic phenotypic effects. For convenience, I shall distinguish between evolutionary transformations at the biochemical level, which affect the form of an organism slightly or not at all, and transformations of form and structure.

BIOCHEMICAL TRANSFORMATIONS

The products of many genes are proteins, including enzymes. Thus biochemical changes, with physiological consequences, can stem either from a change in the structure and therefore the function of a protein, or from changes in its amount or in the pattern of its distribution in different cell types or at different times in development. Changes in the concentration of an enzyme (or other protein) are effected by alteration of a variety of regulatory factors, including transcription of the DNA. In Chapter 10 I describe instances of biochemical adaptation in bacteria that entail mutations in regulatory genes.

Changes in the structure of a protein can greatly affect its properties, as is graphically illustrated by the sickle-cell hemoglobin Hb^s, in which a single amino-acid substitution reduces its capacity to bind oxygen. Adaptation to different environments is often achieved at the biochemical level by variation in the reaction rates of enzymes and in the conditions of temperature and pH at which their activity is greatest (Hochachka and Somero 1973). Seldom is an enzyme modified during evolution to serve an entirely new function, for its previous function remains necessary. But very often a locus is duplicated (*Chapter 10*), and the two cistrons diverge by mutation and selection to take on different functions (Ohno 1970), as in the divergence of α hemoglobin from myoglobin and of the several hemoglobin chains from each other.

In some instances changes in enzyme systems (probably in the structure of the enzyme) may be prerequisite to success in a new adaptive zone. The adaptation of insects and reptiles to terrestrial life is facilitated by the excretion of uric acid rather than urea. Bumblebees are more prevalent in cold regions than most other bees, probably because of their biochemically based homeothermy. They maintain a high level of fructose diphosphate that is split by fructose-diphosphatase in a reaction that is coupled to the reaction $ATP \rightarrow ADP + P_i$ + heat. AMP is generated from ADP by a further cleavage of P_i; but in bumblebees (unlike most species) fructose diphosphatase is not feedback-inhibited by AMP, so the production of heat proceeds apace (Hochachka and Somero 1973).

Physiological adaptation is often accomplished merely by changes in the levels of enzyme maintained. Detoxification of DDT in some resistant strains of houseflies is due to an increased titer of DDT-dehydrochlorinase, which is present at low levels in susceptible flies and presumably catalyzes other reactions under natural conditions (O'Brien 1967). Changes in enzyme regulation affect the distribution and amount of pigments, lipids, lignins, and other major tissue constituents. Thus desert plants have a thick layer of waxes on the epidermis, but this is largely a quantitative change in materials widespread among plants.

Major biochemical changes, however, such as elaborate new biosynthetic pathways that must entail the evolution of many new enzymes with mutually compatible catalytic properties, are complex adaptations that seldom evolve. As every biochemist knows, the biochemical characteristics of organisms are far less diverse than morphological features. Many physiological adaptations entail not biochemical but structural or behavioral changes. The thermoregulatory problems of different mammals are solved largely by differences in behavior, body size, shape (e.g., the length of appendages), and thickness of fur. Evolution of these characteristics consists of changes in the mysterious processes of growth and differentiation.

GROWTH AND DIFFERENTIATION

Evolutionary changes in morphology are based on alterations of two ontogenetic processes: growth, the increase in size by cell division and enlargement, and differentiation, the assumption of distinctive com-

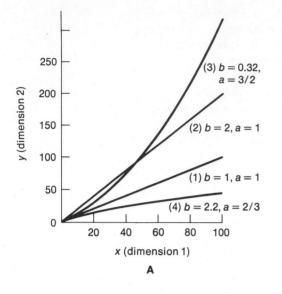

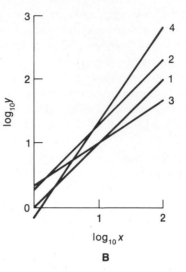

A

B

3 Allometric growth. (A) Arithmetic plot of the lengths y and x of two structures or dimensions. Curves 1 and 2 show isometric growth ($a = 1$); structure 2 equals structure 1 in curve 1 ($b = 1$) and is twice as long in curve 2 ($b = 2$). Curve 3 shows positive allometry ($a > 1$); curve 4 shows negative allometry ($a < 1$). (B) Logarithmic plot of the same curves. The slope equals 1 in curves 1 and 2, which differ only in intercept (cf. Figure 6); it is greater than 1 in curve 3 ($a > 1$), less than 1 in curve 4 ($a < 1$). (C) Illustration of hypothetical evolution of leaf shape, according to these graphs. As the ancestral species doubles in length (x) during its growth from juvenile (J) to the mature form (A_1); it doubles in width (y); that is, $a = 1, b = 1$. Descendant A_2 is twice as wide

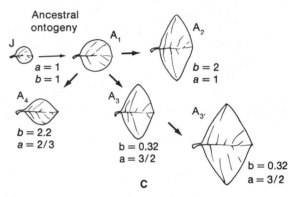

C

when mature, for b has evolved from 1 to 2. If a evolves from 1 to 2/3, a narrower leaf (A_4) results. If a evolves from 1 to 3/2, a wider leaf (A_3) is formed. With this allometric relation, evolution of greater length results in disproportionately greater width ($A_{3'}$).

position and form by different groups of cells. The shape of a structure and its size relative to the whole organism are determined by the time during development at which its growth begins and ends and by its rates of growth in different dimensions. The factors that determine these processes are no doubt complex; but some can be altered simply. For example, the growth of lateral buds in plants is inhibited by hormones secreted by the terminal meristems and is released when the hormones are reduced (say, by pruning).

The differentiation of a group of cells likewise is a response, presumably the activation of genes hitherto in a repressed state, to stimuli from other cells or the external environment. The cells may be competent to respond at some times in ontogeny and not others; and the stimuli likewise may vary ontogenetically, as do hormones, like juvenile hormone in insects, that induce differentiation. Thus fairly simple changes in the timing of stimuli or in the response of cells to these stimuli can have important phenotypic consequences. A low level of thyroxin late in the development of a salamander, for example, can result in the retention of larval characteristics in a large, sexually mature animal. Changes of timing in differentiation can have especially dramatic effects if they are magnified by growth.

CHANGES IN GROWTH RATES: THE EVOLUTION OF SIZE AND SHAPE

During ontogeny the parts of an organism, or of a single structure, grow ALLOMETRICALLY, that is, at different rates. In a series of specimens of different ages one dimension (with magnitude y) may be disproportionately large or small compared with another dimension (with magnitude x). The relation between them may be written $y = bx^a$ or, in logarithmic terms, $\log y = \log b + a \log x$. Thus a is the slope of the relation between y and x on a log log plot, and the constant b describes the y-intercept. If a is the same for two species, a difference in b means that the species differ in the value of y even when they have the same value of x. If the constant a, called the allometric coefficient, equals 1, then y (say the length of a limb) is a constant proportion of x (say body length). If y becomes disproportionately greater as x increases, then $a > 1$; if lesser, then $a < 1$ (*Figure 3*). For example, the head grows during human ontogeny more slowly ($a < 0.3$) than the body (*Figure 4*); if it kept pace with the body, we would be megacephalic monsters as adults.

Many allometric relations are adaptive. Organs like the intestine, whose function is surface-dependent, often grow disproportionately faster than the body as a whole. This happens because the area of an organ increases as the square of its length y, while the volume (and

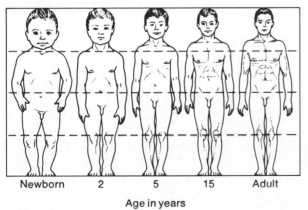

4

Allometric growth in humans. Individuals of different ages, drawn at equal heights, show proportionately less rapid growth of the head and more rapid growth of the legs than of the body as a whole. (Redrawn from Sinclair 1969)

Newborn   2   5   15   Adult

Age in years

weight) of the body increases roughly as the cube of body length x. Thus the ratio of the organ's surface area to the volume of body it serves is constant only if $y = bx^{3/2}$.

Evolutionary changes in adult body size can occur for many adaptive reasons. A specific body size may be adaptive in itself (small animals can hide in small spaces; large ones are often susceptible to fewer predators), or it may be a corollary of the length of time it takes to reach sexual maturity, which itself is adaptive (*Chapter 12*; S. J. Gould 1977). If the ontogenetic relation $y = bx^a$ remains unchanged as body size x evolves, the structure measured by y will be disproportionately larger or smaller (if $a \neq 1$) in the adult descendant than in the ancestor. Thus the plot of y against x for the adults of various species may form the same curve as for the different developmental stages of an ancestral species, extrapolated to newly evolved body sizes. For example, the length of a baboon's face increases disproportionately with body size during growth, and adults of larger species have disproportionately longer faces than smaller species (Freedman 1962); the coefficient a has apparently remained genetically constant during evolution. In the largest species of deer, the extinct Irish elk (*Figure 5*), allometric growth resulted in monstrously large antlers (this case is rather complex, however; see Gould 1974). Thus a change of body size (or of any organ whose parts or dimensions

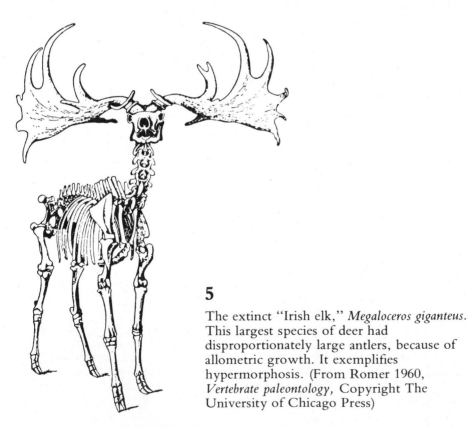

5

The extinct "Irish elk," *Megaloceros giganteus.* This largest species of deer had disproportionately large antlers, because of allometric growth. It exemplifies hypermorphosis. (From Romer 1960, *Vertebrate paleontology,* Copyright The University of Chicago Press)

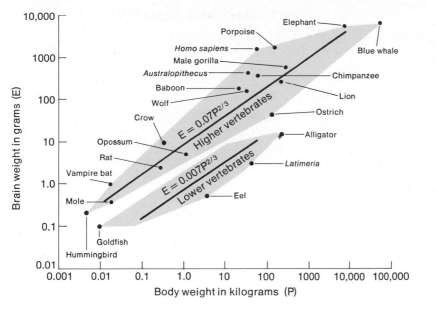

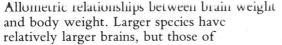

Allometric relationships between brain weight and body weight. Larger species have relatively larger brains, but those of endotherms are 10 times as large as those of ectotherms of comparable size. (Redrawn from Jerison 1973)

6

grow at different rates) results in a change of shape.

Often the ontogenetic relationship *a* changes in evolution, so that the proportions in a larger species are not a simple extrapolation of allometric growth in smaller species. If the length of the face of living and extinct adult horses is plotted against cranial length, the allometric coefficient is 1.8. During the ontogeny of the modern horse (*Equus*), however, *a* = 1.5. If the eohippus *Hyracotherium* had had this growth coefficient and had simply evolved to the size of *Equus,* modern horses would have had relatively shorter faces than they actually have. Thus *a* has become greater during evolution.

Finally, evolution can entail change in the value of *b* in the allometric equation (*Figure 3*). For example, brain weight is proportionately less in larger than in smaller species of both ectothermic (cold-blooded) and endothermic (warm-blooded) vertebrates; *a* = ⅔ in both groups. But the brain of an endotherm weighs 10 times that of an ectotherm of comparable body weight: *b* equals 0.07 vs. 0.007 (Jerison 1973; *Figure 6*).

The topic of allometric growth (Huxley 1932; Cock 1966; Gould 1966, 1977) has many consequences in evolution. For example, many evolutionary changes in shape that are not adaptive in themselves may be allometric by-products of other adaptive changes (*Chapter 12*). It also sheds some light on the question of whether ontogeny recapitulates phylogeny.

ONTOGENY AND PHYLOGENY

Suppose two structures 1 and 2 that have lengths y and x develop according to the equation $y = bx^a$. If a evolves to a smaller value, structure 1 will be retarded with respect to structure 2 in the descendant, compared with the ancestor; at comparable times in ontogeny structure 1 will be smaller, more like the juvenile condition (*Figure 7*; Gould 1977). If a evolves to a larger value, structure 1 will be accelerated with respect to 2 and will be larger in the descendant than in the ancestor at any stage of development; it will have more of the ancestor's adult form in the juvenile stages of the descendant. Thus, if the descendant attains the same adult body size as the ancestor, it will have passed through the ancestor's adult form while still juvenile; its ontogeny will recapitulate its phylogeny.

By considering such shifts in the timing of developmental events, Gould (1977) concludes that two major phenomena, PAEDOMORPHOSIS and RECAPITULATION, can result from each of two processes, RETARDATION and ACCELERATION. Paedomorphosis is the retention of an ancestor's juvenile characteristics in the adult stage of a descendant; recapitulation is the attainment of the ancestor's adult characteristics during the juvenile stages of the descendant. Gould suggests that the following evolutionary events can occur.

Acceleration can produce recapitulation if the development of somatic organs is accelerated relative to the time of reproductive maturation (adulthood). This can occur if natural selection favors the earlier expression of genes which in the ancestor are expressed only late in development (Stebbins 1974). For example, the complexity of the sutures in the shell of ammonites (extinct cephalopods) was as great in the juveniles of descendant species as in the adults of their ancestors and greater in adult descendants than in adult ancestors (*Figure 8*).

Acceleration can produce paedomorphosis if maturation becomes speeded up relative to somatic development; the ontogeny of the descendant is therefore halted at an earlier stage. This kind of paedomor-

7 Evolutionary changes in shape. The allometric coefficient a is smaller in descendant 1 than in the ancestor; the development of the organ is retarded relative to body size, compared to the ancestor. In descendant 2, a has evolved to be larger; the development of the organ is accelerated, compared to the ancestor. Descendant 3 has the same pattern of allometric growth as its ancestor, but it has a proportionately larger organ when adult because it attains greater body size. It exemplifies hypermorphosis and recapitulation. x_A = adult body size of ancestor and descendants 1 and 2; x_3 = adult body size of descendant 3.

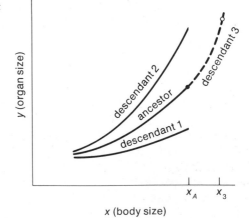

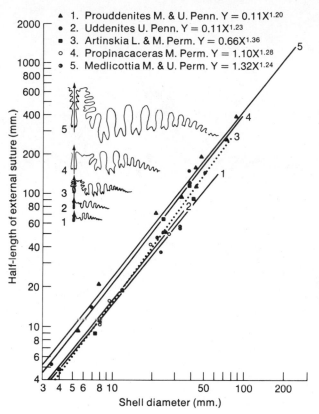

1. Prouddenites M. & U. Penn. $Y = 0.11X^{1.20}$
2. Uddenites U. Penn. $Y = 0.11X^{1.23}$
3. Artinskia L. & M. Perm. $Y = 0.66X^{1.36}$
4. Propinacaceras M. Perm. $Y = 1.10X^{1.28}$
5. Medlicottia M. & U. Perm. $Y = 1.32X^{1.24}$

8

An example of recapitulation and acceleration. The sutures on the shell of ancestral (e.g., number 1) and descendant (e.g., number 5) species of ammonites are illustrated. The older part of the suture is the small tail to the right. The suture of number 5 when young resembles that in the adult of number 1, exemplifying recapitulation; but at any given shell size the length of the suture is greater in the descendant than in the ancestor, as the regression lines indicate. In the descendant, therefore, suture growth is accelerated with respect to shell size. (From Newell 1949)

phosis, called PROGENESIS, seems especially common when natural selection favors rapid maturation or small size. In parasitic crustaceans, for example, sexual maturity is attained at a small size, when the animal still has a larval form beyond which it does not progress.

Retardation can produce recapitulation if maturation is delayed relative to somatic development. This is especially common when there is evolution of larger body size, as in the Irish elk or in the titanotheres (*Chapter 7, Figure 7*). The descendant passes through the morphological forms that its smaller ancestors had at comparable body sizes and attains a more exaggerated form when adult, as its ancestors might have if their allometric growth had been extended to larger sizes. This form of recapitulation is called HYPERMORPHOSIS.

Retardation results in paedomorphosis if somatic development is slowed relative to the time of maturation. The adult descendant retains the juvenile features of its ancestor at the same (or greater) body size that the ancestor attained. The classic example of such paedomorphosis, termed NEOTENY, is in certain salamanders, in which the gills and other larval features are retained in adulthood (*Figure 9*). Gould believes that neoteny is very common and agrees with many previous authors that humans are neotenic, retaining in adult life many of the body proportions (e.g., the short face and relatively large cranium)

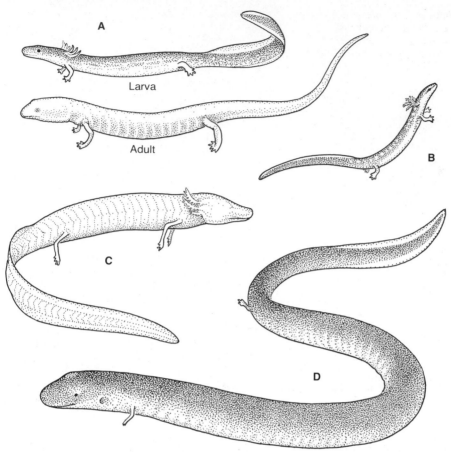

9 Neoteny in salamanders. (A) Larva and adult of *Typhlotriton spelaeus,* showing larva's gills. (B) Adult *Eurycea neotenes,* with gills. (C) A highly modified neotenic derivative of *Eurycea,* the cave salamander *Typhlomolge rathbuni.* (D) An unrelated neotenic form, *Amphiuma means,* in which the gills are retained but are internal. (A and B are redrawn from Conant 1958, C and D redrawn from Noble 1931)

and other features typical of juvenile apes. Systematists have long speculated that many major groups of organisms stem from neotenic ancestors — that insects arose from the larvae of millipede–like ancestors (juvenile millipedes have six legs) or that chordates had their origin in the tadpole larva of tunicates. These speculations, like the neotenic origin of the hominids, are by no means universally accepted.

GROWTH FIELDS AND GRADIENTS

The course of growth of an organism's parts often follows a coherent pattern, as if one or a few factors regulated the growth of the whole. In the house wren there is a proximal-distal gradient of growth rates (Huxley 1932, Reeve and Huxley 1945); relative to body size the thigh grows at the allometric rate $a = 1.29$, the tibiotarsus at $a = 1.41$, the tarsometatarsus at 1.58, the middle toe at 1.57, and the claw at 2.22. Thus if the gradient of *a*'s remained constant in the evolution of greater body size, the leg of the larger bird would consist of a series of disproportionately longer elements from the proximal to the distal region.

When such gradients exist, the separate elements are effectively integrated into a single character, often consisting of functionally related parts (Olson and Miller 1958). Although the mechanistic bases for such integration are not known, one might hypothesize that they stem from a gradient in the concentration of a growth-promoting substance. Such a mechanism establishes a GROWTH FIELD, defined by Van Valen (1962) as "any influence during the development of an organism that produces a detectable pattern among the elements of any set of structures."

Another mechanism that may produce a growth field is competition between developing structures for building materials. The fibula in birds is so reduced that only its proximal end remains; but if a barrier is placed between the developing tibia and fibula, they grow to equal length. In the normal course of development the tibia seems to compete more successfully for cells (Hampé 1960). Rensch (1959) believes that such competition is important in evolution and that it accounts especially for the evolution of vestigial structures. Wingless female moths, for example, have enlarged ovaries, as if the materials used in other moths to build wings were reallocated to egg production.

The outcome of development, however, can be affected simultaneously by positive and negative integrating factors. For example, Van Valen (1962) found that the length of the first molar in the deer mouse *Peromyscus leucopus* is positively correlated with that of the second, but the third molar is not well correlated with the first two (*Figure 10*). He

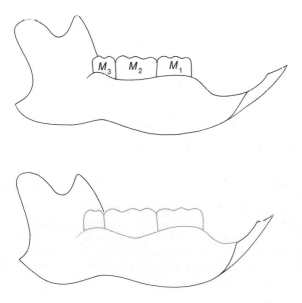

10

Developmental interactions among mouse teeth. As the length of the tooth row increases, M_1 and M_2 remain proportionately more similar to each other than to M_3.

11

D'Arcy Thompson's method of transformations. Replacing the coordinates at left by those at right transforms the shape of a puffer (*Diodon*) into that of an ocean sunfish (*Orthagoriscus = Mola*). (From Thompson 1917)

suggests that the total length of the tooth row varies with body size, yielding positive correlations, but that the total length is held within the limits necessary for proper function, so that negative correlations among the individual teeth are caused when one tooth enlarges at the expense of the others.

Because integrating mechanisms, whatever they may be, apparently exist, complex changes in the geometric form of organisms may stem from rather simple evolutionary changes in development. D'Arcy Thompson (1917) was the first to illustrate this, by showing that rather simple mathematical transformations of the shape of one species could transmute it into the shape of another. If a figure of, say, a porcupine fish (*Diodon*) is inscribed in the Cartesian coordinates *x, y,* it can be

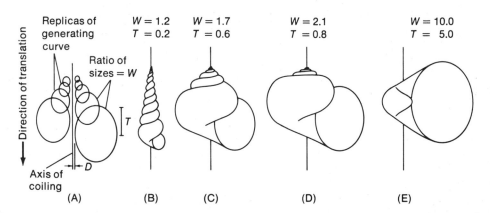

12 Hypothetical snail forms (B–E) drawn from computer-made plots. They were generated from the information in A, which shows the form of the generating curve, the axis of coiling, the size ratio *W* of successive generating curves, the distance *D* of the generating curve from the axis, and the proportion *T* of the height of one generating curve that is covered by the succeeding generating curve in one full revolution. These hypothetical forms closely resemble known species. (After Raup 1962)

transformed into the shape of an ocean sunfish (*Orthagoriscus* = *Mola*) by replacing the *x* coordinates by approximate hyperbolas and the *y* coordinates by a system of concentric circles centered near the head (*Figure 11*).

More recently Raup (1962, 1966; see also Raup and Stanley 1971) has shown that diverse shapes of snail shells, including most of the forms that actually exist, can be generated theoretically by variations in only four basic parameters: (1) the shape of the generating curve, the cross-sectional outline of the shell as it grows spirally about the axis of coiling; (2) the distance *D* between the generating curve and the axis of coiling; (3) the rate of increase *W* of the size of the generating curve as it coils about the axis; and (4) the rate *T* of translation of the generating curve along the axis of coiling (*Figure 12*). The mathematical simplicity of such transformations suggests that evolution can proceed by rather simple changes in the mechanisms of development.

ALTERATIONS IN THE NUMBERS OF PARTS

Extension of the principles of acceleration and retardation may conceptually account for a very common event in evolution, the increase or decrease in the number of SERIALLY HOMOLOGOUS PARTS. The number of identical elements may increase in evolution, as has the number of vertebrae in snakes, body segments in millipedes, or ovules in such plants as lilies. A reduction in number is even more common; reduction in the number of vertebrae, aortic arches, digits, and teeth is one of the most common themes in the comparative anatomy of the vertebrates. Most advanced families of plants have fewer stamens and carpels than the primitive angiosperms such as magnolias. Evolutionary changes in number commonly occur when the number of parts is large and variable (as in the many stamens in such mimosaceous legumes as *Acacia*); but when the number is smaller (generally a derived condition), the parts are much less variable in both number and position, both within and among species. The papilionaceous legumes such as peas, for example, almost always have 10 stamens (*Figure 13*). Stebbins (1974) refers to the greater evolutionary lability of variable multiple features as "modification along the lines of least resistance."

Changes in the numbers of parts may be due to several kinds of developmental event. If the factors responsible for their differentiation remain constant through development, increasing the period of development increases the number; truncation of development decreases it. Conversely the number of parts can decrease if the stimuli responsible for their differentiation or the competence of undifferentiated tissue to respond to such stimuli ceases early in development. For example, the number of legs in millipedes could hypothetically be reduced to the six in insects if leg differentiation stopped at the six-leg stage. Birds and frogs have the embryonic potential to develop teeth, but their development is suppressed, perhaps because the requisite stimuli are lacking.

Some reductions are attributable not to failure of development, but

A   **B**

13   Indeterminate and determinate numbers of
parts. Stamens are numerous and variable in
number in mimosaceous legumes (A,
Enterolobium cyclocarpum), but are usually fixed
at 10 in the papilionaceous legumes (B,
Sarothamnus scoparius). (A from Hutchinson
1969, B from Proctor and Yeo 1972)

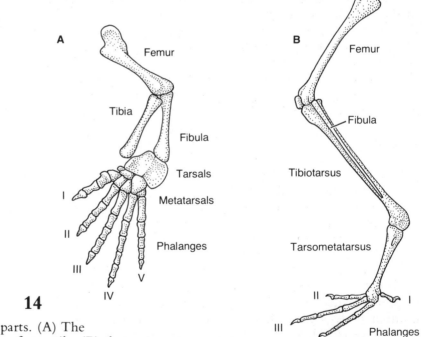

14

Fusion and reduction of parts. (A) The
skeleton of the hind limb of a reptile; (B) that
of a bird. In the bird some of the tarsals have
become fused with the tibia and others with
the metatarsals; the number of digits and
phalanges is reduced; and the fibula is reduced
in size. (A after Romer 1956, *Osteology of the
reptiles,* Copyright The University of Chicago
Press)

to fusion with adjacent structures. A bird has far fewer bones than a reptile, largely because of fusions (*Figure 14*). Families of plants differ largely in patterns of fusion among floral structures. Such patterns are also attributable to changes in developmental timing. For example, petals may be fused into a tube if the tissues (intercalary meristems) between the petal primordia grow when the petal primordia grow; if they do not, the petals are separate (*Figure 15*).

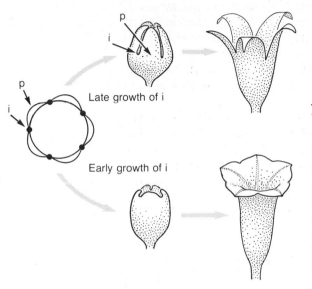

15

Effects of developmental timing on form. The developing flower, shown from above at left, has petal primordia (*p*) separated by intercalary cells (*i*). If these begin dividing only after the petal primordia grow, the petal lobes of the mature flower (side view, at right) are well separated. If the intercalary cells develop along with the petal primordia, a largely fused (sympetalous) corolla is formed. (Adapted from Stebbins 1974, after Payer 1857)

PATTERNS

The number of serially homologous parts also depends, perhaps most frequently, on the pattern by which cells do or do not differentiate into a particular kind of tissue (Sondhi 1963; see also Goldschmidt 1938, Waddington 1956*a*, 1962). The number and geometric arrangement of petals in a flower, digits on a foot, or scales on a snake depend on the mechanisms that dictate that cells in certain sites rather than others differentiate into petals, digits, or scales. It seems probable (Waddington 1962, Sondhi 1963) that alterations in patterns can stem from rather simple changes in competitive interactions among cells, in the spatial pattern of the stimuli that evoke differentiaton, and in the relative times at which the stimuli are present and the cells are competent to respond to them.

Imagine that cells compete for a homogeneously distributed substance required for differentiation. A cell that happens to gain an advantage in growth over its neighbors depletes the growth substance in its immediate neighborhood and forms a territory in which other cells cannot differentiate. Alternatively the cell may chemically inhibit the differentiation of its neighbors. In *Drosophila* it is possible to identify masses of tissue that have different genotypes because of somatic crossing over. The mutant *Hairy wing* (*Hw*) produces an abnormally high density of bristles; and wild-type tissue, if adjacent to *Hw* tissue, develops the same phenotype. The mutant *Hw* must produce an abnormally high level of bristle morphogen or a low level of bristle inhibitor that diffuses into adjacent tissue. Thus genetic variation in the production of an inhibitor can alter the spacing and number of bristles.

Complex patterns arise if tissues respond to an underlying PREPATTERN of an inducing substance (Stern 1968). For example, the tarsus of a female *Drosophila* has an underlying prepattern for the development of a male sex comb (a group of modified bristles), since male epidermal cells on an otherwise female tarsus develop a sex comb. If a mutant is incompetent to respond to a prepattern, the phenotype is altered. The mutant *achaete*, for instance, does not form bristles for this reason (Postlethwait and Schneiderman 1973).

Such prepatterns can conceivably arise from rather elementary principles. The mathematician Turing showed that if two substances that react to give an inducing substance diffuse across an area or field, the reaction can generate a standing wave pattern in the concentration of inducer. The pattern depends on the size and shape of the field, on the rates of diffusion of the precursors, and on the kinetics of their reaction.

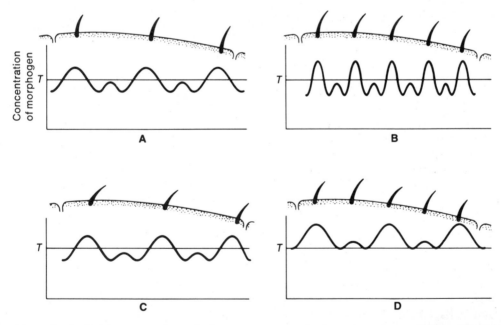

16 Hypothetical changes in an underlying prepattern (A) of a morphogen, or material that induces bristle formation. If the concentration of morphogen exceeds a threshold *T,* a bristle is formed. Change in the kinetics of morphogen synthesis can change the spacing pattern (B) or the position of the morphogen peaks (C); increasing morphogen concentration can also change the number of bristles (D).

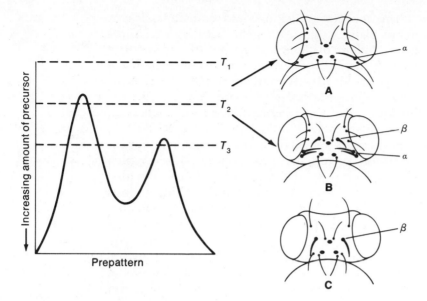

Sondhi's model of origin of a neomorphic ("new") pattern. If the first peak of the prepattern is between threshold levels T_1 and T_2, it produces the wild-type pattern of bristles and ocelli (A) in *Drosophila subobscura*. If the second peak exceeds threshold T_3, additional bristles are formed (B): bristle α is doubled, and new bristles (β) arise. These new bristles are unknown in normal drosophilids, but have a counterpart in *Aulacigaster leucopeza* (C), a member of a related family. (After Sondhi 1962)

Changes in the arrangement of the wave pattern or in the total concentration of inducer (*Figure 16*) can change the pattern of the structures that develop in response to the inducer.

Sondhi (1963) invokes this model to explain some interesting alterations in the pattern of head bristles in *Drosophila* stocks selected for higher bristle number. The new bristles in the selected stocks appeared consistently at specific positions (*Figure 17*), as if selection had raised the concentration of an inducer that is present in a specific pattern in wild flies but which is insufficient to induce bristles in certain positions. In most stocks the normal bristles retained their usual positions, but in others whole groups of bristles were shifted to new sites as if the prepattern had been shifted, as Turing's model predicts it can be. A most interesting aspect of Sondhi's work was that certain new patterns, although never observed in normal *Drosophila*, resembled those of different, related families of flies (*Figure 17*).

DIFFERENTIATION OF SERIALLY HOMOLOGOUS STRUCTURES

Among the most important regularities in evolution is the transformation of homogeneous repeated structures in an ancestor into diverse structures serving different functions in the descendant. In plants such as *Poinsettia* the leaves near the inflorescences are red, serving the function of petals; seeing such a plant makes it easy to imagine how further alterations in form and composition could transform leaves into flower parts. In many plants (e.g., asters) the basal and terminal leaves have different shapes and are bridged by a gamut of intermediate forms. The diversification of the limbs in Crustacea and of teeth in the mammals is the foundation of the adaptive radiation of these groups. Evolutionary homogenization, although less common, does occur; the teeth of the toothed whales are not differentiated into the incisors, molars, and so on of other mammals.

The diverse serially homologous structures of an organism appear to share an initially common pattern of development, after which differentiating influences switch development into one of several paths. This is evident in the homeotic mutants of *Drosophila,* in which various structures are transformed into each other. The homeotic transformations follow a regular sequence (Postlethwait and Schneiderman 1973). The imaginal discs (masses of tissue in the larvae that give rise to adult structures) are not transformed at random; rather the genital disc is most readily transformed to a leg or antenna, the leg disc to an antenna, the antennal disc to a leg or wing, and so on. Moreover, after differentiation has proceeded into an abnormal path, genes that act later in development to finish the structure of a normal organ act on the homeotic organ in the same fashion. Mutations that affect legs but not normal antennae have the same effect on the antennal legs of mutant *aristapedia* flies. Thus each path that a developing organ may take seems to facilitate the subsequent developmental steps appropriate to that path. But the mechanisms by which development is switched to one path or another and by which novel changes are added in evolution to the initial common developmental plan are mysterious.

EVOLUTION AND DEVELOPMENTAL BIOLOGY

Molecular and developmental biology leaves not the slightest doubt that development is exceedingly complex. But evidence from developmental genetics (e.g., the effects of single mutations), the models of developmental changes considered in this chapter (e.g., allometry), and the study of relationships among organisms (e.g., diverse phenotypes in closely related species) together suggest that this very complexity can help magnify slight changes in the biochemical or mechanical factors of development into large phenotypic differences. It may be possible, as Goldschmidt, D'Arcy Thompson, and Raup have theorized, that humdrum ancestors can be transformed into a variety of bizarre descendants by the coaction of a few alterations. Changes in the relative rates of growth of different groups of cells, in body size or the length of development, in the prepattern of inducing substances or the time at which cells react to them, or in the developmental switches that trigger one or another pathway can have novel evolutionary consequences.

Unfortunately we are largely ignorant of the biochemical nature of these events, the mechanisms by which they occur, and especially the control processes that govern them. Development is the greatest mystery in biology, but we may need to understand its complexity in biochemical detail before we can understand the alterations of ontogeny that are the history of evolution.

SUMMARY

The evolution of new phenotypes poses two questions. Why did they evolve — what selective factors, if any, guided their evolution? And how did they originate — what differences in biochemistry and development are there between the ancestral and derived conditions? The adaptive *why* is largely answered; evolutionary innovations often occur when the function of a structure is enhanced, reduced, or changed by an alteration of its adaptive context.

The developmental *how* of evolution is largely unanswered because the mechanisms of development are so poorly understood. It seems likely that great changes in phenotypes may be caused by only slight alterations in relative growth rates, the rapidity of maturation, the temporal and spatial pattern of inducing substances within the developing organism, and the reactions of tissues to such substances — in sum, by slight changes in growth and in the differentiation of tissues. It seems probable that some macroevolutionary changes, between ancestors and their very different descendants, can have arisen from just a few genetic changes. Very often, however, the great phenotypic differences among taxa are the summation of many smaller differences, like the variations within individual species.

FOR DISCUSSION AND THOUGHT

1 In Chapter 6 of the *Origin of Species,* Darwin remarked, "If it could be demonstrated that any complex organ existed, which could not possibly have been formed by numerous successive, slight modifications, my theory [of evolution by natural selection] would absolutely break down." Discuss this statement in light of this chapter. Darwin's chapters 6 and 7 make fascinating reading in this connection.

2 By reference to comparative anatomy and embryology, trace the antecedents of the distinctive characteristics of a major taxon such as the Mammalia as far back as possible. Do all structures arise from antecedent structures?

3 The mutation *bithorax* may transform dipteran halteres into wings like those of other insects; but is it likely that a single such mutation could have transformed wings into halteres?

4 How can the phenotypic effect of a mutation be measured? Is it possible to define a macromutation operationally?

5 Discuss the possible reasons for the degeneration and loss of a complex structure that no longer has any function.

6 Discuss the proposition that structural change in animals is usually accompanied or preceded by a change in behavior. Are there many instances of alteration of behavior that have not (yet) been accompanied by morphological changes? If so, what are the implications of this observation?

7 How can we recognize preadaptations? For example, what preadaptations can organisms have to the toxins that industrial society releases into the environment?

8 Review (or learn) the process by which the somites of a vertebrate develop, and speculate on how the number of vertebrae might evolve.

9 Why should the likelihood that the number of serially homologous parts will evolve depend on whether their number is large and variable or

small and fixed? Why should there be a correlation between their number and fixity?

10 Discuss the proposition that a structure that grows with positive allometry (i.e., $a > 1$) can take on a new function when body size evolves.

11 Evaluate Gould's (1977) distinction between acceleration and retardation in development, and discuss ways in which paedomorphosis by acceleration (progenesis) might be distinguished from paedomorphosis by retardation (neoteny).

12 Discuss the possible implications for the differentiation of serially homologous parts of Stebbins' (1974) principle of allochronic deviation, that is, a change in the timing of differentiation processes relative to each other. For example, double flowers might arise if stamen-differentiating genes are not activated in time to prevent stamen primordia from yielding to the influence of petal-differentiating genes.

13 Provide a developmental speculation for Maynard Smith and Sondhi's (1960) observation that it is difficult to select in *Drosophila* for a loss of the right or left ocellus (simple eye), although the median ocellus is easily lost.

14 Explain how the differentiation of structures can be shifted forward in ontogeny, so that they appear before they can have any function. For example, human embryos have calluses on their feet.

15 Are developmental systems adaptively integrated, as Olson and Miller (1958) suggested? Are they so constructed that variation in one element is accompanied by an adaptive covariation in other elements? Pigment cells and autonomic nerve ganglia are both derived from neural crest cells in vertebrates. Will they vary adaptively in concert? See Chapter 15 for an interesting study of character correlations.

16 Microbial geneticists distinguish structural genes, that produce enzymes and other proteins, from regulatory genes (*Chapter 3*). It is generally presumed that higher organisms also have regulatory genes, although little is known about them. How could phenotypic changes of the kind discussed in this chapter be caused by mutations in regulatory genes? Is it possible that these are responsible for most of the phenotypic mutations (e.g., *vestigial* wing or *bithorax* in *Drosophila*) that we see? See Chapters 10 and 17.

MAJOR REFERENCES

Mayr, E. 1960. The emergence of evolutionary novelties. In *The evolution of life,* edited by S. Tax. University of Chicago Press, Chicago; pp. 349–380. A classic exposition of the subject; see also Chapter 19 in Mayr (1963).

Rensch, B. 1959. *Evolution above the species level.* Columbia University Press, New York. 419 pages. An old but excellent treatment of the evolution of new features, discussed in part from a developmental viewpoint.

Waddington, C. H. 1956. *Principles of embryology.* Allen and Unwin, London. 510 pages. See comment, Chapter 3.

Waddington, C. H. 1962. *New patterns in genetics and development.* Columbia University Press, New York. 271 pages. A speculative treatment of mechanisms of development that could prove useful to evolutionists.

Stebbins, G. L. 1974. *Flowering plants: Evolution above the species level.* Harvard University Press, Cambridge, Mass. 399 pages. A discussion of evolutionary patterns in the angiosperms, with speculations on the developmental bases of new features.

Frazzetta, T. H. 1975. *Complex adaptations in evolving populations.* Sinauer, Sunderland, Mass. 267 pages. Discussion of the possible origins of major phenotypic changes from slight alteration of developmental patterns.

Gould, S. J. 1977. *Ontogeny and phylogeny.* Harvard University Press, Cambridge, Mass. 501 pages. An essay on the role of changes of the tempo of development in evolution, together with a history of the biogenetic law.

PART

THE
MECHANISMS
OF EVOLUTION

Variation

Evolutionary biology, like any science, seeks to understand the complexity of nature by formulating generalizations and rules. We attempt to simplify for the sake of greater comprehension. Yet the beauty and excitement of biology lies in large part in the diversity of living beings. To appreciate fully this diversity and to understand fully the evolutionary processes that give rise to it, we must apprehend the nature of the variation that is both the product and the foundation of evolutionary change.

SPECIES

The higher categories into which organisms are classified have somewhat arbitrary limits. But one taxonomic category, the species, has been thought by many to be real and nonarbitrary. The species is an important concept; it is the fundamental unit of classification, it plays an important role in ecological thought, and it has long dominated the stage in evolutionary theory. Mayr (1963), for example, says that "the species are the real units of evolution, as the temporary incarnation of harmonious, well-integrated gene complexes."

The objective reality of species was not questioned by Linnaeus, to whom they were immutable units created in the beginning by God. Variation within species represented mere imperfections in creatures that, but for the faults of our subcelestial world, would conform to the TYPE, the Platonic "idea" or εἶδος in the mind of God. The conception of variation as imperfection carried into the thinking of early taxonomists, who established a system of taxonomic practice in which specimens were assigned to a species if they conformed to the type specimen, or holotype, on the basis of which the species was originally described. Species, then, were recognized by preservable morphological characteristics and by their conformity to a set standard.

To dismiss variation as unimportant or abnormal and to classify specimens into discrete categories, each represented by a

189

normal type, is a reflection of a whole worldview, a *Weltanschauung* that Mayr (1963) has called TYPOLOGICAL THINKING. It finds its way into other dimensions of life, sometimes with pernicious effects. It is the kind of thinking that dichotomizes: either/or, black/white, good/evil, normal/abnormal. Mayr has shown that the replacement of typological thinking by the recognition and acceptance of variation was pivotal in the development of the modern view of evolution. It is a conceptual change that could be applied with profit to some other realms of thought.

The concept of purely morphological, typological species was never applied absolutely. Very different forms were considered conspecific if they were simply different sexes or developmental stages or if they could be shown to interbreed despite their striking differences. POLY-MORPHISM, the existence of two or more reasonably common geno-types within an interbreeding population (*Figure 1*), makes a purely typological species concept untenable. So do SIBLING SPECIES — those that despite their great morphological similarity differ in other biolog-ical characteristics and do not interbreed. *Drosophila pseudoobscura* and *D. persimilis,* for example, were first recognized as different species by differences in their chromosomes and in ecological characteristics; only later were slight morphological differences in the male genitalia dis-covered. Thus the generally accepted criterion for species is not mor-phological degree of difference, but evidence that two forms do not interbreed in nature. This criterion has led to widespread acceptance of the BIOLOGICAL SPECIES CONCEPT, enunciated by Mayr (1942) as follows: "Species are groups of actually or potentially interbreeding natural populations, which are reproductively isolated from other such groups."

Populations are not assigned to different species if they are merely isolated by topographic barriers such as bodies of water; if they resem-ble one another closely, they are presumed to be potentially capable of interbreeding. Nor is sterility the criterion of species; in both ducks and orchids, species hybridize in captivity and produce fertile offspring, yet do not interbreed in nature. Conceptually individuals are members of the same species if their genes can descend through the generations to unite in the same individual under natural conditions. By their passage from generation to generation, genes from white-tailed deer in Colorado might ultimately unite with those from deer in Florida, so these populations are conspecific. Yet white-tailed deer and mule deer in Colorado have separate gene pools and do not interbreed; their evolutionary paths through time are distinct.

In practice, biological species are usually recognized by phenotypic differences that are taken as evidence of genetic distinctness, not by genetic criteria *per se.* Two species may each be very variable and have overlapping ranges of variation in many characteristics, yet discrete differences in one or more characteristics are taken to imply genetic isolation. Because phenotypic rather than genetic criteria are so often used, the biological species concept is not universally accepted as useful

Polymorphic variation. (A) Different color forms ("snow goose" and "blue goose") of *Chen caerulescens*. (B) The two patterns of the king snake *Lampropeltis getulus* in California. In both cases the two forms are often found in the same litter. (A redrawn from Pough 1951, B redrawn from *Amphibians and reptiles of western North America,* by R. C. Stebbins. Copyright 1954 by McGraw-Hill Book Company. Used with permission)

1

(see, e.g., Heslop-Harrison 1964, Sokal and Crovello 1970, Sokal 1973).

It is undeniable that some populations do interbreed and others do not. If the biological species concept is accepted, species are then best recognized by the characteristics that prevent their interbreeding. These are called isolating mechanisms.

ISOLATING MECHANISMS

Two species are said to be SYMPATRIC if they occur in the same geographic area and so have the opportunity to interbreed. ALLOPATRIC populations occupy different geographic areas and so do not encounter each other. Populations are PARAPATRIC if they meet only along the borders of their ranges. In some cases sympatric populations inhabit different habitats and are to some extent ecologically isolated; for example, species of spadefoot toads occupy different soil types (Wasserman 1957), and species of parasites are often restricted to different hosts. Ecological isolation is one of several PREMATING ISOLATING MECHANISMS (*Table I*) that inhibit interbreeding. Species may be temporally isolated, as are plants that have different flowering seasons (see Grant and Grant 1964) or insects that mate at different times of night (e.g., fireflies; Lloyd 1966).

Although the ecological or temporal isolation between sympatric species is usually incomplete, they usually do not interbreed because of physiological or behavioral features. ETHOLOGICAL ISOLATING MECHANISMS in animals include differences in the courtship behavior of males, in the vocalizations or chemical signals (pheromones) by which one sex attracts the other, or in the color patterns by which an individual recognizes a potential mate. Female fireflies, for example, respond to the light pattern emitted by males of their own species (*Figure 2*). Altering species-specific characters sometimes makes it possible to induce hybridization, proving that the characteristics are indeed isolating mechanisms. Smith (1966) induced species of gulls to interbreed by changing the contrast between the eye and the feathers of the face.

Species often differ most in sexual characteristics if they run a high risk of making a mistake. For example, in birds of paradise, ducks, and hummingbirds (Sibley 1957, Selander 1965), which generally do not form a pair bond, sexual dimorphism is extreme and the species differ greatly in the male plumage. In the few species that do form a pair bond and thus have a more extended courtship and more time to rectify a mistake in mate choice, sexual dimorphism is reduced and the species are more similar to one another. Sympatric species are often more different in reproductive characteristics than allopatric species that are not faced with the threat of hybridization. The mallard (*Anas platyrhynchos*), sympatric in northern North America and Eurasia with many other species of ducks, has highly distinctive male plumage; but the males of related species and races to the south that are sympatric

TABLE I. A classification of isolating mechanisms in animals

1. Mechanisms that prevent interspecific crosses (premating mechanisms)
   a. Potential mates do not meet (seasonal and habitat isolation)
   b. Potential mates meet but do not mate (ethological isolation)
   c. Copulation attempted but no transfer of sperm takes place (mechanical isolation)

2. Mechanisms that reduce full success of interspecific crosses (postmating mechanisms)
   a. Sperm transfer takes place but egg is not fertilized (gametic mortality)
   b. Egg is fertilized but zygote dies (zygote mortality)
   c. Zygote produces an F_1 hybrid of reduced viability (hybrid inviability)
   d. F_1 hybrid zygote is fully viable but partially or completely sterile, or produces deficient F_2 (hybrid sterility)

(From Mayr 1963)

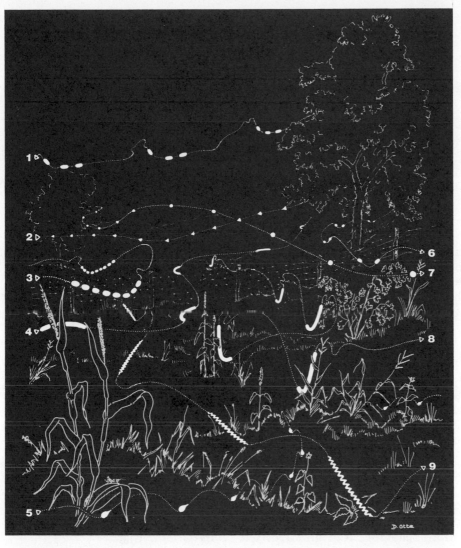

2

The flight paths and flash patterns of male fireflies (Lampyridae) of nine species. Females respond to the patterns of their own species and not to others. (From Lloyd 1966)

with few or no other species have female-like color patterns (*Figure 3*).

In vascular plants the union of pollen nuclei of one species with egg nuclei of another is prevented by ecological and temporal isolation and by physiological and "ethological" barriers. These barriers include physiological incompatibility between the pistil of one species and the pollen of another, and differences in flower color and form that induce different pollinating animals to serve the different species or prevent them from transferring pollen from one species to another.

Under natural conditions premating isolating mechanisms usually prevent the formation of hybrid zygotes. But under laboratory con-

3 The possible effect of sympatric species on plumage characteristics in birds. The male mallard (A), where sympatric with other species of ducks, has distinctive plumage, facilitating recognition by the female. Closely related mallardlike forms, like the mottled duck in Florida (B), are sexually monomorphic in coloration where no other duck species occur with which they might be confused. (Redrawn from Pough 1951)

ditions, and sometimes in nature as well, species that ordinarily do not interbreed do so, revealing any of several POSTMATING BARRIERS to gene exchange. The F_1 hybrid may fail to develop, as in the case of hybrids between the frogs *Rana pipiens* and *R. sylvatica,* which do not develop past the early gastrula stage (Moore 1961). Or it may develop normally but fail to survive in nature. For example, hybrids between the buttercups *Ranunculus millanii* and *R. dissectifolius,* which occupy wet and dry habitats respectively, cannot compete successfully with either parent in the parent's habitat and so in nature occur only in intermediate, disturbed habitats (Briggs 1962). If the hybrids do survive they may not reproduce, either because they are sterile or because they have inappropriate mating behavior. Rao and DeBach (1969) found that male hybrids between two parasitic wasps were accepted by hybrid females but not by females of either parent form.

Hybrid sterility can have either a *chromosomal* or a *genic* basis. If the species' chromosomes differ in structure, they may pair improperly in the hybrid, whose gametes then have incomplete complements of genes (*Chapter 10*). For example, the diploid hybrid between *Primula verticillata* and *P. floribunda,* two primroses, is largely sterile. But tetraploid hybrids called *P. kewensis* have been formed; in these each chromosome pairs properly with its newly made partner, so that meiosis proceeds normally and the plant is fully fertile (Stebbins 1950). So *verticillata* genes are compatible with *floribunda* genes; only the differences in chromosome structure make the diploid hybrid sterile.

In the F_1 hybrid between *Drosophila pseudoobscura* and *D. persimilis,* the chromosomes pair properly but the males are nevertheless sterile.

By a series of elaborate crosses Dobzhansky (1936) showed that sterility was due to genes, scattered over all the chromosomes, that differ in the two species. This genic sterility, implying deep-seated differences in the genetic makeup of the species, appears to be prevalent in *Drosophila* (Ehrman 1962) and probably in many other animal groups as well.

LIMITATIONS OF THE BIOLOGICAL SPECIES CONCEPT, AND HYBRIDIZATION BETWEEN SPECIES

A concept cannot be stretched past the domain in which it applies. Thus the biological species concept does not embrace quite a few organisms. Asexual reproduction is the norm in prokaryotes, in the *fungi imperfecti* like *Penicillium,* in blackberries (*Rubus*) and many other plants, and in many animal groups such as the bdelloid rotifers. In such groups species are defined not by reproductive isolation, but rather arbitrarily by phenotypic differences. In some instances, such as the genus *Rubus,* so many intermediate forms are known that taxonomists differ greatly in the number of species they recognize. In other cases, such as some rotifers (Hutchinson 1968), the asexual genotypes fall into phenotypic classes as distinct as those of sexual species. Each is perhaps adapted to a discrete ecological niche. Some authors (e.g., Sokal 1973) have suggested that the phenotypic distinctness of sexual species could likewise be shaped by the discreteness of their niches and that such species could as well be considered occupants of different niches as reproductively isolated populations.

The biological species concept cannot be used to categorize populations in intermediate states of speciation. This process (*Chapter 16*) is believed to follow most commonly from the geographic isolation of populations that diverge genetically to such a degree that they cannot or do not interbreed when they reestablish contact. Accordingly the genetic divergence realized by the time the populations come into contact is sometimes insufficient to fully prevent interbreeding; such populations are not full species, but incipient species, or SEMISPECIES. Their geographic ranges abut along a HYBRID ZONE that sometimes coincides for many pairs of semispecies, indicating the site of a former barrier. In North America, for example, Remington (1968) has described several such zones, like the one in northern Florida where continental populations of many species interbreed with morphologically distinguishable peninsular populations (*Figure 4*). During the Pleistocene interglacials a strait across northern Florida isolated the peninsular populations long enough for them to become differentiated.

Often the hybrid zone is quite narrow, so that the main population of each semispecies has remained free of the genetic effects of hybridization even though secondary contact between the formerly isolated populations occurred long ago. In northeastern South America a hybrid zone between *Heliconius* butterflies has persisted for at least 200 years (2000 generations), but is only about 50 kilometers wide (Turner 1971). In some instances some characteristics have filtered farther than

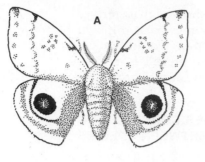

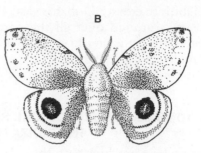

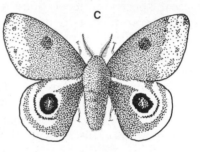

4

Hybridization between partly differentiated forms (subspecies) that meet in northern Florida. (A) The moth *Automeris io io*. (B) A hybrid between A and C, the Floridian form *A. io lilith*. (Redrawn from Remington 1968)

others from one semispecies into the range of the other; this is INTRO-GRESSIVE HYBRIDIZATION (Anderson 1949). Yang and Selander (1968) describe a hybrid zone between northern and southern grackles (*Quiscalus*) in Louisiana that is only 15 to 40 miles wide; but some individuals several hundred miles distant show some characteristics of the other semispecies (*Figure 5*). In such cases it seems likely that advantageous genes acquired through hybridization are retained while disadvantageous genes are eliminated.

Hybridization is not restricted to narrow borders between allopatric forms. Especially among plants, two species may form hybrid swarms in some localities, yet be entirely distinct and show no signs of interbreeding in others. Such hybridization is prevalent in disturbed habitats where the ecological component of reproductive isolation has broken down. *Iris hexagona,* which grows in the exposed tidal marshes of the Mississippi delta, and *I. fulva,* which occupies the shady margins of streams, produce an F_1 hybrid that is only partially fertile; yet populations of hybrids with various combinations of characteristics occur in lumbered, drained swampy areas (Riley 1938). Two finches, the rufous-sided and collared towhees, interbreed in disturbed habitats throughout Mexico yet coexist in some localities without crossing (Sibley 1954). Local hybrid populations can occur without obvious disturbance of the habitat, though; the oaks are considered a taxonomically difficult group because of the prevalence of hybrids (Palmer 1948).

Thus there is a continuum of phenotypic differentiation and reproductive isolation between populations; they may be isolated completely, not at all, or to any intermediate degree. Perhaps we identify hybrids between species only because our typological thinking focuses first on the "pure" forms and then interprets intermediate phenotypes as secondary products. If we focused on the continuum of phenotypes as a unit, we might not worry as much about which are the species and which the hybrids.

GEOGRAPHIC VARIATION

Hybridization between parapatric populations is one form of geographic variation that provides clues to the mechanisms by which new

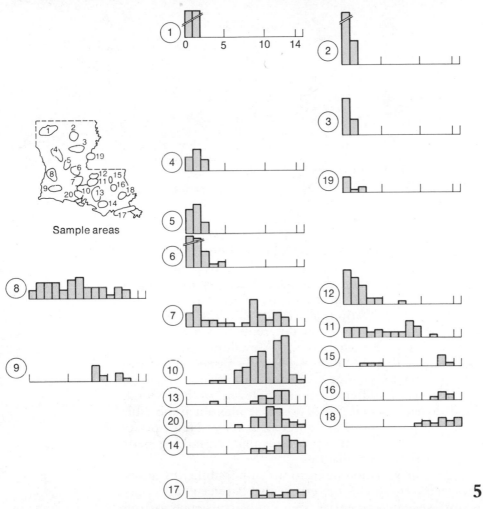

5

Geographic variation in the grackle, *Quiscalus quiscala*. The horizontal axis of each histogram is the score for an individual's back color, ranging from bronze to purple. Predominantly bronze populations in the north give way, over a very narrow region in which there is mixture, to purple populations in the south. The distribution of other differences is similar, but not identical. A few specimens with northern characteristics occur in the south (e.g. locality 15), perhaps exemplifying introgressive hybridization. (Modified from Yang and Selander 1968, in *Systematic Zoology*)

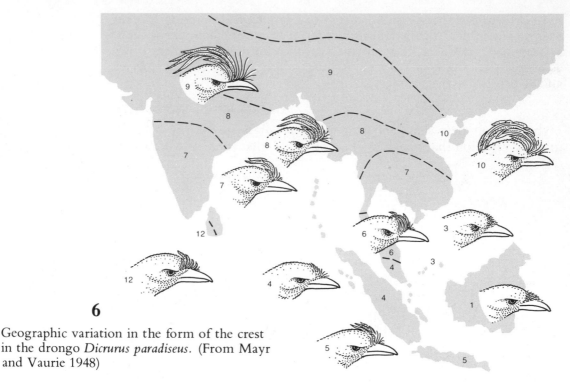

6

Geographic variation in the form of the crest
in the drongo *Dicrurus paradiseus*. (From Mayr
and Vaurie 1948)

species are formed. Because so many historical evolutionary events are
inaccessible to direct observation, many evolutionary hypotheses can
be tested only by examining extant organisms. Much as an ecologist
can trace the historical course of succession by piecing together the
individual stages that exist in various places at the present time, an
evolutionist can use the varying levels of differentiation among populations and species to infer the time course of evolutionary change.
Such observations suggest that evolution is generally a gradual process;
differences among populations range from the immeasurably small,
through varying degrees of differentiation, to levels of behavioral,
chromosomal, and developmental distinction that make it impossible
to specify whether the populations are different species. New species
must therefore be formed by the same processes that engender genetic
differences among conspecific populations.

Patterns of geographic variation point to the impossibility of drawing clear-cut lines between species or, especially, between subspecies
or races. These infraspecific categories are simply constructs of our
imagination, erected for the sake of convenience; we can recognize as
few or as many races as we find convenient, for they have no independent biological reality.

Finally we find that all characteristics are subject to geographic
variation. In some cases the differences among populations seem clearly
adaptive; in others, it is hard to say whether they are adaptive or not.

GEOGRAPHIC RACES

Similar species sometimes have parapatric distributions. For example, most of the species of chickadees in North America meet only along very narrow geographical or altitudinal borders, where competition apparently excludes each from the other's range (Lack 1969). Only by studying them in the area of overlap is it possible to tell that they are reproductively isolated, because they are very similar in appearance. Such a group of closely related ecological replacements, or vicarious species, is called a superspecies or an *Artenkreis*.

Often, however, the ecologically and phenotypically similar forms are totally allopatric and differ to various degrees. For example, island populations of drongos (*Figure 6*), crow-like birds of the Old World tropics, differ in the form of the crest, which may be used in courtship displays, but it is hard to know whether they would interbreed given the opportunity. They might be reproductively isolated (hence different species) or not. Mayr and Vaurie (1948), in fact, treated them as geographic races, or subspecies, of a single species. In zoological taxonomy the term SUBSPECIES means a recognizably different geographic population, or set of populations, of a species. A species that is divided into subspecies (geographic races) is often called a POLYTYPIC species, or a *Rassenkreis*. In instances like the drongos the difference between

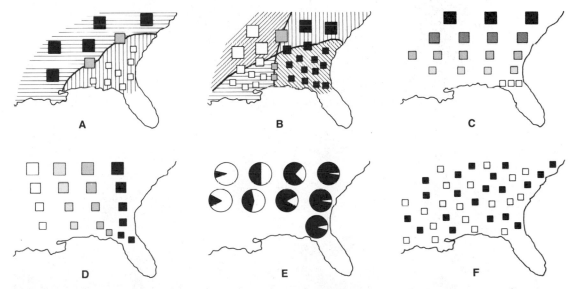

Highly diagrammatic representations of some common patterns of geographic variation. (A) Two classical subspecies that interbreed along a narrow border. Size and color are correlated. (B) Abrupt transition in each of two characters that have discordant distributions. (C) Concordant clines in each of two characters. (D) Discordant clines in each of two characters. (E) An east-west cline in the frequency of black and white individuals; each "pie diagram" represents proportions in a sample from a single locality. (F) A mosaic distribution of two phenotypes, as might be observed if one (black) were a wetland ecotype and the other (white) an upland ecotype.

7

an *Artenkreis* and a *Rassenkreis,* between allopatric species and allopatric subspecies, is arbitrary.

The naming of subspecies is a formal recognition of the geographic variation that almost every species displays, variation that may take many forms (*Figure 7*). A typical example of the way in which subspecies have been described is the rat snake *Elaphe obsoleta* (*Figure 8*). In this instance the subspecies meet along zones of intergradation, within which rather homogeneous populations of intermediate forms are found.

Body color, the blotched pattern, and the possession of stripes in

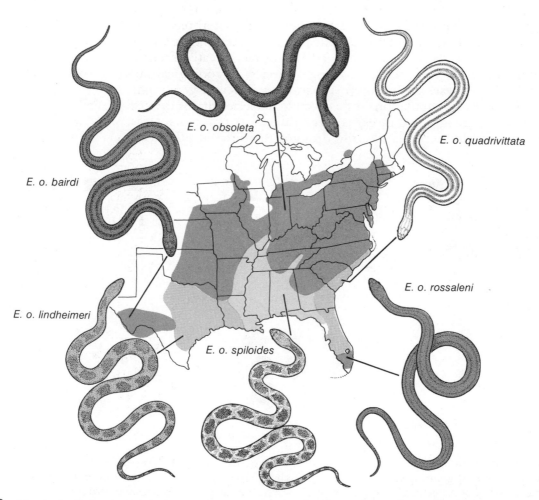

8 Classical subspecies in the rat snake *Elaphe obsoleta;* allopatric geographical "races" that interbreed where their ranges meet. The distinctions among the subspecies are based on several characters with different distributions, and the distinctions in some cases (e.g., *lindheimeri* vs. *spiloides*) are very subtle — some would say trivial. (Redrawn from Conant 1958)

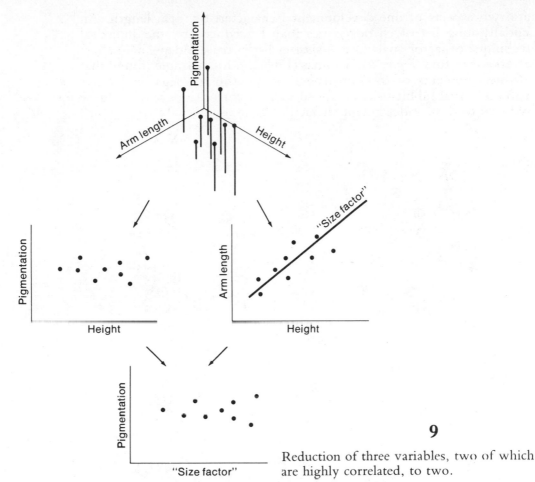

9

Reduction of three variables, two of which are highly correlated, to two.

Elaphe are independently distributed; if subspecies were described only on the basis of body color, they would have a different distribution than if they were defined on the basis of stripes. Thus, unlike the grackles, in which different characters vary concordantly, the characteristics of the rat snakes appear to have differentiated independently, perhaps in response to different environmental factors. The subspecies of rat snakes do not appear to be incipient species; they appear to differ only in single characters rather than in whole character complexes.

CLINES

Different characteristics may vary concordantly (in parallel) or discordantly (independently). But to determine whether the geographic variation of different characteristics is concordant, it is important first to eliminate the correlation *within* populations caused by developmental patterns (see Gould and Johnston 1971). The lengths of the upper arm and the forearm, for example, vary concordantly simply because they are two aspects of one developmental characteristic, arm length. An initially long list of characteristics may be reduced by the statistical technique of factor analysis to a shorter list of truly independent characters, or factors (*Figure 9*). Thomas (1968a,b), for example, found that 16 measurements of the dimensions of mouthparts, legs, and other parts of larval rabbit ticks exhibited various correlations and could be reduced to three independent characters: factor I, a "body size factor"

10 A complex analysis of geographic variation. A and B represent variation in factors I and III in the rabbit tick; each factor is a complex of highly correlated characteristics (I is a body size factor, III an appendage factor). The mean score of individuals from a locality increases with intensity of shading of a circle. There is a general cline from south to north in factor I, and from southwest toward the north and east in factor III. (Modified from Thomas 1968a,b)

reflected in the dimensions of the scutum, the anal plate, and so forth; factor II, an "appendage factor" that accounted for variation in most of the leg segments; and factor III, a "capitulum factor" that measured the mouthparts and some of the leg segments.

Each factor varies geographically, not discretely from one large region to another, but gradually. Mean body size, for instance, decreases gradually from north to south. Such gradual changes along a transect are called CLINES. In the rabbit tick, as in many species, the clines are independent for different characters (*Figure 10*); factor I varies latitudinally, whereas factor III varies from east to west. These patterns suggest that the geographic variation is adaptive; large size is advantageous in cold northern areas where ticks need to store fat during hibernation, and short appendages are probably advantageous in the arid west where a low surface-volume ratio reduces dessication.

Very often populations vary clinally in the frequency of different alleles at one or more loci. In the clover *Trifolium repens,* for example, the proportion of plants that produce cyanide increases from north to south (*Figure 11*). The frequency of cyanogenic plants is apparently determined by a balance between the advantage they derive from being unpalatable to herbivores and the disadvantage they suffer when frost disrupts the cell membranes, releasing cyanide within the plants' tissues (Jones 1973).

ECOTYPES

Clinal variation can occur over very short transects. In many species of plants one or more characteristics change gradually with altitude, and the same pattern of variation occurs on each of several mountain ranges. On mountain tops plants are usually short and differ genetically from lowland populations in other characteristics as well (Clausen, Keck, and Hiesey 1940).

Such ecotypes can have a mosaic rather than a clinal distribution. Sometimes given a subspecific name by botanists, an ecotype frequently occurs throughout the species' range wherever the habitat to which it is adapted occurs. Turesson (1922), for example, described in the hawkweed *Hieracium umbellatum* four ecotypes that occur throughout the range of the species: one adapted to sand dunes, one to sandy fields, another to seaside cliff faces, and another to forests. Ecotypic variation is often more physiological than morphological; for example, metal-impregnated soils near mines are occupied by metal-tolerant populations of grasses that are genetically distinct from conspecific plants on normal soils a few feet away (Antonovics, Bradshaw, and Turner 1971).

POLYTOPIC ecotypes, those that recur in widely separated localities, are thought to have evolved independently from different local populations of the species. In at least some cases the distinctive character of a polytopic ecotype has the same genetic basis in different populations (Clausen 1951), but it seems likely that the genetic basis sometimes varies.

CHAPTER NINE

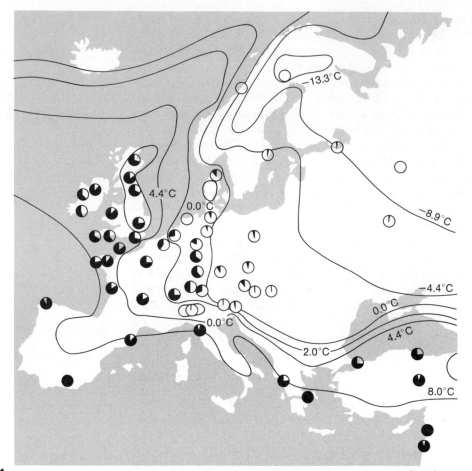

11 Frequency of the cyanide-producing form in populations of white clover (*Trifolium repens*), represented by the black section of each circle. The cyanogenic form is more common in warmer regions. Thin lines are January isotherms. (Modified from Jones 1973, after Daday 1954)

SUBSPECIES AND TYPOLOGY

Many of a species' characteristics vary geographically, and different characters usually have different patterns of variation. Some combination of characters is almost certain to distinguish almost any local population from most other populations, at least on a statistical basis, so in principle there is no clear limit to the number of subspecies that could be recognized. In some cases astronomical numbers of trivially different subspecies have been described; mammalogists have described more than 250 subspecies of the gopher *Thomomys bottae* in the western United States, most with very restricted geographic ranges. Taxonomists used to argue at length about how many subspecies a species consisted of; in some quarters there is still argument over how many human races (which are simply subspecies, or geographic races) should be recognized. Depending on how finely you discriminate differences in blood type, skin color, hair form, and other physical measures, you can name as many human races as you like.

Nowadays many evolutionists accept Wilson and Brown's (1953) argument that the subspecies is so arbitrary a concept that it should be abandoned. Deciding that there are, say, six human subspecies recognized on the basis of three or four characteristics leads easily to thinking of these as real entities and ignoring the elaborate patterns of variation in other characteristics that these typologically conceived categories mask. Once the categories Caucasians, Negroes, and Orientals have been defined, it is easy to suppose that these groups differ in blood type, bone structure, mental qualities, and features other than the few superficial traits by which they are defined, and to assume that all Caucasians are more similar to one another in all features than they are to Negroes or Orientals. In fact the variation in human blood groups and enzymes is so much greater within populations than among "races" that the differences among races account for only about 15 per cent of the genetic diversity in the entire species (Lewontin 1972, Nei and Roychoudhury 1972). The concept of race diverts attention from the overwhelming genetic similarity of all peoples and from the mosaic patterns of variation that do not correspond to racial divisions; it focuses attention instead on the few differences that are readily apparent to the eye. The concept of human races is not only socially dysfunctional; it is biologically indefensible as well. Subspecific names in biological taxonomy may be convenient labels, but they usually obscure the patterns of greatest biological interest.

GEOGRAPHIC VARIATION IN ECOLOGICAL AND REPRODUCTIVE CHARACTERISTICS

Geographic variation is displayed by almost all characteristics, including those most closely associated with a species' ecological role and those which by evolution may result in speciation. Among ecological characteristics, for example, the capacity for physiological acclimation

Terrestrial   Aquatic   Terrestrial   Aquatic

Hidden Lake   Cleawox Lake

12

Ecotypic variation in developmental plasticity. *Ranunculus flammula* plants from Cleawox Lake differ genetically from those at Hidden Lake in the degree to which leaf shape varies when grown under terrestrial or aquatic conditions. (From Cook and Johnson 1968)

varies ecotypically in the goldenrod *Solidago virgaurea*, in which the scope for photosynthesis is greater in plants from exposed habitats than in those from shaded habitats (Björkman and Holmgren 1963). Developmental flexibility, too, varies among populations (Cook and Johnson 1968; *Figure 12*).

Species vary in characteristics that affect their interaction with other species. This sometimes takes the form of character displacement (*Chapter 4*), in which species differ more where they are sympatric than where they are allopatric. Differences in beak shape among conspecific populations of Galápagos finches (*Figure 13*) may be adaptations to reduce interspecific competition.

Character displacement is often seen in characteristics that serve as premating isolating mechanisms. Where the opportunity for interspe-

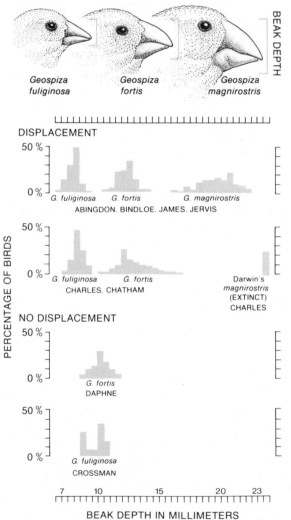

13

Character displacement in Galápagos finches. *Geospiza fortis* and *G. fuliginosa* differ more where they are sympatric than where they are allopatric. (Modified from Lack 1947)

cific hybridization arises, an individual whose sexual features readily identify it as an appropriate mate will be more successful in passing on genes, as will an individual who is highly capable of discriminating between appropriate and inappropriate mates. Males and females of two of the semispecies of the *Drosophila paulistorum* complex taken from allopatric populations interbreed in the laboratory more readily than those from sympatric populations; those with a history of exposure to one another have evolved greater mating discrimination (Ehrman 1965). In Australia allopatric populations of the tree frogs *Litoria (Hyla) ewingi* and *L. verreauxi* have very similar vocalizations, while sympatric populations display quite different vocalizations (Littlejohn 1965). In at least some frogs the auditory system of the female is tuned to the dialect of the males in her local population (Capranica, Frishkoff, and Nevo 1973).

Allopatric populations of *Phlox pilosa* and *P. glaberrima* have pink flowers, but in regions of sympatry *P. pilosa* is frequently white. Pollen from *P. glaberrima* is carried by insects less frequently to white than to pink *P. pilosa* (Levin and Kerster 1967), illustrating clearly the advantage of character displacement.

Not all geographic variation in reproductive characteristics, however, is due to character displacement. The flowering time of a plant can vary due to differences in climate; the color pattern of a bird or fish is affected by many factors besides the preferences of sexual partners; the properties of an effective vocal communication are affected by the entire sound environment.

Postmating barriers to gene exchange also vary geographically. The form and number of chromosomes often vary among populations. For example, isolated populations of the plant *Clarkia unguiculata* differ from one another by three to seven chromosome translocations, and hybrids between the populations are sterile (Vasek 1964).

Interpopulation hybrid infertility may also be due to genic differences. Populations of the plant *Streptanthus glandulosus* (Cruciferae) in California do not appear to differ in chromosome structure, but they vary greatly in the extent to which they form fertile F_1 offspring in the laboratory. The more geographically distant two populations are from each other, the lower the fertility of their hybrid (Kruckeberg 1957). In eastern Asia the gypsy moth (*Porthetria dispar*) consists of "weak" and "strong" geographic populations that Goldschmidt (1940) termed *sex races*. A cross between a weak female and a strong male produces daughters that are phenotypically male; if the female is from a half-weak race, the daughters are sterile intersexes. There is a general north–south cline from strong to weak, but some regions (e.g., Hokkaido) do not fit the clinal pattern (*Figure 14*). Clearly many local populations of gypsy moths would be incapable of exchanging genes if they were to meet. Thus the very properties that define biological species, premating and postmating barriers to gene exchange, vary geographically. This is evidence that speciation is a gradual process in which geographic populations diverge into different species.

VARIATION WITHIN POPULATIONS

Species vary geographically on different levels; variation encompasses variation, from widespread subspecies to local ecotypes that may be interspersed mosaically in such proximity to each other that the variants form a single interbreeding population. Variation among populations therefore blends into, and indeed stems from, variation within populations. The genetic variation within populations is the stuff of evolution.

Most characteristics vary much as does body size; individuals may have almost any value of the characteristic, within limits. Most characteristics have a bell-shaped, *normal* frequency distribution (*Appendix I, Figure 1; Chapter 3, Figure 1*), although some have lognormal distributions (Kerfoot and Kluge 1971). Such distributions arise from the independent action of many sources of variation: several or many gene loci and usually a variety of environmental factors. The amount of variation is measured by the variance (*Appendix I*).

That much of this variation is genetic can be shown by applying selection to the character, for if the population mean can be changed by selecting extreme phenotypes, some of the variation must have been inheritable. Artificial selection, applied to all sorts of traits in a great variety of plants and animals, almost always elicits a response, often quite a dramatic one (*Chapter 15, Figure 2*). Even characteristics that ordinarily show no phenotypic variation are genetically variable! For example, vibrissa (whisker) number is invariant in mice; yet by introducing into a mouse population a mutant gene that disrupts the

14   The distribution of the "sex races" of the gypsy moth *Porthetria dispar* in eastern Asia. (Modified from Goldschmidt 1940)

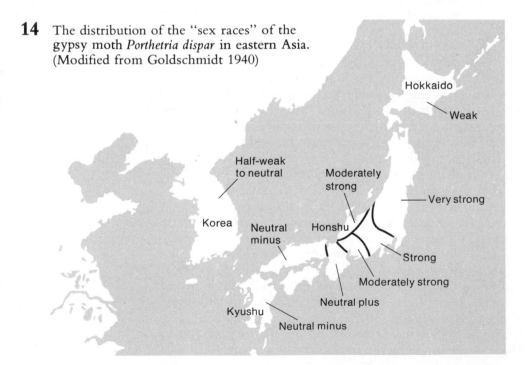

usual developmental pathway, latent variation in vibrissa number caused by genes other than the mutant allele becomes visible. Artificial selection in such a population changes the mean vibrissa number. Thus there are genetic variants that are prevented by homeostatic developmental pathways from exhibiting their effects (Kindred 1967).

Genetic variation is not limited to morphological traits. In *Drosophila* artificial selection has altered geotactic behavior, sexual behavior, developmental rate, fecundity, dispersal ability, feeding preferences, resistance to insecticides and other toxins, crossover rates, and many other characteristics; Lewontin (1974a) has summarized much of this literature. Indeed the whole history of breeding improved strains of domestic plants and animals testifies to the ubiquity of genetic variation in almost every characteristic imaginable. The response to selection is due to preexisting genetic variation, not to the occurrence of new mutations as selection proceeds, for highly inbred homozygous populations hardly respond to selection at all (Clayton and Robertson 1955).

Some characteristics, to be sure, do not respond greatly to selection. The ability to compete with related species seems to be a character of this kind; Park and Lloyd (1955) and Futuyma (1970), among others, found little evidence of a consistent genetic response to competition between species of *Tribolium* flour beetles and of *Drosophila*. Selectable genetic variation is often least in those very characteristics most intimately associated with a species' survival and reproductive success (*Chapter 14*). Still the ubiquity of genetic variation leads one to wonder whether its amount ever limits a population's rate of evolution or its ability to adapt to environmental changes.

Much of the theory of evolutionary genetics cannot be easily tested with quantitatively varying characters, for the theory requires that we be able to measure the frequencies of specific genes. This is difficult to do with continuously varying traits. Hence most of the study of the genetics of natural populations has sought to identify the alleles at individual loci, to determine how many loci are variable, and to assess the relative importance of the factors that effect gene frequency change. The history of these studies is summarized by Lewontin (1974a).

POLYMORPHISMS

In classical genetics it is impossible to know whether a gene exists, much less to be able to study its properties, unless alternative alleles at that locus can be identified. So the study of genes in populations focused on discrete variants such as the A–B–O blood types or white eyes in *Drosophila*. But in populations of most species we seldom find a characteristic with two or more discrete phenotypes that would tempt us to perform a Mendelian cross. Populations of eastern gray squirrels, for example, contain occasional black, melanistic individuals, but such "mutants" are rare. Thus classical geneticists distinguished the prevalent wild type from mutants and assumed that populations were genetically homogeneous. In this classical view of population structure, a wild-type allele is prevalent at most loci; the only variation is due to occasional mutations which, since they seem rare, must be deleterious. In this view evolution is a very slow process, since it must await rare advantageous mutations to provide adaptation to environmental change.

Occasionally, however, polymorphic loci were discovered in which two or more discrete phenotypes each make up at least 1 per cent of the population. Some polymorphisms are TRANSITIONAL; we come on the population while one allele is replacing another. The most famous case of such a polymorphism is the industrial melanic form of the moth *Biston betularia,* which was almost unknown before the Industrial Revolution but by 1895 constituted about 98 percent in some populations. In this and most other cases, though, the one allele does not completely replace the other; it arrives at a stable intermediate frequency and forms a BALANCED POLYMORPHISM.

A complex balanced polymorphism has been studied in the land snail *Cepaea nemoralis.* Six alleles affect the color of the shell, which can be several shades of brown, pink, or yellow. A closely linked locus governs the presence or absence of dark bands; other, unlinked, loci control the intensity and the number of bands (0–5). The relative numbers of the various forms differ greatly from one place to another, even between localities less than a mile apart. Fossils show that the polymorphism has persisted at least since the Pleistocene (Diver 1929). That the gene frequencies are affected by natural selection is suggested by their correlation with habitat (*Figure 15*). Open fields and hedgerows are predominantly populated by yellow unbanded snails, while brown banded individuals are more common in woods. Cain and Sheppard (1954a) and Murray (1962) showed that this correlation is at least partly due to predation by thrushes, which are less likely to find yellow snails than other forms in grassy areas where the yellow coloration is more cryptic. Moreover, the genotypes differ in their susceptibility to extreme temperatures (Lamotte 1959).

The marine copepod *Tisbe reticulata* has a color polymorphism (Battaglia 1958). In laboratory crosses the offspring consistently show an excess of heterozygotes, which have higher survival than either homozygote. The magnitude of the difference in viability depends on the culture conditions; the excess of heterozygotes is greater in highly crowded than in less dense cultures and is also influenced by salinity. Chapter 13 demonstrates that if the fitness of the heterozygote is greater than that of either homozygote, both alleles remain in a balanced polymorphic state. The polymorphism in *Tisbe* is one of the better documented cases of such heterozygote superiority. It also illustrates that the relative viabilities of genotypes vary with environmental conditions and that the effect of a gene on survival may not be due to the gene's obvious morphological manifestation. It seems implausible that the coloration of these copepods was itself responsible for their ability to survive in the laboratory; rather the coloration seems to be a pleiotropic effect, perhaps not adaptive in itself, of a gene whose unknown physiological effects are responsible for the existence of the poly-morphism.

CRYPTIC VARIATION

Although a fair number of such obvious polymorphisms are known, they are more the exception than the rule. Thus early geneticists were rather surprised to find that many populations actually harbor a great many mutants. Most of these are recessive and are brought to light by inbreeding stocks of, say, *Drosophila,* thus making them homozygous. From an extensive study of *Drosophila mulleri* Spencer (1957) estimated that every wild fly somewhere in its genome carries an average of one mutant gene that would cause some morphological abnormality in homozygous condition. Each of these mutants is individually quite rare, having an average frequency of about 10^{-3} (Lewontin 1974a).

Such morphological mutations are known at only a small fraction of the loci that *Drosophila* is thought to have; moreover it may be easy to miss seeing many of them by simply looking at flies. A better estimate of the true prevalence of mutations comes from the study of more unambiguously recognizable genes: modifiers of viability. The study of such genes requires special techniques that can be applied only in some species of *Drosophila*. Through a series of crosses between a wild fly and a laboratory stock that carries a marker gene and an inversion that prevents crossing over (*Chapter 10*), it is possible to bring a single chromosome from a wild population into homozygous

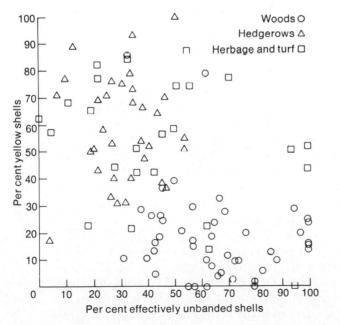

15

Microgeographic variation in the land snail *Cepea nemoralis*. Coloration and banding vary with habitat and are loosely correlated with each other. (From Cain and Sheppard 1954a)

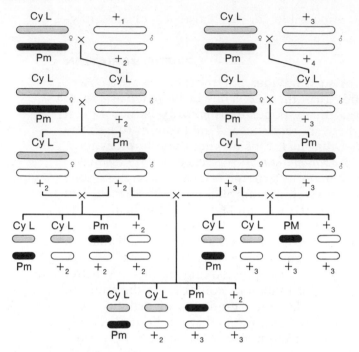

16 Crossing technique for "extracting" a chromosome from a male *Drosophila melanogaster* and making it homozygous to detect recessive alleles. The chromosomes marked *Cy L* and *Pm* carry dominant mutant markers, as well as inversions that prevent them from crossing over with other chromosomes. The crosses shown produce flies that are homozygous for either of two wild-type chromosomes ($+_2$ and $+_3$), as well as flies heterozygous for these same two chromosomes. (From Dobzhansky 1970, with permission of the publisher)

condition (*Figure 16*). The crosses ultimately produce a family of flies made up of some that are genetically identical for the laboratory chromosome and some identical for the wild chromosome. These two kinds of fly may be expected in, say, a 2:1 ratio. The degree of deviation from this ratio is a measure of the inviability of the flies that are homozygous for the wild chromosome; if no wild-type flies appear, the wild chromosome must carry a recessive lethal gene. Performing such crosses with many different wild flies makes it easy to determine what fraction of wild chromosomes are lethal, semilethal, subvital, quasinormal, or supervital in homozygous condition.

Studies of this kind (reviewed by Wallace 1968*a*, Dobzhansky 1970, and Lewontin 1974*a*) have been performed on many *Drosophila* populations; some typical results are illustrated in Figure 17. Most heterozygotes for randomly chosen wild chromosomes have quasinormal viability, but homozygotes are in general less viable. In fact, about 10 per cent of the chromosomes are virtually lethal in homozygous condition. Almost every individual carries abnormal genes. This also appears to be the case in corn (Crumpacker 1967) and Douglas fir (Sorensen 1969) and probably in human populations as well (Morton, Crow, and Muller 1956).

Moreover, any two quasinormal, seemingly identical chromosomes from different flies are genetically different. When they are allowed to recombine, and the recombined chromosomes are then made homozygous, the newly synthesized chromosomes vary in their viability and are sometimes even lethal (*Figure 18*). So the original chromosomes must have had different genes, which in new combinations interact to affect viability.

The revelation of all this variation was something of a shock in the 1920s and 1930s, when genetic uniformity was taken for granted. It resulted in a swell of new opinion among many geneticists, led by Theodosius Dobzhansky: a population is an immensely diverse assortment of genotypes, and there is no such thing as a wild-type, or normal, genotype. Rather the norm *is* diversity. The words *normal* and *abnormal* begin to lose their meaning.

However, a fly with two lethal chromosomes is usually healthy as long as the chromosomes were derived from different wild individuals. Therefore most of the lethal genes in a wild population are not allelic to one another (Wright, Dobzhansky, and Hovanitz 1942), so lethal genes must occur at many different loci. And the lethal allele at any one locus is actually very rare; its frequency is generally about 10^{-3} to 10^{-2}. Thus based on this evidence, any two wild *Drosophila* could be genetically identical, save for the one or two lethal genes they carry.

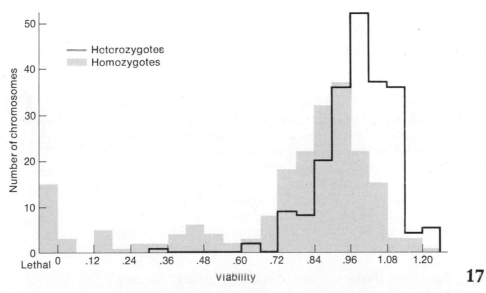

17

The distribution of relative viabilities of wild-type chromosomes in *Drosophila pseudoobscura,* in homozygous and heterozygous condition. The chromosomes vary greatly in the viability they confer, especially when homozygous. (From Lewontin 1974a, by permission of the publisher)

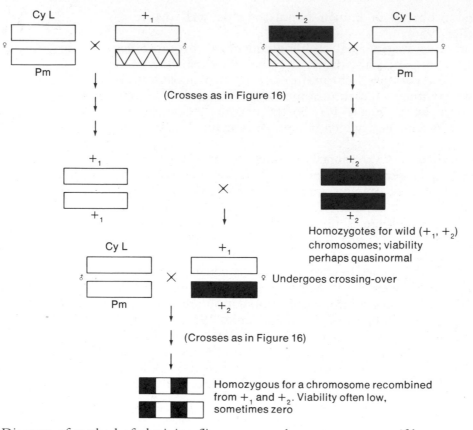

18 Diagram of method of obtaining flies homozygous for a chromosome formed by recombination of two wild chromosomes. Such a chromosome is sometimes lethal when homozygous, even if homozygotes for each of the constituent wild chromosomes are highly viable.

Just why these genes exist in wild populations has long been debated. Some geneticists (e.g., Hiraizumi and Crow 1960, Crow and Temin 1964) hold that many lethals are somewhat deleterious when heterozygous as well as when homozygous and are thus eliminated by natural selection; they exist only because of continuing mutation. Others (e.g., Dobzhansky, Krimbas, and Krimbas 1960; Wallace and Dobzhansky 1962; Band and Ives 1963; Wallace 1966b) are inclined to argue that these lethals may actually be slightly beneficial when heterozygous and that they are kept in the population by heterozygote superiority. The evidence for each of these points of view is many-faceted and too intricate to go into here.

In any case these rare alleles do not contribute greatly to any evolutionarily meaningful measure of genetic diversity. In Chapter 13 I show that rare recessive alleles cannot increase in frequency very rapidly by natural selection, so they cannot contribute much to immediate evolution. To assess the variation that permits evolutionary change, we want to know what fraction of loci have two or more common alleles and whether or not these alleles confer different adaptations and hence the capacity to adapt to environmental changes. A useful measure of genetic variation that takes into account the number of alleles at a locus and their relative frequencies is the frequency H of heterozygotes in a randomly mating population. For two alleles with frequencies p and q, $H = 2pq$, which is greatest when both alleles are equally common.

VARIATION IN PROTEINS

It was impossible to tell what fraction of loci is polymorphic or what the heterozygosity of an average locus is as long as it was impossible to count the number of loci that have no genetic variation. But it may be possible to do so if most loci code for proteins (especially enzymes). Different forms of, say, lactic dehydrogenase are coded by different alleles; but two individuals with the same form of the enzyme are presumably genetically the same. So it is possible to identify an invariant gene locus by finding an invariant enzyme. By taking a random sample of all kinds of enzyme and determining what fraction of them is genetically variable, one can estimate the fraction of polymorphic loci. Knowing which individuals are homozygous and which heterozygous for each enzyme, one can calculate the heterozygosity at each locus.

The most common technique for distinguishing different genetic forms (allozymes) of the same enzyme is gel electrophoresis. Extracts from several individuals are placed in a porous gel, often of starch, across which an electric potential is applied. If the amino acid composition of a given enzyme varies and if the amino acid substitutions carry different charges, the enzymes differ in net charge and move at different rates; so they are separated in the gel. Their positions are then found by flooding the gel with a substrate on which the enzyme acts,

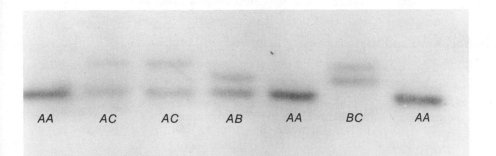

19

Polymorphism in an enzyme, leucine aminopeptidase, in the mussel *Mytilus edulis*. A starch gel with the enzyme profile of seven individual mussels representing four genotypes. The enzyme is a monomer; hence homozygotes have one band and heterozygotes have two. (Gel courtesy of Richard K. Koehn)

together with a dye that reacts with the product of the enzyme-substrate reaction to yield a colored band that indicates the position of the enzymes. Thus different genotypes are identified by differences in the position of the bands on the gel (*Figure 19*). These differences usually prove to be genetic, so an experienced worker can recognize different genotypes and be quite sure that they are genetically different, even in species that cannot be crossed in the laboratory.

The first assays of genetic variation by protein electrophoresis were published in 1966, when Harris reported on variation at 10 enzyme loci in humans and Lewontin and Hubby investigated 18 loci in each of several populations of *Drosophila pseudoobscura.* Lewontin and Hubby's data (*Table II*) were a surprise to the classical school of geneticists, for they led to the conclusion that an average *Drosophila* population is polymorphic at no less than 30 per cent of its loci and that the polymorphic loci have so many alleles (2–6) at such high frequencies that an average fly is likely to be heterozygous at about 12 per cent of its loci. If these loci are truly a random sample of the roughly 10,000 loci in *Drosophila,* there must be about 2000 to 3000 polymorphic loci in a local population, and an average fly has 700 to 1200 heterozygous loci. If so, any two flies from a single population differ at about a quarter of their loci. This amount of variation is truly staggering. Moreover, Harris' data indicated that humans, with 30 per cent polymorphic loci and 10 per cent average heterozygosity, were similarly highly variable.

Since these studies, more extensive work on *Drosophila* and *Homo* has revealed that these early estimates of genetic variation were, if anything, underestimates. Many of the apparent alleles revealed by electrophoresis are shown by detailed analysis to consist of several alleles whose enzyme products have similar electrophoretic mobility when first examined (Singh, Lewontin, and Felton 1976; Coyne 1976). A class of protein variants with the same electrophoretic mobility is an ELECTROMORPH.

Surveys on a great variety of species have revealed that almost all

TABLE II. Polymorphism and heterozygosity at 18 enzyme loci in *Drosophila pseudoobscura*

Population	Number of loci polymorphic	Proportion of loci polymorphic	Proportion of genome heterozygous per individual
Strawberry Canyon	6	0.33	0.148
Wildrose	5	0.28	0.106
Cimarron	5	0.28	0.099
Mather	6	0.33	0.143
Flagstaff	5	0.28	0.081
Average		0.30	0.115

(After Lewontin and Hubby 1966)

TABLE III. Genetic variation at allozyme loci in animals and plants

	Number of species examined	Average number of loci per species	Average proportion of loci	
			Polymorphic per population	Heterozygous per individual
Insects				
Drosophila	28	24	0.529	0.150
Others	4	18	0.531	0.151
Haplodiploid wasps[1]	6	15	0.243	0.062
Marine invertebrates	9	26	0.587	0.147
Marine snails	5	17	0.175	0.083
Land snails	5	18	0.437	0.150
Fish	14	21	0.306	0.078
Amphibians	11	22	0.336	0.082
Reptiles	9	21	0.231	0.047
Birds	4	19	0.145	0.042
Rodents	26	26	0.202	0.054
Large mammals[2]	4	40	0.233	0.037
Plants[3]	8	8	0.404	0.170

(After Selander 1976)
[1] Females are diploid, males haploid
[2] Human, chimpanzee, pigtailed macaque, and southern elephant seal
[3] Predominantly outcrossing species

species have comparably high levels of variation (Selander 1976; *Table III*). There are some consistent differences among species. Vertebrates tend to be less highly polymorphic than invertebrates, and species that form small local populations or are otherwise known to be inbred have reduced levels of heterozygosity. But even inbreeding plants such as wild oats are highly polymorphic; they consist of a variety of different homozygous genotypes. Haploid organisms are also polymorphic; Milkman (1973) found extensive variation at five enzyme loci in samples of *Escherichia coli* taken from single fecal samples. Whether they are inbreeders or outbreeders, inhabitants of constant or fluctuating environments, short-lived or long-lived, species harbor an enormous amount of variation within their populations.

There is reason to believe that the loci typically sampled are not a truly random sample of the genome, for some classes of enzymes seem consistently more polymorphic than others (see Selander 1976, Johnson 1976); but even if the proportion of polymorphic loci is less than 30 per cent, there must still be hundreds or thousands of polymorphisms in almost every population. Accounting for this variation is the subject of a major controversy. But whatever the reasons for its existence, the genetic variation revealed at the molecular level, like that revealed at the phenotypic level by artificial selection, is so great that the concept of a wild-type, normal genotype must be abandoned. So much variation exists that the capacity for rapid genetic response to environmental changes must be immanent in almost every population.

INTRASPECIFIC VARIATION AND HIGHER CATEGORIES

A description of patterns of variation within species would be incomplete without asking whether the diversity within species is of the kind that can give rise to new species, genera, or higher categories. This subject is also treated in Chapters 8 and 10.

The multitude of cases in which it is impossible, indeed meaningless, to specify whether two forms are conspecific should suffice to show that the differences among closely related species are of the same kind as those that distinguish well-differentiated populations of the same species. This is as true of characters essential to the biological species concept, such as intersterility and sexual isolation, as of the features by which we most easily recognize species. In many instances (*Figure 20*) a characteristic polymorphic within one species is fixed in a related, indisputably distinct species. Moreover it is possible by artificial selection to transcend the limits of variation that ordinarily exist within the species, as Kindred's work on vibrissa number in mice shows. Darwin noted that artificial selection had yielded breeds of pigeons with characteristics that distinguish families or orders of birds; the fantail breed has 30 to 40 tail feathers instead of the 12 or 14 that all species of pigeons naturally possess.

Of course there exist differences between some species, and especially between genera or higher categories, that are not included within the variation of any one of the species. But occasionally the characteristics that define major taxa do vary intraspecifically as well. In the family Compositae the sunflower tribe Heliantheae has bracts and ray florets that are lacking in the Helenieae. Clausen, Keck, and Hiesey (1947) discovered a small population of plants that lacked ray florets and bracts but was unlike any known member of the Helenieae. Distinctive enough to merit recognition as a new genus, it was provisionally named *Roxira serpentina*. It could readily be crossed, however, with *Layia glandulosa* in the Heliantheae and formed a vigorous and fertile F_1 hybrid (*Figure 21*). The authors concluded that the new genus was just an aberrant ecotype of *L. glandulosa* in which two loci control the presence or absence of ray florets and bracts.

The kinds of characteristic that define certain higher categories often

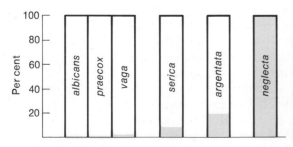

20

The second cubital crossvein is a rare "mutation" in three species of the bee genus *Andrena,* occurs as a polymorphic character in two species, and is a species-typical character in *Andrena neglecta.* (After Timofeeff-Ressovsky 1940)

21
The "new genus" *Roxira* (P_2, above) differs by but a few genes from *Layia glandulosa* (P_1), as illustrated by segregation in the F_2 generation. "*Roxira*" is actually conspecific with *L. glandulosa*, but this example illustrates that some of the characteristics that distinguish genera could be caused by a few mutations. (From Clausen, Keck, and Hiesey 1947, *The American Naturalist*, Copyright The University of Chicago Press)

vary within species in other taxa in which these characters are not diagnostic. The number and form of the teeth define various taxa of mammals, but they are highly variable within species of most other vertebrates. Even as fundamental a character as the number of cotyledons, which distinguishes the two great classes of angiosperms, varies intraspecifically in the shrub *Pittosporum,* and it has proved possible to select tricotyledonous strains of snapdragons (Stebbins 1974).

THE ORGANIZATION OF GENETIC VARIATION

The variation within populations, like that among populations, sometimes forms patterns. These may suggest the existence of organizing factors whose operation will be discussed in subsequent chapters. Patterns in the genetic structure of a population can be nonrandom associations of alleles either within or among loci.

If the frequency of each allele at a locus is known, the expected frequency of each genotype can be calculated on the assumption that the alleles are combined at random (*the Hardy-Weinberg theorem, Chapter 10*). Sometimes, however, heterozygotes are more common relative to homozygotes than expected. For example, there is an excess of heterozygotes for sickle-cell hemoglobin in West African human populations. In the bluegill sunfish (*Lepomis macrochirus*) heterozygotes for the enzyme glutamate oxalate transaminase were present in excess (Avise and Smith 1974). An excess of heterozygotes is generally known or presumed to be caused by their superior survival.

Conversely, heterozygotes are often less common than they

TABLE IV. Genotype frequencies and inbreeding coefficients for three loci in the self-fertilizing wild oat *Avena fatua*

Genotype	Frequency	Frequency of recessive allele	Inbreeding coefficient (*F*)
BB	0.712		
Bb	0.138	0.219	0.597
bb	0.150		
HH	0.583		
Hh	0.167	0.334	0.625
hh	0.250		
LsLs	0.775		
Lsls	0.125	0.162	0.539
lsls	0.100		

(After Jain and Marshall 1967)

"should" be. Heterozygotes for a color polymorphism in the South American fly *Drosophila polymorpha* made up only 39 per cent of a wild population, whereas the expected frequency was 42 per cent (da Cunha 1949). Heterozygote deficiencies of this kind are usually ascribed to some inbreeding effect (*Chapter 11*). Self-fertilizing species of plants, although genetically variable, are almost entirely homozygous (*Table IV*). The variation in such species is organized in a very nonrandom way.

Alleles at different loci can also be organized into nonrandom patterns. Because of recombination between loci, each of the alleles at one locus (*A* and *A'* for example) should become randomly distributed with respect to the alleles at another (*B* and *B'*), even if the loci are on the same chromosome. Thus if the frequencies of *B* and *B'* are 60 per cent and 40 per cent respectively, *A* should be associated with *B* 60 per cent of the time and with *B'* 40 per cent of the time, and similarly for *A'*. This condition of LINKAGE EQUILIBRIUM is violated, however, if *A* is more consistently associated with *B*, and *A'* with *B'* (or *vice*

22 Random and nonrandom association of linked genes. Alleles *A* and *B* have frequency 0.7; alleles *a* and *b*, 0.3. The expected frequencies of the four types of chromosome are as indicated by the white columns; the population is in linkage equilibrium (*D* = 0, in the notation of Chapter 10). If *D* = 0.05, the coupling chromosomes *AB* and *ab* are in excess (black columns); if *D* = −0.05, the repulsion chromosomes *Ab* and *aB* are in excess (gray columns).

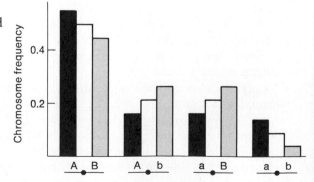

versa) than chance alone would dictate. The loci are then in a state of LINKAGE DISEQUILIBRIUM (*Figure 22; Chapters 10, 14*).

For example (Ford 1971), the flowers of the primrose *Primula vulgaris* are of two forms. The "pin" form (*ss*), with long style and low-set anthers, is inherited as if it were recessive to the "thrum" form (*Ss*), which has a short style and elevated anthers. This polymorphism, called heterostyly, favors outcrossing, since thrum pollen can readily be placed on a pin style, and *vice versa*. Because thrum (*Ss*) is self-incompatible, the genotype *SS* is not ordinarily produced. Occasionally, however, plants with both a long (pin) style and elevated (thrum) anthers are found (*Figure 23*). These, it turns out, are recombinant genotypes: pin and thrum plants actually differ not at one locus, but at two very closely linked loci, of which one (*G, g*) controls the style

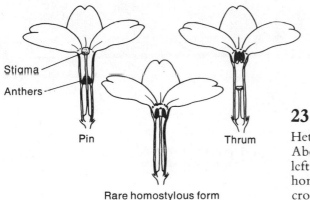

Stigma

Anthers

Pin   Thrum

Rare homostylous form

23

Heterostyly in the primrose *Primula vulgaris*. Above, the "pin" and "thrum" phenotypes at left and right respectively. Below, the homostyled phenotype formed by occasional crossing-over. (From Ford 1971)

and the other (*A, a*) the anthers. *G* is almost always associated with *A*, and *g* with *a*; thrum plants are *GA/ga* and pin plants *ga/ga*. The exceptional plants with the combination of pin and thrum characteristics are *gA/ga*. Thus *G* and *A* are inherited together as if they were one locus, a SUPERGENE that is only rarely broken up by crossing-over. The prevalent combinations, *GA* and *ga*, are advantageous compared to the rare combinations *Ga* and *gA* and are kept together by natural selection.

We might expect to find such supergenes whenever alleles at different loci interact to form especially favorable combinations. But natural selection, which would maintain the integrity of such combinations, is opposed by recombination, so loosely linked loci are ordinarily randomly associated unless recombination is reduced.

For reasons illustrated in Figure 24 (see also *Chapter 10*), paracentric chromosome inversions suppress crossing-over. Many populations of *Drosophila* are polymorphic for different inverted sequences. Dobzhansky (1970), who with his colleagues studied these polymorphisms extensively, long ago postulated that each inversion type carried a particularly adaptive combination of alleles at various loci, COADAPTED

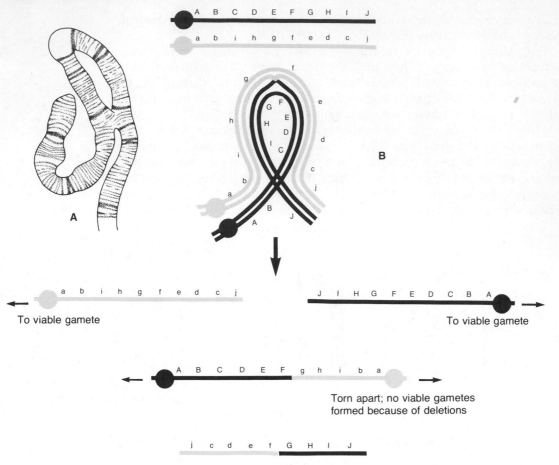

24 Chromosomal inversions in *Drosophila*.
(A) Synapsed chromosomes in a *Drosophila pseudoobscura* heterozygous for *Standard* and *Arrowhead* sequences. (B) The behavior of inversions in heterozygous (heterokaryotypic)

form. Crossing-over is suppressed because crossover products lack centromeres or substantial blocks of genes. (A redrawn from Strickberger 1968)

to function well together. His predictions were verified by Prakash and Lewontin (1968), who found that different groups of inversions in *Drosophila pseudoobscura* and *D. persimilis* carried different alleles at two electrophoretic loci.

It is possible to tell which inverted chromosome sequences have been derived from which others (*Figure 25*). Prakash and Lewontin found that related inversions in the *Santa Cruz* group (*SC, CH, TL*) shared alleles that were different from those prevalent in the *Standard* group (*ST, PP, AR*). The latter were shared by both *D. pseudoobscura* and *D. persimilis*. Thus the association of the alleles with the inversion types and with each other must antedate the speciation event. The association is not due simply to different mutational events in the different chromosomes, for each allele is found in low frequency in the "wrong" chromosomes. Thus it seems likely that natural selection has maintained these combinations of genes intact for at least several million generations.

Inversions are not the only way to keep gene combinations intact. Ameiotic parthenogenesis and vegetative reproduction reduce recombination to zero. Populations of the moth *Alsophila pometaria* consist largely of asexual females and, in the minority, sexual females and males. Alleles at several enzyme loci are not associated with each other in the sexual form, but the parthenogenetic moths consist of several clones, each with a distinct multilocus genotype (Mitter and Futuyma 1977).

A fully inbred population can consist of a variety of homozygous genotypes. Hamrick and Allard (1972) found two alleles at each of six loci in the self-fertilizing wild oat *Avena barbata*. Of the 2^6, or 64, possible homozygous genotypes, only two were common, in moist and dry microhabitats, respectively. Hamrick and Allard postulated that these are the two highly advantageous gene combinations, which can persist unbroken because of the infrequency of recombination in this species.

In randomly mating sexual populations, loosely linked genes should generally be randomly associated with one another (but see *Chapter 14*). Yet in some instances they are not. Charlesworth and Charlesworth (1973) found that alleles at each of five polymorphic enzyme loci were correlated with alleles of at least one other locus, in populations of *D. melanogaster*. The pairs of loci that showed consistently strong associations were not necessarily closely linked. So despite the free recombination that must randomize these alleles with respect to one another, they tend to come in packages.

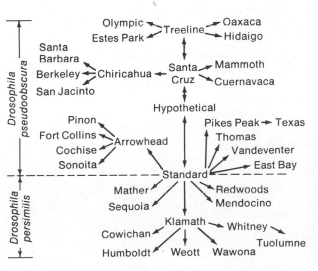

25

The phylogeny of arrangements of the third chromosome in *Drosophila pseudoobscura* and *D. persimilis*. The species share the *Standard* sequence, which may therefore be ancestral to the others. (From Anderson et al. 1975)

SUMMARY

Many or most of a species' characteristics vary from one population to another; even the characteristics that define species vary geographically, so that species are often hard to delimit. A characteristic's pattern of variation is often far from random; it may give evidence of the history of the species or of the adaptive properties of the characteristic. Because different characteristics often vary independently and gradually within a species, it is usually difficult to divide a species into unambiguously meaningful divisions, or subspecies. Variation is nested within variation, down to the level of variation within populations, which is so luxuriant that it often makes the variation among populations seem meager by comparison. The genetic variation within populations, like that in the species as a whole, is often organized into nonrandom patterns. These patterns and other properties of the variation suggest the ways in which variation arises, persists, and changes — the mechanisms of evolutionary change.

FOR DISCUSSION AND THOUGHT

1 Does the continuum from population to species to genus necessarily imply that evolution is gradual?

2 Discuss the prevalence and implications of typological thinking as it applies to evolutionary theory, to scientific thought, to psychology, to sociology, to human affairs.

3 How can morphological or physiological evidence be used to determine whether two sympatric forms are different species? How about allopatric forms?

4 Are Springer spaniels and Irish setters different species? St. Bernards and Chihuahuas? sterile worker bees and queen bees? black and white South Africans? tetraploid and diploid plants of the same "kind"?

5 Discuss the essential differences among geographic isolation of populations, premating (e.g., ethological) isolating mechanisms, and postmating (e.g., hybrid sterility) barriers to gene flow. Are these all evolved properties? Do they all evolve in the same way? Should they all be called isolating mechanisms?

6 Some species have a two-year life cycle, so that individuals who mature in even years cannot interbreed with those that mature in odd years. Can this temporal analogue of geographic isolation lead to genetic divergence? Should the two "populations" be considered different species?

7 In some polymorphic species a certain amount of assortative mating occurs. The blue and white forms of the blue goose, for example, mate among themselves more frequently than they do with one another (Cooke and Cooch 1968). If such assortative mating is less than 100 per cent perfect, should the two forms be considered different species?

8 Provide an alternative explanation for observations interpreted as introgressive hybridization, such as those on the grackles.

9 Is it possible, or likely, that two species that have long coexisted without interbreeding will merge by hybridization into one? How would such hybridization differ from that between partially differentiated allopatric populations, like grackles?

10 In cases of circular overlap the last in a series of populations that interbreed along a circular distribution overlaps with the first, without interbreeding (*Chapter 16*). How many species are there? Analyze the cases

cited by Mayr (1963) and determine whether such a phenomenon has ever been unequivocally demonstrated. Why are such cases interesting?

11 If a characteristic of a species changes abruptly at a border between two subspecies, how could one tell whether the discontinuity was due to adaptation to a discontinuity in the environment or to the rejoining of two formerly isolated, genetically differentiated populations? See Endler (1977).

12 The concept of the ecotype has been applied primarily to plants. Do animals also display ecotypic variation?

13 Early geneticists were familiar with the continuous variation in morphological characters and with the effects of artificial selection. Yet they were surprised at and reluctant to accept the notion of extensive genetic variation, as revealed by the study of lethal genes and enzyme polymorphisms. Why?

14 Evolution by natural selection is commonly supposed to entail the "survival of the fittest," whereby the best genotype prevails over those that are inferior. How does our recognition of extensive genetic variation modify this view?

15 Distinguish between chromosomal variation (e.g., inversion polymorphism) and genic variation. What differences between these kinds of variation are important for evolution? Can a species have one without the other?

16 Would you expect the amount of genetic variation within a population that inhabits a very changeable environment to be different from that within a population in a more constant environment? Why?

17 Mayr (1963), among others, has argued that the populations along the periphery of a species' geographic range harbor less genetic variation than those in the central part of the range. Why might you expect this? Why is the argument important?

18 Would you expect asexual or sexual species to be more finely adapted to their environment? Why? How could you tell?

19 Early in this chapter I stressed that geographic populations may have different genetic compositions. Later I stressed an apparently opposing point of view, that most of the genetic diversity of a species may lie within rather than among populations. Can these statements be reconciled?

20 What are the implications of this chapter for the use of the terms *normal* and *abnormal* in reference to human beings? What are the implications for the classification of humans into races and the further implications for social thought and policy?

MAJOR REFERENCES

Mayr, E. 1963. *Animal species and evolution.* Harvard University Press, Cambridge, Mass. 797 pages. One of the most important books on evolution. Extensive discussion of geographic variation and of genetic variation within populations, the latter now outdated.

Mayr, E. 1969. *Populations, species, and evolution.* Harvard University Press, Cambridge, Mass. 453 pages. An abridgment of Mayr 1963.

Dobzhansky, Th. 1970. *Genetics of the evolutionary process.* Columbia University Press, New York. 505 pages. This and its predecessors, the several editions of *Genetics and the origin of species,* are among the most influential books on evolution from the viewpoint of population genetics.

Lewontin, R. C. 1974. *The genetic basis of evolutionary change*. Columbia University Press, New York. 346 pages. A description of genetic variation as revealed by classical methods and by electrophoresis, together with a review of the history of the subject and a thorough analysis of the application of population genetic theory to data.

Genetic Variation and its Origins

The preceding chapters have treated primarily the products of evolution, the phenomena that evolutionary theory means to explain. This theory, the statement of the mechanisms of evolutionary change, is largely the subject of population genetics, the analysis of genetic changes in populations. This chapter deals with a fundamental property of genetic variation and with the origins of that variation.

THE HARDY-WEINBERG THEOREM

In a laboratory cross between a homozygous dominant (AA) stock and a homozygous recessive (aa) stock, the F_2 generation shows a 3:1 ratio of dominant to recessive phenotypes. Isn't it reasonable, asked a naive geneticist in 1908, to suppose that by its numerical predominance the dominant gene will completely hold sway as the recessive becomes less and less common and finally extinct? Won't a population's phenotypes, then, be entirely due to dominant genes, and won't recessive genes exist only as rare mutants? If this supposition were true, phenotypes would exist not because of their advantageous properties, but solely because of their Mendelian ratios. It is not true, of course, as G. H. Hardy and W. Weinberg independently proved in 1908.

In a diploid population of N individuals there are $2N$ genes at locus A. The three genotypes AA, AA', and $A'A'$ number n_{AA}, $n_{AA'}$, and $n_{A'A'}$ respectively. The proportion of AA, n_{AA}/N, will be denoted D. Similarly H is the proportion of heterozygotes and R the proportion of $A'A'$ individuals. Notice that $D + H + R = 1$. The proportions D, H, and R are the GENOTYPE FREQUENCIES. We may also calculate the GENE FREQUENCIES of the alleles A and A'. If the total number of A genes is n_A and that of A' genes is $n_{A'}$, then the frequency p of the A allele is $n_A/2N$, and the frequency q of the A' allele is $n_{A'}/2N$. Note that $p + q = 1$. The number of A genes equals twice that of AA individuals plus the number of heterozygotes, or $n_A = 2n_{AA} + n_{AA'}$. Thus

$n_A/2N = (2n_{AA} + n_{AA'})/N$, so $p = D + H/2$. Similarly $q = R + H/2$.

If mating occurs at random, the probability of an $AA \times AA$ mating is D^2, the product of the probabilities that each mate has the genotype AA. Similarly the probability of a mating between AA and AA' is $(D \times H) + (H \times D)$, or $2DH$ (since there are two such crosses, $AA \times AA'$ and $AA' \times AA$).

The probability that an offspring from the $AA \times AA$ mating will have the genotype AA is 1; but from an $AA \times AA'$ mating two genotypes arise, each with a probability of $\frac{1}{2}$. The matings and their outcomes are listed in Table I. The frequency of AA progeny is $D^2 + DH + H^2/4$, or $(D + H/2)^2$; the frequency of heterozygous progeny is $2(D + H/2)(H/2 + R)$, and that of $A'A'$ offspring $(H/2 + R)^2$. But $D + H/2 = p$, and $H/2 + R = q$. Thus after one generation of random mating, the frequencies of the three genotypes are $p^2:2pq:q^2$, no matter what the initial genotype frequencies D, H, and R were. That is, the proportions of genotypes are found by the binomial expansion $(p + q)^2$. Among these progeny, moreover, the frequency of A is $p^2 + \frac{1}{2}(2pq) = p(p + q) = p$, which is the same gene frequency that the parents had.

Thus the relative abundance of the alleles A and A' does not change from one generation to the next; the only change in the genetic complexion of the population is a redistribution of the genotypes into frequencies that they will retain in all subsequent generations. Whether either of the alleles is dominant over the other is irrelevant; dominance describes the phenotypic effect, not the abundance, of an allele. This then is the Hardy-Weinberg theorem, the foundation of the entire genetical theory of evolution: *under the conditions we have implicitly assumed, a single generation of random mating establishes binomial genotype*

TABLE I. Derivation of the Hardy-Weinberg theorem

Mating	Frequency of mating	Progeny		
		AA	AA'	A'A'
$AA \times AA$	D^2	D^2		
$AA \times AA'$	$2DH$	DH	DH	
$AA \times A'A'$	$2DR$		$2DR$	
$AA' \times AA'$	H^2	$H^2/4$	$H^2/2$	$H^2/4$
$AA' \times A'A'$	$2HR$		HR	HR
$A'A' \times A'A'$	R^2			R^2
Total	$(D + H + R)^2 = 1$	$(D + \frac{1}{2}H)^2 = p^2$	$2(D + \frac{1}{2}H)(\frac{1}{2}H + R) = 2pq$	$(\frac{1}{2}H + R)^2 = q^2$
		(because $p = D + \frac{1}{2}H$, $q = \frac{1}{2}H + R$)		

NOTE: D = initial frequency of AA, H = that of AA', R = that of $A'A'$; $D + H + R = 1$.

frequencies, and neither these frequencies nor the gene frequencies p *and* q *will change in subsequent generations.*

In both natural and laboratory populations genotype frequencies are often very close to these theoretical expectations. For example, of 1612 specimens of the scarlet tiger moth *Panaxia dominula* that Ford and his associates collected (Ford 1971, p. 136), 1469 had white spotting (*AA*), 5 had little spotting (*A'A'*), and 138 were intermediate (*AA'*). The frequency *p* of the gene *A* was thus $D + H/2$, or $1469/1612 + \frac{1}{2}(138)/1612 = 0.954$. The frequency of *A'* was 0.046. The genotype frequencies, according to the Hardy-Weinberg theorem, should be $(0.954)^2 = 0.9101$ *AA*, $2(0.954)(0.046) = 0.0878$ *AA'*, and $(0.046)^2 = 0.0021$ *A'A'*. The numbers of the three genotypes that we should expect in a sample of 1612 moths are, then, 1467 *AA* (approximately 0.9101×1612), 142 *AA'*, and 3 *A'A'*. Because the expected and observed numbers are very similar, we can conclude that this locus was in Hardy-Weinberg equilibrium.

EXTENSIONS OF THE HARDY-WEINBERG THEOREM

The Hardy-Weinberg theorem can be extended in several ways. Let the three alleles *A, A', and A''* have gene frequencies *p, q,* and *r*, where $p + q + r = 1$. Then the genotype frequencies after one generation of random mating are given by the binomial expansion $(p + q + r)^2$:

AA	AA'	AA''	A'A'	A'A''	A''A''
p^2	$2pq$	$2pr$	q^2	$2qr$	r^2

If there are many, say *k*, alleles with frequencies $p_1, p_2, \ldots, p_k$, the genotype frequencies are given by $(p_1 + p_2 + \ldots + p_k)^2$.

The Hardy-Weinberg theorem can be extended to cases of polyploidy also. In a tetraploid plant two alleles *A* and *A'* with frequencies *p* and *q* assort into five genotypes (*AAAA, AAAA', . . ., A'A'A'A'*). If the chromosomes assort randomly in meiosis to form diploid gametes, these genotypes are formed in accord with the expansion $(p + q)^4$. In general, for an *n*-ploid organism, the genotypes are given by $(p + q)^n$. In some cases, though, meiosis in polyploids is not all that simple, and the theorem must be modified accordingly.

If a locus is sex-linked rather than autosomal, two-thirds of the genes in a population with a 1:1 sex ratio are carried by the homogametic sex (e.g., the human female), and one-third by the heterogametic sex (e.g., the human male). Since males derive their sex-linked genes only from their mothers, the gene frequency among the males in generation *t* must equal the frequency among the females in the previous generation; and the gene frequency in females must be the average of that of their mothers and fathers. As a result the gene frequency of the population as a whole remains constant, but within each sex it oscillates toward an equilibrium that is the same for both sexes (*Figure 1*).

The extension to two (or more) loci is complicated, but it is very important. Let p_1 and q_1 be the frequencies of the alleles A and A' at one locus; p_2 and q_2 the frequencies of B and B' at a second locus. The loci recombine by crossing over and segregating in the double heterozygote $AA'BB'$ at a rate R which ranges from 0 (for loci that do not recombine) to ½ (for independently segregating loci). If the loci are on the same chromosome, then R is the crossover distance between the loci.

The critical concept in multiple-locus theory is GAMETE FREQUENCY. There are nine possible zygotic genotypes ($AABB$, $AABB'$, ..., $A'A'B'B'$), but only four possible gametes (AB, AB', $A'B$, $A'B'$) that unite at random to form zygotes. Denote the frequencies of these gametes by the respective symbols g_{00}, g_{01}, g_{10}, and g_{11}, and note that $g_{00} + g_{01} + g_{10} + g_{11} = 1$. The frequency p_1 of the allele A is $g_{00} + g_{01}$, since these are the only gametes that carry A. The frequency p_2 of B is $g_{00} + g_{10}$, and similarly for the other two alleles.

If the probability that a gamete contains allele A is independent of its carrying allele B, the frequency (g_{00}) of the AB gamete is then the product of the probabilities, $p_1 p_2$. Similarly in this case $g_{01} = p_1 q_2$, $g_{10} = q_1 p_2$, and $g_{11} = q_1 q_2$. The gametes AB and $A'B'$ are COUPLING GAMETES, and AB' and $A'B$ are REPULSION GAMETES. The term D, defined as $D = g_{00} g_{11} - g_{01} g_{10}$, expresses the degree to which coupling and repulsion gametes differ in frequency. If the alleles at the two loci are randomly combined into gametes, coupling and repulsion gametes are equally common, and $D = (p_1 p_2 \cdot q_1 q_2) - (p_1 q_2 \cdot q_1 p_2) = 0$.

However, the alleles may not be combined at random. For example, if a population has just been formed by mixing two homozygous stocks $AABB$ and $A'A'B'B'$ in equal numbers, the gene frequencies p_1, p_2, q_1, and q_2 all equal ½, but only two kinds of gametes are formed, AB and $A'B'$, each with a frequency of ½. Thus $D = (½ \cdot ½) - (0 \cdot 0) = ¼$. The population is said to be in a state of GAMETIC EXCESS, or LINKAGE DISEQUILIBRIUM. Often termed the COEF-

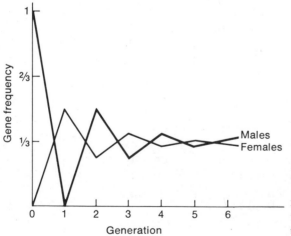

1
Approach to equilibrium at a sex-linked locus, in a species in which females have two X chromosomes, males one. The allele has an initial frequency of zero in females, one in males. Its frequency in the population is ⅓ throughout. (From Crow and Kimura 1970)

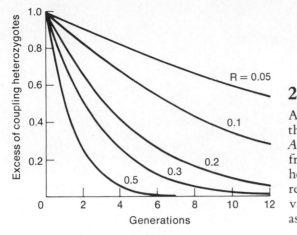

2

Approach to linkage equilibrium for two loci that begin in coupling phase (all heterozygotes $AB/A'B'$, none $AB'/A'B$). The difference in frequency between coupling and repulsion heterozygotes declines faster if R, the recombination frequency, is greater. The value $R = 0.5$ represents independent assortment. (After Falconer 1960)

FICIENT OF LINKAGE DISEQUILIBRIUM, D ranges from $+\frac{1}{4}$ when only coupling gametes exist, to $-\frac{1}{4}$ when all gametes are in the repulsion phase.

Several points of interest are proven in Box A. For each of the two loci viewed separately, a single generation of random mating establishes the genotype frequencies p^2, $2pq$, and q^2 just as if the other locus did not exist; the Hardy-Weinberg theorem holds true. However, the degree of association between alleles A and B and between A' and B' persists for a while. There is a correlation between alleles at the two loci, that is visible among the zygotes as an excess of some genotypes (those carrying AB and $A'B'$ combinations) and a deficiency of others (those with AB' and $A'B$ combinations in their genetic makeup). The strength of the correlation is measured by D, which only gradually declines toward zero as recombination assorts the genes into random combinations. The rate at which D approaches zero (a state of linkage equilibrium) depends on how tightly the loci are linked; the value of D in the nth generation is $D_n = D_0 (1 - R)^n$ (*Figure 2*). Thus even if two loci are linked, the characteristics they determine are not necessarily correlated with one another in the population. Conversely two characteristics that are correlated with one another may be determined by unlinked genes, for even independently segregating loci can show some degree of gametic excess for a number of generations.

ASSUMPTIONS OF THE HARDY-WEINBERG THEOREM

The Hardy-Weinberg theorem and its extensions are based on certain assumptions; violations of these assumptions cause changes in the frequencies of genes, of genotypes, or of both. The discrepancies between an "ideal" Hardy-Weinberg population and real populations are the ingredients of evolution. The assumptions underlying this theorem are the following:

1. The size of the population is infinite, or effectively infinite. But

A Two Loci

The initial gamete frequencies are g_{00}, g_{01}, g_{10}, and g_{11} for AB, AB', $A'B$, and $A'B'$ respectively. The gene frequencies of A, A', B, and B' are, respectively,

$p_1 = g_{00} + g_{01}$, $q_1 = g_{10} + g_{11}$, $p_2 = g_{00} + g_{10}$, and $q_2 = g_{01} + g_{11}$. These gametes combine at random to form zygotes in the frequencies shown in the table at right. These produce gametes with new frequencies g'_{00}, and so on. The coupling and repulsion heterozygotes $AB/A'B'$ and $AB'/A'B$ are written separately; each yields a given gamete type either by recombination (with probability R) or by not recombining (with probability $1 - R$):

Parent	AA			AA'			A'A'		
	BB	BB'	B'B'	BB	BB'	B'B'	BB	BB'	B'B'
	g_{00}^2	$2g_{00}g_{01}$	g_{01}^2	$2g_{00}g_{10}$	$2g_{00}g_{11} + 2g_{10}g_{01}$	$2g_{01}g_{11}$	g_{10}^2	$2g_{10}g_{11}$	g_{11}^2
	$\dfrac{AB}{AB}$	$\dfrac{AB}{AB'}$	$\dfrac{AB'}{AB'}$	$\dfrac{AB}{A'B}$	$\dfrac{AB}{A'B'}\;\;\dfrac{AB'}{A'B}$	$\dfrac{AB'}{A'B'}$	$\dfrac{A'B}{A'B}$	$\dfrac{A'B}{A'B'}$	$\dfrac{A'B'}{A'B'}$

New gamete frequency

$$g'_{00} = g_{00}^2 + (\tfrac12)2g_{00}g_{01} + (\tfrac12)2g_{00}g_{10} + (\tfrac12)(1 - R)2g_{00}g_{11} + (\tfrac12)(R)2g_{10}g_{01}$$

$$g'_{01} = (\tfrac12)2g_{00}g_{01} + g_{01}^2 + (\tfrac12)(R)2g_{00}g_{11} + (\tfrac12)(1 - R)2g_{10}g_{01} + (\tfrac12)2g_{01}g_{11}$$

$$g'_{10} = (\tfrac12)2g_{00}g_{10} + (\tfrac12)(R)2g_{00}g_{11} + (\tfrac12)(1 - R)2g_{10}g_{01} + g_{10}^2 + (\tfrac12)2g_{10}g_{11}$$

$$g'_{11} = (\tfrac12)(1 - R)2g_{00}g_{11} + (\tfrac12)(R)2g_{10}g_{01} + (\tfrac12)2g_{01}g_{11} + (\tfrac12)2g_{10}g_{11} + g_{11}^2$$

Each of these expressions may be simplified. For example,

$$g'_{00} = g_{00}^2 + g_{00}g_{01} + g_{00}g_{10} + (1 - R)g_{00}g_{11} + (R)g_{10}g_{01} = g_{00}^2 + g_{00}g_{01} + g_{00}g_{10} + g_{00}g_{11} - Rg_{00}g_{11} + Rg_{10}g_{01}$$

$$= g_{00}(g_{00} + g_{01} + g_{10} + g_{11}) - R(g_{00}g_{11} - g_{10}g_{01}) = g_{00} - RD$$

because $g_{00} + g_{01} + g_{10} + g_{11} = 1$ and $g_{00}g_{11} - g_{10}g_{01} = D$, the coefficient of linkage disequilibrium. Similarly $g'_{01} = g_{01} + RD$, $g'_{10} = g_{10} + RD$, $g'_{11} = g_{11} - RD$.

Thus gamete frequencies change. But gene frequencies do not; for example, $p'_1 = g'_{00} + g'_{01} = (g_{00} - RD) + (g_{01} + RD) = g_{00} + g_{01} = p_1$, its initial frequency. Moreover each locus viewed separately is in Hardy-Weinberg equilibrium. The frequency of AA, for example, is $(g'_{00})^2 + 2g'_{00}g'_{01} + (g'_{01})^2 = (g'_{00} + g'_{01})^2 = p_1^2$, as it should be under Hardy-Weinberg equilibrium, and the same is true of each other genotype. But the two-locus genotypes deviate from randomness if $D \neq 0$. If $D > 0$, for example, the frequency of $AABB$, g_{00}^2, exceeds $p_1^2 p_2^2$, while that of $AAB'B'$, g_{01}^2, is less than $p_1^2 q_2^2$. Hence the traits controlled by the two loci will be correlated.

in a finite (real) population the frequency of an allele may vary by chance.

2. Individuals mate with one another at random. But the pattern of breeding in a population is often not random, and this has important consequences.

3. All the alleles are equally competent in making copies of themselves, which enter the gene pool in the gametes. If alleles differ in their replacement rates, their frequencies may change. This phenomenon is called selection. A corollary assumption is that the alleles segregate in a one-to-one ratio into the gametes produced by heterozygotes.

4. There is no input of new copies of any allele from any extraneous source. If there were, and if one allele entered the population at a greater rate than another, the gene frequencies would change. There are two possible sources of new copies: migration of genes from another population (gene flow), and the transformation of one allele into another (mutation). These are the sources of genetic variation.

MUTATION

Heredity is, as Dobzhansky (1970, p. 30) says, a conservative force conferring stability on biological systems. Yet no mechanism composed of molecules and subject to the impact of the physical world can be perfect. Mistakes in the copying of the genetic message produce mutations: new messages, themselves capable of self-replication. A mutation is conceived to be a change in the genetic information itself rather than the formation of new combinations of preexisting genetic messages through recombination. The conceptual distinction between mutation and recombination, though, can be vague at times.

It is convenient to distinguish two large classes of mutations: macrolesions, which affect more than one cistron; and microlesions, or POINT MUTATIONS, which are confined to a single genetic locus. In eukaryotes macrolesions are generally changes in the structure of the chromosomes.

CHANGES IN THE KARYOTYPE

Because of the immense variety of meiotic mechanisms and chromosomal behavior that animals and plants display, the subject of mutational changes in chromosome structure is most complicated. It is treated extensively by Swanson (1957), Stebbins (1950, 1971b), and White (1973).

Understanding the origins of chromosome rearrangements and their prospects of propagation requires a thorough understanding of meiosis. Two other salient facts must be kept in mind. First, the loss of a significant amount of genetic material, a whole chromosome or a major piece of a chromosome, is almost always highly deleterious.

Second, such deficiencies often occur when a chromosome arm loses its centromere. Some organisms, such as nematodes and the hempiteran and homopteran insects, have holocentric chromosomes; each part of the chromosome has centromeric activity, so that even small chromosome fragments travel to the poles of the meiotic spindle. But in most organisms the chromosome has a localized centromere and is literally lost without it.

We may distinguish two classes of changes of the karyotype: events that alter the amount of genetic material and those that simply rearrange the genes into new sequences.

CHANGES IN THE COMPLEMENT OF GENES
Polyploidy

If the first meiotic division fails to occur, unreduced gametes are formed: diploid ($2N$) gametes from diploid plants, for example. The union of such a gamete with a haploid (N) gamete yields a triploid ($3N$) zygote, which is usually sterile because each of its gametes receives a highly imbalanced complement of genes (*Figure 3a*).

Two unreduced gametes may unite to form a tetraploid ($4N$) zygote, which is also usually sterile (*Figure 3b*). But a few naturally occurring tetraploids are known that are thought to have arisen within a single diploid species (AUTOTETRAPLOIDS). The fireweeds (*Epilobium*), for example, include autotetraploid forms (Mosquin 1967). Autotetraploidy is most common in parthenogenetic animals and plants, such as the brine shrimp *Artemia salina* which consists of not only a sexual diploid form but also asexual diploids, triploids, tetraploids, pentaploids, octaploids, and decaploids.

Polyploidy is nevertheless widespread in nature. ALLOPOLYPLOIDY, the formation of a polyploid from the diploid hybrid between two genetically differentiated populations (e.g., two species), is common, especially in plants. If the genetic content of the chromosomes of the two parental forms is sufficiently different, the tetraploid hybrid is fertile because its chromosomes form $2N$ sets of bivalents rather than N sets of irregularly segregating quadrivalents (*Figure 3c*).

Such tetraploids behave chromosomally as if they were diploid, but at many gene loci they are tetraploid. Perhaps because of the doubled gene dosage, tetraploids are often larger and more robust than diploids, have higher enzyme and hormone levels, and differ from diploids in many physiological and ecological respects.

Over evolutionary time the identical genes that a tetraploid has derived from its diploid ancestors can experience different mutations, so that initially identical loci may diverge and the amount of genetic information may be augmented. As this process continues, the tetraploid may evolve to become diploid not only in its chromosomal behavior, but in its genetic makeup as well (Stebbins 1950).

Aneuploidy

If one chromosome in a diploid organism is represented by other than

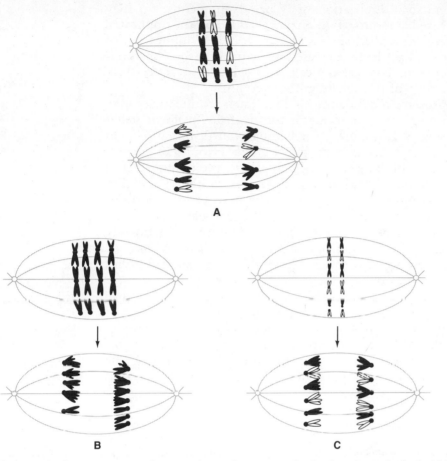

Diagrams of meiotic metaphase and anaphase. (A) A triploid typically produces gametes with unbalanced chromosome complements, as does an autotetraploid (B). An allotetraploid (C), if formed by polyploidization of a hybrid between two diploid species (with "white" and "black" chromosomes respectively), forms gametes with balanced chromosome complements.

two copies, the individual is termed aneuploid. Absence of a chromosome often causes lethality or sterility; a well-known example in humans is Turner's syndrome, entailing a sex chromosome complement denoted XO. Similarly individuals polysomic for the X chromosome (having the complement XXY, XXXY, XXXXY, or XXXXXY) are sterile and are described as having Klinefelter's syndrome. In general, aneuploidy is deleterious and seems unlikely to have much long-term evolutionary significance.

Duplications and deficiencies

If two homologous chromosomes in meiotic synapsis are not perfectly aligned, locus for locus, crossing-over between them is unequal and gives rise to one chromosome with an interstitial deficiency and one with a duplication. Such duplications may be advantageous, for they may provide increased amounts of the enzymes coded by the duplicated loci. Hansche (1975) described such a case in yeast. A duplication of the acid monophosphatase locus increased in frequency, since the doubled amount of enzyme enabled the cells to exploit more effectively the low concentrations of phosphate in the medium.

Duplications produce redundant gene loci. Such redundancies may be typical of some species; for instance, the hagfish *Eptatretus stoutii* appears to have several loci coding for hemoglobin (Ohno 1970). In the long run gene duplication may increase the amount of genetic information. As in the diploidization of polyploids, the redundant loci may experience different mutations and diverge. Another possibility is that unequal crossing-over in a heterozygote may yield a chromosome that carries both alleles in tandem. As this chromosome increases in frequency, the species becomes monomorphic for two functionally related genes rather than polymorphic for the two alleles (*Figure 4*). Such evolution seems to have occurred in some catostomid fishes (suckers) (Koehn and Rasmussen 1967).

Unequal crossing-over and gene duplication explain the existence of isozymes and of paralogous polypeptides such as the subunits of vertebrate hemoglobin (*Figure 4b*). Multiple forms of the same protein are more the rule than the exception among enzymes and other proteins and bespeak an important role for gene duplication in amplifying the information content of the genome (Ohno 1970).

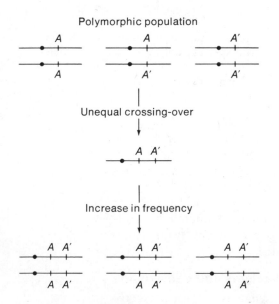

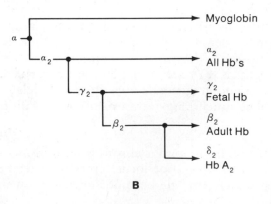

4

Effects of gene duplication. (A) Attainment of permanent "heterozygosity." (B) Evolution of the hemoglobin chains by gene duplication, indicated by the solid circles. (B from Ingram 1963, with permission of the publisher)

REARRANGEMENTS OF THE GENES

Alterations of the karyotype may simply reorganize the positions of genes relative to one another without changing their number. Such processes may change the number of chromosomes.

Inversions

If two breaks occur simultaneously in one chromosome, the segment between them may be rotated 180°, and be rejoined to the terminal pieces. Such an inverted chromosome first occurs in heterozygous condition (forming a HETEROKARYOTYPE), since it is rare; and a heterozygous inversion has important effects. A single crossover within the inversion yields a dicentric chromosome, with two centromeres, some duplicated loci, and some loci lacking, and an acentric chromosome

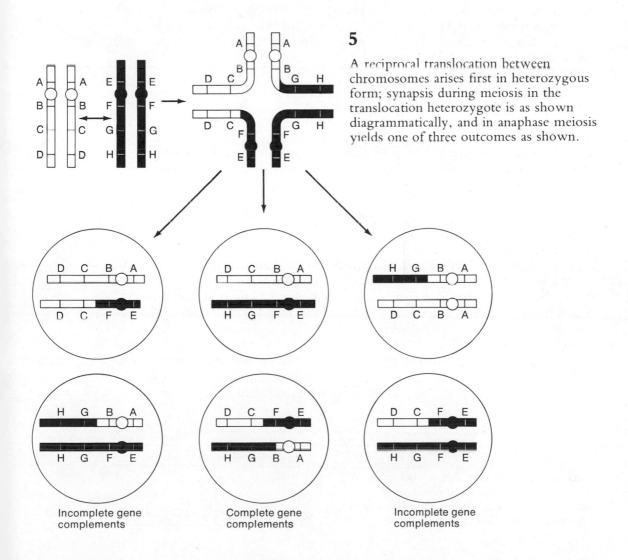

5

A reciprocal translocation between chromosomes arises first in heterozygous form; synapsis during meiosis in the translocation heterozygote is as shown diagrammatically, and in anaphase meiosis yields one of three outcomes as shown.

Incomplete gene complements

Complete gene complements

Incomplete gene complements

that is lost during anaphase (*Figure 24, Chapter 9*). Neither daughter cell gets a full complement of genes, so the only viable gametes formed by a heterokaryotype come from cells in which no single crossovers occurred within the inversion. An inversion therefore acts as a suppressor of crossing-over.

Because inversions cause partial sterility, they probably seldom become established in nature, except in some important instances. In many of the true flies of the order Diptera paracentric inversions do not lower fertility. There is no crossing-over in males, so the output of sperm does not suffer; and meiosis in females is peculiarly arranged so that the dicentric chromosomes pass into the polar bodies and only the intact chromatid is incorporated into the egg nucleus.

Translocations

When two breaks occur simultaneously in two nonhomologous chromosomes, a translocation may occur (*Figure 5*), forming a chromosome from two nonhomologous chromosomes. In the heterozygous condition, in which translocations first occur, the chromosomes achieve gene-for-gene synapsis by assuming rather complicated attitudes. Of the three ways in which the four members of this aggregate can segregate two by two, only one yields viable gametes, so translocation heterozygotes should be about two-thirds sterile.

Still, some species are heterozygous for translocations; one remarkable case will be described in Chapter 14. And quite often, closely related species are each chromosomally monomorphic but differ from each other by one or more reciprocal translocations. Thus translocations become established despite their initial disadvantage.

Fusion and fission of chromosomes

Metacentric chromosomes have the centromere somewhere in the middle; acrocentric chromosomes have the centromere very near the end. The two arms of a metacentric chromosome in one organism are often homologous to two different acrocentric chromosomes in another. White (1973) suggests that metacentrics arise by a reciprocal translocation between two acrocentrics and that a metacentric may dissociate into two acrocentrics by a reciprocal translocation with a minute donor chromosome (*Figure 6*).

Fusions and fissions often give rise to variation in chromosome number. For instance, Staiger (1954, cited in White 1973) found that the intertidal gastropod *Thais lapillus* varied in chromosome number

6

A possible mechanism for increase in chromosome number: dissociation of a metacentric chromosome into two acrocentrics by translocation with a minute "donor" chromosome. (From White 1973)

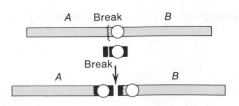

from $2N = 26$ to $2N = 36$ and that different populations had different mean numbers.

Many chromosomal mutations have little phenotypic effect; it is usually impossible to tell just by looking at a specimen whether or not any of its chromosomes are inverted, translocated, or fused. The significance of chromosome rearrangements probably lies less in their immediate effects on the phenotype than in their effects on recombination, on the sterility of heterozygous individuals, and on the sterility of hybrids between closely related species. To account for most changes in the biochemical, physiological, or morphological characteristics of species, we must look to changes within the genes themselves, the changes that are usually called point mutations.

POINT MUTATIONS

Most point mutations are thought to be changes in the sequence of nucleotide bases of single cistrons, changes that in the case of structural genes alter the amino acid sequence of the polypeptide chain for which the cistron codes (*Figure 7*). For example, a transversion from the RNA triplet GAA (or GAG) to GUA (or GUG) substitutes valine for glutamic acid. This is the mutational event that caused the abnormal beta chain in sickle-cell hemoglobin (Hb^S). Because of the degeneracy of the genetic code, many (about 24 per cent) codon substitutions do not change amino acid sequences and thus do not affect the phenotype.

More drastic changes in amino acid sequences are caused by frameshift mutations (*Figure 7*). Because the RNA sequence is translated into an amino acid sequence by triplets, a deletion or addition of base pairs shifts the reading frame, thereby altering the amino acid sequence throughout the whole chain, distal to the lesion. A second insertion or deletion at a subsequent site reestablishes the original reading frame, so that only a short nucleotide sequence is read in altered triplets. Whether such "small" frameshift mutations have been important in

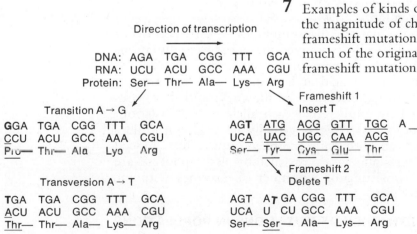

7 Examples of kinds of point mutation. Notice the magnitude of change caused by a frameshift mutation and the reconstitution of much of the original message by a second frameshift mutation.

Direction of transcription

DNA: AGA TGA CGG TTT GCA
RNA: UCU ACU GCC AAA CGU
Protein: Ser— Thr— Ala— Lys— Arg

Transition A → G
GGA TGA CGG TTT GCA
CCU ACU GCC AAA CGU
Pro— Thr— Ala Lys Arg

Transversion A → T
TGA TGA CGG TTT GCA
ACU ACU GCC AAA CGU
Thr— Thr— Ala— Lys— Arg

Frameshift 1
Insert T
AGT ATG ACG GTT TGC A _ _
UCA UAC UGC CAA ACG
Ser— Tyr— Cys— Glu— Thr

Frameshift 2
Delete T
AGT ATGA CGG TTT GCA
UCA U CU GCC AAA CGU
Ser— Ser— Ala— Lys— Arg

evolution is unknown. The available evidence suggests that frameshift mutations usually produce nonfunctional products; in yeast such mutations are completely recessive, whereas mutations caused by base pair substitutions are usually partially dominant, evincing some activity of the altered gene (Wills 1968).

In addition to base pair substitutions and frameshift mutations, several kinds of "pseudomutational" events are known. In bacteria plasmids can become integrated into the chromosome and affect the activity of the bacterial genome in ways that look like the effects of mutation (Drake 1970). In higher organisms such as fungi and *Drosophila,* recombination between two very closely linked markers seems to "convert" an allele at one site into the allelic form represented on the other chromosome. Although the exact mechanism of such gene conversion is unknown (Stahl 1969), it is generally agreed to entail some kind of recombination within the limits of a single cistron. Such INTRAGENIC RECOMBINATION can give rise to new gene products.

For example, cistrons coding for polypeptide sequences Val-Thr-Arg-Leu and Glu-Thr-Arg-Gly, if recombined, can yield a new sequence Val-Thr-Arg-Gly. Ohno and his colleagues (1969) believe that a polymorphism of 6-phosphoglycerate dehydrogenase in the Japanese quail arose in this manner. They point out that by intragenic recombination, variation begets variation; the more alleles there are, the more new alleles can come into being. New alleles could arise at rates several orders of magnitude greater than the mutation rates that are usually observed (Watt 1972), but whether intragenic recombination is important in evolution is unknown.

At the phenotypic level wild type genes mutate into mutant alleles, and mutants revert, by back mutation, to the wild type. At the molecular level, however, a true back mutation is most unlikely. Suppose A is replaced by G at one site in a cistron of 423 nucleotides (the size of the gene for human hemoglobin α). The probability that the next mutation will occur at this site is $1/423$, and the probability that G will mutate back to A rather than to C or G is $1/3$ (assuming that transitions and transversions are equally likely, which is not quite true). So the probability of a true reversion is $1/1269$.

Microbial studies show that true back mutation is infrequent. For example, Allen and Yanofsky (1963) showed that most back mutations of the tryptophan synthetase enzyme of *E. coli* result not from a reversion to the original amino acid at a mutated site, but from a second amino acid substitution elsewhere in the enzyme that restores the enzyme's function. This implies that each allele in a population, recognized by its phenotypic effect, may actually be a class of molecularly different isoalleles, all with the same phenotypic effect. Indeed many of the allelic enzymes (allozymes) revealed by electrophoresis are heterogeneous entities, consisting of a number of enzymes with the same electrophoretic mobility.

THE DYNAMICS OF MUTATIONS IN POPULATIONS

The frequency of each of the alleles (or chromosome types, for that matter) that may exist in a large population changes as the allele mutates into other alleles and as it is generated by mutation of other alleles. For simplicity, assume that there are only two alleles A and A', with frequencies p $(= 1 - q)$ and q. Then the rate at which A mutates to A' is u, and the rate of back mutation from A' to A is v.

In each generation (Δt), q changes to an extent Δq by the addition and loss of A' alleles through mutation. The absolute increment in q is $u(1 - q)$, since the frequency of A is $1 - q$, and the probability of mutation to A' is u. Similarly q declines by the amount vq by back mutation to A. The net change in frequency is

$$\Delta q/\Delta t = u(1 - q) - vq.$$

If u and v are very small, as they generally seem to be, the change in gene frequency will be very small in each generation. For example, if u and v are 10^{-5} and 10^{-7} respectively and $q = 0.5$ at time t, q will have changed to only 0.50000455 by the next generation. Crow and Kimura (1970) estimate that whatever the gene frequency happens to be, it will take about 70,000 generations for it to get halfway to the equilibrium gene frequency by mutation alone. Thus mutation *by itself* is a very weak evolutionary force.

The gene frequency reaches equilibrium and is no longer changing when $\Delta q/\Delta t = 0$. The equilibrium frequency $\hat{q}$ is found by solving for q in the equation

$$\Delta q/\Delta t = 0 = u(1 - q) - vq$$

and is

$$\hat{q} = u/(u + v).$$

Thus if $u = 10^{-5}$ and $v = 10^{-7}$, the gene frequency should ultimately come to rest at $\hat{q} = 0.909$, although it can take a very long time to arrive at this point.

This is a STABLE EQUILIBRIUM; and since the concept of a stable equilibrium recurs, I pause here to define the term. An equilibrium value of some system is one from which the system will not move unless acted on by external factors (*Figure 8*). A ball at rest on a perfectly level, flat table is at equilibrium. But it is a NEUTRALLY STABLE

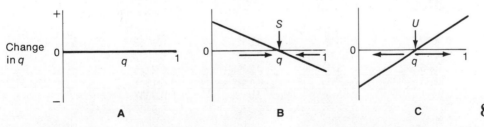

Kinds of gene frequency equilibria. (A) All gene frequencies are neutrally stable; there is no tendency either to increase or decrease. (B) $q = S$ is a stable equilibrium toward which q returns if displaced. (C) $q = U$ is an unstable equilibrium, from which q moves away in either direction if displaced.

position, since the ball will not return to this point if moved. The gene frequency at a locus that perfectly meets the assumptions of the Hardy-Weinberg theorem is a neutrally stable gene frequency. It will remain at value q indefinitely until some force changes it, but then it will remain at its new value q', its new neutrally stable position, thereafter.

A ball balanced on the tip of a pool cue is at equilibrium, but the slightest disturbance will move it to a new position, so the cue tip is an UNSTABLE EQUILIBRIUM point to which the ball will not return. Some genetic equilibria are unstable.

Conversely a ball in a perfectly concave bowl comes to rest at a point to which dynamic forces will return it should it be dislodged. This equilibrium is stable in that the ball moves back to it. There may be several or many stable equilibrium points in a system, for example the pockets of a pool table. The equilibrium between forward and back mutation is stable. If $q > u/(u + v)$, then $\Delta q / \Delta t$ is negative and q declines back to its equilibrium. If $q < u/(u + v)$, then $\Delta q / \Delta t$ is positive and q increases.

In fact, though, gene frequencies are probably seldom at their mutational equilibria. Since the equilibria are approached so slowly, other events in the meantime are likely to have more important effects. Selection and gene flow are likely to be important; even chance is important.

RATES OF MUTATION

If a population is very large, a mutation rate of 10^{-5} is a meaningful figure. If the population size is, say, 10^6, about 10 progeny in every generation will carry a new mutation at a given locus. More realistically, if the population size is about 1000 to 10,000, a mutant offspring will appear only once in a hundred or so generations and is thus effectively a unique event.

The fate of this unique mutation is very greatly influenced by

TABLE II. Probability of loss or survival of a single mutation

Number of generations	Probability of extinction		Probability of survival	
	No advantage	Advantage of 1 per cent	No advantage	Advantage of 1 per cent
1	0.3679	0.3642	0.6321	0.6358
3	0.6259	0.6197	0.3741	0.3803
7	0.7905	0.7825	0.2095	0.2175
15	0.8873	0.8783	0.1127	0.1217
31	0.9411	0.9313	0.0589	0.0687
63	0.9698	0.9591	0.0302	0.0409
127	0.9847	0.9729	0.0153	0.0271
Limit	1.0000	0.9803	0.0000	0.0197

(After Fisher 1930)

TABLE III. Spontaneous mutation rates of specific genes

Species and Locus	Mutations per 100,000 cells or gametes	
	Forward	Back
Escherichia coli		
streptomycin resistance	0.00004	
resistance to T1 phage	0.003	
arginine independence	0.0004	
Salmonella typhimurium		
tryptophan independence	0.005	
Neurospora crassa		
adenine independence	0.0008–0.029	
Drosophila melanogaster		
yellow body	12	
brown eyes	3	
eyeless	6	
Zea mays (corn)		
sugary seed	0.24	
I to *i*	10.60	
Homo sapiens		
retinoblastinoma	1.2–2.3	
achondroplasia	4.2–14.3	
Huntington's chorea	0.5	
Mus musculus (house mouse)		
a (coat color)	7.1	0.047
c (coat color)	0.97	0
d (coat color)	1.92	0.04
ln (coat color)	1.51	0

(After Dobzhansky 1970, with permission of the publisher)

chance. The mutation appears first in heterozygous form and so will be propagated, if at all, by the mating $AA \times AA'$. If the population is stable in size, the average number of surviving offspring from any mating is two. The probability that any one offspring will receive the A' gene from its heterozygous parent is ½, so if this mating yields two progeny, the probability that neither will receive the gene is $(½)^2$. Taking into account the likelihood that this mating could yield fewer or more than two offspring, Fisher (1930) calculated that the probability is about 0.37 that the new mutation will immediately be lost from the population. There is a comparable chance of loss in subsequent generations, so by the fifteenth generation, the chance that the mutant is still present is only about 0.11 (*Table II*). The same principle holds even for mutations that are advantageous in heterozygous condition.

The likelihood that a given kind of mutation will persist is greater if it recurs often: if mutation rates are high and if the population is large. Some mutation rates are listed in Table III. They vary greatly from locus to locus, probably to a greater extent than these figures

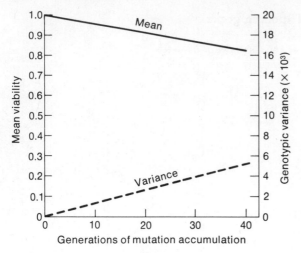

9

Effects of the accumulation of spontaneous mutations on viability. Mean viability decreases, and variation increases. (After Mukai et al. 1972)

indicate, since very low mutation rates are unlikely to be observed. The average rate of mutation seems to vary among organisms as well, being approximately 10^{-5} per locus per generation for higher animals and plants and 10^{-9} to 10^{-8} for microorganisms such as bacteria. The rate of back mutation from mutant to wild-type alleles is less than the rate of forward mutation, both in microorganisms (Drake 1970) and in eukaryotes (*Table III*). This is probably because a greater number of different isoallelic mutations impairs the function of an enzyme, yielding the same mutant phenotype, than restores a mutated enzyme to its wild-type function.

Although mutation rates are too low to affect the gene frequencies at any one locus very greatly, the sum of these small effects over many loci may appreciably alter the genetic complexion of a population. For example, Mukai (1964) and Mukai et al. (1972) isolated a single wild chromosome from a population of *Drosophila melanogaster* and for 40 generations maintained many lines of flies with this chromosome in heterozygous condition, so that recessive mutants could accumulate without being eliminated because of their deleterious effects. Periodically some of these chromosomes were brought into homozygous condition so that the viability of homozygous flies could be measured. Over the 40 generations the average viability of homozygous flies declined, and the variation among the replicate lines increased (*Figure 9*), because of the accumulation of deleterious mutations. From this and other work it appears that in *Drosophila* an average of about one new mutation may occur somewhere in the whole diploid genome, per fly per generation — and this underestimates the true total mutation rate, since mutations that do not measurably affect viability are not taken into account.

Total mutation rates per genome have also been calculated by determining the mutation rate for some specific loci and extrapolating

TABLE IV. Comparative spontaneous mutation rates

Species	Base pairs per genome	Mutation rate per base pair replication	Mutation rate per genome per generation
Bacteriophage lambda	4.7×10^4	2.4×10^{-8}	0.001
Bacteriophage T4	1.8×10^5	1.1×10^{-8}	0.002
Salmonella typhimurium	3.8×10^6	2.0×10^{-10}	0.001
Escherichia coli	3.8×10^6	4.0×10^{-10}	0.002
Neurospora crassa	4.5×10^7	5.8×10^{-11}	0.003
Drosophila melanogaster[1]	4.0×10^8	8.4×10^{-11}	0.93

(After Drake 1974)
[1] The *Drosophila* values are per diploid genome per generation of flies, not per generation of cells as in the other species.

to the whole genome. On this basis it appears that species of micro-organisms with larger genomes have a lower rate of mutation per base pair of DNA than species with smaller genomes (Drake 1974), so that the total rate of mutation for the whole genome is approximately the same for these very different species (*Table IV*). But among higher organisms, this conservation of total mutation rate breaks down: *Drosophila* has almost 10 times as much DNA as *Neurospora,* but about the same rate of mutation per generation per base pair, resulting in a far higher total input of mutations.

THE FACTORS THAT AFFECT MUTATION RATES

Have organisms evolved an optimal mutation rate? It might be supposed that a reasonably high mutation rate is adaptive, for if there were no mutation, the genetic variation of a population would ultimately be exhausted and further adaptation to environmental changes would be impossible. But because most mutations are deleterious, individuals with the lowest mutation rates should have the greatest number of surviving offspring. For this reason, genes that suppress mutation tend to have an advantage over those that allow mutations, so the mutation rate should evolve toward zero (*Figure 10*).

Mutation rates are influenced by environmental factors such as heat, ultraviolet light, ionizing radiation, and chemical mutagens. Thus any gene that influences an organism's susceptibility to those factors

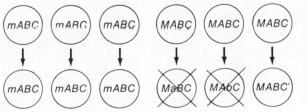

10

An allele *M* that raises the mutation rate at other loci suffers a selective disadvantage if most of the new mutations with which it is associated are deleterious (*a, b*), even if some (*C'*) are neutral or advantageous.

affects the mutation rate. Moreover in *Drosophila* there are mutator genes that cause exceptionally high mutation rates at a great many loci (e.g., Ives 1950). At least 30 loci affect the sensitivity of yeast to mutation by ultraviolet light, and at least two loci enhance spontaneous mutation rates (von Borstel et al. 1973). In *Escherichia coli* mutator genes include Treffer's mutator, *mutT,* which greatly increases mutation rates at many other loci by enhancing the frequency of A:T to C:G transversions.

But mutator alleles do not seem to be prevalent in natural populations, so mutation rates are not as high in nature as they could be if the mutator alleles were more common. A mutator gene that promotes transversions throughout the genome causes an immense number of deleterious alleles and is eliminated with them.

But mutator alleles in asexual populations can sometimes increase in frequency (Leigh 1970, 1973). Asexual populations sometimes evolve in a rather peculiar manner called periodic selection (Atwood, Schneider, and Ryan 1951). For example, the frequency of a gene for phage resistance may fluctuate dramatically in a culture of bacteria, even though no phage are present at any time (*Figure 11*). This can occur if an advantageous mutation at some locus, say *A,* happens to arise in one of the few cells that carries an allele for phage resistance (r^+) at another locus. Because A and r^+ are not dissociated by recombination r^+ increases with A. But if a still more advantageous mutation B occurs in a phage-sensitive (r^-) cell, Br^- has an advantage over Ar^+, so phage resistance declines. Because there is no recombination, a newly arisen advantageous mutation can cause an increase of neutral or even disadvantageous alleles at linked loci, such as mutator alleles. This may explain cases in which the mutator allele *mutT* increases in

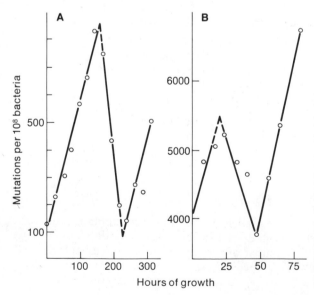

11

Periodic selection in *E. coli.* The number of mutant cells resistant to phage becomes greater in a strain carrying a mutator allele (B) than in a wild-type strain (A). In both, however, the proportion of mutant cells fluctuates drastically as clones carrying advantageous mutations, but not the allele for phage resistance, displace the phage-resistant genotypes. These cultures are not grown with phage; samples from the culture are tested separately for phage resistance. (From Nestmann and Hill 1973)

cultures of *E. coli* (Gibson, Scheppe, and Cox 1970; Nestmann and Hill 1973).

This mechanism cannot operate in a sexual population, though, because the mutator gene is often associated with some of the many deleterious mutations it has caused and dissociated by recombination from its occasional advantageous production. Thus there is a conflict; a certain amount of mutation is certainly "good" for a sexual species, yet the species cannot evolve a higher mutation rate from a lower one because mutations are usually harmful to the individuals that carry them. This is one of many instances in which the species in the long run does not profit from genetic changes caused by natural selection.

THE ADAPTIVE VALUE OF NEW MUTATIONS

Are most mutations harmful? The great majority of the classical mutations in *Drosophila,* such as *white* eye and *cut* wing, are deleterious, and in caged populations they are rapidly replaced by wild-type alleles (see, e.g., Ludwin 1951, Rendel 1951). But these alleles cause gross alterations of the phenotype and are perhaps the very kinds of mutation we most expect to be harmful.

It seems likely that at least some mutations have neutral effects. For example, Yamazaki (1971) carefully examined the fitness of flies carrying two allelic forms of an esterase and was unable to find any difference in viability, developmental rate, or fecundity; the gene frequencies in several experimental populations did not change for two years, suggesting that neither allele had an advantage over the other. If these alleles differed in their effect on fitness, it was too slight a difference to be discerned.

Whether an allele is detrimental, neutral, or salubrious may depend on the environmental circumstances. Even the most seemingly harmful mutations can be advantageous in certain environments. The ability to synthesize essential amino acids is a most important adaptation, yet Zamenhof and Eichhorn (1967) showed that *E. coli* mutants incapable of such syntheses are competitively superior to wild-type bacteria if these amino acids are supplied to them. For human populations in a malaria-free environment, hemoglobin *S,* which in heterozygous form causes sickle-cell anemia and in homozygous form causes sickling disease, is unquestionably harmful. Who, *a priori,* would have supposed that such a disease could confer protection against malaria?

Moreover the genetic background in which a mutation arises may well determine its fate. The simplest instance is that of a genetic priority effect; a new, possibly beneficial mutation is not advantageous if its function is already served by an established gene. Thus, for instance, several different hemoglobin mutants reduce human susceptibility to malaria, and certain of these have complementary distributions among the human populations of the Old World tropics (Livingstone 1964); none has increased within the range of the others.

One of the most important aspects of the genetic environment seems to be the total degree of heterozygosity of the genome. The remarkable experiments of Bruce Wallace (1958; also Mukai, Yoshi-kawa, and Sano 1966) illustrate this effect.

Wallace isolated a single wild-type chromosome (+) from a highly homozygous stock of *D. melanogaster* by the crossing techniques used in such work (*Chapter 9*) and set up the mating

$$\frac{Pm}{+} \times \frac{Cy\ L}{+}$$

which yields genotypes

$$\frac{Cy\ L}{Pm} \quad \frac{Cy\ L}{+} \quad \frac{Pm}{+} \quad \frac{+}{+}$$

The *CyL* and *Pm* chromosomes carry slightly deleterious dominant markers and inversions that prevent crossing-over. The phenotypes appear in the ratio $1:a:b:c,$ so the value of c measures the relative viability of the $+/+$ flies.

The crux of Wallace's experiment was to expose some of the males bearing the + chromosome to X rays and thus induce a small number of new mutations. Let an irradiated wild-type chromosome be denoted by $+^*$; then the cross

$$\frac{Pm}{+^*} \times \frac{Cy\ L}{+}$$

yields the genotypes

$$\frac{Cy\ L}{Pm} \quad \frac{Cy\ L}{+^*} \quad \frac{Pm}{+} \quad \frac{+}{+^*}$$

in the ratios $1:a':b:c',$ where c' measures the viability of flies heterozygous for random new mutations. Because $+/+^*$ and $+/+$ flies were genetically identical except for their new mutations, the relative values of c' and c indicate whether random mutations affect viability in heterozygous condition.

To almost everyone's surprise, Wallace found that on average c' was greater than c; that is, $+/+^*$ flies had a higher viability than $+/+$! Thus although the radiation may induce both harmful and beneficial mutations, the average effect of a new mutation is an improvement of viability *if* the fly is heterozygous for the mutation and *if* it is predominantly homozygous throughout the remainder of its genome. In later experiments (Wallace 1963) the new mutations were induced in a more outbred, heterozygous background genotype, and under these conditions the advantage of the new mutations was no longer evident. A mutation that is advantageous to an "unhealthy," highly homozygous fly has little effect on a "healthy," highly heterozygous individual.

It is likely, then, that few new mutations will enhance adaptation in outbred populations that inhabit environments to which they have long since become adapted, but that they will more often be advantageous in novel environments or in inbred populations. Perhaps it is meaningless to ask whether most mutations have positive, neutral, or negative effects without specifying the context in which they occur.

THE RANDOMNESS OF MUTATIONS

Mutations occur at random. This does not mean that all conceivable mutations are equally likely, nor that they are completely independent of the environment; after all, mutation rates are affected by mutagens. Mutation is random in that *the chance that a specific mutation will occur is not affected by how useful that mutation would be.* We would profit from a gene that rendered us less susceptible to the effects of air pollution, but such a mutation is no more likely to arise now than it was a thousand years ago. As Dobzhansky (1970) says, "It may seem a deplorable imperfection of nature that mutability is not restricted to changes that enhance the adaptedness of their carriers. However, only a vitalist Pangloss could imagine that the genes know how and when it is good for them to mutate."

Resistance to a source of mortality, such as a toxin, is based on alleles that are already in the population or come about through mutation, irrespective of the presence of the factor. Two elegant experiments illustrate this. Luria and Delbrück (1943) used the technique of replica plating in bacteria to show that mutations for resistance to an

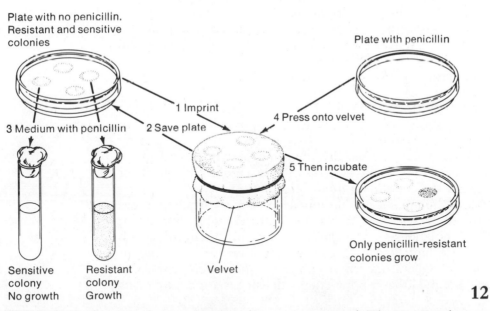

The method of replica plating, showing that mutations for penicillin resistance arise spontaneously and are not induced by penicillin. In step 3 the number of resistant colonies from a plate without penicillin is scored; in step 5 the number of colonies capable of growth when exposed to penicillin is measured. The two numbers are the same; thus penicillin (in step 5) does not induce resistant mutations. (Modified from *General genetics,* 2nd ed., by Adrian M. Srb, Ray D. Owen, and Robert S. Edgar. W. H. Freeman and Company. Copyright © 1965)

12

antibiotic occur independently of exposure to the drug and are not induced by it (*Figure 12*). In a conceptually similar experiment Bennett (1960) showed that the evolution of DDT resistance in *Drosophila* was due to mutations already present in the population. He demonstrated this by breeding from the siblings of the most resistant flies in each generation; but the breeding flies had not been exposed to DDT. After 15 generations the stock was highly resistant to DDT even though none of its ancestors had ever encountered DDT. Thus the genes for DDT resistance were present in the population, irrespective of the population's experience of DDT.

As far as we know, then, the environment does not evoke the appearance of favorable mutations; nor can a population accumulate mutations in anticipation of a change in its environment just because they might some day be useful.

THE IMPORTANCE OF NEW MUTATIONS

How important is the process of mutation? If there were no mutation, there would be no evolution; but the question is more meaningful if we ask whether mutation rates, or the kinds of mutations that occur, commonly limit the rate or direction of evolution. This is not the same as asking whether the amount and kind of genetic variation in a population, only a small fraction of which is the product of immediate mutational events, limits the rate and direction of evolution. The question is instead whether populations that already contain so much genetic variation would evolve any faster if mutation rates were higher.

If the response of populations to new selective pressures depended entirely on normal rates of spontaneous mutation, they would adapt very slowly. Clayton and Robertson (1955) subjected highly inbred (and hence homozygous) lines of *Drosophila* to selection for increased and for decreased bristle number for 17 generations. They observed such slight changes that they estimated that spontaneous mutation gives rise in each generation to about $1/1000$ of the amount of genetic variation in this characteristic usually observed in wild populations. Thus any immediate response to selection for change in this character hardly depends on the continuing input of mutations.

In a few cases, though, increasing mutation rates by irradiation has improved responses to selection. Ayala (1966), for example, found that large, outbred populations of *Drosophila serrata* and *D. birchii,* which presumably carried high levels of genetic variation, attained higher density in a novel laboratory environment if they were irradiated (*Figure 13*). He assumed that a population's density is a measure of its adaptedness.

There must certainly be natural instances as well in which the genetic variation in a population is inadequate for the demands of the environment, and the population must await the right mutations. Kettlewell (1973) notes that while some species of moths have evolved melanic forms adapted to the soot-blackened British landscape, other species have not; in some species no aberrant melanistic specimens have ever been captured. On the other hand, studies of artificial selection indicate that populations have enough genetic variation to respond to almost any new selection pressure; they do not seem to have to wait for new mutations. How often a population has taken one evolutionary path rather than another, or has failed to adapt to new ecological opportunities, or has become extinct, because the right mutations did not occur at the right time, is one of the weightier questions in evolutionary biology, and one that will not soon be answered.

THE SUFFICIENCY OF MUTATION

We may ask, finally, whether mutation provides the kinds of variations that, preserved by natural selection, are the foundations of organic diversity.

The phenotypic effects of mutations vary enormously. Individual mutations may have only slight effects on a character such as bristle number, as in Clayton and Robertson's work, or they may have the drastic effects typical of the many mutants studied in *Drosophila*. A single mutation such as *singed* changes the shape of all the bristles, *Curly* the shape of the wing. Homeotic mutants, by acting on organs with fundamentally similar developmental patterns, convert one into another: the antenna into a leglike structure (*aristapedia*), the metathorax into a mesothorax (*bithorax*). These drastic mutations so disrupt development that they may never have been, in themselves, the material of adaptive evolutionary change; but they demonstrate that profound phenotypic changes of the kind that distinguish genera or families can in principle be caused by single mutations. Still, the work of such people as Clausen, Keck, and Hiesey (1940) demonstrates that many

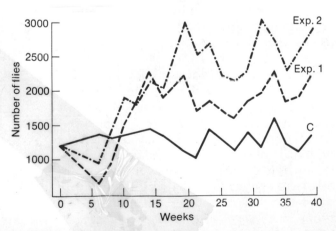

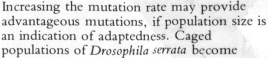

13

Increasing the mutation rate may provide advantageous mutations, if population size is an indication of adaptedness. Caged populations of *Drosophila serrata* become denser if irradiated in the early generations (Exp. 1, Exp. 2) than if not (C). (From Ayala 1968; copyright 1968 by the American Association for the Advancement of Science)

of the genetic differences between species are not single drastic mutations, but the summation of slight effects at many loci.

It is important to ask, though, whether true evolutionary novelties
actually arise by mutation. For example, can both a new enzyme and
the regulatory system that modulates its production arise by mutation?
Clarke (1974) describes several possible pathways whereby the ability
to metabolize a novel substrate may evolve. (1) A change in enzyme
structure may allow metabolism of a novel substrate that is chemically
similar to the normal substrate. (2) The new substrate may be unusable
simply because it cannot enter the cell; a mutation in a permease may
serve to transport it across the cell membrane. (3) Many enzymes are
subject to end-product inhibition, perhaps by an interaction between
the end product of a biochemical reaction and a "promoter" gene. If
the promoter gene mutates so that massive quantities of enzyme are
produced, an enzyme with only a slight affinity for the novel substrate
may metabolize enough of it to enhance growth. (4) Constitutive
mutations of operator genes (*Chapter 3*), resulting in permanent derepression of enzyme synthesis, have the same effect. Clarke cites cases
of mutational changes in the metabolic capacities of bacteria that appear
to represent each of these possibilities.

Several of these are illustrated by Clarke's work on the bacterium
Pseudomonas aeruginosa, in which the wild-type form of an amidase
metabolizes acetamide and propionamide, yielding sources of both
carbon and nitrogen. It metabolizes butyramide at very low efficiency,
but butyramide inhibits its synthesis, as do valeramide and phenylacetamide.

Clarke was able to obtain a mutant that was not repressed by
butyramide and could thus grow on it, as well as mutants with structurally different enzymes that metabolized butyramide more efficiently. From these strains, moreover, she selected further mutants
with different enzyme structures that were able to grow on valeramide
and phenylacetamide, a very different compound from the normal
substrate metabolized by this enzyme. She notes that species related to
P. aeruginosa vary in the amides they grow on. *P. cepacia,* for instance,
has two different amidases, one that metabolizes acetamide and another
that attacks valeramide and phenylacetamide. It is tempting, as Clarke
says, to suppose that a single amidase locus was duplicated in the past
and that the duplicate loci then diverged in their structure and function.

Such experiments with microorganisms show that drastic, yet
adaptive, changes in intricate biochemical pathways can arise by mutation. Most morphological changes in evolution, however, have probably entailed alterations not of enzyme structure, but of regulatory
genes that affect patterns and rates of development (*Chapters 8, 17*).
Many of the classical mutations in *Drosophila* — *Curly* wing or *scute*
bristles, for example — may well be mutations of regulatory genes,
but we are entirely ignorant of how such genes operate. We know
only that mutations in developmental patterns exist.

Do mutations happen often enough to account for evolution? A numerical example may suffice. The ratio of the paracone and the ectoloph of the third molar in horses almost doubled over a period of 16 million years; this is a fairly typical rate of evolution. Postulate a mutation that changes the ratio by only 1 per cent; the substitution of 100 such alleles would double the ratio. If only one such mutation were substituted at a time, the observed rate of evolution would be accountable by one substitution per 160,000 years, or one per 32,000 generations if we assume a generation time of five years. If, conservatively, the mutation rate at each locus is only 10^{-6} per gamete per generation, and the population size is only 10^5 individuals, each of the 100 mutants will arise 3200 times during each period of single gene substitution. Thus the new alleles are certain to be available when their time has come.

Of course, many evolutionary changes have been far more rapid; indeed, the evolution of horse teeth may have occurred in a few very rapid bursts, rather than as a continual gradual process, if Eldredge and Gould (1972) are correct (*Chapters 7, 16*). But these calculations were unrealistically conservative; the population size in most species is greater than 100,000: the mutations in such a population are carried into daughter populations in which they can increase rapidly in frequency (*Chapter 16*); mutations are not selected sequentially but in concert, since it is the total phenotype, not the individual genes, that is subject to natural selection. And, most important, populations have a large "standing crop" of genetic variation that has accumulated over past generations and is often at hand when environments change.

RECOMBINATION: THE AMPLIFICATION OF VARIATION

Genetic variation arises not only from mutation, but from recombination. In eukaryotes this results from two processes that are often but not always associated with each other. These are sexual reproduction and the formation of gametes genetically different from those that united to form the individual that produces them. The latter too entails two processes that are not inextricably associated: the independent segregation of nonhomologous chromosomes and crossing-over between homologous chromosomes.

While all these events contribute to recombination in organisms that we think typical (e.g., ourselves), there are many variations. Some uniparental organisms have meiosis, independent segregation, and crossing-over. For example, in a parthenogenetic race of the moth *Solenobia triquetrella* a normal meiosis occurs and the haploid egg begins to develop normally, but after four nuclei are formed by mitosis, they unite in pairs, restoring the diploid chromosome number. Some organisms have normal meiosis but no crossing-over, as do male *Drosophila* and many other Diptera. In males of the fungus gnat *Sciara* meiosis is bizarre; only the chromosomes that the male inherited from his mother are transmitted into the sperm cells, so segregation of chromosomes is anything but independent.

These variations in the meiotic machinery illustrate that recombi-

14 Comparison of the relative positions of various loci on the X chromosome of *Drosophila melanogaster,* based on the chromosome itself (C) and on recombination values among the loci (G). Recombination frequency is not directly proportional to the linear distance between the loci on the chromosome. (After *Principles of genetics,* by Sinnott, Dunn, and Dobzhansky 1958.

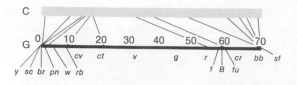

Copyright McGraw-Hill Book Company. Used with permission)

nation rates evolve. Even the frequency of crossing-over is subject to precise genetic control. The genetic map of *Drosophila,* constructed on the basis of crossover frequency between loci, differs substantially from the chromosome map, constructed by determining which of the visible bands carry these same loci (*Figure 14*). Chinnici (1971) and others have shown that the crossover frequency between a pair of loci is under genetic control and can be altered by artificial selection (*Figure 15*) without affecting the frequency of crossing-over elsewhere on the chromosome.

THE SIGNIFICANCE OF RECOMBINATION

Recombination results in new combinations of genes that may confer properties transcending those of either parent. If, for instance, + and − alleles at each of several loci affect body size, two parents with genotypes +−−+−/−+−−+ and −−+−+/+−+−− are much the same size, but they can produce offspring with genotypes +++++/+++++ and −−−−−/−−−−−, spanning the range of possible body size. Thus variation can arise at a far higher rate than it could by mutation alone.

But these extreme homozygous genotypes arise only infrequently. For example, if these five loci segregate independently, the probability is only $1/64$ that an offspring from this mating will have the genotype +++++/+++++. Most individuals in a population have a mix of + and − alleles and are intermediate in body size. Latent variation for extreme phenotypes exists, but these phenotypes are rare, unless alleles with like effects are bound together into nonrandom sets. An inversion, for example, could hold the + alleles at each of several loci in linkage disequilibrium.

Dobzhansky and his colleagues have extensively studied the effect of recombination on variation in viability (reviewed by Lewontin 1974). Single chromosomes isolated from a wild population (*Figure 16, Chapter 9*) vary greatly in the viability they confer in homozygous condition. Heterozygotes formed by mating these homozygotes at random are less variable in viability (*Figure 17, Chapter 9*). Crossing-over in these heterozygotes produces recombinant chromosomes that can be made homozygous (*Figure 18, Chapter 9*). The viability of such homozygotes varies greatly, even when the chromosomes from which they are derived have similar, normal viability. Thus superficially similar wild-type chromosomes carry different genes, which by interacting with one another in new combinations have different effects. Recombination among only 10 "normal" chromosomes can give rise to as much as 75 per cent of the total variance in viability displayed by the homozygotes for a much larger sample of chromosomes extracted from a wild population. This implies that a population founded by just a few individuals can have almost as much genetic variation as the entire population from which they came.

THE IMPACT OF RECOMBINATION ON EVOLUTION

One way of assessing the role of recombination is to see what effect a restriction on recombination has on a population's response to selection. Carson (1959) did this by selecting for motility in populations of *Drosophila robusta* from two localities. The strains from Nebraska entirely lacked chromosome inversions, while those from Missouri were highly polymorphic for inversions in all their chromosomes and so must have been less capable of crossing-over. The chromosomally polymorphic populations showed less response to selection than the monomorphic ones, suggesting an important effect of recombination.

Carson's work is consonant with the long-standing dogma that recombination, especially sexual reproduction, enables a population to adapt rapidly to changing environmental conditions by bringing together advantageous alleles at different loci (*Figure 16*). Asexual orga-

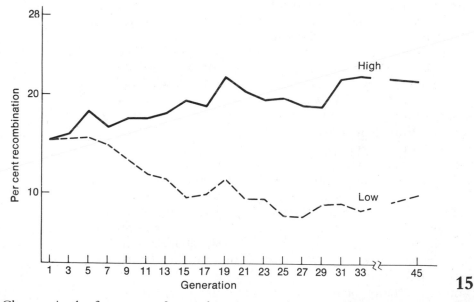

15

Changes in the frequency of recombination between the loci *sc* and *cv* in laboratory stocks of *Drosophila melanogaster* selected for high and low recombination between these loci. (After Chinnici 1971)

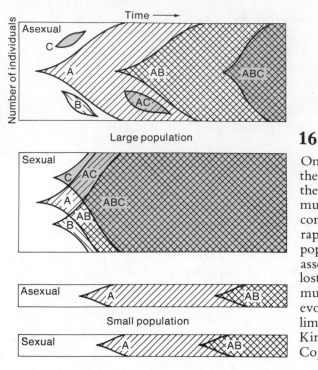

16

One view of the effect of recombination on the rate of evolution. If evolutionary rates and the precision of adaptation depend on new mutations (A, B, C) that are advantageous in concert, then adaptedness is achieved more rapidly in a sexual than in an asexual population because the mutations are assembled by recombination. This effect is lost in small populations, however, because mutations occur so rarely. This view of evolution assumes that favorable mutations limit the rate of evolution. (From Crow and Kimura 1965, *The American Naturalist,* Copyright The University of Chicago Press)

nisms lack this capacity because mutations at different loci occur together only when one mutation arises in a clone descended from an individual that previously experienced the other. Thus, as Crow and Kimura (1965) have argued on mathematical grounds, evolution should be faster in sexual than in asexual populations.

But recombination may not accelerate evolution that much. The problem is that while recombination may give rise to wonderfully advantageous genotypes, it breaks them apart just as quickly (Eshel and Feldman 1970, Felsenstein 1974, Williams 1975). Only if a newly arisen advantageous combination can be held intact by some restriction on recombination can the advantageous combination be perpetuated and increased. Thus most genotypes in a sexual population are inferior to the best possible genotype, which only rarely (if ever) appears and then immediately vanishes. Thus Williams (1975) argues that sex reduces the level of a population's adaptation to the environment and can actually *retard* the rate of evolution (*Figure 17*). Felsenstein (1974), however, presents an intricate model from which he concludes that in a changing environment, the accelerating effect of recombination generally outweighs its retarding effect; the rate of evolution, he claims, is enhanced by recombination.

Recombination can certainly be instrumental in saving a population from extinction when the environment changes drastically. An asexual population may be made up of a few genotypes superbly adapted to prevailing conditions, but a sexual population includes rare recombinant genotypes that can survive in the face of new adversity. Under this hypothesis the extinction rate of sexual forms is less than that of their asexual counterparts, accounting for the prevalence of sex among the eukaryotes. In support of this proposition many authors have pointed out that asexual reproduction is only sporadically distributed among higher taxa of plants and animals and that most asexual "species" have "all the earmarks of recency" (Mayr 1963). Many asexual forms are extremely similar to the sexual species from which they are thought to have been derived and often still have all the accoutrements of sex — as in the dandelion (*Taraxacum officinale*), which is asexual but has a fully developed inflorescence that attracts pollinators.

The effect of recombination on the rate of evolution is thus open to debate; indeed its effect on preserving species from extinction has also been questioned (Williams 1975). Sex, in fact, is a thoroughly controversial subject, even in evolutionary biology. There are good reasons for arguing that it should never have evolved at all, an argument that I defer to Chapter 12.

EXTERNAL SOURCES OF VARIATION

Variation arises within a population by mutation and recombination, but it may also be imported from other, genetically different populations. Utterly novel combinations of genes may stem from the transfer

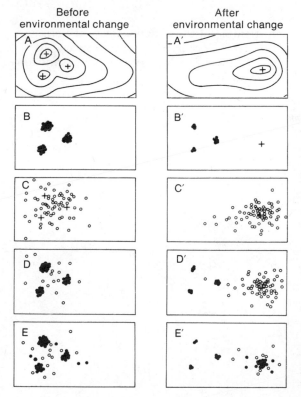

Before environmental change | After environmental change

17

The effect of sex on the adaptedness of a population. (A) and (A') Levels of fitness of different multilocus genotypes, the most highly adapted of which are marked by + (*Figure 8, Chapter 13*). A new genotype is favored after the environment changes (A'). Asexual genotypes are denoted by solid circles, sexual genotypes by clear circles. A population of asexual clones consists exclusively of highly adapted genotypes (B), whereas most sexual genotypes have lower fitness (C), especially if they are competing with more highly adapted clones (D). After an environmental change, however, the fitness of an asexual population is far lower (B') than that of a sexual population (C', D'). Eventually (E, E') new clones may arise at the new adaptive peak. (From Williams 1975)

of genetic material from one species to another. Such transfers may play an important role in the evolution of bacteria, where bacterio-phages and plasmids carry bits of chromosome from one bacterial genus to another; and there is even evidence, from the similarity of protein structure, that bacteria may sometimes derive genes from their mammalian hosts (Reanney 1976).

Interspecific hybridization among the higher plants and animals can spawn variation that transcends that of either parent species (*Figure 18*). Some species seem to make a habit of hybridizing with others. From morphological evidence, Harlan and deWet (1963) concluded that the grass *Bothriochloa intermedia* has incorporated genes from *B. ischaemum* in Pakistan, from *Dichanthium annulatum* in Pakistan and India, from *Capillipedium parviflorum* in northern Australia, and from *B. insculpta* in East Africa.

Because hybrids seem often to be adapted to disturbed environ-ments, hybridization may sometimes provide enough genetic variation to invade environments that would otherwise be closed to the species. Laboratory populations of hybrids between the fruit flies *Dacus tryoni* and *D. neohumeralis* were better able to adapt to high temperatures than were populations of either "pure" species, leading Lewontin and Birch (1966) to postulate that the hybridization known to occur nat-urally between these species might have been responsible for the ex-

Viola pedatifida   *Viola sagittata*

18

An example of the variation that may arise from hybridization between two species of violets. (After Stebbins 1950, with permission of the publisher)

F_2 hybrids

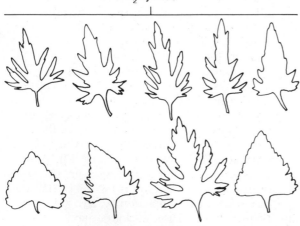

pansion of the species' range over the last hundred years.

Allopolyploid plants are most prevalent in stringent habitats (Stebbins 1970) and abundantly demonstrate the evolutionary success that hybrid genotypes may enjoy. Such hybrid genotypes, moreover, may prove to be the founding stocks of phyletic lines that ultimately achieve great taxonomic distinction. Most of the families and other higher taxa of plants have large chromosome numbers that are thought to have originated by polyploidy. For example, although most members of the tribe Phaseolae (beans) have gametic numbers of 10 or 11, the species of *Erythrina,* a genus of tropical trees in this tribe, have gametic numbers of 21 or 42. Hybridization between two species of Phaseolae, one with 10 and one with 11 chromosomes in the haploid set, could have produced a tetraploid (with 21 gametic chromosomes) that later developed the distinctive features of the genus *Erythrina.* A later allopolyploid between two *Erythrina* species would again double the chromosome number to 42. (Note that saying that a group of plants that we now call a genus or family originated from an allopolyploid is *not* the same as saying that allopolyploidy causes the sudden appearance of a distinctive new type. Polyploidy does not create new genera.)

Hybridization between species is but a special case of the admixture of genes from genetically differentiated populations. Far more common than interspecific hybridization is the simple passage of genes from one population to another of a single species. Such GENE FLOW or GENE MIGRATION may well be the most important source of genetic variation for a population — and, paradoxically, a most important barrier to adaptation. But this brings up a complicated new topic, the pattern of mating within a species.

SUMMARY

Within a population the frequencies of alleles remain constant if the population is very large and if there is no mutation, gene flow from other populations, or selection. Moreover, if mating occurs at random, the frequency of each genotype remains fixed at a value specified by the binomial theorem. This is the Hardy-Weinberg law on which all of evolutionary genetics rests.

Among the forces that change gene frequencies are mutation, which in itself changes the genetic composition of a population only very slowly, since mutation rates are low. But it is the ultimate source of all variation. Both changes in the DNA sequence of single genes and alterations in the number and form of the chromosomes have significant evolutionary consequences.

Most mutations seem to be harmful, but a good many are beneficial, especially if the population is in a stressful environment to which it is not fully adapted. There is no evidence, though, that the environment ever induces the occurrence of the specific mutations that would be adaptive in that environment. It is generally thought that beneficial mutations only slightly modify the phenotype. At least some mutations are clearly the stuff of true evolutionary novelty.

Within populations, variation that arose by mutation is recombined into a great spectrum of genotypic variants. The formation of new combinations of old genes is a major proximal source of variation. Some of these combinations owe their origin to gene flow from other populations of the same or other species. Recombination may in some instances accelerate and in other cases retard the rate of evolution.

FOR DISCUSSION AND THOUGHT

1 One generation of random mating establishes the genotype frequencies p^2, $2pq$, q^2, no matter what the genotype frequencies were previously. Postulate long-term changes in genetic composition that would take many generations to build up and might then be annihilated by one generation of random mating.

2 People often object, "If the Hardy-Weinberg theorem only holds when there is no mutation, gene flow, or selection, and only in a randomly mating population, what good is it? There are no gene loci that really meet these assumptions." How do you answer this objection?

3 Evaluate the pros and cons of constructing a mathematical model of genetic changes in populations rather than a simple verbal description. Apply your evaluation either to the model of linkage disequilibrium between loci or to the effects of mutation rates on gene frequencies.

4 Suppose you find that in a wild population of beetles, black-winged specimens usually (but not always) have yellow tarsi, while brown-winged specimens usually have brown tarsi. What can you say about the genetic basis for these traits?

5 Discuss the validity of viewing mutations as "mistakes" in gene replication, rather than adaptive events that are programmed into the genes.

6 How would our understanding of evolutionary changes in the number and form of chromosomes be improved if we knew, at a molecular level, what a centromere is and what causes chromosomes to pair in meiosis?

7 Geneticists have long realized that polyploids cannot be classified simply as autopolyploids or allopolyploids. Why not?

8 Would you expect similar gene loci, arisen through duplication and subsequent divergence, often to be closely linked? How could you account for their not being linked?

9 Discuss the implications that the widespread existence of isozymes (not allozymes) might have for the hypothesis that heterozygotes are more fit than homozygotes.

10 Several kinds of chromosome aberrations, such as inversions and translocations, usually cause the heterozygote to have diminished fertility. But they are heterozygous when they first appear and so should never increase in frequency. Nevertheless, populations and species differ from one another in chromosome sequence. How can you account for this paradox?

11 Compare the definition of *stability* in this chapter to the usual uses of this word. Of what value is the concept of a stable equilibrium? How applicable is it to a natural population if it takes more than 70,000 generations to arrive at the equilibrium point?

12 Discuss the differences in the consequences of an increased mutation rate for a population of flies vs. a human population.

13 In light of the material in Chapter 8, discuss whether mutations with miniscule phenotypic effects, rather than large effects, are most likely to be adaptive. May Goldschmidt's idea of the "hopeful monster" have been right?

14 Do the experiments by Luria and Delbrück and by Bennett really prove that an environmental change does not induce mutations specifically

adapted to that environment? Evolutionists almost universally reject the Lamarckian notion that the environment can induce the appearance of adaptive mutations. On what evidence is this conviction based?

15 Consider some of the deviations from "normal" meiosis: holocentric chromosomes, the lack of crossing-over in male *Drosophila,* the peculiarities of *Solenobia* and *Sciara.* What advantages could each of these systems have? How could they have evolved?

MAJOR REFERENCES

Mettler, L. E., and T. G. Gregg. 1969. *Population genetics and evolution.* Prentice-Hall, Englewood Cliffs, N. J. 212 pages. An introduction to population genetics.

Wilson, E. O., and W. H. Bossert. 1971. *A primer of population biology.* Sinauer, Sunderland, Mass. 192 pages. A clear, simple introduction to the elements of basic population genetic theory.

Spiess, E. B. 1977. *Genes in populations.* Wiley, New York. 799 pages. A comprehensive treatment of the theory and data of population genetics.

Stebbins, G. L. 1971. *Chromosomal evolution in higher plants.* Arnold Press, London. 208 pages. A short but comprehensive treatment of chromosome changes.

The Hardy-Weinberg theorem assumes that all the individuals in an infinite population mate at random. But in real life populations are finite in size, and not all the individuals of a species have an equal chance of mating with one another. The breeding pattern of the species may be structured in any of several ways that have important consequences for evolution.

Mating can be nonrandom in two major ways. Individuals may preferentially mate with one another or avoid each other on the basis of one or more phenotypic characteristics. Some birds, for example, prefer others of the same coloration as mates. This is ASSORTATIVE MATING, a topic treated at the end of this chapter and again in Chapter 16, which discusses the origin of new species.

The other major deviation from randomness occurs when mating is more likely among related than among unrelated individuals. One major reason for this pattern is that the capacity for dispersal is quite limited in many species. Hence mates are often born near each other and are likely to be related. A PANMICTIC species, one in which all members over the entire geographic range of the population are equally likely to mate, is a rarity in nature. Most species are fragmented into local populations, which may or may not be discrete (*Figure 1*) and among which there may be much or little dispersal of individuals, resulting in much or little gene flow. If there is very little dispersal, each individual can mate only with its relatives. This form of inbreeding is a convenient concept with which to begin the analysis of population structure.

THE INBREEDING COEFFICIENT

The topics that follow are treated by Wright (1921, 1931, 1969), who developed much of the theory, and by Crow and Kimura (1970), who provide exceptionally comprehensible derivations.

In Figure 2*a* a brother and sister, the offspring of unrelated parents, have a

CHAPTER 11

Population Structure

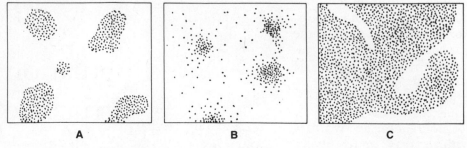

1 Some patterns of distribution. (A) Discrete populations, corresponding to the "island model." (B) Perhaps the most common pattern, ill-defined populations between which density is low. (C) A more or less uniform distribution, corresponding to the "isolation by distance" model, except that patches of unfavorable habitat are acknowledged.

daughter I. Although both grandparents carry allele A_1, both of I's A_1 genes are descended from her grandfather's A_1 gene rather than from her grandmother's. Thus I's two A_1 genes are IDENTICAL BY DESCENT. The daughter I is homozygous at this locus, but it is possible to be homozygous (as is I's mother) and yet have genes that are not identical by descent. (Notice a problem that I shall ignore: the two A_1 genes in I's mother are identical by descent if they stem from the first A_1 allele that arose by mutation.)

An individual with two alleles identical by descent is AUTOZYGOUS at that locus; otherwise it is ALLOZYGOUS. The INBREEDING COEFFICIENT f of an individual I is defined as the probability that I is autozygous. The daughter I is autozygous if she has the genotype A_1A_1, A_2A_2, $A_1'A_1'$, or A_3A_3. She is A_1A_1 only if (a) her grandfather passed his A_1 gene to her father ($p = \frac{1}{2}$) and (b) her father in turn passed it on to her ($p = \frac{1}{2}$) and (c) her mother also inherited the A_1 gene from I's grandfather ($p = \frac{1}{2}$) and (d) her mother in turn passed it on to I ($p = \frac{1}{2}$). Thus the probability that I is autozygous for A_1 is $(\frac{1}{2})^4 = \frac{1}{16}$. But she may similarly be autozygous for any of the other three genes, so

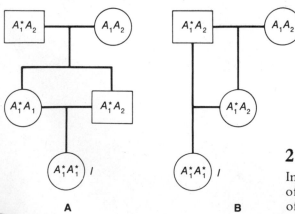

2 Inbreeding due to (A) sib mating, (B) parent-offspring mating. The inbreeding coefficient of I is $\frac{1}{4}$ in both cases.

the probability that she is autozygous for any of the four genes is $4(^1/_{16}) = \frac{1}{4}$. This, then, is the inbreeding coefficient of an individual derived from the incestuous union of brother and sister. And since the probability that two mates will have an autozygous offspring is a measure of their relationship, ¼ is also the value of the COEFFICIENT OF CONSANGUINITY of two siblings.

Similarly the inbreeding coefficient of the child of a coupling between father and daughter or mother and son (*Figure 2b*) is also ¼. Self-fertilization constitutes more intense inbreeding, for if a plant with the genotype A_1A_1' fertilizes itself, the probability is ¼ that the offspring will be A_1A_1 and the same for $A_1'A_1'$, so $f = \frac{1}{2}$.

If some matings in a population are consanguineous, the population has an inbreeding coefficient F, defined as the probability that an average individual is autozygous. In a very large population with completely random mating, the probability that relatives will mate approaches zero, as does F; conversely $F = 1$ if every individual is autozygous.

THE EFFECTS OF INBREEDING ON A POPULATION

A population inbred to some extent F has two kinds of homozygotes. Of the proportion $1 - F$ of the population that is allozygous, p^2 have the genotype AA. Of the fraction F that is autozygous, and hence necessarily homozygous, the probability that an individual carries allele A is p. Hence the total frequency of AA is $p^2(1 - F) + pF$. Since $1 - p = q$, this expression becomes $p^2(1 - F) + pF = p^2 + F(p - p^2) = p^2 + Fpq$. Similarly the frequency of $A'A'$ is $q^2(1 - F) + qF$, or $q^2 + Fpq$. Thus in an inbred population, both homozygotes are more prevalent than in a randomly mating population, in which their frequencies are p^2 and q^2 (*Figure 3*).

Conversely a fraction $2pq$ of the allozygous moiety $1 - F$ is heterozygous. Thus in an inbred population the frequency of heterozygotes H_F is $2pq(1 - F)$. So inbreeding decreases the proportion of heterozygotes and increases both homozygous classes. Inbreeding

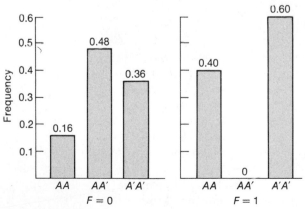

3

Genotype frequencies in a population with allele frequencies $p = 0.4$ and $q = 0.6$, when mating is random ($F = 0$) and when completely inbred ($F = 1$).

4 Decrease in heterozygosity due to (A) inbreeding systems of mating and (B) finite population size. In (A) systems of mating are (curve A) exclusive self-fertilization, (curve B) sib mating, (curve C) double first-cousin mating. In (B) N equals population size. (A from Crow and Kimura 1970, B from Strickberger 1968)

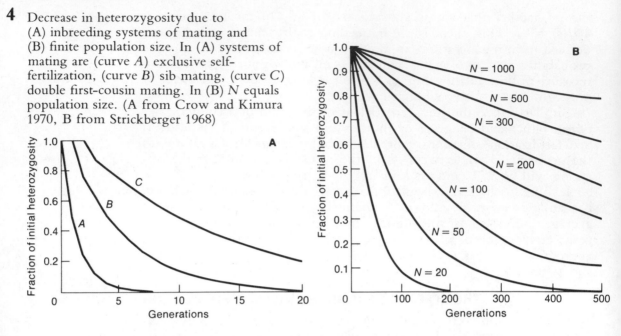

changes the genotype frequencies, but not the gene frequencies. The frequency of A, for example, is $p^2 + Fpq + \frac{1}{2}(2pq)(1 - F) = p^2 + Fpq + pq - Fpq = p(p + q) = p$, just as in a randomly mating population.

It is possible to estimate F by comparing the expected Hardy-Weinberg proportion of heterozygotes $H_E = 2pq$ with the observed proportion of heterozygotes H_F. Since $H_F = 2pq(1 - F)$, or $H_E(1 - F)$, we have $F = (H_E - H_F)/H_E$; it is the proportion by which heterozygotes are lower in frequency than they would be in a randomly mating population. For example, Selander (1970) found that among the house mice (*Mus musculus*) in a barn in Texas, the genotypes AA, AA', and A'A' at an esterase locus had the frequencies 0.226, 0.400, and 0.374 respectively. Since $p = 0.226 + 0.200 = 0.426$, the Hardy-Weinberg proportions should be 0.181, 0.489, and 0.329. The heterozygotes were less prevalent than expected, and $F = (0.489 - 0.400)/0.489 = 0.18$.

This implies that the population consists to some extent of family groups. We may deduce that mice tend not to wander far from their birthplace before reproducing or are prevented from mating except with their relatives. Indeed behavioral studies indicate that house mice form small groups that repel outsiders, especially males.

As long as the pattern of mating remains constant, F is likely to increase toward 1. This is most clearly seen by considering the most extreme form of inbreeding, exclusive self-fertilization. Only half of a heterozygote's offspring are heterozygous, so in each generation the frequency of heterozygotes is halved. If mating entails other kinds of

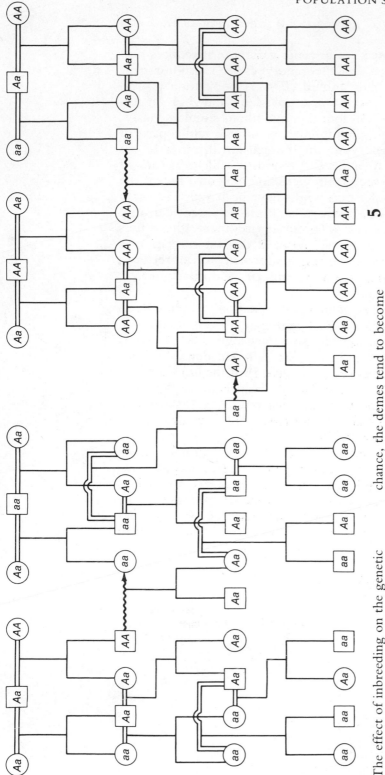

5

The effect of inbreeding on the genetic composition of a population composed of four inbred demes among which there is only occasional gene exchange (indicated by wavy lines). Most matings (denoted by double lines) are between sibs or half-sibs. By chance, the demes tend to become monomorphic for different alleles. In this example each deme can have up to one male (square) and two females (circle) at the time of reproduction.

consanguineous unions, then H approaches 0 and F approaches 1 at a lesser rate (*Figure 4*).

Inbreeding does not in itself erode genic variation; it reduces heterozygosity. Variation persists, but the genotypes are more homozygous than heterozygous. A fully inbred ($F = 1$) population has no heterozygotes but consists of the homozygotes AA and $A'A'$ in frequencies p and q respectively. All loci become inbred simultaneously. If k loci each have two alleles, the inbred population could consist of 2^k different homozygous genotypes. In the self-fertilizing wild oat *Avena barbata,* Hamrick and Allard (1972) found two alleles at each of five enzyme loci. Almost all the plants were homozygous, but of the $2^5 = 32$ homozygous genotypes that might have been expected, only two were common. Hamrick and Allard took this to mean that only these genotypes were well adapted to the environment, the others having been eliminated by natural selection. If this interpretation is correct, *Avena* exemplifies an interesting consequence of inbreeding: advantageous homozygous gene combinations are held intact and are not broken up by recombination.

Inbreeding actually increases phenotypic variation. A population composed of inbred family groups is an aggregate of subpopulations, or DEMES, among which there is little gene exchange. Each deme is becoming more and more homozygous, but different demes are becoming homozygous for different alleles (*Figure 5*). If the homozygous genotypes are phenotypically very different, the genetic and phenotypic variation *within* each subpopulation is declining, but the variation in the entire population is actually augmented by the increased differentiation *among* subpopulations (*Box A*).

Box A also demonstrates that if one allele is dominant or if the heterozygote has a more extreme phenotype than either homozygote (is OVERDOMINANT), the mean phenotype will decline as recessive alleles are revealed. One of the consequences is INBREEDING DEPRESSION: the decline in such characteristics as fecundity (or yield in corn and other crops) and viability as a strain becomes inbred (*Table I*). In human

TABLE I. Inbreeding depression in rats

Year	Nonproductive matings (per cent)	Average litter size	Mortality from birth to 4 weeks (per cent)
1887	0	7.50	3.9
1888	2.6	7.14	4.4
1889	5.6	7.71	5.0
1890	17.4	6.58	8.7
1891	50.0	4.58	36.4
1892	41.2	3.20	45.5

(After Lerner 1954. Data from Ritzema Bos)
NOTE: The years 1887–1892 span about 30 generations of parent × offspring and sib matings.

A The Effect of Inbreeding on a Metric (Gradually Variable) Character

Assume that the genotype AA increases some character such as height by an amount a over the background level; that $A'A'$ decreases height by an amount $-a$; and that AA' increases height by an amount d. If $d = a$, the allele A is dominant; if $d > a$, there is overdominance; if $d = 0$, there is no dominance, since the heterozygote is exactly intermediate between the homozygotes. Thus we have

	AA	AA'	$A'A'$
Phenotype	a	d	$-a$
Frequency	$p^2(1 - F) + pF$	$2pq(1 - F)$	$q^2(1 - F) + qF$.

If $F = 0$, the mean $\bar{x}_0$ equals $p^2a + 2pqd - q^2a$. In a population inbred to the degree F, the mean $\bar{x}_F$ equals

$$a[p^2(1 - F) + pF] + d[2pq(1 - F)] - a[q^2(1 - F) + qF]$$
$$= ap^2 - ap^2F + apF + 2pqd - 2pqFd - aq^2 + aFq^2 - aqF$$
$$= (p^2a + 2pqd - q^2a) + apqF - apqF - 2pqdF$$
$$= \bar{x}_0 - 2pqdF.$$

Thus if there is no dominance ($d = 0$), inbreeding does not change the mean. If A is dominant or overdominant ($d > 0$), the mean declines linearly with the degree of inbreeding (this is called inbreeding depression).

Now consider the effect of inbreeding on the variance (*Appendix* I), making the simplifying assumption that $d = 0$.

The variance V_F is the sum of the squared deviations of the phenotypes from the mean, weighted by the frequency of each phenotype, or

$$V_F = [p^2(1 - F) + pF](a - \bar{x}_F)^2 + [2pq(1 - F)](0 - \bar{x}_F)^2 + [q^2(1 - F) + qF](-a - \bar{x}_F)^2$$
$$= p^2(a - \bar{x}_F)^2 - Fp^2(a - \bar{x}_F)^2 + Fp(a - \bar{x}_F)^2 + 2pq\bar{x}_F^2 - F2pq\bar{x}_F^2 + q^2(-a - \bar{x}_F)^2 - Fq^2(-a - \bar{x}_F)^2 + Fq(-a - \bar{x}_F)^2.$$

When $F = 0$, the variance V_0 is

$$V_0 = p^2(a - \bar{x}_0)^2 + 2pq(0 - \bar{x}_0)^2 + q^2(-a - \bar{x}_0)^2,$$

and in a totally inbred population ($F = 1$) the variance V_1 is

$$V_1 = p(a - \bar{x}_1)^2 + q(-a - \bar{x}_1)^2.$$

Since $d = 0$, we have $\bar{x}_0 = \bar{x}_1 = \bar{x}_F$, and then the equation for V_F can be written

$$p^2(a - \bar{x})^2(1 - F) + 2pq\bar{x}^2(1 - F) + q^2(-a - \bar{x})^2(1 - F) + pF(a - \bar{x})^2 + qF(-a - \bar{x})^2.$$

Thus by substitution $V_F = (1 - F)V_0 + FV_1$, so the total variance is decreased within $[(1 - F)V_0]$, and increased among (FV_1), lines.

We may simplify this by noting from Appendix I that the variance $V = \Sigma_i p_i(x_i - \bar{x})^2$ can also be written $V = \Sigma_i p_i x_i^2 - \bar{x}^2$. Then V_0 becomes $p^2a^2 + q^2(-a)^2 - \bar{x}^2$, and V_1 becomes $pa^2 + q(-a)^2 - \bar{x}^2$. Now note that $\bar{x}_0$ (hence also $\bar{x}_1$) is $p^2a + 2pq(0) - q^2a = a(p^2 - q^2) = a(p - q)$. Hence $\bar{x}^2 = a^2(p^2 - 2pq + q^2)$. Substituting this into the expressions for V_0 and V_1, we obtain $V_0 = 2pqa^2$ and $V_1 = 4pqa^2$. Hence $V_1 = 2V_0$: the variance in the inbred population is twice that in the randomly mating population. By substituting $V_1 = 2V_0$ in the expression $V_F = (1 - F)V_0 + FV_1$, we obtain $V_F = V_0(1 + F)$.

Thus, for additively acting genes, the phenotypic variance of the whole population increases linearly with the degree of inbreeding.

populations the deleterious consequences of inbreeding are well known; they include a higher incidence of mortality, mental retardation, albinism, and other physical abnormalities (see, for example, Schull and Neel 1965, Cavalli-Sforza and Bodmer 1971, Stern 1973). On the basis of studies of inbreeding Morton, Crow, and Muller (1956) estimated that each of us, on average, carries the equivalent of three to five lethal recessive genes.

The outbreeding mechanisms of many plants may have evolved to reduce the chance of producing inferior, highly homozygous offspring. Outbreeding, the reverse of inbreeding, increases heterozygosity, which in species that normally are not inbred is associated with higher fitness and/or greater physical robustness. These effects are often referred to as HETEROSIS, perhaps most outstandingly illustrated by hybrid corn varieties (*Figure 6*).

INBREEDING AND OUTBREEDING IN ANIMALS AND PLANTS

The incidence of inbreeding in natural populations varies greatly among species. Self-fertilization is unusual in animals, even among hermaphroditic species such as snails in which courtship commonly leads to mutual insemination. Hence inbreeding in animals depends on mating among close relatives. The likelihood that this will occur depends on how far individuals have dispersed from their birthplace when they reproduce. The inbreeding coefficient of a population of

6 Heterosis (hybrid vigor) and its converse, inbreeding depression. The two corn plants at left are two inbred strains; their F_1 is to their right, followed by successive self-fertilized generations from the hybrid. (From Jones 1924)

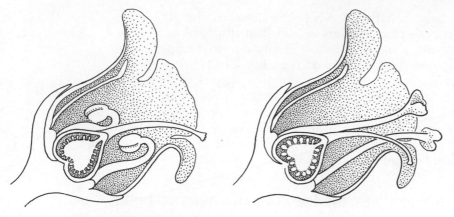

7

One of many mechanisms that promote outcrossing in plants. In the flower of *Salvia nodosa* the pistil is at first exposed to pollinators and the stamens are reflexed (*left*).

Later (*right*) the pistil becomes less accessible, and the stamens are exposed. (After Faegri and van der Pijl 1971)

great tits was estimated at only $F = 0.0036$ (Bulmer 1973), but inbreeding can be more intense, as in certain parasitic Hymenoptera, in which males hardly disperse at all and mate with their sisters almost as soon as they emerge (Askew 1968). At the other extreme, all the eels (*Anguilla rostrata*) of the Atlantic coast of North America (and perhaps of Europe as well) are believed to travel to the Sargasso Sea to mate, where they probably form one huge, panmictic population.

In the higher plants the distance to which pollen and seeds are dispersed affects the frequency of consanguineous matings. In addition, plants are frequently hermaphroditic and often capable of self-pollination. A single plant may have separate male and female flowers (e.g., in ragweed, *Ambrosia,* or corn, *Zea*), but transferring pollen from one flower to another on the same plant is genetically the same as doing it within a single flower.

But even in such plants there are many ways of reducing the likelihood of selfing. Stigmas commonly are receptive to pollen only after the plant has shed its own pollen. In other cases the flower is so constructed that a pollinator is unlikely to transfer pollen from the anthers to the stigma of the same flower (*Figure 7*). Heterostyly (*Chapter 9*) is one such mechanism. Many plants, moreover, are self-incompatible because of several or many alleles at a locus that affects the stigma's receptivity to pollen. In such species, genotype S_iS_j is receptive only to pollen with a haploid genotype other than S_i or S_j. Because S_i pollen will not grow on an S_iS_j stigma, plants are necessarily heterozygous at this locus, so outbreeding is enforced. The number of alleles can be very great. The entire evening primrose species *Oenothera organensis* consisted of only about 500 individuals when Emerson (1939) studied it, yet it had 45 self-incompatible alleles.

Many plants make a habit of self-fertilization. Within the grass genus *Bromus,* for example, there are self-incompatible, obligate out-crossers, self-compatible species that ordinarily outcross nonetheless, and species that are self-compatible and typically fertilize themselves. Stebbins (1957, 1974) argues that self-pollinating species evolve from outcrossing species under circumstances in which cross-pollination is difficult to accomplish. Thus selfing is especially prevalent among weedy colonizing species, among plants of the Arctic or desert, and on islands, which may be colonized by so few immigrants that the incipient population becomes extinct unless the colonists happen to be self-compatible.

POPULATION SIZE, INBREEDING, AND GENETIC DRIFT

Suppose that a population consists of three males and three females. If their offspring mate at random, one-third of the unions, on average, will be between sisters and brothers and so produce inbred offspring. Thus the small size of a population in itself causes inbreeding. In general, if a population consists of N diploid individuals among whom all matings, including self-fertilization, are equally likely, the likelihood that a given gene will unite with an identical gene from the same parent is $1/2N$; then the zygote is autozygous. The probability of uniting with a different gene is necessarily $1 - (1/2N)$. But the probability that this gene is identical by descent to the first is F_{t-1}, the average inbreeding coefficient of the parents. Thus among the offspring the probability of autozygosity is

$$F_t = 1/2N + (1 - 1/2N)F_{t-1}.$$

Thus any population of finite size becomes more inbred in time, and loses heterozygosity. By substituting $(H_E - H_t)/H_E$ for F_t and $(H_E - H_{t-1})/H_E$ for F_{t-1}, one can easily show that $H_t = (1 - 1/2N)H_{t-1}$. Thus the proportion of heterozygotes declines by a fraction of $1/2N$ in each generation. (This is approximately true also of populations in which self-fertilization does not occur, but the algebra is more complicated.)

A more intuitive way of viewing the process is this. The population alive today is descended from only some of the potential grandparents that existed two generations ago, since by chance some did not have grandchildren. Extended backward in time, this reasoning shows that the present population's grandparents must have been descended from an even smaller number of ancestors in the more distant past. Therefore the coefficient of relationship among the members of the population must increase with time. Because successively fewer of the original genes are represented in successive generations, the population must become more and more homozygous.

This process is essentially a matter of sampling. If we randomly take $2N$ beans (genes) from a bag of red and white beans (A and A' alleles) in which the proportion of reds is p, the probability of getting i reds is given by the ith term of the expansion of $[p + (1 - p)]^{2N}$, or

$$P(i) = \frac{2N(2N - 1) \ldots (2N - i + 1)}{i!} p^i (1 - p)^{2N-i}.$$

The mean of this binomial distribution is $\bar{x} = p$, and the variance is $V = p(1 - p)/2N$ (see *Appendix I*). Thus among a series of initially identical colonies (bean bags) that all started with the same gene frequency p, the mean gene frequency after one generation is still p, but the colonies vary to the extent $p(1 - p)/2N$; and the smaller the colonies, the greater the variation. These random changes in gene frequency are termed GENETIC DRIFT.

The variation among colonies increases in subsequent generations. For example, colonies that have drifted from $p = 0.5$ to 0.55 (the same number, approximately, will have drifted to 0.45) will drift both up and down in the next generation, with a variance of $(0.55)(0.45)/2N$, so that the span of gene frequencies is extended. After t generations the variance in gene frequency will be $V_t = p(1 - p)[1 - (1 - 1/2N)^t]$, and there will come a time when all gene frequencies are equally represented among the colonies (*Figure 8*).

Among these frequencies are 0 and 1: complete fixation (monomorphism) of one allele or the other. But once one of the alleles has been lost, it cannot reappear (except by mutation or gene flow). The number of subpopulations in each monomorphic class increases as more of the populations lose whichever allele is rarer. Eventually all the subpopulations will have lost genetic variation entirely; a fraction

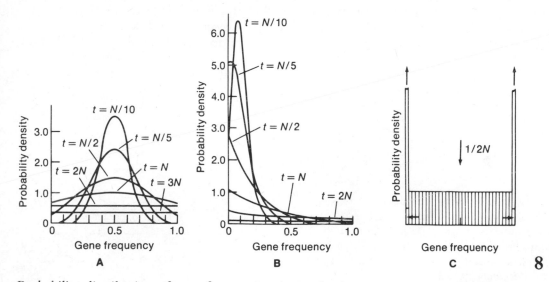

Probability distribution of gene frequencies under genetic drift when the initial gene frequency is (A) 0.5 or (B) 0.1. After $t = 2N$ generations, where N is effective population size, all gene frequencies between 0 and 1 are equally likely. At this time the allele is fixed or lost at a constant rate of $1/4N$ per generation, and each gene frequency class between 0 and 1 is decreasing at rate $1/2N$ per generation (C). (A, B after Kimura 1955, C after Wright 1931)

p of them will be monomorphic AA, and a fraction q will be monomorphic $A'A'$.

During this process the proportion of heterozygotes H in the "population" as a whole decreases — at the rate $H_t = H_0(1 - 1/2N)^t$. After all gene frequencies have become equally likely, a fraction $1/2N$ of the polymorphic subpopulations become monomorphic per generation, so that the proportion of each monomorphic class ($p = 0$ or 1) increases by $1/4N$ per generation. For example (Crow and Kimura 1970), if 1000 subpopulations of 100 individuals each all begin with $p = 0.5$, then 250 of these colonies will consist entirely of AA and 250 entirely of $A'A'$ after about 200 generations. Among the other 500 groups, all the gene frequencies between 0 and 1 will be approximately equally common. In each generation $1/400$, or (conceptually) $1\frac{1}{4}$ colonies will lose the A' allele and so become monomorphic for AA, and $1\frac{1}{4}$ colonies will similarly lose the A allele.

This process does not depend on the initial gene frequency. Thus a rare allele with a frequency of perhaps $p = 0.01$ in all the subpopulations will be lost from about 99 per cent of the colonies, but it has a probability of 0.01 of replacing the common allele — entirely by chance (*Figure 9*).

THE EVOLUTIONARY SIGNIFICANCE OF POPULATION SUBDIVISION

Random genetic drift and "systematic" forces such as natural selection act simultaneously to change gene frequencies. The ultimate fate of genes in a population depends on which has the greater effect: the systematic forces (selection, recurrent mutation, migration) that work toward a particular genetic composition, or the random action of genetic drift (*Figure 10*).

Division of a species into small populations has many implications.

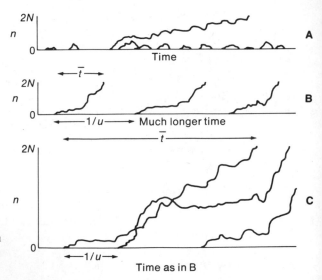

9 Gene substitution by genetic drift. The number of copies n of a mutant fluctuates at random; most mutations are lost, but an occasional mutant drifts to fixation (A). For those that do reach fixation, in B and C, the interval between mutation events ($1/u$) is shorter in a larger population (C) than in a smaller population (B), and it takes longer ($\bar{t}$) for the mutation to be fixed. Hence at any time more selectively neutral alleles will be in the larger population. (From Crow and Kimura 1970)

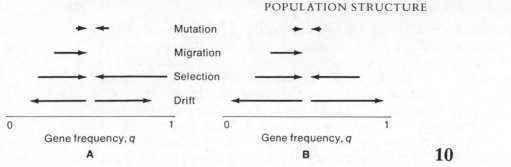

A **B** **10**

The joint action of mutation, migration, selection, and genetic drift on the gene frequency in (A) a larger population and (B) a smaller population. Selection is assumed to favor heterozygotes (*Chapter 13*). The length of an arrow is a measure of the strength of a factor.

Small populations are more susceptible to extinction than large populations (MacArthur and Wilson 1967), and the genetic complexion of a species is affected by population extinction if the populations that survive are genetically different from those that are extinguished (*Chapter 12*). The members of a small population are more closely related to one another on average than members of a large population. A high degree of consanguinity permits the evolution of certain characteristics, such as some social behaviors, that are unlikely to evolve in large panmictic units (*Chapter 12*).

The most far-reaching consequences of population size, though, are on genetic variation. If the genotypes *AA, AA',* and *A'A'* are equal in their ability to survive and reproduce, the alleles *A* and *A'* are NEUTRAL. Among a series of small populations the alleles differ in frequency because of random drift. Some populations are monomorphic for different alleles, while others are still genetically variable. Thus some individual variation in such characteristics as fingerprint patterns, leaf shape, or various enzymes may well be caused by neutral alleles, and many of the genetic differences among populations — or among closely related species, for that matter — may be caused by drift and fixation of different neutral alleles rather than by adaptation to different environments.

Because selection and genetic drift operate simultaneously, an advantageous allele can be lost from a population by chance. Conversely even a highly deleterious allele may become fixed by drift if the population is small enough. For example, heterozygotes for a chromosome translocation have low fertility (*Chapter 10*), so newly arisen translocations are eliminated by selection. If the translocation is common, though ($p > 0.5$; see *Chapter 13*), it replaces the old karyotype from which it arose. It can achieve this critical frequency only by chance and only if the population is small enough. Such a chromosome could be advantageous once established, even if it is disadvantageous when rare; thus the evolution of adaptive characteristics can sometimes depend on genetic drift.

Historically the balance of opinion on the role of genetic drift in evolution has oscillated from one stand to the other. Genetic drift was one of the most popular explanations of variation and evolution in the 1930's, after Sewall Wright's (1921) enunciation of the principle became recognized. A contending school of thought, led by R. A. Fisher (1930), believed that all alleles must have different adaptive properties under at least some environmental conditions and asserted that all variation is adaptive. In the last few decades this has been the most widespread view. But the discovery that most populations are highly polymorphic for enzymes that often seem to differ at most trivially in their properties has once again raised genetic drift to a position of respectability and controversy (King and Jukes 1969, Kimura and Ohta 1971, Lewontin 1974).

EFFECTIVE POPULATION SIZE

The rate of gene frequency change by genetic drift depends on the size of a population, not the total number of individuals that make up the species. But measuring the size of such local populations is not an easy task. If individuals are distributed not in discrete colonies but more or less evenly, each individual mates with others within a local NEIGH-BORHOOD. The population size, the number of individuals in the neighborhood, is $4s^2D$, where D is the density of the population and s is the standard deviation of the distances between the birth sites of individuals and their offspring. These distances are known for few organisms.

But there are further complications. An ecologist may determine that the size of a population is N, but if some of those N individuals do not reproduce, the population is actually smaller from a genetic point of view. For if only half the individuals in a population of 50 breed, $1/50$ rather than $1/100$ of the heterozygosity is lost per generation. The population has an EFFECTIVE SIZE N_e of 25; it is N_e that determines the rate of genetic drift. One factor that can lower the effective population size is an unbalanced sex ratio. If, for example, each male guards a harem of females against other males, the few males that reproduce contribute disproportionately to the ancestry of subsequent generations, so the rate of genetic drift is inflated. If the breeding population consists of N_m males and N_f females, the effective population size is $N_e = 4N_mN_f/(N_m + N_f)$. Thus 100 tribes of mice, each with one male and four females, constitutes an effective population of 320 rather than 500.

This is just a special instance of the more general case. To the extent that some individuals leave more offspring than others, the future population will trace its ancestry to relatively few individuals. Thus, if σ_k^2 is the variance in the number of offspring per parent that survive to reproduce, the effective population size is $N_e = (4N - 2)/(\sigma_k^2 + 2)$. Anything that causes such variation reduces population size, including selection, which by definition constitutes inequality of reproductive success. For example, Eisen (1975) found that the genetic variation for growth rate in mice was reduced virtually to zero after

14 generations of strong artificial selection that reduced population size to 4 or 8 pairs; but there was still variation in selected lines that had been kept at 16 pairs.

The effective population size is further lowered if the generations overlap, so that offspring can mate with their parents (Felsenstein 1971, Giesel 1971), or if the population fluctuates in size from generation to generation. Fluctuations in population size put the population through bottlenecks, during which genetic variation is reduced. The effective size N_e in this instance is approximately found by the relation

$$\frac{1}{N_e} = \frac{1}{t} \sum_{i=0}^{t} \frac{1}{N_i}.$$

For example, if in five successive generations a population consists of 100, 150, 25, 150, and 125 individuals, N_e is approximately 70 rather than the arithmetic mean 110; it is more strongly affected by the lower than by the higher population sizes.

All this means that it is incredibly difficult, perhaps impossible, to measure genetically effective population sizes in nature, so the role of genetic drift is difficult to assess. For an example of the complexities involved in estimating effective population sizes, see Greenwood's (1974) analysis of the snail *Cepaea nemoralis*.

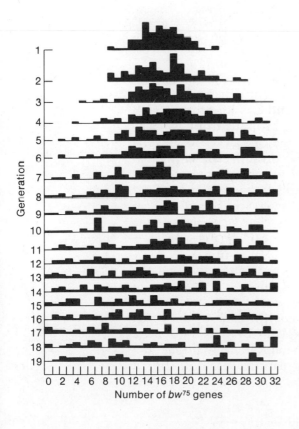

11

Experimental demonstration of genetic drift: the number of bw^{75} genes in each of many small replicate populations of *Drosophila*. After about 12 generations all gene frequency classes have become about equally frequent. (From Buri 1956)

GENETIC DRIFT IN NATURAL POPULATIONS

Several authors have attempted to determine whether the theory of drift is sufficient to explain observations of the rate of genetic change in populations. Some (e.g., Kerr and Wright 1954, Buri 1956) have found that the frequencies of mutant alleles in small laboratory populations of *Drosophila* changed in rather close accord with theoretical predictions (*Figure 11*) if both population size and the known intensity of natural selection against the mutants are taken into account. Some observations of natural populations are at least qualitatively consonant with the predictions of the theory of drift. For example, Selander (1970) found that the mean gene frequency at two protein loci (esterase, *Es-3*[b], and hemoglobin, *Hbb*) was much the same in small and large populations of house mice, but the variation in gene frequency was greater among small than among large populations (*Table II*).

TABLE II. Frequency of alleles at two loci relative to population size of house mice

Estimated population size	Number of populations sampled	Mean allele frequency		Variance of allele frequency	
		Es - 3[b]	Hbb	Es - 3[b]	Hbb
Small (median size 10)	29	0.418	0.849	0.0506	0.1883
Large (median size 200)	13	0.372	0.843	0.0125	0.0083

(After Selander 1970)

In some cases gene frequency differences among populations correspond quantitatively to what one would expect if genetic drift were the only important factor. For example (Kidd and Cavalli-Sforza 1974), the cattle of Iceland are descended from those brought from Norway by the Vikings about one thousand years ago. For each of several blood group loci the genetic difference between the Icelandic population and each of the Norwegian populations was calculated as $F_{ST} = V/\bar{p}(1 - \bar{p})$, where V is the variance in gene frequency among the populations and $\bar{p}$ is the mean gene frequency. (See Box B for an explanation of F_{ST} and why it is useful in this context.) Because the variance among populations increases at the rate $V_t = \bar{p}(1 - \bar{p})[1 - (1 - 1/2N)^t]$, the difference $F_{ST} = 1 - (1 - 1/2N)^t$ at time t.

From historical records it was possible to estimate how long the Icelandic cattle had been isolated, the sex ratio, and the fluctuations in population size. The theoretical value of F_{ST} calculated from these estimates was quite close to the observed value for each of the seven loci. Thus these genetic differences can be explained purely by genetic drift. This analysis does not prove that the differences among the populations are due to drift rather than to natural selection, but it suggests that there is no compelling reason for believing that the differences are due to anything but chance.

B The Wahlund Effect: Genotype Frequencies in a Subdivided Population

Suppose a "population" consists of k subpopulations, each with a different gene frequency p_i. Among the subpopulations the mean gene frequency is $\bar{p} = \Sigma_i p_i/k$, and the variance in gene frequency is $V_p = \Sigma_i(p_i - \bar{p})^2/k$. Now note that V_p is

$$\frac{\Sigma_i(p_i^2 - 2p_i\bar{p} + \bar{p}^2)}{k} \quad \text{or} \quad \frac{\Sigma_i p_i^2 - 2\bar{p}\Sigma_i p_i + \bar{p}^2}{k}.$$

But $\Sigma p_i = k\bar{p}$, so $V_p = \Sigma_i p_i^2/k - \bar{p}^2$. Thus $\Sigma_i p_i^2/k = \bar{p}^2 + V_p$. But $\Sigma_i p_i^2/k$ is the average proportion of AA homozygotes among the subpopulations, the proportion of AA in the population as a whole. Thus the frequency of this homozygous class exceeds the Hardy-Weinberg frequency ($\bar{p}^2$) by an amount V_p. Similarly the frequency of $A'A'$ in the entire population is $\bar{q}^2 + V_p$, and the frequency of heterozygotes is, by subtraction, $2\bar{p}\bar{q} - 2V_p$. This disparity between observed frequencies and Hardy-Weinberg frequencies is termed the Wahlund effect.

This result implies that an investigator who samples from what appears to be a single panmictic population, but is actually an aggregate of subpopulations that vary in gene frequency, will find an unexpected deficiency of heterozygotes. The magnitude of this deficiency is indeed a measure of the degree to which the "population" is actually structured into subpopulations (or, a measure of the variance in gene frequency among the subpopulations). Another such measure, of course, is F; the frequency of heterozygotes may be written either $2\bar{p}\bar{q} - 2V_p$ or $2\bar{p}\bar{q}(1 - F)$. Equating these, we find that $2(\bar{p}\bar{q} - V_p) = 2(\bar{p}\bar{q} - \bar{p}\bar{q}F)$, or $F = V_p/\bar{p}\bar{q}$.

This F, denoted F_{ST}, is different from the F that represents the average inbreeding coefficient of individuals derived from consanguineous matings within a subpopulation. It is useful to recognize, as Wright (1965) does, several levels of F:

F_{IS} the probability that two gametes taken at random within an average subpopulation yield an autozygous individual

F_{ST} the probability that two gametes taken at random from two different subpopulations yield an autozygote

F_{IT} the probability that two gametes taken at random from the entire "population" yield an autozygote

Wright shows that the relationship among these can be written $F_{ST} = (F_{IT} - F_{IS})/(1 - F_{IS})$.

THE EFFECT OF GENE FLOW

Probably few populations are completely isolated. The greater the amount of gene exchange among populations, the more similar their genetic composition will be, unless other factors counteract migration's homogenizing influence.

One such factor is natural selection, which maintains a permanent disparity in the gene frequencies of different populations if different alleles are favored in the various populations (*Box C*). This is reflected in many patterns of adaptive geographic variation. But if migration is strong enough, it can counteract selection to at least some extent, preventing a population from becoming fully adapted to its environment. For example, adult water snakes (*Natrix sipedon*) on the Lake Erie Islands are uniformly grayish in color, whereas mainland adults are strongly banded (*Figure 12*). Among young island snakes, however,

C   Gene Flow among Populations

THE HOMOGENIZING EFFECT OF GENE FLOW

Assume that a population i has a gene frequency q_i but is exchanging genes with other populations in which the mean gene frequency is $\bar{q}$. The population is so large that genetic drift is negligible. In each generation a proportion m of the genes in population i has been brought in by migration from the other populations. In the next generation, then, the gene frequency in population i will be $q_i' = (1 - m)q_i + m\bar{q}$. The change in gene frequency per generation is thus

$$\begin{aligned}\Delta q &= q_i' - q_i \\ &= (1 - m)q_i + m\bar{q} - q_i \\ &= m(\bar{q} - q_i).\end{aligned}$$

Equilibrium is reached when $\Delta q = 0$, or $\hat{q}_i = \bar{q}$. Thus population i loses its genetic distinctness.

MIGRATION AMONG SMALL POPULATIONS

Consider now the effect of a low level of gene flow on the inbreeding coefficient of a small population of size N_e. In generation t the probability of autozygosity is

$$F_t = 1/2N_e + (1 - 1/2N_e)F_{t-1},$$

but this is true only if neither of the gametes that form this individual has come from outside the population. If m is the probability that a gamete has come from outside, then $(1 - m)^2$ is the probability that neither of the two gametes has been imported by migration. Thus

$$F_t = 1/2N_e + (1 - 1/2N_e)F_{t-1}(1 - m)^2.$$

The balance between inbreeding and the outbreeding due to migration establishes an equilibrium value of F. At this time $F_t = F_{t-1}$. Appropriate algebraic manipulation (setting $F_t = F_{t-1} = F$) gives the value of F:

$$F = \frac{(1 - m)^2}{2N_e - (2N_e - 1)(1 - m)^2}.$$

If m is small so that terms in m^2 can be dropped, this reduces to

$$F = \frac{1 - 2m}{4N_e m + 1 - 2m} \cong \frac{1}{4N_e m + 1}.$$

Since $F_{ST} = V_p/\bar{p}(1 - \bar{p})$, this result implies that the variance in gene frequency among populations should be approximately

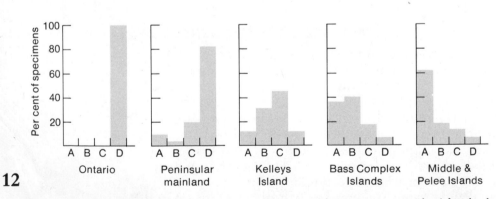

12

Maintenance of genetic variation by gene flow. Water snakes of four pattern types ranging from unbanded (A) to strongly banded (D). The unbanded pattern is advantageous on the islands, but variation persists due to migration from the two mainland areas. (Adapted from Camin and Ehrlich 1958)

$$V_p = \frac{\bar{p}(1 - \bar{p})}{4N_e m + 1}.$$

Thus migration reduces variation among populations.

THE INTERPLAY BETWEEN GENE FLOW AND SELECTION

I anticipate material in Chapter 13 by showing that gene flow can maintain an allele in a population where it is disadvantageous. Assume that allele A' is advantageous in populations other than population i, where because of a difference in the environment the allele is disadvantageous. Let the frequency of A' be $\bar{q}$ in the other populations and q_i in population i. Assume that in population i, the relative number of offspring produced by genotypes AA, AA', and $A'A'$ are 1, $1-s$, and $1-2s$, respectively. As long as there is no migration the frequency of A' will decline in each generation by the amount

$$\Delta q_i = \frac{-sq_i(1 - q_i)}{1 - 2sq_i}.$$

When s is small, this is approximately $\Delta q = -sq_i(1 - q_i)$.

But q_i is augmented in each generation by an amount $m(\bar{q} - q_i)$. Thus the net change in gene frequency is

$$\begin{aligned} \Delta q_i &= m(\bar{q} - q_i) - sq_i(1 - q_i) \\ &= m\bar{q} - q_i(m + s) + sq_i^2. \end{aligned}$$

At equilibrium, $\Delta q = 0$ and the expression is quadratic:

$$\hat{q}_i = \frac{(m + s) \pm [(m + s)^2 - 4sm\bar{q}]^{1/2}}{2s}.$$

Li (1955) explains the algebraic steps leading to the following simplifications of this expression.

If the gene's immigration rate about equals the rate at which it is eliminated by selection ($m = s$), then $\hat{q}_i \cong \bar{q}$.

If selection is much stronger than gene flow ($s \gg m$), then $\hat{q}_i = m\bar{q}/s$, so the deleterious allele has a much lower frequency in i than elsewhere.

---

a full gamut of patterns from uniformity to strong banding is evident (Camin and Ehrlich 1958). The variation is apparently reduced in each generation by selection; gulls and other predators are thought to prey more heavily on young banded snakes than on the more cryptic uniformly patterned individuals. The variation in banding pattern is maintained, however, by migration; biologists have often observed these snakes swimming several miles from the nearest land.

Migration opposes the effect of genetic drift. Migration increases N_e, so N_e and the migration rate m are not really distinguishable variables. But if they are treated as distinct, it can be shown (*Box C*) that remarkably little gene flow reduces the rate at which a population diverges by drift from other populations. If m is small, the variation in gene frequency among subpopulations reaches an equilibrium level of $F_{ST} = 1/(4N_e m + 1)$. If only one of the reproducing adults in the population, per generation, is a migrant, then $m = 1/N_e$ and $F = 1/5$. A single migrant per generation will largely counteract genetic drift.

The same principle holds if the species is distributed not in discrete populations, but uniformly over the landscape. In this isolation-by-distance model, the variation in gene frequency from place to place is great if the neighborhood size of each individual is small; but as the average distance of migration increases, variations in genetic composition become less pronounced (*Figure 13*).

THE NATURAL HISTORY OF DISPERSAL

Although the magnitude of genetically effective dispersal within a species has a profound effect on its genetic characteristics, little is known of these magnitudes in nature. A knowledge of the natural history of species should help to indicate which are more likely than others to have extensive gene flow. For example, gene flow might be greater in marine fishes and invertebrates with planktonic eggs and larvae that are transported long distances by currents than in those whose eggs are laid on the substrate or are guarded by the parent. Rather sedentary animals like snails and wingless beetles probably disperse less than flying insects or spiders that "balloon" by silk threads and are carried long distances by wind.

Gene flow in plants depends on the dispersal of seeds and pollen. Species with bird-dispersed fruits may well be dispersed farther than those that explode the seeds from a capsule for a short distance, but the average distance of dispersal of bird-dispersed seeds depends on the behavior and digestive physiology of the birds. Similarly the behavior of pollinators determines the distance to which pollen is transported; in Costa Rica, for example, early successional species of *Heliconia,* a

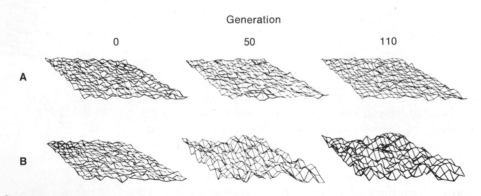

Generation

0   50   110

A

B

13 Effect of neighborhood size on variation in gene frequencies in the isolation-by-distance model. Random changes in gene frequency are simulated by a computer for a 100×100 array of evenly spaced individuals. In series A each had an equal probability of mating with any other; in series B each mates equiprobably within a neighborhood of nine individuals. Local gene frequency differences from place to place within the region are shown by different heights above the horizontal plane, after 0, 50, and 110 generations of mating. Genetic differentiation is more pronounced when neighborhood size is small. (From Rohlf and Schnell 1971, *The American Naturalist,* Copyright The University of Chicago Press)

relative of the banana, are pollinated by territorial hummingbirds that travel only short distances, while the pollen of the forest species is carried farther by hummingbirds that fly long distances between isolated plants (Linhart 1973).

But a casual familiarity with the biology of a species can lead to wrong conclusions about its level of gene flow. Dispersal often does not accomplish gene flow. In many animal species most of the individuals that disperse die before they get a chance to reproduce. In birds, salmon, and many other species that exhibit homing behavior, the dispersal of genes can be limited despite the ability to move great distances, for the animals typically return very near their birthplace to breed. In many species the level of dispersal depends on population densities, for emigration very commonly increases in response to crowding or to a shortage of resources. Some species shift into a qualitatively different mode of dispersal under high-density conditions. Some aphids, for example, produce wingless offspring when uncrowded, but winged offspring when crowded. Migratory locusts (*Schistocerca*) have short wings when population densities are low; as the density increases, later generations develop long wings and launch themselves into the air, and huge flocks are carried by the wind for long distances.

In plants, too, the dynamics of dispersal are often complex (Levin and Kerster 1974). Gene flow in wind-pollinated species can be highly directional where there are prevailing winds. In animal-pollinated species the average distance that pollen is transported depends on the pollinators' behavior, which is affected by the density of the plant population and by factors that determine how many flowers a pollinator must visit before it is satisfied (such as the average amount of nectar per flower), the density of pollinators, and the abundance of other nectar sources.

Because of these complexities, judging how fast the populations of a species will diverge by genetic drift, how precisely the genetic composition of local populations will match the differences in their environments, or how rapidly a favorable mutation might spread through the range of a species requires detailed measurements of migration rates and population sizes.

MEASUREMENTS OF MIGRATION RATES

The few measurements of migration rates and population sizes that have been made suggest that gene flow in many species is more limited than is frequently supposed (Ehrlich and Raven 1969). By planting recessive homozygotes at various distances from a strain marked with a dominant gene, Bateman (1947*a, b*) was able to measure pollen dispersal in both wind-pollinated (maize) and insect-pollinated (radish) crops. The distribution of heterozygous progeny reflected the pollen flow from the dominant stock. The proportion of maize seeds carrying the dominant gene was reduced to 1 per cent at only 40–50 feet from the pollen source and decreased exponentially with distance (*Figure 14*).

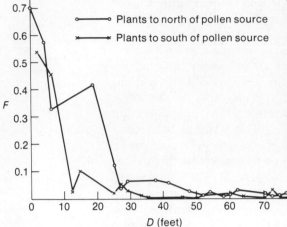

14

Gene flow in corn, a wind-pollinated plant. The vertical axis F gives the proportion of seeds fertilized by a different genetic strain, in plants set at distances D to the north and south of that strain. Prevailing winds affect the amount of gene flow, but in both cases it is quite restricted. Gene flow may be more extensive in other wind-pollinated plants that have lighter pollen than corn. (After Bateman 1947b, *Heredity 1,* 240)

Thus even though wind-carried pollen can be found over the middle of the Atlantic Ocean, most plants must be pollinated almost exclusively by their close neighbors. Both Bateman and Levin and Kerster (1968) found that in insect-pollinated plants most pollen is carried only very short distances, but a small percentage is transported to much greater distances than it would be if it simply dropped off exponentially. As Bateman points out, this low incidence of long-distance migration may be quite important in maintaining genetic cohesion of populations over large areas.

In at least some insects gene flow seems quite limited. Dobzhansky and Wright (1943) released genetically marked *Drosophila pseudoobscura* at a single point and found that they moved an average of 133 meters the first day and about 90 meters per day thereafter. Since females mate within two days after emerging, the average distance between the sites of hatching and mating must be less than 250 meters. Wallace (1966a) pointed out that the dispersal rate of these flies seems to have been inflated by their reaction to an artificially high density at the site of release and suggests that the migration rate in nature may be even less than Dobzhansky and Wright estimated. Moreover this experiment did not measure genetically effective dispersal, which requires that the flies not only disperse but reproduce.

Ehrlich and his colleagues have described an interesting population structure in the checkerspot butterfly *Euphydryas editha* in California. One population of butterflies in a region less than 200 × 200 meters actually consisted of three distinct subpopulations that fluctuated independently in size (Ehrlich 1965). Although there are no physical barriers to flight among the three centers of abundance, the butterflies simply do not fly from one center to another (Brussard, Ehrlich, and Singer 1974). Gilbert and Singer (1973) found, however, that what holds for one population does not necessarily hold for others. In other sites *E. editha* often moves distances greater than a mile.

Among vertebrates, some work on lizards and rodents suggests rather limited movement. In the rusty lizard, *Sceloporus olivaceus,* juveniles disperse an average of 78 meters from their hatching site to their final home, and females wander about 29 meters from their home range to lay eggs; the neighborhood size is about 10 hectares (Kerster 1964). In deer mice (*Peromyscus maniculatus*), at least 70 per cent of the males and 85 per cent of the females breed within 150 meters of their birthplace, and consanguineous matings are rather common (Howard 1949).

No doubt it is easier to measure dispersal in species that do not move around much than it is in more widely ranging species like monarch butterflies or great white sharks, so the view of population structure that has arisen in recent years may well be biased toward the conception that organisms tend to be sedentary and gene flow minimal. But whether it is low enough to allow many differences to arise by genetic drift, we cannot readily tell.

THE FOUNDER EFFECT

The genetic accidents inherent in small population size may be important in another context. If an island or a patch of newly available habitat hitherto unoccupied by a species is colonized by one or a few individuals, all the genes in the population to which they give rise stem from the few carried by the founders and from subsequent mutations and immigrants.

A colony founded by a pair of diploid individuals can have at most four alleles at a locus, although there may have been many more alleles in the population from which the colonists came. Although the number of alleles is reduced, the degree of heterozygosity, and hence the genetic variance, is almost as high as in the source population. (On average it is $(1 - 1/2N)H_0$, where H_0 is the proportion of heterozygotes in the source population and N is the number of colonists.) This is simply because rare alleles contribute little to the source population's level of heterozygosity, and these are the very alleles that are likely to be missing from the colony. The heterozygosity of the colony declines rapidly, however, due to genetic drift if the colony remains small, and it only slowly builds back up by mutation and drift unless natural selection increases the frequency of rare genes (Nei, Maruyama, and Chakraborty 1975; *Figure 15*). If the colony grows rapidly, though, the amount of genetic variation will not be greatly reduced.

The same principle applies to any population that undergoes a severe bottleneck in population size. Mayr (1954, 1963) has argued that the genetic constitution of a population may be so affected by this FOUNDER EFFECT that it accelerates the population's development into a new species, a hypothesis discussed in Chapter 16.

An extreme example of the bottleneck effect is provided by the northern elephant seal (*Mirounga angustirostris*). This species was reduced by hunting to about 20 animals in the 1890s; since then the population has grown to more than 30,000. The effective population size must have been even smaller than 20 at the population's nadir, for the species is polygynous; less than 20 per cent of the males service all the females. Among species that have been examined for electrophoretic variation, this seal is unique, for Bonnell and Selander (1974) found absolutely no genetic variation in a sample of 24 electrophoretic loci. Genetic variation has been observed in the southern elephant seal (*M. leonina*), which was never as severely reduced in numbers.

POPULATION STRUCTURE AND THE RATE OF EVOLUTION

Wright (1931) was the first to consider the effect of population structure on the rate of evolution. He argued that the "optimal" population structure — the structure that promotes the most rapid, sustained evolution of a species as a whole — is a complex of small populations among which there is slight but persistent gene flow. Why this structure might maximize the rate of evolution is perhaps best seen by considering the two extreme possible structures.

If a species consists of many small populations that do not exchange genes at all, each population evolves independently. Evolution in each population occurs when the environment changes so that formerly disadvantageous alleles become advantageous or when advantageous mutations arise that the population did not have before. The rate at which advantageous mutations arise is proportional to $N_e u$, to the mutation rate per locus multiplied by the number of individuals in which mutations can occur. Hence evolution is slow in small populations. If evolution depends on the increase of preexisting alleles that were formerly rare, it is also slower in small populations, because genetic drift reduces the level of genetic variation. Even alleles kept in polymorphic condition by natural selection are lost by genetic drift in small populations (*Chapter 13*). Hence evolution should be slow in an isolated population, especially if it is small.

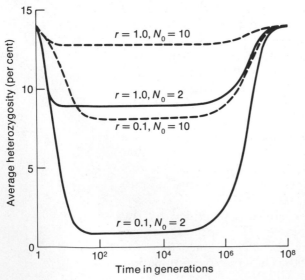

15

Effects on genetic variation of a bottleneck in population size, as in a newly founded population beginning with 2 (solid lines) or 10 (broken lines) individuals. Heterozygosity declines more substantially if the growth rate of the population is low ($r = 0.1$) than if high ($r = 1.0$), because of inbreeding after colonization. (After Nei et al. 1975)

At the other extreme, consider a species like the American eel in which the entire species over a broad geographical range forms a single, panmictic, interbreeding population. Mutations arise in profusion, but few of them are advantageous over the great variety of local conditions that the species faces, so most of them do not persist. The genetic composition is rather static, consisting of those "jack of all trades" alleles that confer reasonably high fitness in each of the many environments into which they are carried by continual gene flow. Moreover genetic variants that, like some chromosome translocations, are advantageous when common but deleterious when rare, cannot increase from their initial low frequency except by genetic drift. (In the terminology of Chapter 14, the population goes by chance from one adaptive peak to the slope of a higher adaptive peak.) Therefore certain highly advantageous genetic changes can happen in small populations, but not in large ones.

In a species consisting of a multitude of populations, some may become extinct, while others — perhaps those that arrived at highly advantageous genetic compositions with the help of genetic drift — persist. Then the evolution of the species as a whole may not be based merely on selection of fit individuals, but may be reinforced by the selection of fit populations. This cannot happen if the entire species is panmictic. For these many reasons, then, evolution is likely to be very slow in an abundant panmictic species.

Evolution should be most rapid, Wright argued, in the intermediate case. Mutations that arise anywhere in the species' range can be transmitted throughout the species if they are advantageous in a wide variety of environments; yet each population can also become highly adapted to its local environment rather than arrive at a "compromise" genotype that must be adapted to other environments as well. The genetic variation useful for adaptation to environmental change is augmented by gene flow from other populations; yet some populations are small enough to arrive by genetic drift at gene combinations that are advantageous only when they are common. Such populations persist while less fortunate populations dwindle, so the genetic complexion of the species changes by selection of populations as well as individuals.

Because most species are subdivided into populations, it might seem that most should fit Wright's model of the optimal structure for rapid evolution. But because only a little gene flow is sufficient to reduce genetic differentiation of populations, it may well be that almost all widespread species have a suboptimal structure; there may be so much gene flow among their populations that they evolve very slowly. This theme is pursued in Chapter 16, in which I expand the argument that most species are at most times genetically quite stagnant and that it is primarily in small, isolated populations, in the process of speciation, that much evolutionary change occurs.

ASSORTATIVE MATING

The discussion of gene flow bears heavily on the topic of speciation.

If new species arise as geographically isolated populations, the likelihood of speciation must depend in part on how isolated they are. So species with reduced gene flow must be more likely to undergo speciation.

In this context a very different basis for nonrandom mating is important. Inbreeding entails mating between individuals that are *genetically* similar; but if *phenotypically* similar individuals tend to mate with one another, the genetic consequences are rather different. For example, there is some tendency toward a correlation in height between mates in the human population of the United States; tall women tend to marry tall men and short women, short men (Spuhler 1968). Now suppose for the sake of argument that height were controlled by a single gene locus (it is not) and that the heterozygote were intermediate between the homozygotes. If individuals were to mate *only* with others of like phenotype, the effects would be the same as inbreeding *with respect to this one locus:* the population would become segregated into two reproductively isolated moieties, one tall (AA) and one short ($A'A'$) in phenotype. They would be two species. But the AA individuals can be genetically variable in all other respects, as can be the $A'A'$ pool. Unlike inbreeding, mating preference based on phenotypic appearance — assortative mating — does not induce homozygosity at all loci, but only at those loci that determine the characteristic that defines mating preference (and at very closely linked loci). At all other loci the AA and $A'A'$ populations should have the same gene frequencies, with Hardy-Weinberg proportions of genotypes.

Assortative mating is already a familiar concept in one context. Allopatric populations that diverge in plumage, voice, or other characteristics important in courtship tend not to interbreed when they meet. They display assortative mating, or ethological reproductive isolation. The more debatable question is how often individuals *within* a population that has never been subdivided mate assortatively on the basis of some polymorphic characteristic. If they mate assortatively and if the assortative mating is almost perfect, a species could split into two without the intervention of any geographic barriers; speciation would not be geographic, but sympatric. Whether this ever occurs is a matter of interminable debate (see, for example, Mayr 1963, Maynard Smith 1966*b*, Bush 1975), which I shall treat in Chapter 16.

SUMMARY

Inbreeding distributes genes from the heterozygous to the homozygous state. A population within which inbreeding occurs becomes subdivided into increasingly homozygous subpopulations, but the genetic variation among subpopulations more than compensates for the homogeneity within each, so that the variation of the population as a whole increases. In each subpopulation gene frequencies can fluctuate by chance, until one allele or another becomes fixed and genetic variation at that locus is lost. This process, genetic drift, happens in any finite population, but is more rapid in small than in large populations. Because almost every species is distributed in semi-isolated populations that are often effectively smaller than they seem, some of the genetic changes in most species are likely to be random.

Migration, the flow of genes among populations, increases genetic variation and counteracts genetic drift; but by the same token it prevents precise adaptation of the populations to their respective environments, just as recombination does within populations. Unless the populations are very small or are subject to strong selection for different gene frequencies, migration maintains a genetic cohesion of the populations of a species and prevents populations from differentiating into different species.

Nonrandom mating within a population may entail not inbreeding, but phenotypic correlation among mates. Positive assortative mating, whereby individuals choose mates that are similar in one or more phenotypic features, can divide a population into two reproductively isolated units — species — if individuals have absolute preferences. Although this is known to occur when two formerly geographically isolated forms come into contact, it is unclear whether it ever happens within the confines of a single population.

FOR DISCUSSION AND THOUGHT

1 Any of many characteristics may reduce the incidence of inbreeding in plants: self-sterility, heterostyly, dioecy, progressive flowering such that few flowers on a plant are open at any one time, protogyny or protandry, dispersal of seeds or pollen. What evidence is required to state confidently that the advantage of outbreeding is the *raison d'être* of any one of these characteristics? In what other ways might each of these be advantageous?

2 If individuals from a normally outcrossing species are inbred, a decrease in fitness is usually observed among their progeny. How, then, can inbreeding mechanisms such as cleistogamy evolve? What advantages are inherent in inbreeding?

3 If effective population sizes and migration rates are so difficult to measure, even difficult to define unambiguously, why are these concepts used?

4 Suppose a population was founded a few thousand generations ago by a pair of individuals but has had a large constant size since then. Does it still have a small effective population size because of the initial bottleneck? What determines when the effect of the bottleneck is so diminished that we can consider it a large population?

5 Given the impact that inbreeding has on phenotypic variation, what effect is it likely to have on the rate at which a population's genetic composition is changed by natural selection?

6 What effects do monogamy vs. polygamy (polygyny or polyandry) have on effective population size? How might these mating systems evolve in birds or mammals?

7 How small does a population have to be before genetic drift has an important effect on the gene frequencies at a locus? Why is this not a very meaningful question?

8 If genetic drift is acting, at equilibrium we expect to find gene frequencies of either 0 or 1 in each of many small isolated populations. So finding a different gene frequency distribution seems to imply that genetic drift is not important. Criticize this argument.

9 In a large, randomly mating population two polymorphic loci should eventually reach linkage equilibrium. What is the effect of population subdivision or of small population size on the degree to which loci depart from linkage equilibrium?

10 Suppose we know that the alleles at a locus are neutral. Then, instead of predicting genetic divergence from a knowledge of population sizes and the length of time two populations have been isolated, we might deter-

mine the length of separation from a knowledge of the population size and the degree of genetic divergence. This procedure has given rise to the idea of a "protein clock." By measuring the difference in amino acid composition of the proteins of several species, we may be able to tell when they arose (see, e.g., Nei 1971). What assumptions must hold true for such analyses to be valid?

11 What evidence would be sufficient to determine whether the geographic variation in some characteristic of a species was primarily adaptive (due to natural selection) or random (due to genetic drift)?

12 Ehrlich and Raven (1969) argue that gene flow among the populations of a species usually proceeds at a very low level. Why then, they ask, don't species show far more geographic variation than they do? How is it possible that we can recognize white-tailed deer from the Atlantic to the Pacific coast as members of one species?

13 As long as widely separated populations remain the same species, the only evolutionary change in the species as a whole must be due to mutations that are advantageous throughout the species, in all the environments it inhabits. Since there must be few such mutations, evolutionary change must necessarily be slower in widespread panmictic species than in isolated local populations. Discuss. Suppose that an adaptation advantageous in many environments (e.g., a general adaptation permitting entry into a new adaptive zone; see Chapter 5) is based on a *combination* of genes at several loci, no one of which is advantageous on its own. Can such a combination spread from one population to another?

14 How does assortative mating differ from sexual selection (Chapter 12)? What are the genetic consequences if individuals tend to mate with dissimilar, rather than similar, individuals? See Lowther (1961) for an example. Assortative mating exists, but why should it? Why should the tendency to mate with like individuals evolve?

15 In all the cases in which sympatric assortative mating has been described, the fidelity of mate preference is imperfect. But if it were perfect, the forms would be called different species. Is it possible, then, to identify cases in which assortative mating has caused sympatric speciation?

MAJOR REFERENCES

The books by Wilson and Bossert, Mettler and Gregg, Dobzhansky, Lewontin, and Spiess treat the material in this and succeeding chapters (12–15). Other major works are the following.

Crow, J. F. and M. Kimura. 1970. *An introduction to population genetics theory*. Harper & Row, New York. 591 pages. A major exposition of mathematical population genetics, containing extensive treatment of inbreeding and stochastic effects.

Kimura, M., and T. Ohta. 1971. *Theoretical aspects of population genetics*. Princeton University Press, Princeton, N.J. 219 pages. Emphasizes the role of genetic drift in evolution, from a largely theoretical standpoint.

Wright, S. 1968–1978. *Evolution and the genetics of populations.* Vol. 1: *Genetic and biometric foundations,* 469 pages; Vol. 2: *The theory of gene frequencies,* 511 pages; Vol. 3: *Experimental results and evolutionary deductions,* 613 pages; Vol. 4: *Variability within and among natural populations,* 580 pages. University of Chicago Press, Chicago. A major, highly technical, treatise on population genetics by one of the founders of the field.

Natural selection is the central principle of evolutionary theory and is indeed one of the most pervasive concepts in all of science. Although exceedingly simple in principle, it can be so diverse in its operations as to be blatant or subtle, simple or intricate. It is an elementary concept; nonetheless it is so commonly misunderstood that a great deal of nonsense has been uttered, even by respected biologists, in its name. It is viewed by some as a dark force, as the inexorable action of an unfeeling, uncaring universe, and by others as the creative agent of nature, instilling progress into history. It is invested by some with the qualities of Providence, as in statements that characteristics have evolved "for the good of the species"; it is invoked as a natural law, with the moral force of an ethical precept from which even human society is not and should not be exempt. Andrew Carnegie, for example, argued that the law of competition "is here; we cannot evade it; no substitutes for it have been found; and while the law may sometimes be hard for the individual, it is best for the race, because it insures the survival of the fittest in every department."

These misconceptions stem largely from the metaphorical ways in which the concept is often used, even by Darwin himself when he writes in such a vein as "Nature, if I may be allowed to personify the natural preservation or survival of the fittest, cares nothing for appearances, except insofar as they are useful to any being. . . . Man selects only for his own good: Nature only for that of the being which she tends." But by such poetical speech, as Darwin noted, natural selection is too easily viewed as "an active power or Deity," omniscient, omnipotent, and, depending on one's point of view, either beneficent, shaping species into perfect form, or malevolent, red in tooth and claw.

DIFFERENTIAL SURVIVAL AND REPRODUCTION

But natural selection is not providential; it is neither moral nor immoral; it carries no

 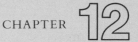

The Nature of Natural Selection

Owing to the struggle [for existence], variations, however slight and from whatever cause proceeding, if they be in any degree profitable to the individuals of a species, in their infinitely complex relations to other organic beings and to their physical conditions of life, will tend to the preservation of such individuals, and will generally be inherited by the offspring. The offspring, also, will thus have a better chance of surviving, for, of the many individuals of any species which are periodically born, but a small number can survive. I have called this principle, by which each slight variation, if useful, is preserved, by the term Natural Selection, in order to mark its relation to man's power of selection. But the expression often used by Mr. Herbert Spencer of the Survival of the Fittest is more accurate, and is sometimes equally convenient.

Darwin, in *On the Origin of Species*

ethical precepts; and it need not entail struggle or competition. The preeminent agent of evolution is, very simply, the differential survival and reproduction of entities that differ from one another in one or more respects. Selection is not *caused* by differential survival and reproduction; it *is* differential survival and reproduction — and no more. It shapes order, form, and wondrous adaptations from the chaos of mutation and recombination, but it is an impersonal, purely mechanical, statistical process nonetheless.

Natural selection in the broadest sense applies to all of nature, not just to gene frequencies. Some isotopes decay more rapidly than others, so the longest-lived are the most prevalent. The form of the universe may well be the product of a selective process; of the possible orbits the celestial bodies might have taken, some are more stable than others and so are more likely to have persisted. Selection of a kind determines the species composition of ecological communities; species that cannot coexist with others have been eliminated.

Selection may be simply a matter of differential birth rates. If two strains of bacteria are kept in a chemostat with unlimited nutrients and flushed out at random to keep the volume of culture constant, the strain with the higher rate of division ultimately replaces the other completely. The difference in growth rate is natural selection; but the system does not change toward a goal, there is no violent struggle, there is no morality, and the state of the system that follows from the selective process is not better than its previous state; it just *is*.

A FORMAL TREATMENT OF THE ELEMENTARY PROCESS OF SELECTION

Selection entails changes in numbers, by births and deaths, and so can be described in terms of the equation for population growth $dN/dt =$

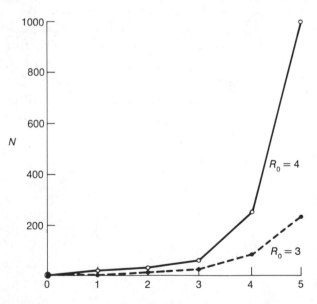

1

The growth of two asexually reproducing genotypes in a population with discrete generations. The proportion of the more prolific genotype ($R_0 = 4$) rapidly becomes far larger than that of the less fecund genotype ($R_0 = 3$). The differential growth rate of the genotypes is an instance of natural selection.

rN from Chapter 4. The instantaneous per capita growth rate r_i of the ith kind of organism measures its fitness. FITNESS, in an evolutionary context, is measured only by growth rate, not by attributes that enter into our everyday uses of the term, such as physical fitness.

If growth occurs in discrete generations, the replacement rate R ($\cong e^r$) may be used to measure fitness. Suppose there are two asexual genotypes A and B in a population, with replacement rates R_A and R_B (*Figure 1*). They number N_A and N_B respectively; $N_A + N_B = N$, the total population size. The proportions of A and B are $N_A/N = p$ and $N_B/N = q$. The growth rate of the entire population is then $pR_A + qR_B = \bar{R}$.

After one generation $N_A' = N_A R_A$ and $N' = N\bar{R}$, so the new proportion of A is $p' = N_A R_A/N\bar{R}$. Because $N_A = pN$ and $\bar{R} = pR_A + qR_B$,

$$p' = \frac{pNR_A}{N(pR_A + qR_B)} = \frac{pR_A}{pR_A + qR_B}.$$

The *change* in the proportion of A is

$$\Delta p = \frac{pR_A}{pR_A + qR_B} - p, \quad \text{or} \quad \frac{pq(R_A - R_B)}{\bar{R}}.$$

As long as $R_A > R_B$, p increases ($\Delta p > 0$) until A has replaced B. The rate of change Δp depends only on the relative values of R_A and R_B, the genotypes' RELATIVE FITNESSES. Whether the absolute fitnesses R_A and R_B are 2 and 1 or 2000 and 1000, the rate of evolution is the same.

The growth rate of a genotype depends on environmental conditions. If genotype A has a replacement rate of R_1 when the environment is cold and R_2 when it is warm, a succession of cold and warm environments for six generations will increase its population by $R_1^3 \times R_2^3$; thus its fitness R_A is the geometric mean of its fitnesses over a series of generations. If, on the other hand, in each generation a fraction x of the population grows in one environment (e.g., a certain nutrient) at rate R_1 and part $(1 - x)$ grows in another environment at rate R_2, the growth rate of genotype A is $R_A = xR_1 + (1 - x)R_2$. In a spatially heterogeneous environment, therefore, a genotype increases at the arithmetic mean of the fitnesses it has in each environmental type.

This viewpoint is one way of analyzing changes in gene frequency in sexually reproducing populations. Assume that the genotypes AA, AA', and $A'A'$ have the Hardy-Weinberg frequencies p^2, $2pq$, and q^2 and that the replacement rates of these genotypes are respectively R_1, R_2, and R_3. Each of the genotypes in which an allele occurs is an environment in which it replicates itself at the rate of replacement of that genotype. Allele A, for example, occurs in the two genotypes AA and AA'. Of the A alleles in the population, a proportion $p^2/(p^2 + pq) = p$ occurs in the AA homozygotes, and a proportion $pq/(p^2 + pq) = q$ occurs in the heterozygotes. Therefore the mean fitness of the A allele is $R_A = pR_1 + qR_2$. Similarly the mean fitness of the A' allele is $R_{A'} = pR_2 + qR_3$. As above, allele A changes in frequency at the rate $\Delta p = pq(R_A - R_{A'})/\bar{R}$. Whether A increases or decreases in frequency depends on the sign of $(R_A - R_{A'})$. Suppose, for example, that R_1, R_2, and R_3 are, respectively, 1, 1, and 0.5 and that $p = 0.1$. Then $(R_A - R_{A'}) = 0.45$, so $\Delta p > 0$ and A increases.

But suppose the fitnesses R_1, R_2, and R_3 are, respectively, 0.5, 1, and 0.5; then the heterozygote is most fit. Then if $p = 0.1$, $(R_A - R_{A'}) = 0.4$, so A increases as before. However, if the initial frequency of A is $p = 0.9$, then $(R_A - R_{A'}) = -0.4$; $\Delta p < 0$, and A declines in frequency. Whether an allele increases or decreases, therefore, depends not only on the external environment but on the genetic composition of the population as well. Changes in gene frequency under selection can become much more complex if we take into account other loci, for each allele occurs in an immensely variable background of genotypes; each may have different capacities for survival and reproduction and so may facilitate the replication of the allele to a different extent.

Three aspects of the use of R (or r) as a measure of fitness together account for much of the intricacy of evolution by natural selection. (1) An allele's growth rate is not constant; it depends on the environment. (2) It is the resultant of several components that can vary independently. (3) The growth rates of genes, genotypes, populations, and other entities may be independent of each other.

THE EFFECT OF ENVIRONMENT ON FITNESS

Some alleles are probably always lethal: those that cause severe developmental abnormalities, for example. But most alleles and most genotypes vary in fitness, depending on the environment. Even such a seemingly harmful allele as that coding for sickle-cell hemoglobin can be advantageous, and so can rise in frequency, under appropriate conditions. In *Drosophila pseudoobscura* different chromosome inversions, which carry different alleles, seem to confer similar fitnesses under some conditions (e.g., 16°C) in the laboratory; but at 25°C the genotypes have very different fitnesses. In one experiment, for example, the relative fitnesses were estimated to be $ST/CH = 1$, $ST/ST = 0.89$, and $CH/CH = 0.41$; the CH homozygote had less than half the survivorship and fecundity of the heterozygote (Dobzhansky 1970).

As a result, an allele may increase in frequency in some places and decline in others; or its frequency may fluctuate as environments fluctuate. The frequencies of ST and CH chromosomes, for example, change seasonally in natural populations (*Figure 2*). Thus changes in gene frequency are determined by immediate conditions and cannot be influenced by future environments. For this reason there can be no goal in evolution, nor predestined direction; if a species changes consistently in one direction for a long time, it must be because the environment changes or because a particular genetic constitution that would at all times be most adaptive is reached only slowly, as mutations arise that more closely approximate the ideal state. Not even avoidance of extinction can be thought of as a goal, for the genetic

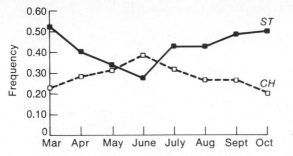

2

Seasonal changes in the frequencies of two chromosome inversions in *Drosophila pseudoobscura*. (After Dobzhansky 1970, with permission of the publisher)

changes impelled by current circumstances may well prove lethal in the future.

Moreover many characteristics that might be "good for the species" — such as those that maintain the genetic variation needed for response to environmental change — do not evolve if they are not beneficial to the individuals who carry them. Genes that cause high mutation rates, for example, might in the long run save the species from extinction, but most such genes are eliminated (*Chapter 10*). Thus evolution does not necessarily happen for the good of the species; it just happens.

SEXUAL SELECTION AND THE T LOCUS

Selection may not act to the species' advantage; sometimes it even has disadvantageous effects. Many species become ecologically highly specialized, for example; if specialized species indeed have higher extinction rates, present adaptedness leads to future disaster.

In many instances alleles with high fitness do not help their bearers cope with such exigencies as enemies or weather but achieve their high fitness, as Darwin put it, by "the advantage which certain individuals have over others of the same sex and species solely in respect of reproduction." Features that enable a female to bear more offspring are such; but Darwin had in mind especially those secondary sexual characteristics of males — the horns of mountain sheep, the bright nuptial colors of sticklebacks and minnows, the train of the peacock — that cannot be considered necessary to survival and may even be deleterious because they increase susceptibility to predation. They simply enable their bearers to father more offspring, by conferring superiority in inter–male combat or display or by rendering the male more attractive to females.

Sexual selection, as Darwin called this aspect of natural selection, raises many interesting problems (see papers in Campbell 1972). Why, for example, are males typically more elaborately adorned than females? A likely answer is that males, who lose little if they copulate with an inferior female, need not be as choosy as females, whose whole reproductive effort may go to naught if they mate with inferior males and have unfit offspring. It is to the female's advantage to exercise careful choice; thus it is the male who courts, with all the elaborate appurtenances pertaining to courtship.

But in what way may a female's offspring be unfit, should she choose an inferior male? Again, simply with respect to reproduction. If her sons have short plumes, inherited from a short-plumed father, when females generally prefer long-plumed males, neither her genes that influence her taste in males, nor her mate's genes for short plumes will be propagated. Thus evolution can occur "by convention," in which the mating preferences of the majority determine what characteristics shall be favored. This self-reinforcing evolution can exaggerate characteristics (as in the peacock) until the disadvantages (e.g., from predation) counter the advantage from mating success.

Since an allele's fitness depends solely on its reproductive success, its increase can sometimes actually harm the species. For example, some recessive alleles at the T locus in house mice cause homozygous males to be sterile. A male $+t$ heterozygote, however, produces 80–95 per cent t-bearing sperm. This distortion of segregation, or MEIOTIC DRIVE, constitutes a selective advantage to the t allele. The increase of t from meiotic drive should be countered by the sterility of the homozygotes, so that it should reach an equilibrium frequency of 0.7. Yet its frequency was found to be only 0.37 in a large mouse colony over a four-year period. To explain the discrepancy, Lewontin (1962) simulated a population on a computer and discovered that a population would have an equilibrium frequency of 0.46 (pretty close to 0.37), if it consisted of inbred demes, each of two males and six females. Many such demes become extinct when t rises in frequency (from meiotic drive) until both males are sterile homozygotes. By genetic drift other demes lose the t allele and so persist. The mean gene frequency, 0.46, is that of the entire ensemble of subpopulations (*Figure 3*), in which the allele can never get very common, because when it does the deme becomes extinct. Thus the gene frequency depends on population subdivision, genetic drift, GENIC SELECTION (in this case, meiotic drive), INDIVIDUAL SELECTION (sterility), and the extinction of populations by a selectively superior gene.

THE DEMOGRAPHY OF SELECTION AND THE EVOLUTION OF LIFE-HISTORY CHARACTERISTICS

An allele seldom has only one effect on the phenotype. Its several

3

Results of a computer simulation of the T locus; the equilibrium distribution of small populations with various frequencies of the t allele. Unshaded portion represents populations in which males, but not females, are all tt homozygotes. (After Lewontin 1962, *The American Naturalist,* Copyright The University of Chicago Press)

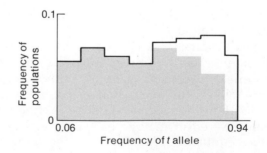

pleiotropic effects may bear different advantages and disadvantages. Hemoglobin *S,* for example, is advantageous insofar as it confers protection against malaria, disadvantageous insofar as it reduces the oxygen-bearing capacity of the blood. The sum of these effects determines the net fitness of the allele and so its fate in a population.

A gene may be expressed at each of several stages of an individual's ontogeny, and it may be advantageous at some age yet disadvantageous at other ages; or it may express itself at only one age. The net fitness of the gene is its value of r, which depends on its bearers' survival l_x and fecundity or mating success m_x at each age x. In a growing population r increases as a function of survival, fecundity, and the rapidity with which the age of reproduction is attained. If an allele's fitness is measured by r, should we not expect all species to have evolved the longevity of redwood trees, the fecundity of codfish, the short generation time of bacteria?

The concept of reproductive value (*Chapter 4*) helps resolve this puzzle. The reproductive value of an individual is its average contribution to the growth rate of the population. Thus an allele expressed at an age of high reproductive value changes in frequency more rapidly than one having a comparable effect at an age of low reproductive value. One consequence is that an allele with seemingly deleterious effects may actually have rather high fitness. For example, Huntington's chorea, a severe disorder of the human nervous system caused by a dominant allele, generally causes death. But it is usually not expressed until the age of 30 or 40, when most people have already borne all the children they will have. Thus the net reproductive rate of choreics, 1.0308, is almost as high as that of the population as a whole, 1.0485 (Cavalli-Sforza and Bodmer 1971), and the gene is not eliminated as fast as it would be if it were expressed at an earlier age.

Thus two genotypes can differ in fitness only if they differ in R (or r); otherwise there is no selection. This means, for example, that a characteristic cannot have evolved by selection if it is a mere convenience. If the form of a horse's tail, used as a fly whisk, has evolved by selection, it is not simply because horses with efficient fly whisks had a more comfortable life; it is because they survived better, perhaps by suffering less from diseases transmitted by flies. If modern medicine has reduced prereproductive mortality from diseases and inborn errors of metabolism, it has reduced the advantage that resistant genotypes had over susceptible genotypes. Natural selection can still act in industrial human populations, but the opportunity is provided more by the variation in birth rates than in death rates (Crow 1958). Even if there is variation in fecundity or a great deal of prereproductive mortality, selection may not be occurring. A great deal of mortality and reproductive failure is nonselective, suffered equally by all genotypes.

In growing populations reproductive value increases from a low value at birth to a peak at the age of first reproduction and declines rapidly thereafter. Therefore an allele that increases fecundity by a given degree increases fitness more if it is expressed earlier in life than later. Reproduction often exposes an organism to risks, so that the chances of subsequent survival and reproduction are reduced (*Figure 4*). But if the increment in fitness from early reproduction exceeds the decrement in later reproduction, the allele for early reproduction increases, and the species' reproductive activity shifts to earlier ages. This explains not only how a species may come to abandon reproduction late in life, but also how a short life span can evolve, why organisms are not immortal. Genes that promote life past the age of last reproduction have no advantage; indeed any genes that have advantageous effects early in life but disadvantageous pleiotropic effects late in life have a net advantage and so contribute to senescence (Williams 1957; see also Williams 1966).

   Conversely species can evolve late reproduction and long life spans. If reproduction is very risky and if fecundity increases with age, *r* can be greater if an individual has many offspring later in life instead of having a few offspring early in life. Gadgil and Bossert (1970) cite the case of migratory salmon. Because fecundity and the energy resources needed for the trip are proportional to body size, *r* is greater for a genotype that delays reproduction until the fish has reached a critical size.

   Most species of migratory salmon put all their energy into reproduction and then die — there is no profit in saving some energy for future survival because the chances of being able to make another reproductive journey are so low. Many species, however, especially if their environment is fairly constant, have a high prospect of future survival and reproduction, if they do not expend all their energy on reproduction at any one time. Moreover they may maximize their chances of leaving surviving offspring by repeated reproduction, for all their progeny could be lost in a bad year. In such species (such as trees), repeated reproduction can be advantageous. Annual plants, on the other hand, inhabit such an ephemeral environment that they are almost certain not to survive long enough to reproduce twice; their best strategy is to put all their energy into seed production and, as a result, to die.

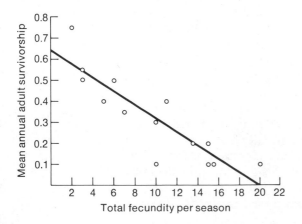

4

An illustration of the cost of reproduction; fecundity *vs.* survivorship in 13 species of lizards (one represented by two populations). (From Tinkle 1969, *The American Naturalist*, Copyright The University of Chicago Press)

How many offspring should a female have when she does reproduce? Energy and nutrients can be divided among many small seeds or eggs or a lesser number of large ones. If the likelihood of an offspring's survival does not depend on how well provided it is with nutrients or defenses, the female will have the greatest number of surviving progeny if she has a great many small eggs or seeds. Such might be the case for an oyster or cod, whose planktonic eggs are unlikely to escape filter-feeding predators no matter how well supplied with yolk. But the seeds of a coconut palm can survive only if they have enough endosperm to grow an extensive root system in the nutrient-poor sands in which they germinate. Hence the seed must be large, and the fecundity of the tree must necessarily be low. Lack (1954) found that starlings, in somewhat similar manner, have the greatest reproductive output if they lay no more than five or six eggs. The parents are unable to feed larger broods adequately, so that the number of surviving offspring from large broods is less than the number from more modest clutches.

The prospects of survival of a young animal or plant may depend on its being well nourished early in life, if it must face competition in a crowded population. The genotypes best adapted to uncrowded conditions are often least fit when crowded. For example, Solbrig (1971) discovered that dandelions from a mowed lawn had high seed output when grown sparsely, but not when subjected to strong competition. Dandelions from an unmowed, weedy site a few feet away were genetically different; they had lesser fecundity at low density but were competitively superior to the lawn dandelions when crowded and produced more seeds.

In resource-limited populations that are often at their carrying capacity (K) and compete for resources, the most fit genotypes are those that most efficiently use resources, even if they grow slowly; that have few offspring, but these well prepared for competition; that live long and reproduce repeatedly. Such K-SELECTED species, then, often evolve characteristics that result in a lower value of r_m, the rate of increase when crowding is least. Because such populations are relatively stable (r, the instantaneous rate of increase, is zero), a genotype with the capacity for rapid increase under uncrowded conditions (r_m large) can be replaced by a genotype with a lower potential rate of increase (r_m small), because the genotypes have no opportunity to express their potential rates of increase. But in populations that frequently grow exponentially — populations of fugitive species, or those decimated by density-independent climatic factors — genotypes with high intrinsic rates of increase are often most fit. This distinction between r-selected and K-selected species, first made by MacArthur and Wilson (1967), has contributed importantly to an extensive, highly mathematical theory of the evolution of demographic characteristics which I have described only superficially (see e.g., Cole 1954; Schaffer 1974; Stearns 1976, 1977).

In K-selected populations evolution may be expected to increase the carrying capacity and hence population density because genotypes that are best able to resist the limiting factor, say by finding food more efficiently, are most fit. This is true in simple models (Anderson 1971); but if the organisms defend resources (e.g., by protecting territories), genotypes that monopolize more resource will outbreed those that monopolize less. Thus population density will decline as such genotypes increase in frequency. A population of birds will be smaller if each holds a larger territory. Thus neither reproductive rates nor population sizes necessarily increase in evolution.

UNITS OF SELECTION

Probably no one definition of selection can satisfy everyone, for the concept embraces a wide variety of phenomena. A reasonably comprehensive definition, however, might be this: selection is the statistically consistent differential survival and/or reproduction of two or more classes of entities. This definition includes three features of selection worthy of note.

All changes in gene frequencies, such as genetic drift, require that one allele replicate itself more than another. Yet genetic drift is not selection, because the differential reproduction is not statistically consistent; it is not correlated with genotype. Conversely a great deal of random mortality occurs, even to individuals with the potentially most advantageous of genotypes. But if individuals with one genotype survive and reproduce better *on average* than those with another, selection occurs.

But if selection is a matter of averages, there must be at least several replicate copies of each gene, several individuals with each genotype, for an average superiority to be expressed; there must be *classes* of genes, or genotypes, or whatever. Because individual genes replicate themselves exactly, each kind of gene exists in many copies, so it is possible to say that one kind has higher fitness on average than another. But in a sexually reproducing species, each individual organism is a unique collection of genes, a multilocus genotype, that does not reproduce itself exactly and will never recur again. It is meaningless to speak of the fitness of Johann Sebastian Bach, because there is no way of telling whether the difference between Bach's reproductive success and that of other people was due to chance or to the properties of his unique genotype.

Selection, then, requires multiple copies and thus heredity. The stronger the degree of heredity, the more efficaciously selection can operate and the longer its effects will last. Thus selection of whole genotypes in a sexual species hardly exists because they do not reproduce themselves; selection of whole chromosomes exists, but is weak, because a superior chromosome ordinarily lasts only a few generations before it is dissolved by crossing over. Short segments of chromosomes (or longer segments held intact by inversions) are so seldom fragmented by crossing over that they can be meaningful selectable entities, meaningful units of selection. But the single gene itself, because of its fidelity of replication, must be the level at which selection has its most enduring effects.

Paradoxically, however, the individual gene is often not the entity to which a fitness value may be ascribed, even though it is the gene, not the genotype, that changes in frequency under selection. A gene does not replicate itself alone and naked, but through the reproduction of the organism that carries it, whose fitness is determined by the whole of the genotype. An allele's average fitness is that of all the individuals that carry it, each with a unique overall genotype. In many cases the phenotypic effect of an allele is so slight, and thus contingent on its interaction with other genes, that it cannot be said to have a fitness independent of the rest of the genotype. I return to this puzzling problem in Chapter 14.

Clearly, then, several kinds of entity in biological systems can be units of selection: genes, chromosomes, anything that replicates itself and retains its identity from generation to generation (*Figure 5*). The

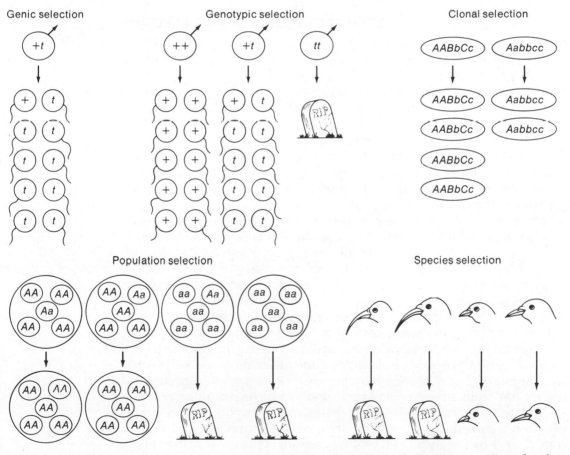

Genic selection   Genotypic selection   Clonal selection

Population selection   Species selection

Diagrammatic illustration of selection at various levels.   **5** units of selection are of many kinds (Lewontin 1970). Cultural patterns, for example, such as words or religious beliefs, have a certain degree of heredity and are subject to a kind of selection; we could say, for example, that Buddhism and Christianity have had higher "fitness" than Manichaeism or Zoroastrianism. Cultural evolution occurs not only in humans, but also in other primates and in many birds in which song dialects develop by learning (Lemon 1975).

Genetic units of selection include genes and to a lesser extent genotypes as well as populations and species, the differences among which are based on the genetic properties of their members. INTERDEMIC SELECTION, or GROUP SELECTION, occurs when populations of one kind originate or die out at a different rate from populations of another kind. The case of the *T* locus in mice exemplifies this process; the genetic composition of an ensemble of populations is partly determined by the higher rate of extinction of demes in which the *t* allele is prevalent. SPECIES SELECTION is the same kind of process. If species with one characteristic speciate more rapidly or become extinct less often than those with another characteristic, the mean phenotype of the species taken as a group shifts.

GROUP SELECTION AND KIN SELECTION

It is difficult at first glance to see how certain characteristics of organisms could have evolved by natural selection. For example, the warning calls that many birds emit when they see predators approaching may benefit the flock, but not the individual who calls and thereby attracts the predator's attention. How, then, could an allele that disposes its bearer to such an altruistic act increase in frequency and not be eliminated from the population? The same problem applies to many of the behavioral traits in social animals that in large part are the fabric of the social organization itself. Perhaps the ultimate act of altruism is found in the worker castes of the social insects, which forego reproduction apparently for the good of the colony.

Other characteristics also pose this problem. Warning (aposematic) coloration is advantageous to a distasteful animal if the coloration is already so widespread that most predators have learned to avoid such animals; but the first aposematic mutant must have been at a disadvantage, drawing the attention of naive predators. Sexual reproduction itself, seemingly, should not have evolved (Maynard Smith 1971, Williams 1975). An allele that codes for sexual reproduction can increase only half as fast as one that causes its bearer to reproduce asexually, because all the offspring of an asexual female are females that will themselves bear offspring, but only half the offspring of a sexual female are daughters (*Figure 6*). Sex may be advantageous to the species, but it does not seem advantageous to the individual organism or gene.

Wynne-Edwards (1962) was, perhaps, most responsible for bringing these problems under careful scrutiny. He raised a good deal of controversy by suggesting that the population densities of many animal species are limited not by extrinsic factors but by altruistic behavioral acts: restraining reproduction or emigrating when increasing popula-

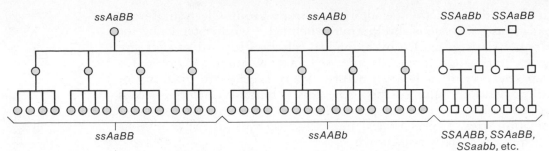

The advantages and disadvantages of an allele *S*, which codes for sexual reproduction, compared to those of an allele *s* coding for asexual reproduction. Each female produces four equally fit offspring, but the frequency of the sexual allele rapidly drops from ½ in the first generation to ⅕ by the third generation. If, however, the environment then changes, so that only *aabb* genotypes survive, allele *S* will persist but *s* will be eliminated. Circles represent females, squares males.

6

tion densities signal impending depletion of resources. Recognizing that a genotype that refrained from reproducing or emigrated for the good of the population would be at a disadvantage, he suggested that such behavior must have evolved by interdemic, or group, selection. Populations that by chance consist of altruistic genotypes and refrain from excessive reproduction or overexploitation of their resources persist; populations that overexploit their resources become extinct (*Figure 5*).

Williams (1966) and others have argued, however, that group selection is very unlikely to be important. "Altruistic" alleles, if such exist, can increase within populations only by genetic drift, against the strong force of natural selection. A population of altruists is genetically unstable in that a single "selfish" allele that enters by mutation or gene flow must rapidly replace the altruistic allele; and this process can be countered by group selection only if populations of nonaltruists become extinct at a very high rate. Williams argued that because populations originate and are extinguished at a much lower rate than the individuals that compose them, group selection must be a weaker force than individual selection.

The question, then, is whether genes that promote the survival of a population, but lower the survival and reproduction of the individuals that bear them, can increase in frequency. In the last few years several theoretical analyses of this problem have been carried out (see Wilson 1975a and references therein). Levin and Kilmer (1974), using computer simulation, determined that an altruistic allele could be maintained in a fixed or polymorphic state by group selection, but only under highly restrictive conditions. The altruistic allele must not be too deleterious to its bearers, but it must very appreciably enhance the survival of the population; there must be very little gene flow (less than 5 per cent, generally) among demes; and the population must be very small (10–25 individuals).

In laboratory populations group selection has been shown to counteract individual selection (Wade 1977). Strong group selection of large populations of *Tribolium* beetles overwhelmed a weaker tendency for the populations to evolve by individual selection toward smaller size (*Figure 7*). But not enough is known to say whether conditions for group selection often hold in nature. Many species are made up of small populations among which gene flow is limited (*Chapter 11*), and the extinction rates of such populations may be quite high; but whether population sizes, dispersal rates, and extinction rates fit the conditions for group selection is not known.

Group selection may well happen in nature, but most of the phenomena it is invoked to account for can be at least as plausibly explained in other ways. Despite Wynne-Edwards' argument, predator populations are probably more often limited by extrinsic factors over which they have no control than by self-restraint (Slobodkin 1974). And many seemingly altruistic traits can be explained by a special form of selection at the level of the gene, KIN SELECTION.

The first mutation for warning coloration in a population of cryptic butterflies, if it arises in only a single individual, is probably eliminated by a naive predator. But if it arises in the germ line of the parent and is transmitted to a whole brood of butterflies, the aposematic allele that the siblings carry may have a higher chance of survival than cryptic alleles in the population at large, because a predator that kills one of the brood may on the basis of this experience avoid the rest (*Figure 8*). That is, the allele accrues high fitness not directly by enhancing the survival of its bearer, but indirectly by enhancing that of its bearer's relatives (kin) who carry other copies of the allele. (Compare Bennett's indirect selection of DDT resistance by sib selection, *Chapter 10*.)

This indirect selection, or kin selection, is actually a very obvious concept; it is the basis of parental care. A female mammal benefits herself nothing by suckling her young, but she is enhancing the survival of her own genes, half of which (on average) are carried by each

7

Mean numbers of adult *Tribolium castaneum* beetles in populations evolving by individual selection only (C) and by group selection for high (A) and for low (B) population size. Individual selection (C) causes a decline in population size, an interesting effect in itself. This can be enhanced (A) or counteracted (B) by group selection, which was imposed by propagating new demes in each generation only from those demes that had the highest (A) or lowest (B) densities in the previous generation. In series C all demes were propagated in each generation; there were no extinctions. (After Wade 1977)

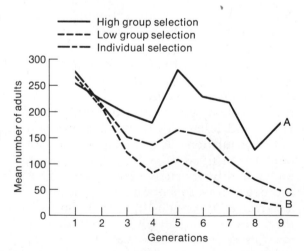

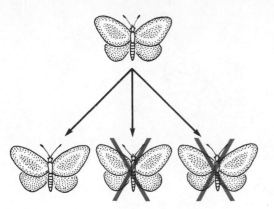

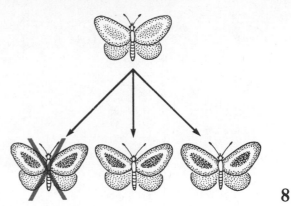

8

An example of kin selection. Cryptically colored individuals fall prey to predators. A newly arisen mutation carried by the members of one family confers higher survival if the predators that kill part of the family avoid those that look like it. The mutation responsible for the warning coloration is thus propagated.

of her offspring. The rate of increase of a gene, then, depends on its INCLUSIVE FITNESS (Hamilton 1964): the reproductive rate of its bearer plus that of the copies of the gene carried by relatives to whom the bearer may extend help. The higher the coefficient of consanguinity between two relatives, the more a given degree of sacrifice increases the fitness of a gene. Thus, hypothetically, since half my genes are carried, on average, by a son or daughter, half by a sibling, and one-fourth by a niece or nephew, my genes would enjoy a higher rate of increase if I sacrificed my life and future reproduction on behalf of more than two offspring or siblings, or more than four nieces or nephews, than if I refrained from self-sacrifice. Who "should" sacrifice how much for whom is a rather intricate theory; for example, parents might be expected to sacrifice more for their offspring than *vice versa,* since the expected reproductive output of an adult who has already reproduced is less than that of one of its offspring (again compare the concept of reproductive value).

Altruism can evolve only if the behavior benefits other bearers of the altruistic allele more than the individuals who do not carry it (*Figure 9*). Otherwise there is no differential reproduction of altruists and nonaltruists, except for the loss of reproduction that the altruist suffers. It seems, then, that behavior benefiting the population as a whole — such as "prudent" predation that would avert overexploitation of a resource — should not evolve. If predators seem prudent, it must be either that they cannot achieve higher proficiency of predation (Slobodkin 1974) or that the most proficient populations or species of predators have indeed overexploited their resources and become extinct, leaving us only inefficient predators (Gilpin 1975). This would be evolution by group selection.

But altruism generally bestowed on other members of the popu-

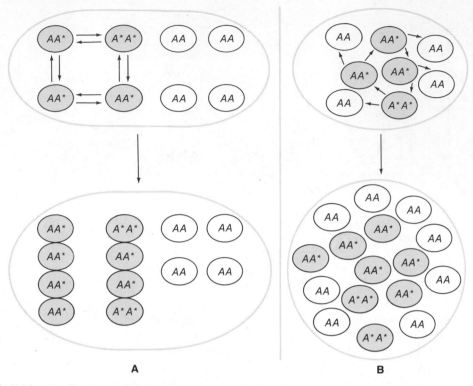

9 "Altruism" can evolve by kin selection only if the benefits of the altruistic act accrue preferentially to other bearers of the alleles (A^* in this figure) that cause altruistic behavior. In (A), altruists raise each others' fitness, so allele A^* increases in frequency; but in (B), bearers of both A^* and A profit equally from the altruists' presence, so A^* does not increase in frequency.

lation can evolve if most members of the population are relatives. This is most likely when the population is highly inbred. The coefficient of inbreeding of a population is likely to be greatest when the population is small and there is little gene flow (*Chapter 11*). But these, the conditions that are favorable for the evolution of social behavior, are also favorable for the operation of group selection, so it is hard to tell whether a given altruistic trait has evolved by kin selection or group selection. The social behavior of animals often knits them together into inbred flocks or tribes and so enhances the opportunity for both kin selection and group selection. Thus social behavior may reinforce its own evolution. But even in non-social species, individuals may interact more often with relatives than non-relatives by chance alone, so opportunities for the evolution of altruism may be common (Wilson 1977).

The theory of kin selection may in principle explain a diverse array of behavioral traits, from cannibalism (or its absence) to sterility in the social insects; the subject is treated in its full intricacy in E. O. Wilson's (1975*a*) comprehensive treatise on sociobiology. Social behavior in animals often fits the predictions of kin-selection theory (e.g., Sherman

1977); but there have been more indirect than direct tests of the theory, for a genetic basis for altruistic behavior has not yet been identified and studied in any species.

GROUP SELECTION AND THE EVOLUTION OF SEX

Of the traits that seem advantageous to a species or population but not to an individual, none is more puzzling than sexual reproduction itself. In large populations at least, recombination can hasten the rate of evolution (Crow and Kimura 1965, Felsenstein 1974; Chapter 10) and so lessen the likelihood of extinction by environmental changes. But this advantage to the species does not explain why sex should have evolved in the first place; genes conferring recombination do not rise in frequency because they will avert future extinction. The lower extinction rate of sexual than of asexual forms may be responsible for the prevalence of sex (as may be the higher speciation rate of sexual forms; Stanley 1975b); but if so, it is group selection or species selection that explains its ubiquity.

Even though recombination is not advantageous to the individual, it is advantageous at the level of the individual gene (Felsenstein and Yokoyama 1976). An allele that promotes recombination finds itself in a greater variety of genetic backgrounds than one that promotes parthenogenesis. If the environment fluctuates so that different gene combinations are favored at different times, the recombination allele is preserved and increased along with the recombined genotypes with which it is associated (Maynard Smith 1971). Even in a constant environment, there is severe competition among genotypes; if enough genotypes are generated by a recombination allele, some are likely to be fitter than the best parthenogenetic genotype and will win the competition (Williams 1975, Williams and Mitton 1973).

Although there is a clear advantage to alleles that promote recombination, there is a serious disadvantage as well. If a parthenogenetic genotype and a sexual genotype have the same fecundity, the intrinsic rate of natural increase of the sexual genotype is only half that of the parthenogenetic form, because all the latter's offspring are female and thereby contribute to the growth of the population. Williams and Mitton (1973) have calculated that if fecundity is great enough, the likelihood of having a winning genotype (better than the parthenogenetic genotypes) among a sexual female's progeny outweighs this enormous disadvantage — so sexual reproduction is favored. But in less fecund species such winning genotypes seldom appear, so sexual reproduction seems disadvantageous, in the balance. Perhaps it is indeed maintained by group selection.

ADAPTATION AND ADAPTATIONS

It is generally agreed that evolution by natural selection is a process of adaptation to the environment whereby species become better adapted. I do not usually dispute this statement, but I am not always sure what it means. The concept of adaptation is rather elusive.

For the sake of clarity I will define *adaptation* as a *process* whereby the members of a population become suited over the generations to survive and reproduce. This process increases the level of *adaptedness* of the population's members. *An adaptation* is a feature of an organism or population whereby its adaptedness is achieved. The feature is adaptive because of its *adaptive function,* the specific means by which it enhances survival or reproduction. Thus, the hydrogen cyanide in white clover leaves is an adaptation; its adaptive function, or adaptive significance, is to reduce herbivory; and it evolved by a process of adaptation that, we may presume, increased the adaptedness of clover populations.

It is not always easy, however, to determine whether a feature is an adaptation, nor to determine its function. It takes little ingenuity to hypothesize a function for a characteristic and thereby to rationalize why it evolved, but it is quite another matter to prove the hypothesis. Evolutionary biology suffers, in fact, from unbridled speculations about adaptive functions — the "idle Darwinizing," as Lewontin (1977b) has called it, that makes amusing party conversation but bad science. In *The Naked Ape,* to pick an egregious example, Morris (1967) "explains" all sorts of human features more or less plausibly; but he provides no strong evidence to support such proposals as that women's breasts are larger than those of other primates because they mimic the buttocks and so act as sexual signals. Some of these facile stories are quite amusing; the exposed mucosa of our lips is supposed to serve in a sexual context because of their resemblance to the vulval labia. The more professional evolutionary literature is full of equally unsupportable tales of adaptation, especially about various aspects of animal behavior.

Anyone can usually think of several or many functions for a feature. But its obvious function is sometimes illusory, and some features almost certainly have no direct adaptive significance at all. Physiologists have long argued, very plausibly, that the scales of reptiles serve (in part) to reduce water loss through the skin, but then they found that an aberrant scaleless gopher snake was as resistant to dessication as its scaled fellows (Licht and Bennett 1972). The traditional explanation of Bergmann's rule (homeotherms are larger in colder regions) is that the lower surface-volume ratio of a large animal reduces heat loss, but the trends in body size may actually be evolutionary responses to interspecific competition (McNab 1971). Distinguishing between these hypotheses may not be easy.

Biological features are not necessarily adaptive. The process of mutation has the *consequence* of increasing genetic variation, but this is almost certainly not its *function (Chapter 10).* Rather, mistakes in replication are probably unavoidable (Williams 1966). And as Dobzhansky (1956) has eloquently argued, many features are the partial products, the side-effects so to speak, of adaptive developmental patterns but are not in themselves adaptive. The bristles on the head of *Drosophila* are part of a pattern that includes the ocelli (Maynard Smith and Sondhi

1960); the number and arrangement of bristles is itself probably not adaptive.

Allometric relations may have nonadaptive consequences. For example, among primates only humans have a prominent nose and a protruding chin. These seem (S. J. Gould 1977 and references therein) to be consequences of humans' lower rate of development of the jaws relative to the cranium. The relative shortness of our jaws may be either adaptive in itself, or it may be merely one of the many aspects of our neotenic development that makes us resemble the young of other primates (which have short jaws). In any case we may have a prominent nose simply because it retains its original growth rate and position while the bones below it do not grow out to meet it. And the chin is formed because the growth of the upper tooth-bearing part of the mandible is more retarded than that of the underlying basal part. Thus adaptive significance may lie more in whole developmental patterns than in their individual components or end products.

The adaptive significance of a feature should be demonstrated, not assumed. The best evidence that a feature evolved by natural selection is derived from the study of polymorphisms; cyanogenic clover plants, for example, suffer less herbivory than acyanogenic plants (Jones 1973). Experimental manipulations can demonstrate adaptive functions. Bentley (1976, 1977) showed that extrafloral nectar glands attract ants, which reduce damage to the plant by herbivorous insects. Functional analyses often provide sound evidence for adaptation. Engineering analyses of stresses, for example, explain the size and shape of skeletal elements (Gans 1974). And an adaptive function can often be inferred if a characteristic in unrelated species is correlated with a specific environmental factor; the purplish color and carrion odor typical of so many fly- and beetle-pollinated flowers can be presumed to attract these insects. But a unique species-specific feature cannot be analyzed in this way. This is one reason why the adaptive significance, if any, of uniquely human traits is less often demonstrated than hypothesized.

WHAT IS MAXIMIZED IN EVOLUTION?

If an adaptation is difficult to demonstrate, adaptedness is even more so. Evolution by natural selection is supposed to increase the adaptedness of species, a notion that is even formalized in Fisher's fundamental theorem of natural selection (*Chapter 13*). But there are no objective criteria by which the adaptedness of populations can universally be measured. Natural selection can increase or decrease a population's potential growth rate, or its abundance (*Figure 7*), or its ecological versatility, or its developmental or behavioral plasticity. Adaptive evolution in some instances produces species with broad ecological tolerances, which can occupy wide geographic ranges and withstand pronounced environmental changes. In other cases, its product is narrow specialization and susceptibility to extinction. Extinction, indeed, seems to be the eventual fate of all species. As Tennyson lamented in *In Memoriam*:

Are God and Nature then at strife,
   That Nature lends such evil dreams?
   So careful of the type she seems,
So careless of the single life. . . .

'So careful of the type?' but no.
   From scarpéd cliff and quarried stone
   She cries 'A thousand types are gone;
I care for nothing, all shall go.'

There are no universal goals nor tendencies in evolution, nothing that is universally maximized — not even the likelihood that the species will continue to survive. Nor can there be, for the direction of evolutionary change at any time depends on the genetic variants that happen to exist and the environmental events that impinge on the population. By the nature of selection, species will not adhere to any single path, much less be directed by final causes to predestined ends. Instead, species are each unique genetic systems, shaped by efficient causes that vary from species to species, in space and in time. And diverse reactions have diverse results. We cannot judge a wasp superior to a clam, a raccoon more adapted than a crocodile, an ape more successful than a mole. To attempt the judgment is to compare the incomparable and to erect arbitrary standards of excellence to which natural selection does not adhere.

SUMMARY

Natural selection consists of differences in survival and reproduction among different hereditary entities. These entities may be genes, genotypes, populations, or species, so selection can act at each of these several levels. The differential survival and reproduction of genes and genotypes within populations, however, is generally considered the most potent force in evolution. The growth rate, or fitness value, of an allele relative to that of other alleles determines how rapidly genetic change occurs by selection. The fitness of an allele is determined, through its effect on the phenotype, by its interaction with the prevailing external environment and with the rest of the genotype in which it is embedded. Because fitness depends on current rather than future circumstances, natural selection does not act for the ultimate good of the species and can sometimes even cause the species' extinction. Natural selection thus does not necessarily improve a species' adaptedness. Indeed adaptedness is a difficult concept to define. Although many (but not all) characteristics of species are adaptations that have evolved by natural selection, no single property of species increases universally during evolution. Thus there is no criterion by which we might compare the adaptedness, or evolutionary "success," of different species.

FOR DISCUSSION AND THOUGHT

1 Gene frequencies track environmental changes with a time lag, so that a population may not be quite up to date in adaptation. How could one determine when organisms are not yet fully adapted? Need we suppose that organisms are invariably as well adapted to their environments as they might be?

2   Discuss the possible meanings of the word *adapted* and how the degree of adaptedness might be measured.

3   If natural selection cannot prepare a population for future events, how can we explain cases of apparent foresight, such as the southward migration of birds before the onset of cold weather?

4   Fisher's fundamental theorem of natural selection (*Chapter 13*) says that a population's fitness increases under natural selection. How can this hypothesis be tested? How can this theorem be reconciled with the fact that extinction is common?

5   Some authors have supposed that a species, like an individual, has a life cycle that is necessarily followed; it is born by speciation, flourishes, and then dies out. Does any mechanism exist that could cause this to happen universally? Criticize the argument.

6   Slobodkin and Rapoport (1974) write that organisms seem to be playing a "game against nature" in which the only "pay-off" is continued survival as a species. Evaluate this analogy and its implications.

7   *Drosophila* with the gene "grandchildless" have a normal number of offspring but have no grandchildren. Thus it takes two generations for their low fitness to become evident. How long do the descendants of different genotypes have to be traced to decide which is more fit? Discuss the implications of this example for definitions of fitness and of selection.

8   In our society many people suffer pain from wisdom teeth; the pattern of our tooth development obviously is not optimal. Suppose, as I imagine is the case, that the development or nondevelopment of wisdom teeth is in part genetically determined, but that our survival and reproduction is not at all affected by this difference. Will we lose wisdom teeth by evolution? Extrapolate the argument to other aspects of the future evolution of technological peoples. What is the general evolutionary implication of this argument?

9   Discuss the roles of individual selection and group selection in the evolution of dispersal (see Van Valen 1971). In what kinds of environments do you expect organisms to evolve a high capacity for dispersal?

10 Although it seems that the ability to kill one's competitors should be adaptive, ritualized fighting rather than lethal combat has evolved in many species. Provide an explanation. (See Maynard Smith 1972.)

11 Discuss the ways in which cultural analogues of selection, gene flow, and genetic drift might shape the historical changes in a language or in other cultural patterns.

12 I have stressed in this chapter that selection acts on genes, genotypes, populations, and so on, but it is often said that selection acts on the phenotype. Integrate these points of view.

13 Behavior that reduces another individual's fitness without increasing that of the perpetrator may be considered spiteful; is such behavior likely to evolve? (See Wilson 1975a, p. 119.)

14 The principle of kin selection enables us to recognize that cooperation can evolve as readily as competition. Does this have any bearing on the common supposition that humans are innately aggressive? If human social behavior evolved largely by kin selection, as Wilson (1975a) and Alexander (1974) suggest, does this mean that we are genetically programmed to be kinder to relatives than to nonrelatives? What predictions about human social behavior can be made from the theory of kin selection or the theory of natural selection in general? See Chapter 19.

15 In much recent evolutionary literature a model is deemed valid if data from the natural world are consistent with the model's predictions. For example, Weinrich (1977) specifically invokes this criterion of validation to conclude that noncoital sexual activities in humans are adaptive, by reinforcing the pair bond. Is this a sufficient ground on which to judge a hypothesis of adaptive significance? What would be necessary to prove Weinrich's hypothesis?

MAJOR REFERENCES

Lewontin, R. C. 1970. The units of selection. *Ann. Rev. Ecol. Syst.* 1:1–18. Treats the essence of natural selection at its several levels.

Williams, G. C. 1966. *Adaptation and natural selection: A critique of some current evolutionary thought.* Princeton University Press, Princeton, N.J. 307 pages. A highly readable essay, dealing mostly with the implications of individual selection and group selection. Perhaps the clearest, most forceful statement of what natural selection is.

Wilson, E. O. 1975. *Sociobiology: The new synthesis.* Harvard University Press, Cambridge, Mass. 697 pages. Chapters 4 and 5 provide a clear encapsulation of individual, kin, and group selection.

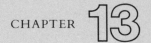

The Effects of Selection on Gene Frequencies

Natural selection is the central concept in evolutionary biology. The diverse attributes of organisms are very largely adaptations shaped by natural selection from the genetic variation that arises by mutation and recombination. Differential survival and reproduction of different genotypes, with different phenotypic properties, changes the genetic composition of a population. This process of individual selection within populations and the differential survival of populations and species that have diverged by this same process constitute evolution by selection. The fundamentally important event in evolution, then, is a change in gene frequency by natural selection.

The simplest, but naive, view of selection is this: a new mutation that arises confers superior survival and reproduction, becomes more and more common, and ultimately replaces the preexisting allele. This is directional evolution, by DIRECTIONAL SELECTION. It is exemplified by the well-known case of the industrial melanic form of the peppered moth (*Biston betularia*) which in about 50 years increased from extremely low frequencies to about 99 per cent in some populations, because melanic moths suffer less predation from birds than the typical gray moths. This process of gene replacement is indeed the essence of evolution by natural selection. But many factors complicate the picture, affecting how fast genetic change occurs and whether the new allele completely replaces the former allele. Some modes of selection, unlike this process of directional selection, do not tend toward an invariant, monomorphic population, but maintain gene frequencies at intermediate values. The variation thus maintained, by various forms of BALANCING SELECTION, is the material on which directional selection is most likely to act if the environment and thus the mode of selection is altered. That is, much of directional evolution consists of increases in the frequency not of new mutations but of alleles that were already rather abundant. This is because the rate of gene frequency change depends on the gene frequency itself. To see this, it is necessary to examine formal models of selection. In this examination both the factors that cause allele replacement and those that maintain variation will come to light.

DIRECTIONAL SELECTION

The rate of gene frequency change by selection depends on the RELATIVE FITNESSES of the various genotypes, their relative rates of increase at any time (*Chapter 12*). Since the fitnesses are relative, it is convenient to set the fitness of the most fit genotype equal to 1. Then the relative fitness of a genotype with 80 per cent as great a capacity for increase is 0.8. The difference, 0.2, is the COEFFICIENT OF SELECTION (usually denoted s) against the inferior genotype.

Assume that selection consists of differential mortality among genotypes that are in Hardy-Weinberg equilibrium in each generation, before the selective mortality occurs. Assume for the moment, moreover, that the relative fitnesses remain constant from generation to generation. Consider the case of an advantageous allele A' that is recessive with respect to fitness. The relative fitnesses must then be $1 - s$, $1 - s$, and 1 for genotypes AA, AA', and $A'A'$ respectively (allele A is dominant with respect to fitness, so AA and AA' must have the same, inferior fitness). Because of differential mortality, A' must increase in frequency in any given generation by an amount $\Delta q = q' - q$, the difference between the frequency q among the zygotes before selection and q' after mortality has occurred.

The new frequency of A' is the fraction of A' genes that survive, divided by the fraction of all genes (A and A') that survive. Of the q^2 A' genes in homozygous condition, the relative proportion that survives is 1; of the $(\frac{1}{2})(2pq)$ A' genes in the heterozygotes, the relative proportion that survives is $1 - s$. Therefore the fraction of surviving A' genes is $(\frac{1}{2})(2pq)(1 - s) + q^2(1)$, or $pq(1 - s) + q^2$. Of all (A and A') genes a proportion p^2 is in AA homozygotes, of which $1 - s$ survives; $2pq$ is in heterozygotes, of which $1 - s$ survives; and q^2 is in $A'A'$ homozygotes, of which a relative fraction 1 survives. Thus $p^2(1 - s) + 2pq(1 - s) + q^2(1) = 1 - s(1-q^2)$ is the surviving fraction of all genes. The frequency of A' after selection must therefore be $q' = [pq(1 - s) + q^2]/[1 - s(1-q^2)]$. The numerator of this equation is, very simply, half the proportion of surviving heterozygotes, plus the surviving $A'A'$ homozygotes. The denominator is the total initial population, $p^2 + 2pq + q^2$ (=1), less the proportion s of inferior genotypes AA and $A'A$ ($p^2 + 2pq$, which equals $1 - q^2$) that died. The denominator, the relative proportion of all genes that survive, is the average fitness of individuals in the population and is often denoted $\bar{w}$.

Substituting this expression for q' into $\Delta q = q' - q$ and after a little algebra, we obtain the change in gene frequency per generation:

$$\Delta q = \frac{sq^2(1 - q)}{\bar{w}} \quad \text{or} \quad \Delta q = \frac{spq^2}{\bar{w}}.$$

This equation is a special instance, for an advantageous recessive allele, of a more general equation derived in Box A.

This equation shows that the advantageous allele increases in frequency (Δq is positive), at a rate that is proportional to the strength of selection s and to both the frequency p of the deleterious allele being replaced and the frequency q of the advantageous allele itself. Thus the rate of genetic change is great only if both alleles are common in the population. For example, suppose $s = 0.1$; then if $q = 0.001$ (as it might be if A' had recently arisen by mutation), $\Delta q = 0.0000001$; if $q = 0.1$ (as it might be if A and A' were a polymorphism), $\Delta q = 0.001$.

The gene frequency therefore has a critical impact on the response of a locus to selection. Newly arisen mutations are likely to contribute little to a population's adaptation to environmental changes, because they increase in frequency very slowly when rare. Complete replacement of one allele by another takes a very long time (*Table I; Figure 1*), because the process is slow both when the favorable allele is very rare and when the deleterious allele has become very rare. In fact, a deleterious recessive allele is seldom completely eliminated by selection, because when it is rare it occurs only in heterozygotes, in which it is "protected" by dominant alleles. Natural populations should therefore have many deleterious recessives in low frequency, as is in fact the case (*Chapter 9*).

TABLE I. Generations required for change in frequency of a deleterious allele

CHANGE IN FREQUENCY Selection coefficient (s):		NUMBER OF GENERATIONS						
		1 (lethal)	0.80	0.50	0.20	0.10	0.01	0.001
No dominance								
From q_0	To q_n							
0.99	0.75	3	4	7	17	35	350	3,496
0.75	0.50	1	1	2	5	11	110	1,099
0.50	0.25	1	1	2	5	11	110	1,099
0.25	0.10	1	1	2	5	11	110	1,099
0.10	0.01	2	3	5	12	24	240	2,398
0.01	0.001	2	3	5	12	23	231	2,314
0.001	0.0001	2	3	5	12	23	230	2,304
Deleterious recessive								
From q_0	To q_n							
0.99	0.75	}1	5	8	21	38	382	3,820
0.75	0.50		2	3	9	18	176	1,765
0.50	0.25	2	4	6	15	31	310	3,099
0.25	0.10	6	9	14	35	71	710	7,099
0.10	0.01	90	115	185	462	924	9,240	92,398
0.01	0.001	900	1,128	1,805	4,512	9,023	90,231	902,314
0.001	0.0001	9,000	11,515	18,005	45,011	90,023	900,230	9,002,304

(After Strickberger 1968)

A   Gene Frequency Change under Selection

Suppose we have these genotypes with these fitnesses:

Genotype	AA	AA'	$A'A'$
Frequency	p^2	$2pq$	q^2
Fitness (w)	1	$1 - s_1$	$1 - s_2$

Then the *average fitness* is

$$\bar{w} = p^2 + 2pq(1 - s_1) + q^2(1 - s_2)$$
$$= 1 - 2pqs_1 - q^2s_2$$
$$= 1 - 2s_1q + 2s_1q^2 - s_2q^2.$$

Also

$$d\bar{w}/dq = -2s_1 + 4s_1q - 2s_2q$$
$$= 2(-s_1 + 2s_1q - s_2q).$$

After one generation the new gene frequency q' will be the proportion of A' genes that survive divided by the proportion of all genes that survive. This latter proportion is the sum of all the genotype frequencies, each multiplied by the fraction that survives, or

$$p^2(1) + 2pq(1 - s_1) + q^2(1 - s_2)$$
$$= 1 - 2pqs_1 - q^2s_2 = \bar{w}.$$

The fraction of A' genes that survives is given by half the fraction of surviving heterozygotes, plus the contribution from surviving $A'A'$ homozygotes: $\frac{1}{2}(2pq)(1 - s_1) + q^2(1 - s_2)$. So

$$q' = \frac{\frac{1}{2}(2pq)(1 - s_1) + q^2(1 - s_2)}{1 - 2pqs_1 - q^2s_2}.$$

Then the change in gene frequency is $\Delta q =$ $q' - q$, which after some algebra becomes

$$\Delta q = \frac{pq(2s_1q - s_1 - s_2q)}{1 - 2pqs_1 - q^2s_2}. \tag{1}$$

This is a general expression if fitnesses are constant. For example, suppose $s_1 = 0$, so that A' is deleterious but completely recessive with respect to fitness. (That is, the heterozygote and the AA homozygote have the same, higher fitness.) Then

$$\Delta q = \frac{pq(-s_2q)}{1 - s_2q^2} = \frac{-s_2pq^2}{\bar{w}}, \tag{2}$$

which is the general expression for selection against a recessive gene.

Note that eq. 1 includes the term

$$2s_1q - s_1 - s_2q,$$

which is $\frac{1}{2}d\bar{w}/dq$.

Substituting this into eq. 1, and substituting $\bar{w}$ for the denominator $1 - 2pqs_1 - q^2s_2$, we obtain

$$\Delta q = \frac{pq}{2\bar{w}} \cdot \frac{d\bar{w}}{dq}, \tag{3}$$

which is the most general expression for gene frequency change under selection. As long as selection coefficients are considered constants, then whatever the selection coefficients are, $\bar{w}$ may be calculated as before, and direct substitution into eq. 3 yields an expression for the rate of gene frequency change that is proper to the particular case.

The rate of evolution at a locus depends not only on gene frequencies and on the strength of selection, but on the phenotypic expression of the trait. An advantageous recessive allele (fitnesses $1 - s$, $1 - s$, 1 for AA, AA', and $A'A'$ respectively) increases at the rate $\Delta q = spq^2/[1 - s(1 - q^2)]$, as we have seen; meanwhile, the deleterious dominant allele is eliminated at the rate $\Delta p = -spq^2/[1 - s(1 - q^2)]$. If the recessive allele A' is deleterious (fitnesses 1, 1, $1 - s$ for AA, AA', and $A'A'$ respectively), it is eliminated at the rate $\Delta q = -spq^2/(1 - sq^2)$, while the advantageous dominant A increases at the rate $\Delta p = spq^2/(1$

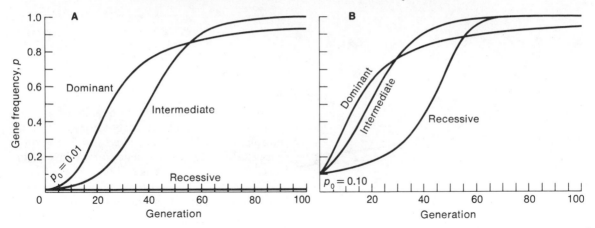

Increase of an advantageous allele from an initial frequency of (A) $p_0 = 0.01$, as if recently arisen by mutation, and (B) $p_0 = 0.10$, as if already present as a polymorphism. Change in gene frequency is shown when the advantageous allele A is dominant with respect to fitness, neither dominant nor recessive ("intermediate") to its allele A', and recessive. Fitness values of AA, AA', and $A'A'$ are respectively 1, 1, 0.8 for the dominant case; 1, 0.9, 0.8 for the intermediate case; 1, 0.8, 0.8 for the recessive case. Note that a newly arisen recessive mutation hardly increases at first because it is so rarely exposed in homozygous form. The intermediate allele reaches fixation before either the dominant or the recessive because the dominant reaches fixation only as the deleterious recessive A' is eliminated — a very slow process. The final approach to fixation of the recessive is very rapid because a deleterious dominant A' is being eliminated, a rapid process. Notice that a new advantageous dominant mutation is fixed more rapidly than an equally advantageous recessive, a possible reason for the prevalence of dominant alleles in natural populations. (Computer simulation courtesy of J. P. W. Young)

1

$- sq^2)$. Thus a rare advantageous dominant increases much more rapidly than a rare advantageous recessive, but it achieves complete fixation more slowly because the last few recessive genes are protected in heterozygous form (*Figure 1*). If neither allele is dominant, the fitnesses of AA, AA', and $A'A'$ may be denoted 1, $1 - hs$, and $1 - s$ respectively; $h = 0.5$ if the heterozygote has precisely intermediate fitness, 1 if A' is fully dominant, 0 if A' is fully recessive. Then the advantageous allele A increases at the rate

$$\Delta p = \frac{spq[q + h(p - q)]}{1 - sq(2hp + q)}.$$

The increase of such an allele is shown in Figure 1.

In at least some instances gene frequency changes follow their predicted courses closely; a recessive lethal allele in experimental populations of the flour beetle *Tribolium castaneum*, for example, decreased as predicted from the high initial frequency that the experimenter (Dawson 1970) had established (*Figure 2*).

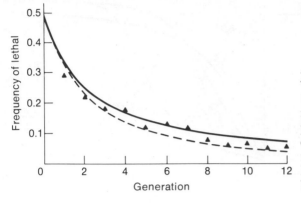

2

Observed decrease in the frequency of a recessive lethal allele in a laboratory population of *Tribolium*. The solid line is the expected change in frequency of a completely recessive lethal; the broken line is that of a lethal that lowers the fitness of heterozygotes by 10 per cent. (From Dawson 1970)

Directional selection exists at a locus if the fittest genotype is a homozygote. The favorable allele increases until it is fixed; the population becomes monomorphic. (The gene frequency at equilibrium is found by setting Δp or $\Delta q = 0$ and solving for p or q.) The population is therefore "cleansed" of deleterious alleles; but these alleles might well be favorable under other environmental conditions. Under directional selection the population loses variation and can adapt to environmental changes only if new mutations occur.

In fact, however, populations are genetically variable at many loci (*Chapter 9*); adaptation to environmental changes is often very rapid, for it entails the increase in frequency not of new mutations or very rare recessive alleles, but of alleles that have persisted in a polymorphic state. Genotypes can differ in fitness, yet genetic variation persists, for several possible reasons.

THE BALANCE OF SELECTION AND RECURRENT MUTATION

A deleterious allele will not be completely eliminated if it arises repeatedly by mutation; its frequency arrives at an equilibrium between input from mutation and loss by selection. The equilibrium frequency of a deleterious recessive allele, for example, is $\hat{q} = (u/s)^{1/2}$, where u is the mutation rate and s is the coefficient of selection against $A'A'$ (*Box B*). Thus, if selection is very weak, the allele will be rather common; if strong, rare. For example, if $u = 10^{-5}$ and s is only 10^{-4}, then $\hat{q} \approx 0.32$ at equilibrium; if $s = 10^{-2}$, then $\hat{q} \approx 0.1$. A recessive lethal ($s = 1$) should have an equilibrium frequency of $\sqrt{u}$, or about 0.003, in large populations. Thus recurrent mutation may explain the existence of rare variants such as albinos or the recessive lethals in *Drosophila* populations. (However, there has been a long argument about these rare variants. One school of thought, including Crow and Temin 1964, has held that these alleles are deleterious in both homozygous and heterozygous form and so persist only because of recurrent mutation. The other, including Dobzhansky, Krimbas, and Krimbas 1960, has argued that they are slightly advantageous in heterozygous condition and are kept in the population by overdominance. The debate, reviewed by Lewontin 1974, is too complex to treat here and has never been fully resolved.)

SELECTION BALANCED BY GENE FLOW

Like mutation, the influx of alleles from other populations can prevent the elimination of an allele by selection. The balance between the influx by migration of an allele favored elsewhere and its elimination from a local population where it is disfavored are treated in Chapter 11. The frequency of the allele will be higher with higher rates of gene flow, and lower with more intense selection. Thus a small population that receives many immigrants from a larger population or a population that is only poorly isolated in a distinctive environment will not become very highly differentiated from neighboring populations, unless the difference in selection regimes is very great. The water snakes on the Lake Erie islands illustrate this principle (*Chapter 11*).

HETEROZYGOUS ADVANTAGE

Selection itself can maintain polymorphism. One form of such balancing selection is HETEROZYGOUS ADVANTAGE, also called overdominance for fitness or heterosis for fitness. If the heterozygote is fitter than either homozygote, both alleles are lost by the death or inferior reproduction of homozygotes, but they are retained by the surviving heterozygote; thus both persist in the population. If the relative fitnesses of the three genotypes AA, AA', and $A'A'$ are $1 - s$, 1, and $1 - t$ respectively, the gene frequencies converge from any initial values (*Figure 3*) to the stable equilibrium $\hat{p} = t/(s + t)$, $\hat{q} = s/(s + t)$ (*Box C*). Thus the position of the equilibrium depends on the balance of fitness of the two homozygotes.

One of the very few (some would say only) well-documented cases of heterozygous superiority at a single locus is that of sickle-cell hemoglobin. Persons homozygous for Hb^S suffer severe anemia and usually die before reproducing; thus $t \approx 1$. Heterozygotes have slight anemia (sickle-cell trait), but in parts of Africa where falciparum malaria is prevalent, their survival rate is higher than that of homozygotes for

---

B   **Balance between Selection and Recurrent Mutation**

Suppose A mutates to A' at a rate u, so the change in the frequency of A' in one generation is $\Delta q = up$. If A' is a deleterious recessive allele, its loss per generation (as shown in Box A) is

$$\Delta q = -spq^2/(1 - sq^2).$$

If q is near 0, then $1 - sq^2 \cong 1$, so the factors operating together yield a change in gene frequency of

$$\Delta q \cong up - spq^2 = p(u - sq^2).$$

At equilibrium, $\Delta q = 0$, so $0 = u - sq^2$ and the equilibrium gene frequency is $\hat{q} = (u/s)^{1/2}$.

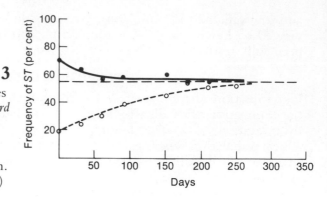

3

Approach to equilibrium when heterozygotes are most fit. Points are frequencies of *Standard* chromosomes in two laboratory populations of *Drosophila pseudoobscura* carrying both *Standard* (*ST*) and *Arrowhead* chromosomes, which are most fit in heterozygous condition. (From Wallace 1968, after Dobzhansky 1948)

normal hemoglobin Hb^A, which suffer a disadvantage of $s \approx 0.15$. The expected frequency of Hb^S is therefore $\hat{q} = 0.15/1.15 = 0.13$, which is about the frequency actually observed in West Africa (Allison 1961, Cavalli-Sforza and Bodmer 1971). Heterozygous superiority seems to explain the persistence of some other natural polymorphisms, such as that for coloration in the copepod *Tisbe reticulata* (*Chapter 9; Table II*).

It is frequently observed that heterozygous genotypes are fitter than homozygous genotypes, but seldom can their superiority be shown to stem from overdominance at a single locus. If two strains are homozygous for different deleterious recessives, the heterozygote will be fitter because each recessive is masked by its dominant counterpart. Suppose, for example, that the relative probabilities of survival are 1, 1, and 1 − s for genotypes *AA, Aa,* and *aa,* and the same holds true for genotypes *BB, Bb,* and *bb* at a second locus. Assume that fitnesses are multiplicative across loci because the probability of survival conferred by one locus is independent of that conferred by the other. The fitness of *aabb* is then $(1 - s)^2$. Then two monomorphic strains *AAbb* and *aaBB* each have fitness 1 − s, but the F_1 heterozygote *AaBb* has a fitness of 1. The heterozygote is heterotic (*Chapter 11*), although neither locus is, individually.

This consideration makes it very difficult to prove the existence of

TABLE II. Viability of genotypes in the copepod *Tisbe reticulata* in the laboratory

Culture condition	Genotype:		
	V^vV^v	V^vV^m	V^mV^m
High crowding	0.660	1	0.615
Medium crowding	0.675	1	0.758
Low crowding	0.893	1	0.901

(After Battaglia 1958)

single-locus overdominance. A visible locus A may appear to be heterotic, yet not be if A is in linkage disequilibrium (*Chapter 10*) with a closely linked locus B whose effect is not evident. If a deleterious recessive a is associated with the advantageous dominant allele B, and the deleterious recessive b with the advantageous dominant A, the genotype Ab/aB (perceived only as a heterozygote Aa) will be most fit. This "pseudo-overdominance" will be a permanent condition if the linkage disequilibrium is maintained by a suppressor of recombination, such as a chromosome inversion; the loci are held together in a super-gene. This may be one reason for the inversion polymorphisms that are prevalent in many species of *Drosophila;* under many conditions chromosome heterozygotes are superior in fitness to chromosome homozygotes. An example of such heterosis is provided in Chapter 12; a full treatment can be found in Dobzhansky (1970). Close linkage of a neutral or even a deleterious allele to a superior allele at a neighboring locus can cause such an allele to increase in frequency (Maynard Smith and Haigh 1974).

It is theoretically possible for a single locus to be overdominant for fitness if the heterozygote is not superior in any one environment but has average superiority (MARGINAL OVERDOMINANCE) over a variety of environmental conditions. Suppose, for example, that AA, AA', and

C   Selection in Favor of Heterozygotes

Let the frequencies and fitnesses at a locus be

Genotype	AA	AA'	$A'A'$
Frequency	p^2	$2pq$	q^2
Fitness	$1 - s$	1	$1 - t$

Over one generation the change in gene frequency is

$$\Delta q = q' - q = \frac{pq + (1 - t)q^2}{(1 - s)p^2 + 2pq + (1 - t)q^2} - q$$

$$= \frac{q - tq^2}{1 - sp^2 - tq^2} - q$$

$$= \frac{-tq^2 + sp^2q + tq^3}{1 - sp^2 - tq^2}.$$

Note that $tq^3 - tq^2 = -tq^2(1 - q) = -tpq^2$, so the numerator becomes $sp^2q - tpq^2$; hence

$$\Delta q = \frac{pq(sp - tq)}{1 - sp^2 - tq^2}.$$

The change Δq can be either positive or negative, depending on the sign of $(sp - tq)$, so q can increase or decrease until $\Delta q = 0$ at equilibrium. Then

$$\Delta q = 0 = \frac{pq(sp - tq)}{1 - sp^2 - tq^2} = sp - tq$$

$$= s(1 - q) + tq = s - q(s + t).$$

Solving for q, we find

$$\hat{q} = \frac{s}{s + t} \quad \text{and} \quad \hat{p} = \frac{t}{s + t}.$$

Note, further, that $\bar{w} = 1 - sp^2 - tq^2 = 1 - s + 2sq - sq^2 - tq^2$. Then $d\bar{w}/dq = 2(s - sq - tq)$. The value of q that maximizes $\bar{w}$ is found by setting $d\bar{w}/dq = 0$ and solving for q. This value is $q = s/(s + t)$. This is one demonstration of the general principle that if selection is constant at a single locus, the equilibrium gene frequency is that gene frequency at which the average fitness is maximal.

$A'A'$ have the fitnesses 1, 0.95, 0.75 in environment 1 and 0.75, 0.95, 1 in environment 2; each homozygote is most fit under different conditions. If environmental variation is fine-grained (*Chapter 4*), the net fitness of a genotype is its average over environments. In this example, if the proportions of environments 1 and 2 are 0.6 and 0.4, the average fitnesses of the three genotypes are 0.90, 0.95, and 0.85, and there is a stable polymorphism.

Lerner (1954) suggested that marginal overdominance might be quite common if heterozygotes are better buffered against the deleterious impact of different environmental conditions than homozygotes so that they maintain higher average fitness in a heterogeneous environment. Illustrating this point, Dobzhansky and Levene (1955) tested the viability at several temperatures and on several food media of *Drosophila* strains homozygous vs. heterozygous for whole chromosomes. Some of the homozygotes were exceptionally fit under one or two of the experimental conditions, but none had uniformly high survival over all the conditions. Heterozygotes for these same chromosomes, however, had a high survival rate that was much more uniform from one environment to another. This experiment does not prove that marginal overdominance exists at single loci, but it does illustrate the principle. Thus, for example, fluctuating temperatures or the heterogeneity of internal cellular conditions in different circumstances could cause heterozygotes for an enzyme to be most fit, even if the enzyme activity of the heterozygote was intermediate between that of the two homozygotes (Gillespie and Langley 1974, Berger 1976).

Berger (1976) has reviewed *in vitro* studies of enzymes that suggest the possibility of single-locus heterosis. In enzymes that form dimers, the heterodimers (combinations of the polypeptides produced by different alleles) that are formed in heterozygotes can show INTERALLELIC COMPLEMENTATION whereby the heterodimers have properties that transcend those of the homozygotes' homodimers. In alcohol dehydrogenase in maize the homodimer of C^mC^m homozygotes has low activity but high stability, the homodimer of *FF* homozygotes has high activity and low stability, but the heterodimer in C^mF heterozygotes has both high activity and high stability (Schwartz and Laughner 1969). But even heterozygotes for monomeric enzymes sometimes have nonintermediate properties; at 25°C enzyme activity of an esterase in the fish *Notropis stramineus* is greater than that of either homozygote (Koehn, Perez, and Merritt 1971). Thus from a biochemical point of view heterozygote superiority could well be common in nature.

The theory of polymorphism by heterozygote advantage is quite complex when there are more than two alleles. If all the heterozygotes have the same fitness and are far superior to all the homozygotes, all the alleles persist; otherwise some are likely to be lost (Wright 1969, pp. 40ff; Lewontin et al. 1978). A particularly interesting case arises if a new mutation, allele A_3, combines the effects of two heterotic alleles A_1 and A_2. If A_3A_3 is as fit as the most fit heterozygote, A_3 replaces the previous alleles and the population becomes monomorphic. Un-

D   Frequency-Dependent Selection

The model presented by Cook (1971) assumes A dominant to A'. Hence there are two phenotypes, and the fitness of each is inversely proportional to its frequency. Thus we have these fitnesses:

Genotype	AA	AA'	$A'A'$
Frequency	p^2	$2pq$	q^2
Fitness	$1 - s(1 - q^2)$	$1 - s(1 - q^2)$	$1 - sq^2$

If a fitness of 1 is assigned to the most fit phenotype at any time, the genotypes AA, AA', and $A'A'$ will have fitnesses 1, 1, and $[1 - sq^2]/[1 - s(1 - q^2)]$. Hence

$$\Delta q = \frac{q(1 - q) + q^2[1 - sq^2]/[1 - s(1 - q^2)]}{p^2 + 2pq + q^2[1 - sq^2]/[1 - s(1 - q^2)]} - q.$$

Since the denominator equals $\bar{w}$, this expression reduces to

$$\Delta q = \left(\frac{sq^2(2q^2 - 1)(q - 1)}{1 - s + sq^2}\right)\Big/ \bar{w}.$$

At equilibrium,

$$\Delta q = 0 = sq^2(2q^2 - 1)(q - 1),$$

which is satisfied by $q = 0$, $q = 1$, or $q^2 = \frac{1}{2}$. Hence there is a stable equilibrium at $\hat{q} = \sqrt{\frac{1}{2}}$.

---

equal crossing over and gene duplication in an A_1A_2 heterozygote can form an A_1A_2 chromosome (a new "allele" A_3) which in homozygous form (A_1A_2/A_1A_2) seems likely to be as advantageous as the A_1A_2 heterozygote. By this argument we would not expect heterotic loci to be common in nature, since gene duplication is probably a fairly common event; it has produced isoenzymes and other paralogous polypeptides, for example (*Chapter 10*). At present, therefore, neither theory nor data can tell how common single-locus heterosis should be in natural populations.

FREQUENCY-DEPENDENT SELECTION

Superior fitness of a heterozygote is not the only pattern of selection that can maintain a polymorphism. Consider the way in which the sex ratio of a population can be kept near 1 (Fisher 1930). If the population consists primarily of females, there is a relative shortage of males; those who are present are almost guaranteed of reproducing, while many females will fail to find mates (unless the males are highly polygamous). If the population has an excess of males, the reverse is true. Whenever the sex ratio deviates from 1, the rarer sex has an advantage and so the population comes to equilibrium at a sex ratio of 1. Variations among species in their breeding systems may modify this scheme; so, for example, sex ratios tend to be highly skewed in favor of females in the mites, in which mating among siblings is common (Hamilton 1967).

The fitness of a genotype can therefore vary with the genetic composition of the population. If a genotype is most advantageous when it is rare, it will be kept in the population (*Box D*). Such FREQUENCY-DEPENDENT SELECTION (known also as apostatic or reflexive selection) may be quite common in nature (Clarke and O'Donald 1964, Murray 1972). For example, Ehrman (1967) has found that female *Drosophila*, when offered a choice of males of two genotypes, often mate with the

4

An example of frequency-dependent selection, predation by a fish on three color forms of the corixid bug *Sigara distincta*. Each suffers disproportionately higher predation when common than when rare. Cf. Figure 8, Chapter 4. (After Clarke 1962, based on data of Popham 1942)

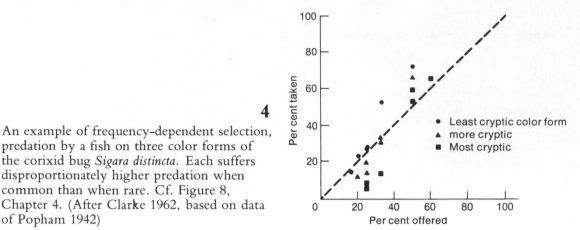

rarer. Predation may also have a frequency-dependent effect; vertebrate predators often form a "searching image" (Tinbergen 1960) for the most common kind of prey item and tend to ignore the rarer phenotypes (*Figure 4*). Self-incompatibility alleles in plants are maintained in this way, for the rarer an allele S_i is, the more likely an S_i pollen grain is to land on a stigma that does not have the genotype S_iS_j (*Chapter 11*, p. 271).

VARIABLE SELECTION IN TIME OR SPACE

Probably the most obvious factor that can potentially maintain genetic variation in a population is the heterogeneity of the environment. But the effect of environmental variation turns out to be much more complicated than might be thought. Suppose, for example, that in a species with two generations per year, different homozygotes are favored in each of the generations and have fitnesses as follows:

	AA	AA'	A'A'
Fitness in spring	1.0	0.8	0.6
Fitness in fall	0.7	0.9	1.0

If the frequency of A' is $q = 0.5$ in the spring, it drops to 0.438 by the end of that generation and then rises again in the fall generation, to 0.481. Over one year there has been a net change in the gene frequency. Selection in the fall is not precisely balanced against selection in the spring (why should it be?), so there is a long-term trend toward the loss of A'. Fluctuating selection may slow the rate at which a gene is lost from the population, but only under exceptional circumstances can it keep both alleles in the population indefinitely (Haldane and Jayakar 1963; *Figure 5*). Thus, for example, although the frequencies of different types of chromosome in *Drosophila pseudoobscura* fluctuate with the seasons, the seasonal variation in the environment does not itself explain the persistence of the polymorphism.

Similarly spatial variation such as a mosaic of microenvironments

E Two Niches with Inferior Heterozygotes

Suppose there are two equally common niches or microhabitats with each homozygote adapted to one of them so that the homozygotes have equal fitnesses ($w = 1$). If the heterozygote is not as well adapted to either, it has an inferior fitness of $1 - s$. Then

$$\bar{w} = p^2(1) + 2pq(1 - s) + q^2(1)$$
$$= 1 - 2sq(1 - q).$$

Note that $d\bar{w}/dq = -2s + 4sq = 2s(2q - 1)$; substituting this into $\Delta q = pq/2\bar{w} \cdot d\bar{w}/dq$ (the general equation from Box A) yields

$$\Delta q = spq(2q - 1)/\bar{w}.$$

Equilibrium ($\Delta q = 0$) is reached when $q = 0$, $q = 1$, or $q = \frac{1}{2}$. But $q = \frac{1}{2}$ is an unstable equilibrium, because $\Delta q > 0$ if $q > \frac{1}{2}$, while $\Delta q < 0$ and q goes to zero if $q < \frac{1}{2}$. Therefore if the slightest accident changes the gene frequency so that it departs from $\frac{1}{2}$, it will continue to change until it comes to rest at either 0 or 1: the population will be homozygous for *either AA or A'A'*. It is easily shown that the same happens if one homozygote has higher fitness than the other, but both have higher fitness than the heterozygote.

can maintain polymorphism, but only under certain conditions (Hedrick et al. 1976, Felsenstein 1976). Imagine, for example, that AA has high fitness in moist sites, $A'A'$ is equally fit in dry places, but that the heterozygote is not quite as well adapted to either. This results in an unstable equilibrium (*Box E*). The gene frequency is unchanging when $q = 0.5$ exactly, but whichever allele by chance becomes even slightly more common will take over the population. There is a drain of alleles in the death of heterozygotes and a conservation of both in homozy-

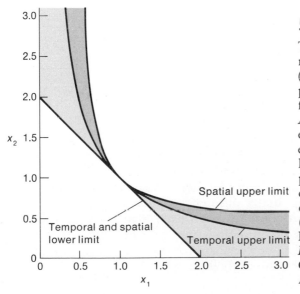

5

The conditions for stable polymorphism maintained by environmental heterogeneity (multiple-niche polymorphism and polymorphism maintained by environmental fluctuations). The relative fitnesses of AA, AA', and $A'A'$ are 1, 1, and x_1 in environment 1; and 1, 1, and x_2 in environment 2. Combinations of x_1 and x_2 lying within the gray regions give polymorphism, but the set of such combinations is more restricted when the environment varies temporally than spatially. (Adapted from Hedrick et al. 1976. Reproduced, with permission, from *Annual Review of Ecology and Systematics,* vol. 7. Copyright © 1976 by Annual Reviews Inc. All rights reserved)

gous form. But because the homozygous "bank" of the rarer allele is smaller than that of the more abundant allele, the rare allele is drained to zero first.

This does not happen if the population is completely inbred, or if the two homozygotes never mate with each other. There is then no heterozygote drain; and the two homozygotes, each superior in its niche, coexist in the same way that two species can (*Chapter 4*). In fact, we might call them different species. This is pretty much what Hamrick and Allard (1972) found in the self-fertilizing wild oat, *Avena barbata*. At each of five enzyme loci one homozygote is prevalent in moist sites, while another is common in more xeric sites. A field of wild oats is a mosaic of genotypes, mapping quite closely the mosaic of environmental conditions.

The conditions under which genotypes adapted to different micro-environments or niches can form a polymorphism were explored first by Levene (1953). He imagined a case in which a proportion c_1 of the adult population inhabits niche 1, and a proportion c_2, niche 2. Suppose mating occurs entirely at random and that the offspring then settle at random into niches, as seeds might. Selection then takes place, with these fitnesses:

	AA	AA'	$A'A'$
Fitness in niche 1	W_1	1	V_1
Fitness in niche 2	W_2	1	V_2

Levene showed that the polymorphism is stable if $\Sigma c_i/V_i > 1$ and $\Sigma c_i/W_i > 1$. For example, the polymorphism is stable if the niches are equally frequent ($c_1 = c_2 = 0.5$) and if the fitnesses are

	AA	AA'	$A'A'$
Niche 1	1.1	1	0.7
Niche 2	0.7	1	1.1

But in this case the polymorphism is stable because of marginal overdominance; the heterozygote has higher average fitness. A polymorphism may be stable without marginal overdominance, but only under special combinations of fitness values (Maynard Smith 1970).

Figure 5 represents the conditions for stable equilibria, for both spatial and temporal variation, when the two niches are equally frequent. A case that might conform to this model is that of the color polymorphism in larvae of the African swallowtail butterfly *Papilio demodocus* (Clarke, Dickson, and Sheppard 1963; *Figure 6*), in which different genotypes are associated with umbelliferous plants and citrus. The association is stronger in older than in younger larvae, probably because of differential predation by birds. If the difference between young and old larvae in the proportions of each form on each plant represents differential survival, the fitness of the citrus type on umbellifers (X_1 in Figure 5) may be calculated to be (very approximately, for the data are sparse) 0.8, relative to 1 for the umbellifer type $A\_$*.

---

* Here and below, the notation $A\_$ denotes any genotype that includes allele A.

On citrus 72 per cent of the young larvae were the umbellifer type, but none of the last-instar larvae were; thus the relative fitness X_2 of the citrus type on citrus is very large, and X_1 and X_2 appear to lie in or near the stability region in Figure 5.

The likelihood of a stable polymorphism is greater if the genotypes tend to disperse into the habitat to which they are best adapted (Maynard Smith 1970). We know little about whether the different genotypes in a species are capable of doing this; it requires some kind of functional association between the trait on which selection acts and the individual's behavior. A genetic polymorphism in niche-seeking behavior (with, say, alleles B and B') would get A_- and $A'A'$ individuals into their appropriate niches only if there were linkage disequilibrium between the A and B loci. Kettlewell and Conn (1977) found that industrial melanic moths prefer to sit on dark surfaces, while pale gray moths prefer light surfaces; Sargent (1969), however, found no such tendency in the moth *Phigalia titea*. Giesel (1970) has reported that light

6 The umbellifer (left) and citrus (right) forms of the larva of *Papilio demodocus*. (Courtesy Dr. C. A. Clarke)

and dark limpets (*Acmaea digitalis*) tend to move toward the environments to which they are best suited. Some authors (e.g., Thorpe 1956) have suggested that female herbivorous insects are "conditioned" to the species of host plant on which they fed as larvae and tend to lay eggs on that species rather than on other potential hosts. There has been little strong evidence for this, the so-called Hopkins' host-selection principle, but if it does occur in some species, it would facilitate the maintenance of polymorphism.

The effect of environmental variation on genetic variation may be even more complicated than I have indicated. For example, there may often be a frequency-dependent component to selection in a patchy environment. A dry-adapted plant genotype in a monomorphic population might occupy dry and moist microenvironments equally well but be displaced from moist microhabitats by a newly arisen wet-adapted genotype. The rate of increase (fitness) of the wet-adapted genotype is higher when it is rare and still has moist sites to expand into than when it has fully occupied the moist sites. This frequency-dependent effect can often stabilize a habitat-associated polymorphism that, perhaps because of heterozygous inferiority, would otherwise be unstable.

DISTRIBUTIONS OF GENE FREQUENCIES

To determine which factors most affect gene frequencies in nature will prove difficult, for they all may operate in concert. Allele frequencies can be simultaneously affected by mutation rates, by gene flow, by genetic drift, and by one or more modes of selection, as well as by linkage and interaction with other genes. For example, overdominant selection that in a large population yields a stable polymorphism may be overwhelmed by the effect of genetic drift if the population is small. In fact, the loss of an allele by genetic drift may even be hastened by selection in favor of the heterozygote if the theoretical equilibrium frequency is near 0 or 1 (Robertson 1962).

Wright (1931) and subsequent authors have found in their theoretical work that as long as a population is finite in size, the gene frequency at a locus is not exactly determined by selection, mutation, and migration. Rather there is some probability that the gene frequency lies anywhere from 0 to 1, because of the variation introduced by genetic drift. Taking into account selection s, forward and back mutation rates u and v, migration m, and population size N, Wright finds the probability $\phi(q)$ that the population has any specified gene frequency q:

$$\phi(q) = Ce^{4Nsq}q^{4N(mq_m+v)-1}(1-q)^{4N[m(1-q_m)+u]-1}.$$

Some of the results are illustrated in Figure 7. In general, the larger the population, the higher the probability that the gene frequency is near the equilibrium determined by selection and mutation alone. The smaller the population, the more variation there is in gene frequency, and the higher the probability that an allele, even if advantageous, will be lost.

SELECTION AND THE FITNESS OF A POPULATION

In Chapter 12 I noted that it is difficult to measure objectively the adaptedness, or fitness, of a population. In some instances, such as the t allele in mice, natural selection seems even to threaten the well-being of a population. But in general it seems likely that evolution by natural selection must make a population better adapted, whatever that may mean.

In most theoretical formulations of selection, $\bar{w}$ increases as gene frequencies change. By $\bar{w}$ we must mean the fitness of an average individual in the population — its capacity for survival and reproduction — compared with that of a reference genotype. Wright (1932) envisioned a population as occupying a position on an ADAPTIVE LANDSCAPE of gene frequencies, in which peaks represent the possible genetic compositions of a population for which $\bar{w}$ is high, and troughs represent the possible compositions for which $\bar{w}$ is low (*Figure 8*). Because the environment is not constant, the positions of peaks and troughs changes; after an environmental change the genetic composition of a population may correspond to an adaptive valley. As the gene frequencies change in the new environment, the population moves uphill until it comes to rest on an adaptive peak. To move downhill, the population would have to become less adapted, and this does not happen ordinarily (but sometimes it does; see Moran 1964, Turner 1970). There may be multiple peaks (different adaptive compositions), but a population cannot move from one to another by natural selection

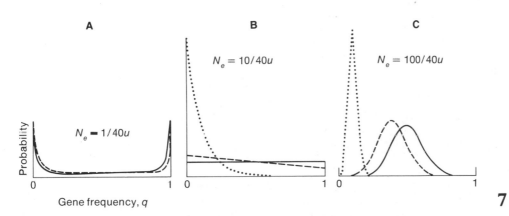

The interaction of selection, mutation, and genetic drift. Each curve is the distribution of probabilities that a population will have gene frequency q of a deleterious gene. Population size is small in A, intermediate in B, large in C. Forward and back mutation rates are assumed to be equal ($u = v$). The selection coefficient against the allele is very small ($s = u/10$), quite small ($s = u$), and moderate ($s = 10u$) for solid, broken, and dotted lines respectively. The gene frequency is likely to be near its deterministic equilibrium (that specified by selection alone) of zero if selection is strong and the population is large. Gene frequencies are much more variable in smaller populations, especially if selection is weak. (From Crow and Kimura 1970, after Wright 1937)

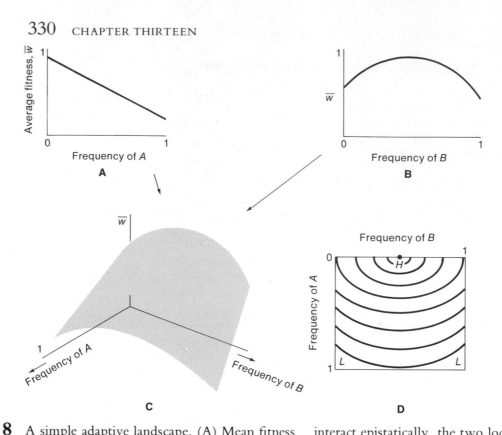

8   A simple adaptive landscape. (A) Mean fitness $\bar{w}$ declines as the frequency of allele A increases, at a locus where fitnesses are $1 - s$, $1 - s/2$, and 1 for genotypes AA, AA', and $A'A'$ respectively. (B) At the heterotic locus B where fitnesses of BB, BB', and $B'B'$ are $1 - t$, 1, $1 - t$ respectively, $\bar{w}$ is maximal when the allele frequency is ½. If they do not interact epistatically, the two loci taken together yield the fitness surface in (C) which can be rendered as the "topographic map" in (D). Here the curves represent gene frequency values of equal fitness. There is a peak at H, toward which the population moves by natural selection; minimal values of $\bar{w}$ are at the points L.

(even if it is higher) because it would have to go downhill through a valley.

Fisher (1930) formalized the notion that selection increases a population's fitness in his fundamental theorem of natural selection (*Box F*), which states that "the rate of increase in fitness of any organism [population] at any time is equal to its genetic variance in fitness at that time." This makes a little intuitive sense if one recognizes that the variance in fitness is determined by the gene frequencies and by the fitness differential among genotypes (the selection coefficient), both of which determine the rate of genetic change.

The theorem is difficult to test or interpret (Turner 1970, Price 1972) because the fitness of a population is so hard to measure. In terms of growth rate or density, a population with an "inferior" genetic composition may flourish as well as one with "superior" alleles (Cain and Sheppard 1954*b*). Some genotypes (such as lethals) may be unconditionally inferior; their death lowers population density, a process that Wallace (1968*b*) has termed HARD SELECTION. Population fitness, as measured by density, in such instances clearly increases as such geno-types are eliminated. But very often, superior genotypes merely replace genotypes that are relatively, not absolutely, inferior; and the popu-lation density, regulated by an extrinsic density-dependent factor, does not change as selection proceeds. In this instance, which Wallace terms

F   Increase of Fitness by Natural Selection

In simple cases the fitness $\bar{w}$ of a population increases as the gene frequencies change due to selection. We can show this easily with two examples.

1. Classical selection at a locus with no dominance

	AA	AA'	$A'A'$
Frequency	p^2	$2pq$	q^2
Fitness	1	$1 - s/2$	$1 - s$

$$\bar{w} = p^2 + 2pq(1 - s/2) + q^2(1 - s) = 1 - sq$$

$$d\bar{w}/dq = -s$$

Therefore $\bar{w}$ is a decreasing function of q and must increase as A' decreases in frequency.

2. Heterozygote superior for fitness

	AA	AA'	$A'A'$
Frequency	p^2	$2pq$	q^2
Fitness	$1 - s$	1	$1 - t$

$$\bar{w} = p^2(1 - s) + 2pq + q^2(1 - t)$$
$$= 1 - s + 2sq - sq^2 - tq^2$$

$$d\bar{w}/dq = 2(s - sq - tq)$$

In this case $\bar{w}$ is maximized when $d\bar{w}/dq = 0$, which is true when $q = s/(s + t)$, the equilib-rium gene frequency.

At what rate does $\bar{w}$ increase? Consider only case 1, the special case of additively acting al-leles, with no dominance. We first calculate the variance in fitness at time t. Since the variance V is $V = \Sigma f_i w_i^2 - \bar{w}^2$ (*Appendix I*), the variance in fitness is

$$V = p^2(1^2) + 2pq(1 - s/2)^2$$
$$+ q^2(1 - s)^2 - (1 - sq)^2.$$

Thus

$$V = p^2 + 2pq(1 - s + s^2/4) + q^2(1 - 2s + s^2)$$
$$- (1 - 2sq + s^2q^2)$$

$$= (p^2 + 2pq + q^2) - 2pqs + pqs^2/2 - 2sq^2$$
$$+ sq^2 - 1 + 2sq - s^2q$$

$$= -2sq(p + q) + 2sq + pqs^2/2$$

$$= pqs^2/2.$$

Now we wish to find $d\bar{w}/dt$, the rate at which $\bar{w}$ changes. This must be

$$\frac{d\bar{w}}{dt} = \frac{dq}{dt} \cdot \frac{d\bar{w}}{dq}.$$

From case 1, $d\bar{w}/dq = -s$. From BOX A

$$\frac{dq}{dt} = \frac{pq}{2\bar{w}} \cdot \frac{d\bar{w}}{dq},$$

so

$$\frac{dq}{dt} = \frac{-spq}{2\bar{w}}.$$

Thus

$$\frac{d\bar{w}}{dt} = \left(\frac{-spq}{2\bar{w}} \right) (-s) = \frac{pqs^2}{2\bar{w}}.$$

If we assume the population is near equilib-rium, $\bar{w} \approx 1$, so

$$\frac{d\bar{w}}{dt} = \frac{pqs^2}{2} = V.$$

Thus the rate of increase in fitness equals the (additive) genetic variance in fitness. This is Fisher's fundamental theorem of natural selec-tion. It does not necessarily hold for more com-plicated genetic models.

SOFT SELECTION, the increase of a population's fitness by selection is not measurable in ecological terms.

GENETIC LOAD AND THE COST OF SELECTION

In 1950 Muller, discussing the deleterious effects of radiation, introduced the concept of GENETIC LOAD, which can be defined as the difference between the average fitness $\bar{w}$ of a population and its maximal possible fitness, that of the fittest genotype (whose fitness is set equal to 1). Thus the load $L = 1 - \bar{w}$. A deleterious recessive, for example, such as might arise by radiation-induced mutation, reaches an equilibrium frequency of $q = (u/s)^{1/2}$; $\bar{w}$ is $1 - sq^2$; thus $\bar{w} = 1 - u$, and $L = u$. If there were n such loci, and the probability of death at each locus were independent of that at other loci, $\bar{w} = (1 - u)^n$ and $L = 1 - (1 - u)^n$, which could be a fairly large number. Much of the population would die in each generation; and if survival were unconditionally dependent on genotype (hard selection), the population density would be reduced.

Every individual with a less-than-maximal fitness is part of the genetic load; thus there must be a mutational load due to deleterious alleles maintained by the balance of selection and migration; an environmental load due to individuals' dispersing into microenvironments inappropriate to their genotypes; a recombinational load due to the proliferation of inferior genotypes by recombination; and a segregational load due to the segregation of inferior homozygotes at heterotic loci.

At an overdominant locus (fitnesses $1 - s$, 1, $1 - t$), $\bar{w} = 1 - sp^2 - tq^2 = 1 - st/(s + t)$ at equilibrium, so $L = st/(s + t)$. This is large if selection is strong. For the case of the sickle-cell hemoglobin polymorphism, $s = 0.15$ and $t = 1$. Thus $L = 0.1$; about 10 per cent of an equilibrium population dies of either anemia or malaria before the age of reproduction (by hard selection). There must be a limit to the number of such loci if they act independently. In a population with 20 such loci, only 0.9^{20}, or 12 per cent, of the offspring of each generation would survive. In a fecund species like *Drosophila,* this leaves enough survivors to maintain a population (*Figure 9*), but in a less fecund species like humans the population would rapidly become extinct. By this argument there cannot be very many overdominant loci in a population (Kimura and Crow 1964), even if selection is weak. For example, if $s = t = 0.02$, $L = 0.01$ and $\bar{w} = 0.99$. If there are 1000 independently acting such loci, $\bar{w} = 0.99^{1000} = 0.0004$.

In the same vein, the replacement by natural selection of one allele by another entails a substitutional load, or COST OF SELECTION. For example, a breeder selecting for higher milk yield in cattle must prevent the lower-yielding individuals from breeding, perhaps by slaughtering them. If the fraction killed in each generation is small, the herd remains large but genetic change is slow. If the fraction killed is large, genetic change is fast but the herd size is reduced. The cost of selection, the number of *genetic deaths* required to change the gene frequency, is about

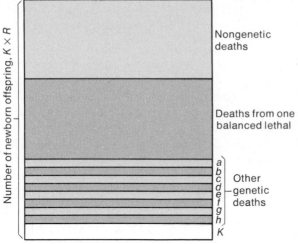

9

The "load space," representing mortality in a new generation of offspring from various factors assumed to act independently. Some mortality, to which all genotypes are equally subject, is nongenetic. A single balanced lethal system may eliminate half those remaining. Segments a–h represent other selective deaths; they might be (a) mutational load; (b–e) mortality at each of four heterotic loci; (f) mortality of inferior genotypes formed by recombination among loci; (g) mortality of genotypes adapted to one niche but without access to it; (h) mortality associated with allele substitutions. K represents surviving adults.

the same whichever alternative is chosen. To bring a favorable allele from a low frequency (10^{-2}) to a very high frequency requires a number of deaths about 30 times the size of the population, if the population is kept at the same size (Haldane 1957). Since the number of genetic deaths per generation is limited by the species' fecundity, there can be rapid genetic change at just a few loci simultaneously or slow change at many loci. Haldane argued that evolution must therefore be slow ordinarily; that it would take, on average, about 300 generations for complete genetic change at one locus, and about 300,000 generations for two populations to diverge completely in gene frequency at 1000 loci. He argued that it might take about 500,000 years for new species to evolve.

But speciation often appears to be much more rapid than this (*Chapter 16*), although it is not known at how many loci species differ. And populations are more polymorphic than genetic load theory suggests they should be, although it is not known whether most of the polymorphisms are maintained by selection. Several authors have therefore argued that the load theory is wrong because it assumes that loci act independently (Sved, Reed, and Bodmer 1967; King 1967; Milkman 1967; Sved 1968). A whole individual lives or dies, so a single death accounts for selection at many loci simultaneously. The survivors may be the individuals heterozygous at the most loci or the ones with the best combinations of newly evolving characteristics. If there is soft selection at most loci, a low $\bar{w}$ does not imply that population size will dwindle to extinction. If ecological factors keep the adult density at K in each generation and each female has R offspring, then $KR - K$ die in each generation, and these may be the most homozygous (*Box G*). Thus many polymorphisms can be maintained, and by a similar argument substitutions can occur at many loci simultaneously under directional selection.

G Multiple Polymorphisms and Genetic Load

I present here the model of Sved, Reed, and Bodmer (1967), with some modifications and deletions for the sake of clarity. Suppose the probability of an organism's survival is related to the number of loci at which it is heterozygous, as in Figure 10b, curve b. The relation cannot be linear as in curve a, since this would imply a far greater disparity in fitness between multiple heterozygotes and multiple homozygotes than is actually observed.

If $s = t$ at each of 10,000 loci, the equilibrium gene frequency at each of these loci should be 0.5. If the loci segregated independently (which in reality they do not), the mean number of heterozygous loci per individual would be 5000. The variance, however, would be quite small: 2500 (*Figure 10a*). Thus the majority of the population (66 per cent) would be heterozygous at between 4950 and 5050 of their loci, so most individuals would have much the same fitness, between the values of f_1 and f_2.

If each individual in a population of size K has R offspring, then KR progeny are produced, of which K survive; suppose the survivors are those that have more than 5050 heterozygous loci (*Figure 10c*). Each of the $KR - K$ individuals that die is homozygous at anywhere from 5049 to 10,000 loci, and its death contributes to the selection that keeps each of these loci polymorphic. What is the selection coefficient s at one such locus?

A survivor has more than 5050 heterozygous loci. Thus the probability that a randomly chosen locus is among these is greater than 5050/10,000. Therefore the probability that this locus is in homozygous condition in a surviving individual is less than 4950/10,000. So the probability of survival of a homozygous locus relative to that of a heterozygous locus is $< 0.495 : > 0.505$, or $< 0.98:1$. Therefore the total selection against homozygotes at this locus is > 0.02, which amounts to $s > 0.01$ against each homozygote, enough to keep the locus polymorphic if the population is reasonably large. This holds for each of the 10,000 hypothetical loci.

Note that fecundity affects the number of loci that can be held polymorphic. If fecundity is less, a larger fraction of progeny survives, if we hold K (the carrying capacity) constant. Suppose K equals half the progeny that are born. Then any individual with more than 5000 heterozygous loci will survive, so selection at each locus will be less intense. A great deal of accidental, nonselective mortality will have the same effect, for this will be distributed over all genotypes equally, as shown by the hatched area in Figure 10c. Thus the K survivors will come from the unhatched area, setting the threshold to the left; more organisms survive that have fewer heterozygous loci. The selection coefficient s against homozygotes at each of these loci will thus be smaller, so genetic drift is more likely to overwhelm the effect of selection and render the locus monomorphic.

But genetic load can nevertheless limit the amount of polymorphism and the rate of evolution. Suppose we postulate that at each of 3000 loci maintained polymorphic by overdominance, $s = t = 0.01$. Then, for every 100 offspring born, two must die because they are homozygous at locus A, two because they are homozygous at locus B, and so on. *If the mortality is not that great, selection is not as strong as stipulated,* and the polymorphisms are more subject to loss by genetic drift. The only escape from this dilemma is strong linkage disequilibrium among loci, so that the homozygote AA is also rather consis-

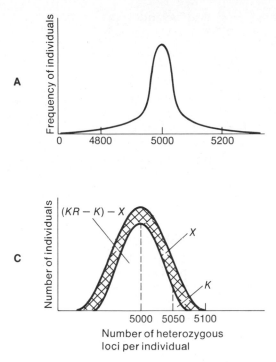

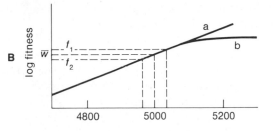

If there are 10,000 segregating loci, most individuals will be heterozygous at about 5,000 loci (A). Postulate that the relation between fitness and heterozygosity is not linear (curve a in B), but asymptotic (curve b). Then if X of the $KR - K$ offspring born per generation die by nonselective causes, $(KR - K) - X$ die of selective causes. The K survivors may be the most heterozygous (C). The smaller K is compared to $(KR - K) - X$, the stronger selection is at each locus. (A and B after Sved, Reed, and Bodmer 1967)

10

tently homozygous for allele B (*Chapter 14*). Selection does require mortality (or differential fecundity, which similarly has limited scope for the amount of selection that can occur). Since the number of potential genetic deaths is determined by fecundity (by $KR - K$), highly fecund species should be more polymorphic and able to evolve more rapidly than less fecund species.

WHY ARE POPULATIONS GENETICALLY VARIABLE?

Until the 1960's, the central debate in evolutionary genetics was whether natural populations are genetically variable. The evidence from electrophoresis and artificial selection is incontrovertible; so the debate now concerns the causes of the variation. The "neutralist" school holds that most genetic variants, especially allozymes, do not differ in fitness and that gene frequencies are determined largely by genetic drift and gene flow. The "selectionist" school affirms that the variants differ in fitness, and that variation is maintained largely by various forms of balancing selection. If the neutralists are right, most molecular evolution (e.g., of protein structure) occurs by genetic drift, and the observable genetic variation is irrelevant to a population's ability to respond to selection. If the selectionists are right, selection is the overriding factor in evolution, and populations have immense evolutionary potential. Papers on this subject appear at such a prodigious rate that my treatment may be obsolete by the time it is published. Nonetheless I shall survey the arguments briefly. More comprehensive treatments are provided by Lewontin (1974) and Selander (1976), and by Kimura and Ohta (1971) and Nei (1975), who summarize the neutralist position.

The neutralist school points out that an entirely neutral mutation has a probability of $1/2N$ of being fixed in a population of effective size N; the rate at which such mutations are fixed is equal to the mutation rate u. Because there are many mutable sites at a locus, we expect to see at any time a number of alleles in various stages of drift toward fixation or loss. If the process has reached a steady state, the level of heterozygosity is $H = 1 - 1/(4Nu + 1)$; for example, if $N = 3 \times 10^4$ and $u = 10^{-6}$, $H = 0.11$ (Lewontin 1974). Under this model populations diverge in gene frequency at a constant rate. The genetic composition varies randomly among populations with no pattern, except for that brought about by gene flow or by linkage to other loci affected by selection. The neutralist model embraces very slightly deleterious mutations (Ohta and Kimura 1975); in large populations these mutations tend not to increase by genetic drift and are eliminated more rapidly than by chance alone, but in small populations they behave like neutral alleles.

If most polymorphisms are attributable to variation in fitness, no precise predictions of the pattern of variation can be made; any nonrandom pattern can be explained by selection, especially because selection and drift can operate jointly, because gene flow can alter expected patterns, and because gene frequencies can be affected by linkage with other loci. Moreover there is no reason to suppose that all, or even most, gene frequencies have reached equilibrium.

We should, I believe, be most interested in which of these explanations accounts for the morphological and physiological variation that is most obviously the stuff of evolutionary change. But because the individual loci that govern such traits cannot be isolated, the fitnesses of known genotypes cannot be measured. Thus it is necessary to study single loci whose pattern of inheritance is known. We know that selection affects the polymorphisms for coloration in *Cepaea nemoralis* and *Tisbe reticulata*, for cyanogenesis in *Lotus*, and for hemoglobin in humans, but these may be biased instances from which we dare not extrapolate to morphological and physiological variation in general. For these reasons and because levels of protein polymorphism are so unexpectedly high, the major debate focuses on variation in enzymes and other proteins.

Rates of protein evolution

From amino acid sequences of homologous proteins in different species, it appears that amino acid substitution rates have been quite high and quite constant (*Chapter* 7). Neutralists argue that the rate of fixation (about 1.6×10^{-9} nucleotide substitutions per year) is about equal to the presumed mutation rate, as expected under the neutral model, that it is too constant to be explained by selection, and that it is too high to be due to selection, for it implies an unbearably high substitutional load. Moreover, proteins or parts of proteins that are functionally less important evolve more rapidly than functionally important enzymes, as expected if a greater proportion of mutations did not affect fitness. Selectionists counter that refined models of substitutional load permit higher rates of allele substitution than had previously been thought possible; they see the constancy as an artifact of averaging substitution rates over long periods of time, during which selection pressures can vary at a stochastically constant rate. But even the divergence among related races and subspecies, which probably became isolated rather recently, seems to have been rather constant (Ayala et al. 1974) and has probably not been accelerated by speciation (Avise and Ayala 1976). The genetic similarity of related species can be interpreted to mean that loci are under similar selection pressures and so have not diverged or that the species share neutral polymorphisms inherited from their common ancestor.

Relations to ecological factors

The average heterozygosity of a population varies rather consistently among major groups of organisms; vertebrates are almost always less variable than invertebrates, for example. This means that it may not be permissible to explain differences in the level of heterozygosity among remotely related species by the differences in their ecology. Many authors (see review by Hedrick et al. 1976) have sought to relate levels of heterozygosity to the degree of environmental heterogeneity that the species are thought to experience. Remarkably few differences emerge, and those few suggest that species in more homogeneous environments can be either more or less heterozygous than those in heterogeneous environments. Among Hawaiian *Drosophila,* habitat specialists are less heterozygous than habitat generalists (Steiner 1977). But specialists are more heterozygous among a series of closely related species of moths that have homogeneous (one host plant) vs. heterogeneous (many hosts) larval diets (Mitter and Futuyma 1979). A possible explanation for such observations is that physiological plasticity evolves in heterogeneous or varying environments and that this buffers biochemical systems from slight environmental variations that in less well buffered species cause polymorphisms to be maintained. The same principle could explain why homeostatic homeotherms are less polymorphic than poikilotherms (Selander and Kaufman 1973). Especially in poorly buffered species, heterozygosity could be due to heterozygous advantage if different allozymes or heterologous enzyme multimers provide greater biochemical regulation over the range of cellular environments in which they operate (Gillespie and Langley 1974, Johnson 1976, Soulé 1976). As Soulé points out, this intrinsic advantage of heterozygosity could be so widespread among species that their level of heterozygosity may be independent of the gross, measurable features of the environment. But the lack of substantial relationships between heterozygosity and ecological factors is at least as favorable to the neutralist as to the selectionist position.

Relations to population size

The ecological property to which heterozygosity of neutral alleles is theoretically most clearly related is population size. But it is difficult if not impossible to measure effective population size N_e in nature (*Chapter 11*). Under the neutralist theory heterozygosity should increase with effective population size. As Lewontin (1974) points out, the observed range of heterozygosity values, if explained entirely by the neutral hypothesis, is so narrow that we would have to imagine N_e lies within a factor of four for all organisms that have been studied (*Figure 11*). This seems unlikely, but it is not utterly inconceivable. Soulé (1976), following Lewontin's approach, has estimated (very roughly) N_e for a variety of species; he concludes that species with small N_e are too heterozygous for the neutral model, while those with large N_e are not heterozygous enough. But low levels of heterozygosity are easily explained by the neutral hypothesis. Heterozygosity will be greatly reduced during population bottlenecks, such as those that probably often accompany the speciation process (*Chapter 17*). It will increase from its nadir only slowly — less than 5 per cent per million generations. A species may simply not be old enough to have reached its equilibrium level of heterozygosity and may never get old enough, since most species do not persist more than about 6 million years. Indeed, as Soulé points out, the difference in heterozygosity between

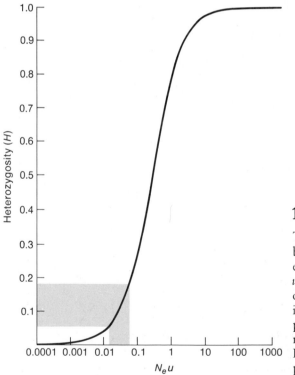

11

The theoretical relation $H = 4N_e u/(4N_e u + 1)$ between heterozygosity H and the product of effective population size N_e and mutation rate u for neutral alleles. The shaded area embraces observed values of H in various species, implying, if most electrophoretic polymorphisms are neutral, a remarkably narrow range of population sizes. (From Lewontin 1974, with permission of the publisher)

vertebrates and invertebrates could be due in part to the shorter evolutionary life span of many vertebrate species.

Patterns of variation within species

Comparisons of heterozygosity in different populations of single species have shown in several cases a positive relation between heterozygosity and local meteorological variability (Bryant 1974, Taylor and Mitton 1974). There is no evidence that the enzyme loci studied, rather than linked genes, are responding directly to the environment. The same objection applies to cases in which allozyme gene frequencies show regular clinal patterns of variation. In some instances, however, strong circumstantial evidence of selection comes from the observation that the enzyme is functionally affected by the environmental factor with which it covaries; Merritt (1972) showed that the temperature kinetics of LDH allozymes vary as predicted by their relative prevalence in minnow populations from cold and warm waters.

Measures of fitness

Differences in genotype frequencies between young and old individuals in the same population can suggest selection; for example, allozyme heterozygotes appeared to have higher survivorship than homozygotes in a mussel population (Koehn, Turano, and Mitton 1973). But, again, this difference could be due to linkage with selected blocks of loci. This reservation holds for most laboratory experiments, such as those of Powell (1971), who found that *Drosophila* populations maintained in heterogeneous environments retained more heterozygosity than those kept under more homogeneous conditions, or those of Yamazaki (1971), who began replicate populations with different esterase allele frequencies to see whether they would converge under selection toward a single equilibrium frequency. That they did not converge may be taken as evidence in favor of the neutral hypothesis or as evidence that the selective factors that act on the esterase locus were absent from the laboratory. If they had converged, however, the experiment could have been interpreted to mean that there are overdominant loci for which a neutral esterase locus is only a marker. Some evidence for this interpretation of similar experiments is provided by Jones and Yamazaki (1974).

Biochemical evidence

Some enzymes are consistently more polymorphic than others. This has been thought to be especially true of those that act on substrates derived from the external environment (such as food) and those that govern the rate-limiting reactions in biochemical pathways (Johnson 1976; but see Selander 1976). One can argue plausibly that both these classes of enzymes should be especially polymorphic due to selection. But it is easy to advance a neutralist explanation by supposing that the less polymorphic enzymes are those that can function only if they have a very precise configuration; variants are deleterious. The polymorphic enzymes are simply those subject to less rigorous functional constraints, and they can exist in any of a variety of equally functional forms.

Although allozymes often show functional differences *in vitro*, many amino acid substitutions have no apparent effect on protein function. Cytochrome *c*'s from different species, for example, can have quite different amino acid sequences yet be indistinguishable in tests of their activity. Neutralists take this as *prima facie* evidence that many amino acid substitutions do not affect fitness. Selectionists counter with the unproven argument that the proteins may behave differently in different cellular milieus.

It seems, then, that convincing arguments are offered by both neutralists and selectionists. Selectionists enjoy a tactical advantage in that they can use statistically significant differences as evidence for their position, while neutralists must defend a null hypothesis (of "no difference") which, like any null hypothesis, can in principle never be proven. They present a strong argument, though, that empirical observations are compatible with the null hypothesis. But almost any observation can be construed as compatible with either theory, for both theories are so immensely flexible that they can explain almost anything, and the critical parameters of the theories, especially effective population size, are almost unmeasurable. Lewontin (1974, especially pp. 266–271) analyzes the reason why the rich theory of population genetics is unable to resolve this argument about the very observations the theory was designed to explain.

The answer to the question, Is enzyme polymorphism due to selection or drift? will, of course, be: Both. But this is really no answer at all. No neutralist would deny that some few enzyme polymorphisms are maintained by selection, and even the staunchest selectionist would not deny that some amino acid substitutions must have such trivial effects that they have vanishingly small effects on fitness and so fluctuate in gene frequency by drift alone. The real question is, What fraction of the variation is attributable to each factor, and what modes of selection act on the fraction that selection does maintain? At this point, it appears that the answer may come only through the painstaking analysis of multitudes of individual loci.

SUMMARY

The evolution of a population consists of changes in gene frequencies. When these changes are due to selection, the rate of genetic change depends on the magnitude of the differences in fitness among genotypes, on the frequencies of the alleles, and on the degree of dominance. Selection may act either to eliminate all but the most fit allele from a population or in various "balancing" modes to maintain stable genetic polymorphisms. The genetic variation that is thus maintained is important in providing genetic flexibility for adapting to changes in the environment. Although selection theory provides many reasons for the persistence of variation, it is not known how much of the very substantial variation that does exist is maintained by these factors and how much is attributable to genetic drift. There are good reasons to suppose that much variation is adaptively neutral, and equally plausible reasons to believe that genotypic variation in fitness is the cause of its existence.

FOR DISCUSSION AND THOUGHT

1   Most of the models discussed in this chapter have implicitly assumed that selection is based on differences in survival. Would any conclusions be changed by considering differences in fecundity?

2   If heterozygotes in a population appear to be less frequent than expected under the Hardy-Weinberg law, can you say anything about the pattern of selection? What if they are more frequent? What if they fit Hardy-Weinberg proportions exactly?

3   It has been postulated that the variation in beak size in a bird population may be due to the adaptation of each beak size to a different range of food types (Van Valen 1965). Discuss the likelihood that this is true and the conditions that would be necessary to stabilize such variation.

4   What examples can you find in the literature on artificial selection in which it has proven more difficult to select two characters simultaneously than either character independently? What relevance do such results have for evolutionary theory?

5   Discuss the effects that inbreeding in a population would have on the rate at which the population becomes adapted to a changed environment.

6   Just as there is a segregational load when the heterozygote at a locus is most fit, there is a recombinational load when we consider two or more loci since some genotypes formed by recombination have a lower fitness than others. What factors affect the magnitude of the recombinational load?

7   Darwin noted that "with all beings there must be much fortuitous destruction, which can have little or no influence on the course of natural selection" (*Origin,* p. 80). For example, the genotype of a small planktonic crustacean must have little influence on whether it is eaten by a baleen whale; it is either lucky or unlucky. Evaluate Darwin's statement, and discuss its implications.

8   Darwin focused on the fact of population limitation as giving rise to the "struggle for existence." Can selection act in a population that is growing exponentially? How does it differ from selection in a stable population? How does the amount of genetic variation in a stable population change if the population is suddenly permitted to grow?

9   It is often stated that the effect of natural selection is to replace poorly adapted genotypes with better adapted ones. How does the theory in this chapter lead us to modify this statement?

10   That allozyme frequencies are sometimes uniform over the geographic range of a widespread species can be taken as evidence of balancing selection, since otherwise the alleles should drift to different frequencies in different local populations. Neutralists counter that just a little gene flow prevents divergence by drift. But what, then, is the effective population size, and how does it compare to the N_e that neutralists must postulate to account for the observed level of heterozygosity?

11   Haldane's estimate of the time needed for speciation assumed complete allele replacement (from $q \approx 0$ to $q \approx 1$) at a great many loci, in popu-

lations constant in size. How will the estimate of speciation rate differ if these assumptions are invalid? How valid are they?

12 Hebert (1974) found that in parthenogenetic populations of the water flea *Daphnia magna* the proportion of heterozygotes at two enzyme loci increased in successive generations and that heterozygotes had a higher fecundity. Does this prove that heterozygotes at these two loci are fitter than homozygotes?

13 Soulé et al. (1973) compared populations and species of lizards; among species of *Anolis* there was a strong correlation between allozyme heterozygosity and morphological variability, but among populations of *Uta* the correlation was weak. Why is the strength of this correlation important to establish? Is the resolution of the neutralist-selectionist argument on enzyme variation important to our understanding of the evolution of morphological and other characters?

MAJOR REFERENCES

Wallace, B. 1968. *Topics in population genetics.* Norton, New York. 481 pages. Extensive discussion of the analysis of gene frequencies and selection in natural populations.

Wilson, E. O., and W. H. Bossert. 1971. *A primer of population biology.* Sinauer, Sunderland, Mass. 192 pages. Elementary selection equations are included.

Lewontin, R. C. 1974. *The genetic basis of evolutionary change.* Columbia University Press, New York. 346 pages. Treats selection theory as it applies to genetic variation.

Cook, L. M. 1971. *Coefficients of natural selection.* Hutchinson and Co., London. 207 pages. A good introduction to selection theory and its application.

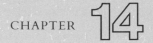

CHAPTER 14

Selection on Polygenic Characters

Although single-locus models of the kind treated in previous chapters suffice for an understanding of many evolutionary events, most characteristics of organisms are affected by more than one locus. The theory of the evolution of such characteristics soon becomes formidably complicated and difficult to test, but it demonstrates that evolution by natural selection can take some surprising turns not predicted by the simple theory described so far.

CONTINUOUS VARIATION

The effect of selection on a characteristic depends on the amount of variation. When only a single locus varies, the rate of genetic change depends on the allele frequencies and on the disparity in fitness among the genotypes, the disparity being dependent on the phenotypic differences among the genotypes. The allele frequencies and the fitness differential together determine the VARIANCE in fitness, on which the rate of genetic change depends (*Chapter 13*).

The same concepts underlie the theory of genetic change when fitness depends on a phenotypic characteristic whose variation is caused by the segregation of alleles at more than one locus. This phenotypic variation, however, has several components. Understanding the nature of this variation is prerequisite to an appreciation of the effect of selection on most characteristics.

Most variation is continuous in kind — body size is an example. Some characteristics, such as the number of scutellar bristles in *Drosophila,* vary discretely, but only because a fly can have four *or* five bristles, but not 4½; the expression of an underlying continuum of bristle-making materials is discrete (Rendel 1967). Thus the models considered in this chapter have broad applicability.

A characteristic such as bristle number or body size commonly has a normal, or bell-shaped, distribution in a population (*Figure 1*). Such a distribution typically arises when the state of a characteristic results from a great many independent factors, each with a relatively small effect. Thus if genotypes AA and $A'A'$ have mean bristle numbers of 18 and 20 respectively, each of these genotypes nonetheless varies in bristle number and may overlap greatly, if environmental variables such as temperature and nutrition influence bristle development or if other loci also contribute to the variation.

The total amount of variation can be thought of as a sum of two components: the variation due to the existence of different genotypes in the sample and the variation caused by the variation in the environment. Measuring variation by the variance (see *Appendix I* for a review if necessary), we may write

$$V_P = V_G + V_E;$$

the VARIANCE IN PHENOTYPE V_P is the sum of the variance due to genetic variation (the GENOTYPIC VARIANCE V_G) and that due to environmental variation (the ENVIRONMENTAL VARIANCE V_E). Notice that V_P can be large either if the population is genetically quite homogeneous and the environment is very variable (V_G small, V_E large) or if the environment is uniform but there is a lot of genetic variation (V_E small, V_G large). Environmental variation V_E tends to be greater for some characteris-

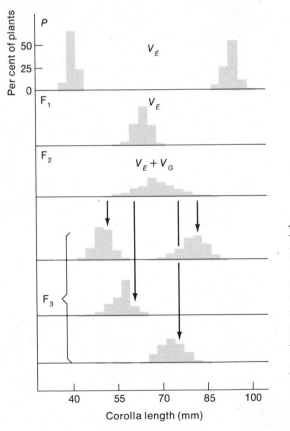

1

Inheritance of a metric, or continuously varying, trait: corolla length in the tobacco species *Nicotiana longiflora*. Variation is environmental (V_E) in the two inbred parent strains and their F_1 offspring, but has a genetic component (V_G) as well in the F_2 and F_3. Arrows show the mean of the parents of each of the F_3 families. (From *The genetics of human populations* by L. L. Cavalli-Sforza and W. F. Bodmer. W. H. Freeman and Company. Copyright © 1971. After Mather 1949)

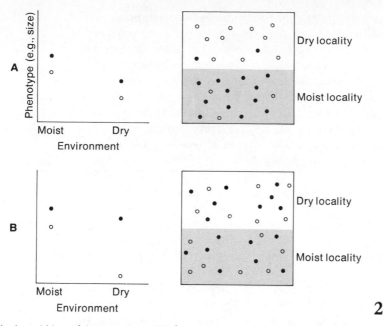

Covariation (A) and interaction (B) between genotype and environment. (A) The effect of environment on the phenotype is the same for two genotypes, but they are differentially distributed over the environments so the variation in the population as a whole is inflated by the covariance [Cov(G, E)] arising from this association. (B) The genotypes are distributed randomly over environments, but the phenotypic variation is affected by their differential phenotypic response to the environmental variation.

2

tics, such as body weight in humans which is affected greatly by nutrition, than for other characters such as eye color. These latter kinds of characters are said to be more highly canalized. Note that two populations containing the same amount and kind of genetic variation may differ greatly in V_E if one has been reared in a more heterogeneous environment than the other. Environmental variation in seed set or height, for instance, is likely to be greater in a collection of wild plants than if the same group of plants had been reared in a uniform garden. Thus we should not extrapolate from one population to another in describing the relative influences of genetic and environmental factors on the phenotypic variation.

Phenotypic variance V_P, however, is not usually the simple sum $V_G + V_E$. If there is a correlation between an individual's genotype and environment, there will be a COVARIANCE between genotype and environment (Cov G, E) that can either inflate or reduce V_P (*Figure 2*). If individuals that are genetically programmed to be larger also tend to develop in environments that contribute to large body size (e.g., those with good nutrition), and similarly for "small" genotypes, V_P is greater than if genotypes are randomly distributed over environments; if the converse occurs, V_P is less. Thus any kind of habitat selection by different genotypes affects the phenotypic variance.

Conceptually quite different from the covariance between genotype and environment is the variance attributable to the INTERACTION between genotype and environment, $V_{G\times E}$. This exists if different genotypes are affected by environmental variations in different ways. For example, some people seem easily affected by their consumption of beer and potato chips, while others, perhaps because of their different genotypes, are consistently slender no matter how much they consume. Both physical characteristics and fitness show a great deal of genotype-environment interaction (*Figure 3*).

So far, then, $V_P = V_G + V_E + 2 \text{ Cov } (G, E) + V_{G\times E}$. In the discussion that follows, I will make the simplifying assumption that both Cov (G, E) and $V_{G\times E}$ equal zero, but this is clearly not always true. The variance due to interaction $V_{G\times E}$ is especially important; the greater the interaction between genotype and environment, the less valid the entire subsequent discussion of variance analysis will be (Lewontin 1974*b*, Feldman and Lewontin 1975).

Consider now the components of the genotypic variance V_G, ignoring V_E for the moment. Assume that one locus A affects body size and that the mean size of the heterozygote is exactly between that of the homozygotes; for example, $AA = 10$ cm, $AA' = 9$ cm, $A'A' = 8$ cm. Because the variance is the average squared deviation of an observation from the population mean, the phenotype of an individual can be measured by its deviation from some arbitrary standard, say the size of the heterozygote. Then $AA = +1$, $AA' = 0$, $A'A' = -1$, or more generally, $+a, 0, -a$. The mean phenotype is $\bar{x} = p^2(a) + 2pq(0) + q^2(-a) = a(p^2 - q^2) = a(1 - 2q)$. The variance (*Appendix I*) is

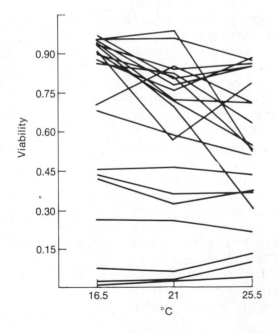

3

Genotype-environment interaction: the viability of each of 21 fourth-chromosome homozygotes in *Drosophila pseudoobscura* at each of three temperatures. Variation in viability in this case cannot be partitioned into genetic and environmental components. Notice that no one genotype is superior under all conditions. (From Lewontin 1974*b*, based on data by Dobzhansky and Spassky 1944; in *The American Journal of Human Genetics,* Copyright The University of Chicago Press)

$$V = \sum_i f_i(x_i - \bar{x})^2 = p^2\{a - [a(1 - 2q)]\}^2$$
$$+ 2pq\{0 - [a(1 - 2q)]\}^2 + q^2\{-a - [a(1 - 2q)]\}^2 = 2pqa^2.$$

The variance thus depends on the gene frequencies and on the phenotypic effect a of an allele substitution. If the character of interest is fitness, this is the genetic variance in fitness, as it appeared in Box F in Chapter 13.

In this example the effect of an allele substitution on the phenotype is the same, no matter what other allele the individual has. Thus replacing an A' allele by A in a homozygote ($A'A' \to AA'$) adds an amount a (say 1 cm) to the phenotype, just as it does if substituted in a heterozygote ($AA' \to AA$). These alleles have additive effects on the phenotype, and the genotypic variance they give rise to is termed the ADDITIVE GENETIC VARIANCE V_A. If two loci each act additively in this fashion, the total additive genetic variance (assuming linkage equilibrium) is the sum of the additive variance contributed by each locus ($V_A = \sum_i V_{Ai}$), as in this example in which each A contributes one unit and each B two units to the phenotype:

	BB	BB'	B'B'	
AA	6	4	2	
AA'	5	3	1	**(Case 1)**
A'A'	4	2	0	

Notice that the average phenotype of a brood of offspring equals that of their parents if alleles act additively. In the cross $AABB \times AA'BB'$, for example, the average parental phenotype (the "midparent") is $(6 + 3)/2 = 4.5$, and the average offspring phenotype (assuming independent segregation) is $\frac{1}{4}(6) + \frac{1}{4}(5) + \frac{1}{4}(4) + \frac{1}{4}(3) = 4.5$. *The additive effects of genes are responsible for the resemblance between parents and their offspring.*

If, however, A is dominant so that both AA and AA' provide an increment of 2 units (say cm), then the interaction between the alleles A and A' produces a different phenotype from that predicted by the additive model; the heterozygote is not intermediate. If there is any degree of dominance or overdominance (the heterozygote more extreme than either homozygote), some nonadditive genetic variation is generated, a component of the variance denoted V_D, the DOMINANCE VARIANCE. Dominance reduces the parent-offspring correlation; the mean of the offspring from the cross $AA \times A'A'$, for example, does not equal the mean phenotype of the parents. Here is a case in which there is dominance at each locus, but the phenotypic value is the sum of the contributions of the individual loci:

	BB	BB'	B'B'	
AA	6	6	2	
AA'	6	6	2	**(Case 2)**
A'A'	4	4	0	

Quite often, however, the phenotypic expression of one locus is modified by the genotype at another locus: epistasis. This interaction between loci gives rise to another component of the phenotypic variance, the EPISTATIC or INTERACTION VARIANCE V_I. Epistasis, like dominance, reduces the phenotypic correlation between parents and offspring, as is found for matings among these hypothetical genotypes:

	BB	BB'	$B'B'$	
AA	2	3	5	
AA'	3	4	1	**(Case 3)**
$A'A'$	4	2	0	

Thus the genotypic variance V_G consists of the components $V_A + V_D + V_I$, and the total phenotypic variance (ignoring Cov (G, E) and $V_{G \times E}$) is $V_P = V_A + V_D + V_I + V_E$.

For a continuously varying trait, it is almost impossible to tell how many loci vary and what the gene frequencies are. But it is possible to determine, from correlations among relatives, the magnitudes of the variance components and thus the degree to which the genes have additive, dominant, or epistatic effects. For example, if genes act additively, the parent-offspring correlation should be high. In fact, the

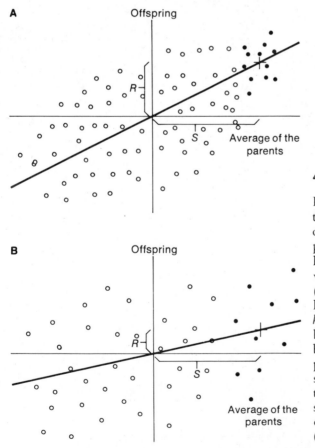

4

Plot of phenotype of offspring against that of their parents. Each point represents the mean of a brood vs. the mean of the brood's two parents (the "midparent" value). The sloping line is the regression of offspring on parent values. It is steeper if the relationship is strong (A) than if it is weak (B). The slope of this line is the "heritability in the narrow sense," h_N^2. Solid points are parents selected for breeding and their progeny. The difference between the mean phenotype of selected parents and that of all potential parents is the selection differential S; the difference between the mean phenotype of the offspring of the selected parents and that of all potential offspring is the response to selection R. (A after Falconer 1960)

fraction of the variance due to the additive effects of segregating alleles is theoretically given by the slope (the least-squares regression coefficient) of the line that relates the phenotypes of offspring to that of their parents (*Figure 4*). This fraction, V_A/V_P, is important enough to merit a special name, the "HERITABILITY in the narrow sense," h^2.

HERITABILITY AND DIRECTIONAL SELECTION

If the only source of phenotypic variation is the segregation of additively acting alleles, then $h^2 = 1$ and the average phenotype of offspring equals that of their parents. If the parents are a selected sample whose mean phenotype differs from that of the whole population by an amount S (the SELECTION DIFFERENTIAL), the population mean will change by just that amount in the next generation; the response to selection R equals the selection differential. If there are other sources of variation, then $h^2 < 1$ and the response to selection will be less marked — for the slope of the parent-offspring relation is lower (*Figure* 4). In general, $R = h^2S$; that is, the response to selection depends on the heritability. Any source of variation that reduces the similarity between offspring and parents — whether dominance, epistasis, or environmental effects — lessens the rate of evolutionary change. Evolution depends on inheritance.

The effect of selection thus depends on a character's mode of inheritance. Directional selection, selection for an extreme state of a characteristic (*Figure 5a*), causes rapid change and ultimately erodes genetic variation, if inheritance is entirely additive. In case 1, for example, directional selection for large size would produce a population monomorphic for the largest genotype *AABB*. If, on the other hand, body size was overdominant, directional selection for large size would yield a polymorphic, variable population, for the fittest genotype would be the heterozygote *AA'BB'*:

	BB	*BB'*	*B'B'*	
AA	3	4	3	
AA'	4	5	4	**(Case 4)**
A'A'	3	4	3	

STABILIZING AND DIVERSIFYING SELECTION

Seldom, though, is the biggest best. Extreme phenotypes almost always suffer some disadvantage; even if there is some directional selection in favor of one extreme, there is counterselection against it. Long fur may be advantageous for a mouse in a cold environment, but fur that is too long is a hindrance. Thus stabilizing, or balancing, selection keeps the mean character state at some intermediate value (*Figure 5b*). For example, Hecht (1952) found that in lizards of the genus *Aristelliger* the mean body size was determined by the balance between the advantage that larger lizards have in defending territories and the disadvantage they suffer in being more susceptible to predation by owls. The reasons for the superiority of intermediate forms are more often

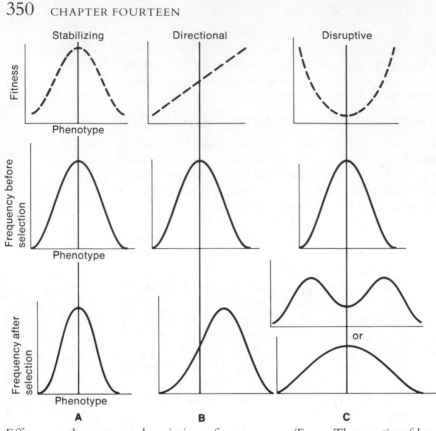

5 Effects on the mean and variation of a quantitative character of three modes of selection. The relation between fitness and phenotype may be due to natural or to artificial selection, or to both acting together.

(From *The genetics of human populations* by L. L. Cavalli-Sforza and W. F. Bodmer. W. H. Freeman and Company. Copyright © 1971)

obscure. For example, for four out of six characteristics (e.g., head width) of the milkweed beetle *Tetraopes tetraophthalmus,* males found mating were less variable and were closer to the population mean than those who were not found *in copulo* (Mason 1964). So it appears that males who were extreme in any of these traits had a lesser chance of reproducing than those closer to the average.

Many such examples can be given. In a classic study Bumpus (1899) found that intermediate-sized house sparrows had higher survivorship during a storm than did those of either extreme size (but see Johnston, Niles, and Rohwer 1972). Karn and Penrose (1951) showed that human infants with intermediate weight at birth have the highest survival rate (*Figure 6*). Eggs of intermediate weight have the highest hatching success in ducks (Rendel 1953) and chickens (Lerner and Gunns 1952). So stabilizing selection must be the usual state of affairs. When the environment changes, directional selection may shift the mean phenotype toward a new optimum, but such selection is only temporary.

Selection on some characteristics is neither directional nor stabilizing. If two or more states of a characteristic are simultaneously favored, DISRUPTIVE or DIVERSIFYING SELECTION is operating (*Figure 5c*). The level of testosterone production in mammals may be such a character; either a high or a low titer is favored, but an intermediate level is disastrous. The polymorphic butterfly *Papilio dardanus* provides an even clearer example. In any given part of Africa, the color pattern mimics one of several distasteful species of butterflies. The dominance relations among the genotypes are such that intermediate butterflies, which would mimic none of the models, do not occur (Clarke and Sheppard 1960; also see Ford 1971).

The topic of disruptive selection, which is close to speciation and multiple-niche polymorphisms, will be considered further in Chapter 16.

THE EFFECT OF SELECTION ON GENETIC VARIATION

The pattern of inheritance determines how each mode of selection affects genetic variation. Suppose, for example, that the character is additively inherited, with the following phenotypic values (e.g., body size):

	BB	*BB'*	*B'B'*
A A	4	3	2
A A'	3	2	1
A'A'	2	1	0

(Case 5)

Directional selection for large size results in monomorphism for the fittest genotype *AABB*. Under stabilizing selection the fitness of a genotype might decline as the square of its deviation from the optimum

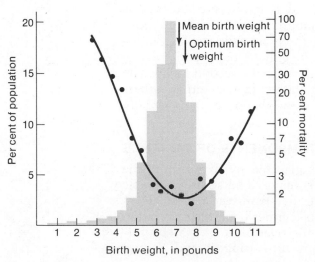

6

Stabilizing selection for birth weight in humans. Early mortality is lowest at a birth weight (optimum birth weight) near the mean for the population. The histogram represents the distribution of birth weights in the population; points represent per cent mortality. (From *The genetics of human populations* by L. L. Cavalli-Sforza and W. F. Bodmer. W. H. Freeman and Company. Copyright © 1971. Based on data from Karn and Penrose 1951)

phenotype (Wright 1969). If the optimum size is 2, the fitnesses of the genotypes in case 5 might be

	BB	BB'	$B'B'$
AA	0.6	0.9	1.0
AA'	0.9	1.0	0.9
$A'A'$	1.0	0.9	0.6

Thus, although the alleles act additively on the phenotypic scale, there is strong epistasis on the scale of fitness. For example, AA confers high fitness if associated with $B'B'$, but low fitness if associated with BB.

When there is epistasis for fitness, there are often several or many genetic equilibria toward which the population may evolve; there are multiple peaks in the adaptive landscape (*Chapter 13*). In this instance (*Figure 7*) the gene frequencies $p = q = 0.5$ (where p and q are the frequencies of A and A'), $r = s = 0.5$ (where r and s are the frequencies of B and B') constitute an equilibrium, as might be guessed from the optimum phenotype of the genotype $AA'BB'$. But it is an unstable equilibrium, at which the average fitness is only $\bar{w} = 0.9$. The adaptive landscape slopes up from this saddle point to two adaptive peaks (*Figure 7*), corresponding to populations monomorphic for $AAB'B'$ or for $A'A'BB$. The population will arrive at one of these peaks, usually the one on whose slope it initially began. Even if it starts at the saddle point, it will drift by chance from this point to an adaptive slope and then climb to a peak. Thus the ultimate genetic composition depends on the history of the population, on its initial gene frequencies. There can be many stable genetic compositions (for an even number n of equivalent additively acting loci, there are $[n!/(n/2)!]^2$ adaptive peaks), so the caprice of history can play a large role in determining a population's genetic character.

If the characteristic is not inherited additively, the outcome of selection can be even more variable. Suppose each locus has an overdominant effect on size, as in case 4, and that intermediate size is favored. All four homozygotes and the double heterozygote are inferior to each of the single heterozygotes; there are four stable compositions, consisting of polymorphism at either locus and monomorphism for either of the two homozygotes at the second locus. The theoretical possibilities become dizzying; rather than entertain others without reason, it is advisable to know how phenotypes are actually inherited.

ESTIMATES OF COMPONENTS OF VARIANCE

Continuously varying characters can be analyzed by partitioning the phenotypic variance into its major components. The additive genetic variance, V_A, can be measured, theoretically, from the relation $h^2 = V_A/V_P$. The heritability h^2 can be determined either from the correlation between parents and offspring or from the relation $R = h^2S$, by imposing artificial selection on a character, measuring R and S, and finding the REALIZED HERITABILITY, *a posteriori*: $h^2 = R/S$.

The environmental variance V_E can be measured if there are a number of genetically identical individuals: a highly inbred line or the F_1 progeny of a cross between two such inbred lines. Any variation within such a brood must be due to the environment. After V_A and V_E are found, the nonadditive genetic variance $(V_D + V_I)$ is found by subtraction. By a sufficiently complicated analysis of resemblances between relatives (see Falconer 1960), the separate components V_D and V_I can be resolved. These methods do not specify exactly how the character is inherited, but they can at least suggest the relative importance of additive gene action, allelic interaction (dominance), and epistasis in affecting phenotypic variation.

None of these values, once measured, is immutable. For example, while it is true that some characteristics, like the growth form in many plants, tend to have a higher environmental variance than others (such as the number of stamens), the magnitude of V_E depends greatly on the magnitude of the difference between the microenvironments from which the plants came. Moreover some genotypes are more sensitive to environmental perturbations than others. For example, Gowen and

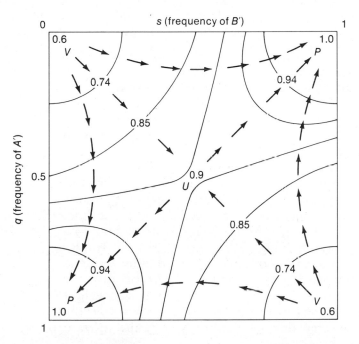

7

A possible adaptive landscape for case 5, described in the text, in which body size is additively inherited at two loci A and B, and stabilizing selection favors intermediate size. Contour lines describe population compositions of equal $\bar{w}$, very approximately. The least fit populations are $q = 0, s = 0$ (pure $AABB$) and $q = 1, s = 1$ (pure $A'A'B'B'$); these are "adaptive valleys" (V). The polymorphic population $q = 0.5, s = 0.5$ is an unstable equilibrium (U), with $\bar{w} = 0.9$. It is a "saddle point" on the landscape, from which a population will evolve toward an "adaptive peak" (P) of composition either $q = 0, s = 1$ (pure $AAB'B'$) or $q = 1, s = 0$ (pure $A'A'BB$).

TABLE I. Relative asymmetry of sternopleural bristle number in *Drosophila melanogaster*

Genotype[1]	Environment[2]	Total asymmetry
Homozygotes	Home	363
	Foreign	451
Intrapopulation heterozygotes	Home	351
	Foreign	342
Extrapopulation heterozygotes	Home	466
	Foreign	442

(After Thoday 1955)

[1] Intrapopulation heterozygotes were heterozygous for two chromosomes taken from the same laboratory population; extrapopulation heterozygotes had chromosomes from two laboratory populations adapted to the same environment.

[2] "Home" is the environment to which the populations from which the chromosomes were taken were adapted; "foreign" indicates tests in another laboratory environment.

Johnson (1946) measured the variation of egg production in two highly inbred, homozygous strains of *Drosophila*. The coefficients of variation were 0.46 and 0.50 in the two strains, but it was only 0.31 for the F_1 cross between them. Table I illustrates another example; homozygous flies tend to be more bilaterally asymmetrical than heterozygous flies. It appears that heterozygotes are less sensitive than homozygotes to environmental disruptions of development (Lerner 1954, Thoday 1955, Jinks and Mather 1955), and they often have higher fitness.

Measurements of the additive genetic variance are frequently unreliable. Even independent estimates of heritability made on the same population in the same environment may vary greatly. For example, since $h^2 = R/S$, the same absolute magnitude of response R should be observed if the same selection differential S is imposed on two samples from the same population, but in opposite directions. Figure 8 shows the result of such selection for body weight in mice. Because the response was much more rapid in the line selected for lower weight than in the upward selected line, h^2 was estimated to be 0.518 on the basis of the downward selected line, but only 0.175 in the upward selected line.

Although estimates of the components of phenotypic variance can be quite unreliable, those that have been made (e.g., *Table II*) show one striking regularity. Characteristics such as bristle number or length of wool, whose effects on fitness are not critical or at least not consistent, tend to have a high heritability; much of the variation is due to the additive effects of genes and very little to dominance or epistasis. Thus many morphological traits seem to correspond, more or less, to

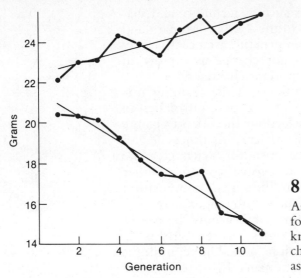

8

Asymmetrical response to artificial selection for six-week body weight in mice. A simple knowledge of the heritability of this characteristic will not predict such asymmetry. (After Falconer 1953)

TABLE II. Heritability and components of variance for some traits of animals

	Heritability $(h^2_N = V_A/V_P)$	$V_D + V_I$	V_E
Cattle			
Amount of white spotting in Friesian breed	0.95		
Milk yield	0.3		
Conception rate	0.01		
Pigs			
Body length	0.5		
Litter size	0.15		
Sheep			
Length of wool	0.55		
Body weight	0.35		
Chickens (White Leghorn)			
Egg weight	0.6		
Age at first laying	0.5		
Egg production	0.2		
Viability	0.1		
Mice			
Tail length	0.6		
Litter size	0.15		
Drosophila melanogaster			
Abdominal bristle number	0.52	0.09	0.39
Length of thorax	0.43	0.06	0.51
Ovary size	0.30	0.40	0.30
Egg production in 4 days	0.18	0.44	0.38

(After Falconer 1960, from various sources)
NOTE: Total phenotypic variance V_P set equal to 1 for all traits

our hypothetical case 1. But characteristics such as fecundity and viability, which are unquestionably related to fitness in an intimate and consistent way, have a low heritability. Their genetic variance is largely nonadditive. These traits, then, often do not change as rapidly in selection experiments as apparently less important characters.

A possible reason for this relationship between the seeming importance of a characteristic and its mode of genetic control lies in Fisher's fundamental theorem of natural selection: the rate of change of mean fitness equals the additive genetic variance in fitness ($d\bar{w}/dt = V_A$). As long as additive genetic variance exists for a character like viability, the mean viability will continue to evolve toward a higher level. But by the time we look at a population, the genetic composition is likely to be at equilibrium and the mean fitness is no longer increasing ($d\bar{w}/dt = 0$). Hence the additive variance for viability must be nearly zero if viability is strongly correlated with fitness, as it clearly is. If there is any genetic variance in fitness, it must be nonadditive; it might be the variance due to dominance or overdominance V_D or to epistatic interactions among loci.

Conversely, if a character such as bristle number or body size shows high additive genetic variance, it may be either that variations in the trait do not affect fitness (the variation is neutral) or that the relation between the phenotype and fitness is not linear (they are not strongly correlated with each other). Such is the case in the hypothetical instance that we have analyzed in which an additively inherited trait is subject to stabilizing selection. But why, given the conclusion that the population should become monomorphic for such a trait, does the variation exist? The answer may lie in the important effects of linkage.

LINKAGE AND POLYGENIC INHERITANCE

In a preceding illustration (case 5), the genotypes $AAB'B'$, $A'A'BB$, and $AA'BB'$ all have equally high fitness, since their phenotype is the intermediate optimum. The point $p = q = \frac{1}{2}$, $r = s = \frac{1}{2}$ is an unstable polymorphic equilibrium, as we noted; if any allele departs slightly, by chance, from 0.5, the population becomes monomorphic. This may happen much less rapidly, though, if the loci are closely linked.

Given two alleles at each of two loci, there are four types of gametes (or chromosomes, as I will call them), AB, AB', $A'B$, and $A'B'$, with gamete frequencies g_{11}, g_{10}, g_{01}, and g_{00}. The gametes AB and $A'B'$, in which alleles with like effects are associated, are coupling gametes; AB' and $A'B$ are repulsion gametes. The term $D = g_{11}g_{00} - g_{10}g_{01}$, the coefficient of linkage disequilibrium (*Chapter 10*), equals zero if the alleles at each locus are randomly combined with those at the other. In this case, $g_{11} = pr$, $g_{10} = ps$, $g_{01} = qr$, $g_{00} = qs$, and $D = pr \times qs - ps \times qr = 0$. The loci are in linkage equilibrium, and the rapidity with which this state is reached (in the absence of selection) is proportional to the rate of recombination R between them in the double heterozygote $AA'BB'$. If the loci are tightly linked (R small), the degree of linkage disequilibrium D will decline only slowly.

If only repulsion chromosomes AB' and $A'B$ exist, and if the gene frequencies are 0.5 at each locus, then $D = (0 \times 0) - (0.5 \times 0.5) = -\frac{1}{4}$; linkage disequilibrium is complete. As long as these are the only chromosome types in the population, the only zygotes will be AB'/AB', $AB'/A'B'$, and $A'B/A'B$, all of which have the optimal phenotype. Linkage disequilibrium has increased the fitness of the population ($\bar{w} = 1$, compared to 0.9 when the loci segregate independently). The population still moves toward monomorphism, though, for recombination slowly produces the other chromosomes AB and $A'B'$, so that the linkage disequilibrium breaks down. Moreover the chromosomes AB' and $A'B$ have equal fitness, so they behave like two neutral alleles subject to genetic drift.

Under some conditions, however, the population remains in a state of permanent linkage disequilibrium, with both loci persisting in a polymorphic state. The mathematical theory (see Bodmer and Parsons 1962, Wright 1969, Lewontin 1974 and references therein) of such systems is quite difficult, so I shall consider only a few special cases. In doing so, I move from effects of two loci on a single character to the more general case of two loci that both affect fitness, whether they act on a single character or on separate traits.

Lewontin and Kojima (1960) analyzed the case in which the genotypes have these fitnesses:

	AA	AA'	$A'A'$
BB	a	b	a
BB'	c	d	c
$B'B'$	a	b	a

They define a coefficient of epistasis as $\epsilon = a + d - b - c$, which is zero if the fitness of each genotype is a simple sum of the fitness conferred by each locus separately (fitnesses additively determined). But such additivity for fitness is most unlikely in nature. For if fitness is measured by the probability of survival, fitnesses may be multiplicative over loci. For example, if the fitnesses at locus A are 0.8, 1, and 0.8 for the genotypes AA, AA', and $A'A'$, and at locus B they are 0.9, 1, and 0.9, then $W(AABB) = W(AA) \times W(BB) = 0.8 \times 0.9 = 0.72$, and similarly for the eight other genotypes. Then $a = 0.72$, $b = 0.9$, $c = 0.8$, $d = 1.0$ and $\epsilon = 0.02$. Thus epistasis for fitness exists.

The behavior of such a system is found by analyzing the change in frequency of each of the four gametes. For example, the change in one generation of the AB gamete is $\Delta g_{11} = [g_{11}(W_{AB} - \bar{w}) - RDW_{11}]/\bar{w}$, where W_{11} is the fitness of the genotype $AA'BB'$, $\bar{w}$ is the mean fitness of the population of zygotes, and W_{AB} is the fitness of the AB gene combination (or chromosome). This is a mean fitness, calculated in the same way in which we found the fitness of an allele (*Chapter 13*): $W_{AB} = \Sigma f_{ij} W_{ij}$, the sum of the fitnesses of the four zygote genotypes (AB/AB, AB/AB', $AB/A'B$, and $AB/A'B'$) in which the chromosome AB occurs, each weighted by its frequency in the population. Thus the genotypic composition of the population determines whether a particular gene combination can increase in frequency.

This equation, then, states that the change in frequency of a chromosome type Δg_{11} depends on its present frequency g_{11}, its selective advantage $W_{AB} - \bar{w}$, the rate R at which it is generated by crossing over in double heterozygotes, the degree D to which these double heterozygotes have an excess of coupling over repulsion chromosomes, and the fitness of the double heterozygotes (W_{11}). This equation and the comparable expressions for the other three gamete types can be used to show for the preceding table of fitnesses that there are three possible genetic equilibria:

1. $g_{11} = g_{00} = \frac{1}{4} + \frac{1}{4}[1 + 4Rd/(b + c - a - d)]^{1/2}$,
   $g_{10} = g_{01} = \frac{1}{4} - g_{11}$,
   $D = \frac{1}{4}[1 + 4Rd/(b + c - a - d)]^{1/2}$;

2. $g_{11} = g_{00} = \frac{1}{4} - \frac{1}{4}[1 + 4Rd/(b + c - a - d)]^{1/2}$,
   $g_{10} = g_{01} = \frac{1}{4} - g_{11}$,
   $D = -\frac{1}{4}[1 + 4Rd(b + c - a - d)]^{1/2}$;

3. $g_{11} = g_{00} = g_{10} = \frac{1}{4}$,
   $D = 0$.

That is, the gene frequencies at each locus are 0.5 because the heterozygotes are most fit and the homozygotes have symmetrical fitnesses; but there can be (1) an excess of coupling over repulsion chromosomes, (2) the reverse, or (3) no excess (no linkage disequilibrium) at all. Lewontin and Kojima show that there is permanent linkage disequilibrium if the loci are sufficiently tightly linked, if $R \leq \epsilon/4d$.

Thus the relation between the recombination rate R and the degree of epistasis ϵ determines whether there will be a persistent association of the alleles at the two loci. Alleles at even tightly linked loci will not be associated unless they interact epistatically in their effect on fitness; and conversely even loci on different chromosomes ($R = \frac{1}{2}$) will show some degree of permanent association if they interact strongly in their effect on fitness. If there is tight linkage ($R \leq \epsilon/4d$), the third equilibrium, in which $D = 0$, exists but is unstable; the genetic composition of the population moves toward a state of linkage disequilibrium.

As a numerical illustration, consider the case when the fitnesses 0.8, 1, 0.8 at the A locus and 0.9, 1, 0.9 at the B locus are multiplied. Then $\epsilon = 0.02$. There will be permanent linkage disequilibrium if $R \leq \epsilon/4d$, or 0.005. If the loci are so tightly linked that $R = 0.001$, then $g_{11} = g_{00} = 0.475$ and $g_{10} = g_{01} = 0.025$ at equilibrium — compared with the gamete frequencies of 0.25 that we would expect if there were neither linkage nor epistasis. (Or g_{11} and g_{00} may be 0.025, and g_{10} and g_{01}, 0.475). With looser linkage ($R = 0.004$) the gamete frequencies are much closer to the unlinked expectation: $g_{11} = g_{00} = 0.285$, $g_{10} = g_{01} = 0.215$.

A particularly instructive example provided by Lewontin and Kojima entails these genotypic fitnesses:

	BB	BB'	$B'B'$
AA	3	4	3
AA'	1	5	1
$A'A'$	3	4	3

Notice that the locus B is overdominant and is therefore stably polymorphic, whatever the genotype at locus A; but the A locus is overdominant only on a BB' background. On a homozygous BB background the AA' heterozygote is inferior. Is there a stable polymorphism at locus A, or is this locus monomorphic, as we expect from single-locus theory? The answer depends once again on history. If the loci are tightly enough linked ($R < 0.11$), a population that from its inception has gene frequencies near 0.5 will remain in a stable polymorphism at the A locus and will be in permanent linkage disequilibrium (e.g., if $R = 0.10$, $g_{11} = g_{00} = 0.394$, $g_{10} = g_{01} = 0.105$). But if the population is initially monomorphic at the A locus, a mutant allele A or A' may or may not rise to a polymorphic equilibrium, depending on linkage. Suppose an A' mutant arises in an AA population that is at gene frequency equilibrium at locus B, and that the loci are not linked. The mean fitness of the AA genotype is $r^2 W_{AABB} + 2rs W_{AABB'} + s^2 W_{AAB'B'} = 3.5$. But the mean fitness of the rare AA' genotype is $r^2 W_{AA'BB} + 2rs W_{AA'BB'} + s^2 W_{AA'B'B'} = 3.0$. Since the heterozygote is not fitter than the prevalent homozygote AA, A' cannot increase. The same holds if A arises as a mutant in an $A'A'$ population.

If, however, locus B is not yet at gene frequency equilibrium, the mutant A may rise toward an equilibrium frequency of 0.5 if it is linked closely enough to the B allele that is rising toward its equilibrium. In the extreme case, for example, imagine that the mutations A and B have both occurred on the same chromosome in an $A'A'B'B'$ population. Because BB' has a higher fitness than $B'B'$, the allele B rises in frequency and carries allele A with it, as long as they remain associated in strong linkage disequilibrium. The locus A, although not overdominant on its own, has overdominance conferred on it by its association with the locus B and so can be brought to a stable polymorphic state. This same principle may apply to quantitative characters of the kind treated earlier. Although stabilizing selection for an intermediate phenotype does not cause stable polymorphism at the loci that control the trait, a single locus that is overdominant for fitness can maintain the polymorphic condition of closely linked loci that govern the characteristic in an additive manner (Wright 1969, p. 76).

MORE CONCLUSIONS FROM TWO-LOCUS THEORY

It will now be apparent that the genetic consequences of selection can be immensely complicated if two linked loci interact. The mathematical analysis of these systems rapidly becomes extremely difficult and is the subject of ever more sophisticated theoretical treatments. Rather than pursue the theory further, I will simply summarize some of the more important conclusions.

If two loci interact epistatically in their effect on fitness, they may be held in permanent linkage disequilibrium, if the rate of recombination between them is sufficiently low.

In general a stable polymorphism at each locus persists only if the locus is overdominant for fitness, but such overdominance is sometimes conferred by association with neighboring loci.

Alleles at each of two loci can sometimes increase in frequency, even if each viewed individually is not advantageous, if they are sufficiently advantageous in combination and are tightly enough linked.

There may be several or many adaptive peaks in the adaptive landscape, corresponding to different equilibrium genetic compositions. Some of these peaks may be higher than others. Which peak a population arrives at depends on its history, especially its initial gene and gamete frequencies. Once on an adaptive peak, the genetic composition remains unchanged, even if there are other genetic compositions that are more fit (higher $\bar{w}$). Populations do not necessarily evolve the best possible genetic compositions.

When there is linkage and epistasis, gene frequencies often move toward their equilibrium values at different rates than when there are no such effects. That is, the rate of response to selection may be slowed or increased because of linkage disequilibrium (for example, the advantageous allele at one locus may be associated with a disadvantageous allele at another locus).

When there is permanent linkage disequilibrium ($D \neq 0$), the average fitness of the population is greater than if the alleles are segregating independently. That is, alleles tend to be packaged in favorable combinations. A corollary is that selection may favor any mechanism that holds these favorable combinations intact. A chromosome inversion is one such mechanism, but the rate of crossing-over between loci, which is subject to genetic control, may also evolve.

Because $\bar{w}$ is increased by the existence of permanent linkage disequilibrium, the genetic load is reduced. That is, two loci may be held in polymorphic condition by fewer "genetic deaths" than if they were segregating independently. For example, if both loci A and B are overdominant for fitness and if allele A is associated with B, and A' with B', the death of a single $AABB$ or $A'A'B'B'$ homozygote contributes to the selection on both loci simultaneously. Both the segregational load required to keep many loci polymorphic, and the substitutional load, or cost of selection required for directional evolution at many loci, may be substantially reduced.

The pattern of association between alleles can change substantially in the course of selection. Lewontin and Kojima describe a theoretical case in which D begins at zero, becomes negative, goes back through zero to become positive, and then returns to its final value of zero.

The heritability (V_A/V_P) measured from parent-offspring resemblances does not predict the long-term response to selection, for linked polygene combinations may have latent genetic variation which is not apparent until it is released by recombination.

If there is epistasis among loci in their effect on a phenotypic characteristic, directional selection for an extreme state of the character will generate linkage disequilibrium (Felsenstein 1965). For example, if the body size of $AABB$ is greater than the simple sum of the contributions of AA and BB taken separately, selection for large body size will build up an excess of AB chromosomes.

By the same token, if a population has an excess of AB chromosomes ($D > 0$), selection for the extreme $AABB$ phenotype will elicit a more rapid response than if there is linkage equilibrium. Conversely the response to directional selection will be slower if there is an excess of repulsion chromosomes ($D < 0$), because the favored AB combination will become available to selection only as it is formed by recombination. Since populations in the same environment may have arrived at different equilibrium compositions (either $D > 0$ or $D < 0$), they may respond very differently to a change in the selective environment.

If a population is in linkage disequilibrium with $D < 0$ before directional selection is imposed, but the new selection regime favors a composition with $D > 0$, tight linkage between loci at first slows the response to selection, but then (when D passes from negative to positive) quickly accelerates the response (Felsenstein 1965). It is thus quite possible for a population experiencing a novel but constant selection pressure to show little or no response for several generations and then to respond rapidly, in a burst of evolutionary change.

EPISTASIS AND LINKAGE DISEQUILIBRIUM IN NATURAL POPULATIONS

Epistatic interactions among loci are very common, as I have stressed in previous chapters. Even when a metric trait such as body size has a predominantly additive mode of inheritance, the loci affect fitness epistatically. But sometimes alleles with distinguishable morphological effects show epistasis at the level of both phenotype and fitness. For example, the capacity to produce hydrogen cyanide in the plants *Lotus corniculatus* (bird's foot trefoil) and *Trifolium repens* (clover) requires the presence of a dominant allele at each of two loci, one coding for an enzyme that synthesizes a cyanogenic glycoside and one for an enzyme that cleaves the glycoside to release HCN. Although these loci are not linked in *Trifolium,* they are rather closely linked in *Lotus,* and the dominant alleles tend to be associated in linkage disequilibrium in polymorphic populations (Ford 1971). A striking case of almost complete linkage disequilibrium is the heterostyle-homostyle system in *Primula vulgaris,* described in Chapter 9. Two loci controlling the male and female characters of the flower are so tightly associated that plants with a disadvantageous combination of pin and thrum features are very seldom found. This is only one of many instances (summarized by Ford 1971) in which a polymorphism that at first appears to be a single locus is actually caused by two or more very tightly linked loci held as a supergene in strong linkage disequilibrium.

In some instances epistasis and/or linkage disequilibrium can be shown to exist, but its adaptive basis is unknown. Levitan (1959) has shown that in *Drosophila robusta,* an inversion in one arm of the X chromosome tends to be associated with an inversion in the other arm, to which it is only loosely linked. The Australian grasshopper *Moraba* (=*Keyacris*) *scurra* is polymorphic for an inversion A on one chromosome, and for an inversion B on another. The nine genotypes do not occur in the frequencies expected from random assortment. From an analysis of the genotype frequencies, Lewontin and White (1960) estimated that their fitnesses showed epistasis:

	BB	BB'	$B'B'$
AA	0.79	1.000	0.83
AA'	0.67	1.006	0.90
$A'A'$	0.66	0.66	1.07

From these fitnesses Lewontin and White calculated what $\bar{w}$ would be for populations with different chromosome frequencies and constructed the adaptive landscape shown in Figure 9. There are two adaptive peaks, at 100 per cent A', 100 per cent B' and 0 per cent A', 55 per cent B'; and an unstable equilibrium, or saddle point, exists at 20 per cent A', 65 per cent B'. Surprisingly the natural population they examined was at the saddle point, rather than on one of the two stable adaptive peaks (but see Turner 1972).

Linkage disequilibrium may exist for more than two loci, of course,

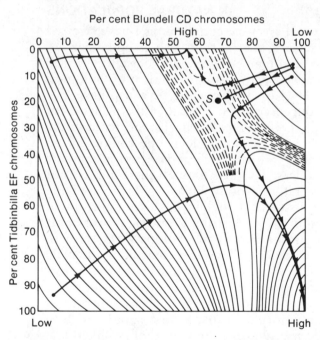

9

The adaptive landscape for a population of the Australian grasshopper *Moraba scurra* that is polymorphic for both the *EF* chromosome and the *CD* chromosome. Based on genotype frequencies in the field, viabilities were calculated for each genotype, and from these the theoretical fitness $\bar{w}$ of populations of each possible chromosomal constitution were calculated. Compositions of equal $\bar{w}$ are indicated by contour lines; the dashed lines indicate finer distinctions of $\bar{w}$ than the solid lines. There are two peaks (high) and a saddle point (*S*). The trajectories are theoretical changes in genetic composition a population would follow from five initial states. (After Lewontin and White 1960)

and may characterize much of the genome. Polymorphic inversions, such as are common in *Drosophila,* hold alleles in nonrandom combinations (*Chapter 9*). The ultimate in linkage disequilibrium is found in evening primroses (*Oenothera biennis* and its relatives). In these plants each of the seven chromosomes has undergone reciprocal translocations with two other chromosomes. In meiosis each chromosome must lie in synapsis with the ends of two other chromosomes, forming a complete ring. Unless the two sets of chromosomes move intact to the poles of the meiotic spindle, the gametes that are formed lack some genetic material and thereby form inviable zygotes. Thus recombination is suppressed, and the entire genome behaves like a single linkage group. Moreover each of the chromosomes bears recessive lethal genes, so that a chromosome set cannot survive in homozygous condition. So the entire genome is permanently heterozygous and experiences no recombination (Cleland 1972, Stebbins 1950).

MORE THAN TWO LOCI

Although single-locus theory is quite adequate for loci that do not interact strongly with the rest of the genome, it becomes increasingly insufficient as the number of interacting loci increases. Some indication of these complexities is found in the theoretical work of Franklin and Lewontin (1970; see Lewontin 1974 for a summary).

Franklin and Lewontin analyzed, by computer simulation, the case in which there are two alleles, denoted 0 and 1, at each of n loci. At each locus the heterozygote has superior fitness, and the probability of survival of an individual is independently affected by each locus. Thus if $W_{00} = 0.9$, $W_{01} = 1$, and $W_{11} = 0.9$ at each locus, an individual homozygous at k loci has the fitness $(0.9)^k(1)^{n-k}$.

Because populations seem to be polymorphic at 30 per cent of their loci, they argue that adjacent pairs must be closely linked, and so they assume a set of 36 loci with the recombination frequency among adjacent pairs equal to $r = 0.0025$. If the population begins with all the alleles randomly assorted with respect to alleles at other loci, there are 2^{36} types of gamete (chromosome) to begin with. But when recombination, mating, and gene frequency changes due to selection are simulated, linkage disequilibrium quickly develops, so that after 300 generations, 85 per cent of the gametes are of only two types! That is, selection favors chromosomes with both 0 and 1 alleles, so that most of the zygotes formed are heterozygous at most of their loci.

A most interesting result arose when Franklin and Lewontin considered asymmetrical fitnesses (*Table III*). If $W_{00} = 0.9375$, $W_{01} = 1$, and $W_{11} = 0.75$, then the equilibrium frequency of the 1 allele should be $\hat{q} = 0.0625/(0.0625 + 0.25) = 0.2$ at each locus. But in their simulation very strong linkage disequilibrium again developed, so that only a few of the 2^{36} possible gamete types were prevalent, and at almost all loci $\hat{q}$ *was closer to 0.5 than to the expected value of 0.2.* To visualize what happened, imagine just five loci in complete linkage disequilibrium, such that the only types of chromosome are 10101 and

TABLE III. Gametic array after 700 generations of computer-simulated selection

	Gamete	Frequency
(1)	111010110101110101001011110101101101	0.265
(2)	000111001010101110110111101111010110	0.134
(3)	000111001010101110110111101111010111	0.108
(4)	111101111101111011111100111010111011	0.111
	Others	0.392

(From Franklin and Lewontin 1970)
NOTE: Fitnesses at each of 36 loci are $W_{11} = 0.9375$, $W_{10} = 1$, $W_{00} = 0.75$.

01010. The fitness of the heterozygote is 1, but that of the homozygote 10101/10101 is $(0.94)^2(0.75)^3 = 0.37$, while that of 01010/01010 is $(0.94)^3(0.75)^2 = 0.47$. Then the frequency of the chromosome 10101 at equilibrium is $\hat{q} = s/(s + t) = 0.63/(0.63 + 0.53) = 0.54$, which is necessarily the frequency of each of its 1 alleles, since they occur only on this chromosome. As more and more 0 and 1 alleles are packed together, their fitnesses become averaged, so that the frequency of an allele becomes determined not by its independent effect on fitness, but by the fitness conferred by the whole chromosome in which it usually resides.

Another striking feature of these results was that two loci may be so far apart on the chromosome that by two-locus theory they should be in linkage equilibrium; but by interacting with the loci between them, they are kept in a state of permanent linkage disequilibrium. Moreover two tightly linked loci can be held in linkage disequilibrium by the loci that surround them, even if the epistatic interaction between them is extremely small. Allele frequencies, and patterns of association among loci, are almost independent of individual locus effects and of pairwise interactions between loci; they depend instead on the relative fitness in the homozygous vs. heterozygous condition of whole segments of chromosomes and on the total map length (amount of recombination) of these segments (Lewontin 1974). Thus, according to this theory, the genetic composition of populations depends not on the fitness of individual alleles viewed in isolation — for they are seldom so isolated — but on the fitness of whole groups of genes, the context in which the alleles function.

Lewontin (1974) therefore argues that even if the fitness differences at a locus are so slight as to be unmeasurable and indeed insufficient to maintain polymorphism by themselves, they can be sufficient to keep the locus polymorphic by interaction with neighboring loci. And if linkage disequilibrium were as prevalent as this theory suggests, the problem of genetic load would almost vanish. Balancing selection could maintain far more variation and evolution under directional selection could be far more rapid than single-locus theory predicts.

DOES THE GENOTYPE CONGEAL?

However, the meager data amassed to date indicate that electrophoretic loci are in strong linkage disequilibrium only when they are associated with inversions (Prakash and Lewontin 1968; *Chapter 9*). Otherwise they tend to show only weak linkage disequilibrium, if any (e.g., Charlesworth and Charlesworth 1973; *Chapter 9*). And such cases do not prove epistatic effects on fitness, because linkage disequilibrium can arise by chance in a finite population; chromosomes with particular combinations of alleles can drift in frequency just as alleles at single loci do.

It appears then that natural gene pools have not coalesced into as few sets of associated alleles as the Franklin-Lewontin model predicts. Why then, as Turner (1967) asks, does the genotype not "congeal," as it has in *Oenothera biennis* or in *Drosophila* inversions? Turner suggests that although many loci interact favorably and so should come into linkage disequilibrium with one another, loci that interact disadvantageously are scattered among them, making it extremely difficult for recombination to dissociate the favorable alleles from the unfavorable alleles in a short chromosome segment. Even when the few favorable gene combinations do arise, they are likely to be quickly dissociated by crossing-over, unless they happen to be preserved by an inversion. Another reason why linkage disequilibrium may not be prevalent is that a particularly advantageous combination of alleles is likely to be advantageous only under certain environmental conditions. When these change, formerly advantageous gene combinations (e.g., a particular inversion) are replaced by more freely recombining genotypes. Thus the amount of recombination in which the genotype engages is likely to be determined by the inconstancy of the environment.

The other possible reason for the relative rarity of linkage disequilibrium (if this turns out to be generally the case) is simply that genes in natural populations may not behave according to the Franklin-Lewontin model; there may be very few overdominant loci and hence little of the overall epistasis required to confer organization on large blocks of chromosomes. This subject requires a good deal more empirical and theoretical exploration before any firm conclusions will emerge.

SUMMARY

Most characteristics vary under the influence of both the environment and several or many loci. For many characteristics these loci have additive effects; the influence of an allele on the phenotype is independent of the rest of the genotype. Such loci are responsible for the resemblance between relatives, and characters that show this mode of inheritance respond rapidly to selection. But many characters, especially fitness itself and characteristics closely correlated with fitness, are controlled by alleles with strong dominance and epistasis rather than additivity. Such characteristics change more slowly under selection.

Nonadditive effects on fitness may be caused by stabilizing selection in favor of intermediate phenotypes. Whether or not stabilizing selection is involved, the combination of epistatic (nonadditive) fitness and close linkage can organize the genome into nonrandom associations of alleles that form especially favorable, coadapted, gene complexes. Such nonrandom association, or linkage disequilibrium, can either accelerate or retard the rate of evolution when selection pressures change; it reduces the genetic load of a population; and it may bring gene frequencies to levels that are determined less by the effects of the individual alleles than by the effects of whole sets of interacting genes.

When genetic change occurs at two or more loci simultaneously, the genetic composition of the population can move to any of several or many different states, or adaptive peaks. The course of evolution by natural selection is then determined in large part by historical accident, and it can be quite unpredictable.

FOR DISCUSSION AND THOUGHT

1   Show that a characteristic like the growth form of a plant, which exhibits a great deal of nongenetic variation due to the environment, responds less rapidly to either artificial or natural selection than a character under "tight" genetic control. What are the evolutionary implications of this fact?

2   Genotypes differ in the degree to which the phenotype is affected by environmental perturbations. Hence Scharloo et al. (1972) were able to select for both increased and decreased sensitivity to temperature of a wing vein in *Drosophila*. What kinds of characteristic should evolve high or low sensitivity to environmental variation? That is, under what selective regimes should a characteristic manifest high vs. low values of V_E?

3   The heritability of a behavioral trait may range from a very high value (when it is usually described as instinctive behavior) to a very low value (when it is called learned). Should we then expect the capacity to learn to vary in any systematic way phylogenetically? What sorts of thing should what sorts of animal be capable of learning? What are the implications of these predictions for research in experimental psychology and other behavioral sciences?

4   Postulate models of biochemical pathways such that allelic variation at the loci coding for the enzymes might result in additive, overdominant, or epistatic variation in the end product of the pathway. Do these models correspond to any known biochemical pathways?

5   Does the prevalence of additive genetic variation seem inconsistent with our view of development as a highly complex pattern of interactions between sets of cells that proliferate and grow at different rates, inducing changes in other sets of cells? How can these two views of development be resolved?

6   What are the consequences of disruptive selection acting on a polygenic trait in which gene action is (a) additive; (b) dominant; (c) overdominant? *A priori,* do you suppose that additively acting or dominant alleles have a selective advantage?

7   Can the pattern of inheritance of a trait evolve? Can a trait evolve from being additively determined to being controlled by dominant or epistatically interacting alleles? Under what conditions might you expect this to occur?

8   At any stage in artificial selection, one may calculate the realized herita-

bility of a character from the relationship $h^2 = R/S$. Usually h^2 decreases as selection proceeds. Why should this be so?

9 Explain Falconer's observation that the response to selection for body weight in mice was different in the upward and downward selected lines.

10 The best-known example of heterosis is in corn (maize). Hybrid corn varieties are taller and have a greater yield than inbred, homozygous strains. Are these characteristics determined by additively acting polygenes or by genes with some other mode of action?

11 Some genotypes of corn manifest general combining ability, meaning that they contribute a predictable increment to seed yield in the F_1 hybrid, no matter what strains they are crossed with. Other genotypes show specific combining ability; they contribute greatly to the yield of the F_1 if crossed with some strains but not with others. How would you characterize such genotypes in terms of additive and nonadditive gene action? Discuss the importance of combining ability for plant breeders and for the response of a population to natural selection.

12 A character with low heritability responds slowly to directional selection on the population. How would you practice artificial selection on such a character if you wanted a rapid response? What relevance does this question have to evolutionary theory?

13 Mayr (1963) has argued that selective changes in gene frequencies at one or several loci may induce gene frequency changes at other loci; thus genetic change begets genetic change, and a "genetic revolution" may transpire. How do the models described in this chapter bear on this hypothesis?

14 Explain how linkage disequilibrium may be caused by genetic drift. What are the implications of this fact for the interpretation of observations of gene frequencies and linkage disequilibrium in natural and experimental populations?

15 Discuss the implications of the models in this chapter for macroevolution. How do the concepts of multiple adaptive peaks, differential genetic responses to similar selection pressures, and associations between alleles at linked loci bear on the question of whether the diverse attributes of different species constitute unique adaptations to different selection pressures? Why, for example, have mammals retained the left aortic arch and birds the right?

MAJOR REFERENCES

Lewontin, R. C. 1974. *The genetic basis of evolutionary change*. Columbia University Press, New York. 346 pages. Treats the theory of polygenic systems in its last chapter.

Falconer, D. S. 1960. *Introduction to quantitative genetics*. Ronald Press, New York. 365 pages. The theory of polygenic traits, and their response to selection.

Lerner, I. M. 1958. *The genetic basis of selection*. Wiley, New York. 298 pages. Same topics as Falconer.

The Response to Selection

Insight into the factors that affect genetic variation should be only a prelude to understanding the essence of evolutionary change: the genetic responses of populations to selection. This chapter examines such responses, in natural populations and especially in laboratory populations subjected to artificial selection.

Under natural selection differences in fitness depend on the totality of characteristics that affect survival and reproduction; under artificial selection, on the other hand, a genotype's reproductive prospects are based largely (but not entirely) on its body size, bristle number, or whatever characteristic or combination of characteristics the experimenter has chosen to select. Studies of artificial selection, moreover, usually deal with small populations, into which gene flow is prohibited. For these reasons it may be risky to extrapolate from studies of artificial selection to the action of natural selection in wild populations. But it is unjustified to dismiss these laboratory studies as "unnatural" and therefore uninteresting. Ever since Darwin, who opened the *Origin of Species* with a chapter on variation and artificial selection in domesticated organisms, evolutionists have used artificial selection to gain insight into the genetic structure and flexibility of populations. And artificial selection is often difficult to distinguish from natural selection, especially when natural selection, in the form of hard rather than soft selection (*Chapter 13*), acts by the survival of only those individuals that possess a specific state of some characteristic. Natural selection for resistance to DDT, toxic soils, or harsh weather is very similar to artificial selection.

STABILIZING AND DISRUPTIVE SELECTION

If artificial stabilizing or disruptive selection is applied to a continuously varying characteristic, one might suppose that after many generations the mean would not change, but that the variance would decrease in the stabilizing selection (S) line and increase in the disruptive selection (D) line. But in actual experiments the outcome of such selection varies. For example, Thoday (1959), selecting in these modes for sternopleural bristles in *Drosophila,* obtained results rather like those we might expect. But Bos and Scharloo (1973) found that after about 30 generations of selection for body size in *Drosophila,* the variance did not change in either S or D lines. Moreover mean body size declined in the S lines and increased in the D lines. Thus the responses of populations to artificial selection may be surprising, and they are likely to be at least as complicated under natural selection.

Bos and Scharloo explain their observations by noting that body size was not symmetrically distributed but skewed toward small sizes. Thus the putatively stabilizing selection for individuals near the mean shifted the mean toward lower levels; selection had a directional component (*Figure 1*). In the D lines both large and small females were selected; but large females produce more eggs, so *natural* selection shifted the mean upward.

The lack of change in the variance follows from the theory of polygenic inheritance (Bulmer 1971). If body size is determined by + and − alleles at each of many loci, an intermediate-sized fly is likely to have a chromosomal constitution like this: $\frac{+-+-+}{-+-+-}$, with + and − alleles in repulsion phase. In the S lines these chromosomes are preserved by selection; but in each generation coupling chromosomes

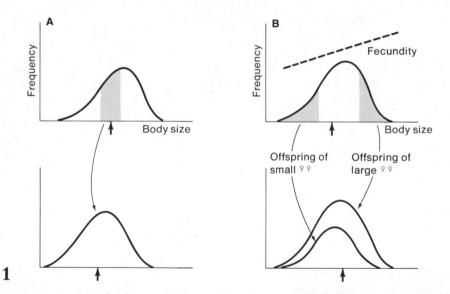

Diagrammatic illustrations of the results of one generation of (A) stabilizing and (B) disruptive selection, by breeding from individuals in the shaded regions of the upper curves. In each case recombination expands variation in the progeny beyond that of the parents to approximately that of the population from which the parents were selected. (A) The skewed distribution causes a decrease in the mean (indicated by the arrow). (B) The higher fecundity of large parents increases the mean.

($+ + + + +$ and $- - - - -$), conferring more extreme phenotypes, are formed by recombination, thus maintaining the high variance. Theoretically the population should become monomorphic for a genotype like $\frac{+ - + - +}{+ - + - +}$ (*Chapter 14*), but only after many generations. In the D lines, on the other hand, coupling chromosomes are preserved by selection, but random mating in each generation forms intermediate heterozygotes. Disruptive selection increases the variance appreciably only if coupled with assortative mating among animals of like phenotype (Bos and Scharloo 1973, Thoday and Boam 1959), or with inbreeding.

DIRECTIONAL SELECTION

It is with directional selection, the primary basis of evolutionary change, that this chapter is most concerned. In the simplest model of directional selection, advantageous alleles replace their predecessors until the genetic variation is used up; the population becomes monomorphic and reaches a plateau, from which further advance is possible only when new advantageous mutations arise.

The actual history of such populations is exemplified by those studied by Mather and Harrison (1949). The main features of their work (*Figure 2*) have been observed in many similar experiments. From a laboratory population of *Drosophila melanogaster,* two lines were taken. Line 1 was selected for a low number of abdominal bristles and line 3 for a high number by choosing the two most extreme individuals of each sex from 20 of each sex on which bristle number was scored in each generation. The main features of the results are these:

Mean bristle number changed dramatically in the first 20 to 30 generations, from 36 to 25 in the "low" line (1), and to 56 in the "high" line (3).

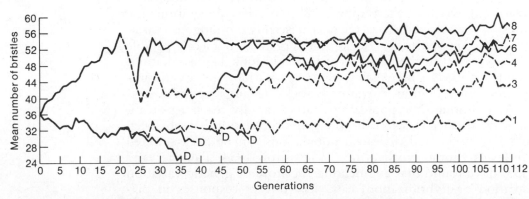

2

Changes in mean bristle number under artificial selection, as practiced by Mather and Harrison. Solid lines denote operation of artificial selection, broken lines relaxation of artificial selection. D indicates that a line died out because of sterility. (After Mather and Harrison 1949)

The mean was still changing when selection was terminated, because the low line died out at generation 34 and the high line, too, was clearly headed toward extinction. Thus fertility and/or viability declined, a CORRELATED RESPONSE that is often (but not always; see Clayton, Morris, and Robertson 1957) observed in selection experiments.

When selection was relaxed in line 3, the bristle number rapidly decreased to its original level.

Bristle number in new lines (e.g., 8) taken from line 3 rapidly increased when selection was resumed; these lines did not suffer reduced fertility. They reached a plateau at lower levels than line 3 had achieved, but further (delayed) responses to selection did occur after as many as 55 generations of no change (e.g., line 8 at generation 82).

When selection was relaxed in lines (e.g., line 7) taken from the plateaued line 8, bristle number did not decline very much.

When line 2, taken from the plateaued line 7, was back selected for lower bristle number, it responded readily, dropping to the level of line 1.

Correlated responses were observed in other characters, such as the number of coxal and sternopleural bristles and the number and form of spermathecae.

Mather and Harrison determined that each of the three major chromosomes carried genes that affect bristle number. In the base population epistasis between chromosomes X and III acted to lower bristle number, but in the high-selected lines these chromosomes exhibited positive epistasis, increasing bristle number nonadditively. Thus selection had increased the frequency of alleles that not only increase bristle number in their own right, but combine synergistically to yield the desired phenotype.

A most striking feature of these data, as in most such studies, is the rapidity and extent of the response. Within a few generations the mean value of the selected character far transcends the range of variation in the original population. The change is gradual, but a comparable change, if preserved in the fossil record, would look like a saltation. The response is due to preexisting genetic variation, not to the immediate effects of mutation, for inbred stocks do not respond in this manner. And although the populations rather rapidly reach a plateau, responding no further to selection, the genetic variation has not been exhausted (*Figure 3*), and plateaued populations can be back selected to lower levels. Some workers (e.g., Scossiroli 1954, Falconer and King 1953) have found that augmenting the store of genetic variation (by irradiation or hybridization) can yield further responses in plateaued populations, suggesting that variation has been exhausted. But Mather and Harrison's observations are more common: the level of genetic variation in itself seldom seems to limit the rate or extent of response to selection. What, then, does limit the response?

GENETIC AND DEVELOPMENTAL HOMEOSTASIS

Several answers to the question of what limits response are suggested by the prevalence of correlated responses, especially the decline in fitness as selection proceeds. Natural selection opposes artificial selection; the genotypes favored by the experimenter are reproductively inferior. The population thus resists change in bristle number and returns to its original state if selection is relaxed. Lerner (1954) referred to this phenomenon as GENETIC HOMEOSTASIS, which he defined as "the property of the population to equilibrate its genetic composition and to resist sudden changes." Lerner's theory is one of two major hypotheses for the results of artificial selection; I shall treat the other later in this chapter.

The basis for genetic homeostasis, Lerner argued, is the superior fitness of heterozygotes, of which the widespread phenomenon of inbreeding depression gives evidence. Strong artificial selection increases homozygosity and thereby reduces fitness. The reason for the low fitness of homozygotes is their reduced capacity for developmental homeostasis.

DEVELOPMENTAL HOMEOSTASIS is the capacity of an individual's genotype to produce a proper, well-formed, adaptive phenotype in the face of the perturbations and insults that occur in the course of development. The development of the normal phenotype is canalized along proper channels and resists diversion into other paths. Thus, for example, the effects of teratogens (agents that cause developmental abnormalities), such as rumplessness in chickens exposed to excess in-

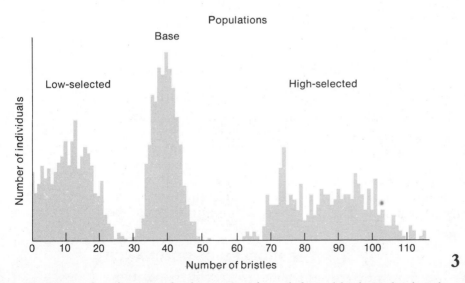

3

Demonstration that directional selection need not diminish variation: the frequency distribution of abdominal bristle number in a stock of *Drosophila melanogaster* (center) from which lines were artifically selected for low (left) and high (right) bristle number for 34 and 35 generations respectively. (After Clayton and Robertson 1957 and Falconer 1960)

sulin, are less drastic in well-canalized than in poorly canalized genotypes.

One measure of the susceptibility of development to environmental perturbations is the magnitude of the environmental variance V_E of a character. In *Drosophila*, in which it is possible to synthesize strains homozygous for 1, 2, or 3 major chromosomes, V_E is greater in more homozygous than in more heterozygous genotypes (Robertson and Reeve 1952). Another measure of sensitivity to the environment is the degree of bilateral asymmetry. The two sides of a fly, having the same genotype, should be identical if it were not for accidental alterations of the developmental pattern. Asymmetry is more pronounced in homozygous than in heterozygous flies, and it increases in populations selected for bristle number (Thoday 1955). Natural populations are also characterized by varying degrees of asymmetry; among island populations of lizards, asymmetry is most pronounced in small populations (perhaps because of inbreeding), especially those most phenotypically distinct from their continental ancestors (perhaps because of directional selection; Soulé 1967).

COADAPTATION

Developmental homeostasis depends not only on the level of heterozygosity, but on the joint action of genes that are coadapted to produce a well-organized pattern of development. For example, bilateral symmetry is greater in heterozygotes than in homozygotes only if the chromosomes come from the same population. Heterozygotes formed by crossing flies from populations adapted to different environments were highly asymmetrical (Tebb and Thoday 1954; Thoday 1955; *Table I, Chapter 14*). Thus alleles at different loci interact epistatically to produce the phenotype. Alleles from different populations, not having been selected to form harmonious combinations with each other, are not coadapted and are more susceptible to perturbations of development (Dobzhansky 1955).

To attribute Mather and Harrison's results to genetic homeostasis, and thus to developmental homeostasis, heterozygosity, canalization, and coadaptation, is not to explain them. These terms merely describe experimentally observed phenomena whose developmental or biochemical bases are quite obscure. But although we do not know what causes canalization and coadaptation, it is important to know how prevalent these phenomena are and how they affect evolution. I shall therefore explore them at some length before returning to the possible explanations of the response to selection.

THE PROBLEM OF DOMINANCE

Coadaptation is also shown by crosses among geographic populations of the swallowtail butterfly *Papilio dardanus* (see Ford 1971 for a succinct summary of work by Clarke and Sheppard). In the Malagasy Republic (Madagascar) the females are monomorphic and resemble the males. On the African continent, however, females are polymorphic, and mimic each of several species of distasteful butterflies. In southern Africa, for example, the genotype $H^T\_$ mimics *Danaus chrysippus,* and has reddish wings; the genotypes $H^C H^C$ and $H^C h$ are almost completely black, resembling *Amauris echeria*; and hh is black with large white spots, like its model *A. niavius*. When these African forms are crossed with one another, the H^T allele is fully dominant over H^C and h, and H^C is fully dominant over h; thus imperfect mimics with intermediate characteristics are not produced. But the progeny of crosses between any of the African forms and the Madagascan race are often intermediate in appearance, because H^T and H^C are not dominant over the Madagascan allele. Thus dominance is not an intrinsic feature of either of these alleles; instead it depends on their epistatic interactions with other elements in the genotype that control their phenotypic expression.

Clarke and Sheppard interpret these results as evidence for Fisher's (1930) theory of the evolution of dominance. Most wild-type alleles are dominant over mutations. This dominance is not a necessary consequence of biochemistry, for even if recessives produce less gene product (e.g., a pigment) than dominants, their less extreme phenotype can often be advantageous, so the recessives should often predominate (be the wild type) in natural populations. Fisher proposed that dominance is not an intrinsic feature of a wild-type allele, but that the degree of dominance evolves as modifier genes at other loci are built into the genotype (*Figure 4*). Thus an allele M at a modifier locus is advantageous if it causes an advantageous allele A at another locus to be dominant over deleterious mutations, such as A'. Thus A' (which, being deleterious, is rare and so exists almost exclusively in heterozygous condition) is not expressed in the heterozygote $AA'MM$.

Clearly, then, the genes A and M interact epistatically in their effects on both the phenotype and on fitness. The gene combination $AAMM$ is coadapted, and dominance breaks down if that genotype is hybridized with members of another population in which the prevalent genotype is $A'A'mm$. Modifier genes such as M canalize the phenotype.

Sved and Mayo (1970) summarize much of the controversial history of Fisher's theory. The major criticism was leveled by Sewall Wright, who pointed out that the modifier allele M would have only a minuscule selective advantage, since it would exert its beneficial effect only when coupled with rare AA' heterozygotes. Since locus M would have other, more important, physiological effects than merely modifying

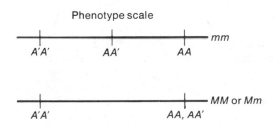

4

The theoretical effect that a modifier allele M will have on the phenotype of a heterozygote AA' at another locus. The allele M, which is assumed to be dominant, increases the dominance of A by conferring the same phenotype on AA' as on AA.

dominance, evolution at the M locus would be determined mostly by these other effects.

It is possible to modify an allele's dominance experimentally by artificial selection. For example, heterozygous currant moths (*Abraxas grossulariata*) are intermediate in color between the homozygous wild-type and lutea morphs (Ford 1940). By breeding from heterozygotes that tended more toward one or the other homozygote, Ford was able, in four generations, to obtain one line in which heterozygotes were almost wild type in appearance and another in which they closely resembled the lutea genotype.

But selection in favor of dominance in this case is much stronger than in the instance Fisher hypothesized. Also the observation that dominance has evolved in *Papilio dardanus* does not support the generality of Fisher's theory, for the same reason. But dominance need not have evolved specifically to suppress the effect of deleterious mutants (Plunkett 1933, cited in Sved and Mayo 1970). Rather dominance may be a particular manifestation of a generalized capacity to resist alterations of the developmental pattern, whether they are caused by genetic (mutational) or environmental perturbations.

PHENOCOPIES, GENETIC ASSIMILATION, AND CANALIZATION

There is indeed no clear distinction between genetic and environmental disruptions of development. The same phenotypic abnormality may be produced by a mutant gene or by an environmental event. For example, the crossveinless condition in *Drosophila* can be produced by any of many alleles or simply by exposing pupae to a heat shock at a critical stage of development (Waddington 1953). Such environmentally produced mimics of mutations have been called PHENOCOPIES, and they give evidence that development, proceeding by the joint action of genotype and environment, can be altered by variation in either.

In fact, the alteration of development by environmental agents, coupled with the action of selection, can result in genetic changes, as

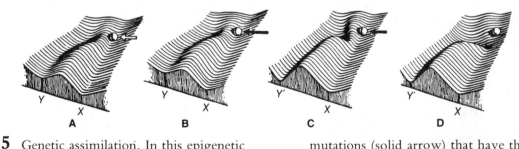

5 Genetic assimilation. In this epigenetic landscape (*cf. Figure 4, Chapter 7*), development is generally canalized into path X, but environmental stimuli (hollow arrow) may push it into path Y. Development along path Y may be genetically canalized by single mutations (solid arrow) that have the same effect as the environmental stimulus (B), or by genetic changes that lower the barrier to path Y (C, D). Path Y becomes the new canalized norm if the channel Y is deepened (Y'). (From Waddington 1957)

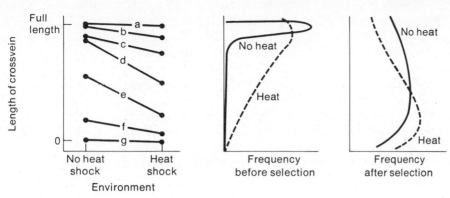

Diagram of the genetic assimilation of crossveinlessness. The population originally consists of a variety of genotypes with different reaction norms; some (such as d, e) are more sensitive to heat shock than others (such as a–c) that are more prevalent in the population. Before selection most flies have full crossveins in the absence of heat shock; but given a shock, a small proportion of the population develops incomplete crossveins. Breeding from these increases the frequency of such genotypes as d–g, so that after selection crossveinlessness appears even in the absence of heat shock.

the crossveinless example illustrates. Although the crossveinless condition is almost never expressed in wild flies, Waddington, by selecting flies that had developed the phenotype when subjected to heat shock, developed a population in which most of the flies were crossveinless when treated with heat; he had selected for increased sensitivity (decreased canalization) to the heat shock. Most remarkably, however, some of these selected flies were crossveinless even without a heat shock, and the crossveinless condition was heritable. An initially acquired, or environmentally induced, characteristic has become genetically determined — a result that no doubt would have pleased Lamarck. But this GENETIC ASSIMILATION of an acquired characteristic simply occurred because genotypes vary in the degree of canalization (*Figures 5, 6*). Some are easily deflected by environmental events into aberrant developmental patterns. When the aberrant phenotypes appear, selection for such weakly canalized genotypes also entails selection for modifier genes that canalize the development into the newly favored channel; as such alleles increase in frequency, less environmental stimulus is required to produce the new phenotype.

Waddington argued that genetic assimilation might have evolutionary importance, in that initially nonheritable responses to the environment could, if adaptive, become genetically controlled. This possibility could explain, he said, why a trait that is not heritable in one species (such as the species-typical song, which in some birds is learned) can be under tight genetic control in others (as in some birds in which the song is innate). That genetic assimilation might occur by natural selection is evident from a study of body size in *Drosophila pseudoobscura,* which is inversely correlated with the temperature at which the flies develop. This ontogenetic response is paralleled by genetic differences among geographic populations exposed to different average temperatures. Populations derived from a single stock were maintained in the laboratory for six years at 16°, 25°, and 27°C. When they were reared at the same temperature, the cold-adapted (16°) flies were larger than the warm-adapted ones (Anderson 1966).

There is strong evidence, then, that genotypes — and characters — vary in developmental homeostasis; some more than others maintain a phenotypic equanimity in the face of fortune's slings and arrows. Even more revealing are experiments of the kind reported by Rendel (1967).

THE GENETIC CONTROL OF CANALIZATION

The recessive sex-linked mutation *scute* (*sc*) in *D. melanogaster* changes the number of bristles on the scutellum from the normal (wild type, denoted +) number of four, to a more variable, lower number. The mutation also reduces the number of abdominal bristles, but in this respect *sc* is not recessive; +*sc* flies are intermediate between ++ and *scsc*.

Rendel supposes that the number of bristles is a discrete expression of an underlying continuous variable that might be called bristle-making ability, or, as Rendel calls it, Make (*M*). It might be, for example, the amount of an enzyme that forms bristle components. Rendel assumes that Make is normally distributed (has a bell-shaped distribution) in a population, but that canalization in ++ flies prevents the expression of bristle numbers other than four. If variation does exist, though, it is possible to determine the underlying distribution of Make and to specify the interval on the Make scale over which a given bristle number is formed. For example (*Figure 7*), if flies with no bristles make up 0.38 per cent of the population, and those with one bristle make up 6.85 per cent, the cutoff points for the one-bristle class lie at 2.67 and 1.49 standard deviations (σ) below the mean level of Make ($\overline{M}$), since 0.38 per cent and 6.85 per cent of the area under a normal curve fall, respectively, beyond 2.67 and 1.49 standard deviations below the mean (*Appendix I*). Thus $2.67 - 1.49 = 1.18\sigma$ is the width of the one-bristle class on the Make scale. By this technique (probit analysis), the relation between bristle number and the amount of the hypothetical substance Make can be determined.

This relation is linear for abdominal bristles. But the number of scutellar bristles has a *canalization plateau*; an increment of 1σ of Make will bring a fly from one to two bristles, but flies with four bristles can vary more than 6σ in their amount of Make (*Figure 8a*). By performing this analysis on ++, +*sc,* and *scsc* flies, Rendel determined that both the +pl and +*sc* genotypes lie within the canalization plateau (*Figure 8b*). Thus ++ does not produce twice as much Make as +*sc,* so the production of Make must be repressed in ++ flies, probably by regulatory genes. The repression results in dominance: the identical phenotype of ++ and +*sc* flies.

By selecting for bristle number in an *scsc* population that varied in bristle number (because of the segregation of minor genes), Rendel could change the mean level of Make, and hence bristle number, without altering the genotype at the *scute* locus. This demonstrates that the regulatory system that represses Make production by the major gene at the *scute* locus does not repress the activity of minor genes. In another experiment Rendel imposed stabilizing selection for low variation in an *scsc* population with a mean of two bristles. The two-bristle phenotype initially spanned 1.5σ on the Make scale and so was not well canalized; but after 50 generations of stabilizing selection,

Example of probit analysis of hypothetical data. The observed distribution of bristle numbers (top) is presumed to result from a normal distribution of "Make" (M), in the middle figure. Bristle class 1 constitutes $6.85 - 0.38 = 6.47$ per cent of the population, the area under the normal curve labeled 1. The cutoff points for 0.38 per cent and 6.85 per cent of a normal curve lie, respectively, at 2.67 and 1.49 standard deviations (σ) below the mean ($\overline{M}$), so bristle class 1 is $2.67 - 1.49 = 1.18$ standard deviations (σ) wide. Similarly bristle class 2 is found to be 3.25σ wide. The relation between bristle number and amount of Make, measured in units of σ, can thereby be plotted as in the lower graph.

7

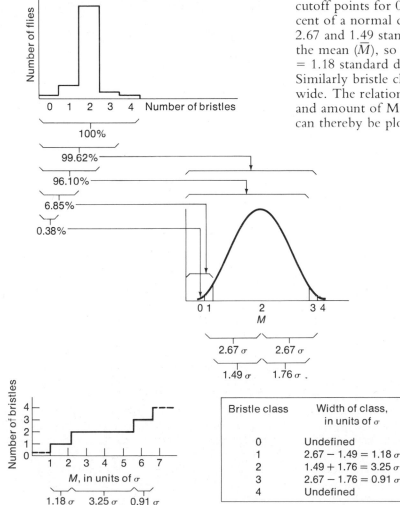

Bristle class	Width of class, in units of σ
0	Undefined
1	$2.67 - 1.49 = 1.18\,\sigma$
2	$1.49 + 1.76 = 3.25\,\sigma$
3	$2.67 - 1.76 = 0.91\,\sigma$
4	Undefined

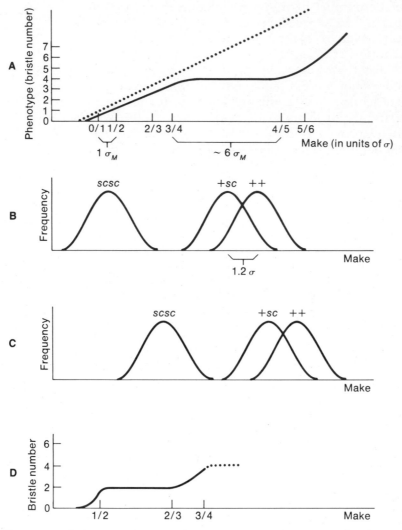

8 Canalization and the response to selection. (A) represents the relationship Rendel found, by the method illustrated in Figure 7, between scutellar bristle number and amount of Make. Because of canalization, the four-bristle phenotype develops over a great range of Make values; the other bristle classes are much narrower. The notations ⁰/₁ and so forth mark the borders, on the Make scale, between the zero-bristle class and the one-bristle class, and similarly for the other classes. Without canalization there would be a linear relation between phenotype and Make, as shown by the dotted line. (B) shows the distribution of Make, and hence of bristle number, in females of each of the three genotypes at the *sc* locus. Since +/*sc* and +/+ flies fall entirely within the canalization plateau, they always have four bristles; *sc*/*sc* flies fall outside the range of canalization, so they vary in bristle number from zero to three. (C) represents the distribution of Make in flies of the three genotypes, after modifier alleles at other loci have been selected to increase mean bristle number. +/*sc* and +/+ have been shifted up out of the canalization plateau; so they have not only more, but more variable, bristles. (D) A new canalization plateau has been shaped by artificial selection; the two-bristle class rather than the four-bristle class is now the stable phenotype over a great range of Make values.

variation was so reduced that the two-bristle class spanned 3.65σ (*Figure 8d*). Thus a new canalization plateau was formed, suggesting that there are genes that set the form of the canalization plateau as well as the major (*sc*) and minor genes that they regulate.

Investigating correlated responses to selection, Rendel found that replacing an *sc* allele by a + allele raises both the number of scutellar and abdominal bristles, while selection for increased scutellar bristles in ++ flies lowers the abdominal bristle number! Rendel postulates (*Figure 9*) that the two kinds of bristles are formed from a common pool of Make, of which a proportion *a* is allocated to scutellar bristles. Substituting + for *sc* increases Make, thus increasing both bristle counts, but selection of minor genes may lower abdominal bristle number by increasing the proportion *a* of Make allocated to scutellars.

Although Rendel's model may not be correct in all details (see Sheldon and Milton 1972), it suggests that dominance is just one manifestation of canalization; that heterozygotes may be better buffered than homozygotes because they lie closer to the middle of the canalization plateau; that epistatic interactions (among major genes, minor genes, and regulatory genes) affect the phenotype strongly; and that gene action can yield rather complex correlated responses to selection. The canalization (dominance) may have evolved, Rendel suggests, not because a particular phenotype is in itself most adaptive, but because of the need to achieve a developmental balance that prevents any one

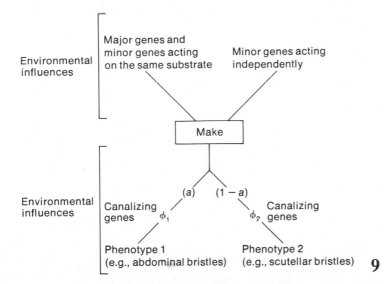

Rendel's model of developmental interactions. Major and minor genes act to produce a pool of Make, of which a fraction *a* is allocated to one part of the phenotype and $1 - a$ to another. ϕ_1 and ϕ_2 are developmental functions, genetically controlled, that regulate the relation between the amount of Make allocated to any part of the phenotype (e.g., scutellar bristles) and the magnitude of the phenotypic end product (e.g., number of scutellar bristles). (After Rendel 1967)

part of the phenotype from monopolizing energy and materials. Selection for increased bristle number can reduce fitness if other organs are deprived of critical biochemical products. Genetic homeostasis is the result. But if selection persists, compensatory changes in other developmental processes may restore fitness, and the phenotype no longer regresses toward its original state when selection is relaxed. Such coordinated changes are the essence of the formation of new coadapted gene complexes.

Genetic changes in Make, or more importantly, in the regulatory system that dictates how the phenotype shall be expressed, depend on rather drastic genetic or environmental perturbations of normally canalized development, and these changes are no doubt unlikely to improve fitness. Adaptive, but major, changes in the control system are perhaps unusual evolutionary events, so that, as Rendel says, "the periods of great evolutionary change will be short-lived and will be interspersed by long periods during which nothing much seems to happen." Well-canalized characters, of the kind that distinguish genera or higher categories, are likely to resist change — but when changes come, they may come not single spies, but in battalions.

INTERPRETATIONS OF THE RESPONSE TO SELECTION

From such concepts as canalization and coadaptation has arisen a highly holistic view of the genetic structure of populations (see, e.g., Mayr 1963, Dobzhansky 1970). In this view most genes are pleiotropic and most characters are affected by many genes. Selection on a character lowers fitness, even if variation in the character is itself selectively neutral, because it increases homozygosity at heterotic loci or because it pleiotropically alters the development of important characteristics. If selection persists, compensatory changes at other loci restore fitness. Correlated responses to selection are due to pleiotropy.

This model can explain the responses to selection that Mather and Harrison observed. Another model, however, which Mather and Harrison themselves favored, provides an equally plausible and indeed simpler interpretation. With the theory outlined in Chapter 14, the responses to selection can very simply be attributed to linkage (*Figure 10*). Postulate that + and − alleles at many loci on a chromosome affect bristle number additively and have little or no effect on fitness. Interspersed among these loci are others that do affect fitness; at some of these loci there are deleterious recessive alleles.

Most chromosomes, because of recombination, have a mixture of + and − alleles in repulsion phase and so confer intermediate bristle number. Strong selection for high bristle number preserves the exceptional chromosomes (e.g., + + + + +) that have many + alleles in coupling. But such chromosomes are likely to carry one or more deleterious recessives. Under intense artificial selection, homozygotes for such chromosomes increase in frequency. There is little opportunity for recombination to produce chromosomes that bear many + alleles but few deleterious recessives, so fitness declines. If artificial selection is relaxed, natural selection eliminates these recessives, the remaining repulsion chromosomes (e.g., $+-++-$) increase in frequency, and bristle number regresses toward the original state. Crossing-over yields more coupling chromosomes, which are less heavily loaded with deleterious alleles because their frequency has been reduced; hence bristle number may increase, without a severe decline in fitness, when artificial selection is resumed. Occasional crossovers in a plateaued population yield delayed responses to selection (*Chapter 14*) from a persistent store of genetic variation that is maintained by linkage disequilibrium. Correlated responses to selection are attributable to linkage, not to pleiotropy, which plays little role in this atomistic view of the genome.

Bristles	Mean number of Homozygous *r* genes
4.4	0.0
5.4	0.4
6.0	1.2
4.8	0.6
4.8	0.2
6.4	0.0

Generation (4.4, 0.0):
```
1+0+1r 1    1r 0+0+1    1+1+0r 1    0+0r 0+1    1+1r 1+0
--------    ---------X  --------X   --------    --------
0+1+0+1     0+1r 0+0    0+1r 1+1    0+0+1+0     1r 0+1+1
```
Artificial selection →

Generation (5.4, 0.4):
```
1+1+0r 1    0+1r 1+1    0+1r 0+1    1r 1r 1+0    1+1+0r 1
--------X   --------    --------    --------X    --------
1r 0+1+1    0r 0+1+1    1+1r 1+0    0+1r 1r 1    1r 0r 1+0
```
Artificial selection →

Generation (6.0, 1.2):
```
1+1+0r 1    1r 1+0r 1    1r 0+1+1    1r 1r 1r 1    1r 0+0r 1
--------X   ---------    --------    ---------    --------X
0+1r 1r 1    1r 1r 1r 1   1r 1r 1+0    1r 1+0r 1    0+1r 1+0
```
Natural selection; artificial selection relaxed →

Generation (4.8, 0.6):
```
1+1+0r 1    1r 0+0+0     0+1+0r 1    1r 0+1+1    0+0+0r 1
--------X   --------X    --------X   --------X   --------X
1r 0+1+1    1r 1r 1+0    1r 0+1+0    1r 0+0r 1    1+1+1r 1
```
Natural selection; artificial selection relaxed →

Generation (4.8, 0.2):
```
1+0+1+1     0+1+0+0     1+1+0r 1    1r 0+1+1    1r 0+1+0
--------X   --------    --------    --------X   --------
1+1+1+0     1r 1r 1+0    0+1+1+0     1+1r 1+1    1r 0+0+0
```
Artificial selection →

Generation (6.4, 0.0):
```
1+1+1+0     1+0+1+1     1+1+1+1     1+0+1+0     1+0+1+1
--------    --------    --------    --------    --------
1r 0+1+1    1+1r 1+1    1r 1+1r 1    1+0+1+1    1r 1+1r 1
```

10

A contrived illustration to show how results in artificial selection may be caused by linkage. Alleles (represented by 1's) that raise bristle number increase under artificial selection, but so do deleterious recessive alleles (*r*) to which they are linked. Viability decreases as flies become homozygous for *r* alleles. When artificial selection is relaxed, *r* alleles decrease in frequency and advantageous dominant alleles (+) at these loci increase; thus new chromosomes carrying combinations of 1 and + alleles are formed. When artificial selection is resumed, these chromosomes and chromosomes derived from them by recombination increase in frequency. Arrows indicate parents of next generation, crosses that crossing–over may occur between chromosomes.

PLEIOTROPY OR LINKAGE? COADAPTATION OR ADDITIVITY?

Whether the pattern of response to selection is best explained by pleiotropy or linkage is as unresolved a problem as the related question, whether genetic variation is influenced more by selection or by genetic drift. The holistic argument for pleiotropy espoused by Falconer (1954, 1960), among others, rests on the known pleiotropic effects of single mutations and on the observation that specific correlated responses often appear consistently in replicated experiments (Clayton et al. 1957). The correlations among characters should be less consistent from one population to another if linkage is responsible.

If linkage is responsible, correlated responses should be less pronounced if selection is weak, because weak selection allows recombination to form advantageous combinations of alleles at both the loci affecting the selected character and those with which they are associated. This expectation is borne out in at least some instances (Eisen, Hanrahan, and Legates 1973). Moreover Davies and Workman (1971) showed that the correlation between the number of sternopleural and abdominal bristles was due more to linkage than to pleiotropy by successfully selecting these characteristics in opposite directions in the same population of flies. A careful mapping of many of the loci that affect these characters showed that they are on different sites on the chromosomes (Davies 1971). Thus at least some correlated responses are due to linkage.

Just how important epistasis is, how coadapted the gene pool may be, is likewise a matter of some debate. It seems, as I noted in Chapter 14, to vary in importance from character to character. In *Drosophila* characteristics like bristle number, wing length, and body weight are mostly additively inherited, while direct components of fitness — viability, fecundity, egg hatchability — show mostly nonadditive genetic variance, much of it due to epistasis (Kearsey and Kojima 1967). Epistasis is apparently most marked in characters that are directionally selected toward the extreme of the phenotypic spectrum.

It is possible that epistatic interactions are more important under natural conditions than in the laboratory. Under conditions of stress genetic variation in fitness will be more evident than in benign environments, and the most fit genotypes may well be those with the best combinations of coadapted alleles. In tobacco, for example, the epistatic component of variation was greater in both "good" and "bad" environments than in "average" environments (Jinks et al. 1973).

Finally the magnitude of epistatic interactions among alleles from a single population is often less than among those taken from different populations of a species. For example, chromosomes from closely related strains of flies contributed a smaller epistatic component of the variance in body size than chromosomes from different geographic populations (Robertson 1954), and there are many instances of genetic incompatibility between populations of a species, as in the intersexes produced in crosses between populations of gypsy moths (*Chapter 9*).

It is in the breakdown of coadaptation that deleterious epistatic interactions become most evident. If an allele's fitness has never been tried in the genetic context of other populations, it is likely to show unpredictable, nonadditive effects when acting in a foreign genetic background. Thus hybrids between populations may be unfit. But within a population alleles that make a good contribution to the phenotype, no matter what background genotype they are embedded in, are likely to persist; they will be, as Mayr (1963) puts it, good mixers, having an additive effect on the phenotype. Thus the epistatic component of the phenotypic variation within populations is not likely to be very great.

THE RESPONSE TO SELECTION IN NATURAL POPULATIONS

Responses to natural selection in wild populations are presumably at least as complex as responses to artificial selection, but the course of directional selection in nature is not well documented. This is partly because natural populations are much more difficult to study, and partly because directional selection is probably rather temporary in nature; most natural environments are not changing much in any consistent way, and the genetic constitution of most populations has probably already reached an equilibrium dictated by local environmental conditions. Indeed most analyses of gene frequencies in natural populations have assumed, rightly or not, that they are in equilibrium.

Nonetheless some instances of evolution in nature have been observed (Chapter 7). Industrial melanic moths (*Biston betularia*) increased from less than 1 per cent to more than 98 per cent within 50 generations because of the lesser predation suffered by dark moths resting on dark tree trunks. In Californian populations of *Drosophila pseudoobscura* (Dobzhansky 1958, Anderson et al. 1975), the *Pike's Peak* (*PP*) inversion, formerly common only east of the Rocky Mountains, rose from less than 0.02 per cent in 1946 to 11 per cent eight years later; since then, *PP* has declined, and another inversion (*Treeline*) has increased. These changes occurred over most of California, in a great variety of habitats. Their selective basis is unknown.

Evolution of physiological adaptations to environmental changes wrought by human activity has occurred in many species but has been studied little. Between 1942, when DDT first came into use, and 1960, evolution of DDT resistance was documented in at least 137 species of insects (O'Brien 1967), although at least some populations of some species do not seem to have the genetic variation required. For example, dieldrin resistance in the mosquito *Anopheles gambiae* has evolved only in certain populations (Brown 1964). A great deal is known about the genetic basis for insecticide resistance, especially in laboratory populations (Crow 1957, Plapp 1976), which exemplify many of the genetic phenomena I have discussed. Resistance can be monogenic or polygenic in different populations of the same species; it is often associated with lower fecundity and other correlated effects that result in a loss of resistance when the populations are no longer subjected to the insecticide; strains resistant to one insecticide can be cross-resistant to others.

The most intensively studied cases of resistance of natural populations to toxins are plants, especially grasses, growing in soils impregnated with heavy metals, especially near mines (Antonovics, Bradshaw, and Turner 1971). Resistance typically has a polygenic basis with high heritability ($h^2 = 0.7$) when measured under uniform conditions. Metal-tolerant genotypes have developed very rapidly (in less than 30 years in some cases), having a selective advantage s of 0.3–0.7 over nonresistant genotypes on toxic soils. On nontoxic soils they are competitively inferior, suffering a selective disadvantage of 0.05–0.3; thus the border between resistant and nonresistant populations is often quite abrupt, just a few feet. Some resistant populations have evolved a degree of reproductive isolation, thus lowering the contamination of the gene pool by nonresistant alleles. They are more highly self-compatible than nonresistant populations and have diverged in flowering time by as much as 25 per cent.

FAILURES IN ADAPTATION

These cases give little evidence of the details of the response to directional selection in natural populations. They prove only that evolution can happen very rapidly. When one considers that populations are so rich in genetic variation and that responses to artificial selection almost invariably occur, the remarkable fact is not that some populations rapidly adapt to changed conditions, but that so few do. All too seldom do biologists point to the overwhelming number of populations that, having failed to adapt to a change of circumstance, have become ecologically or geographically restricted or extinct (*Chapter 5*). Many species of moths have evolved industrial melanism, but many have not, and some seem to have become locally extinct by their failure to do so (Kettlewell 1973). Many plants have not become adapted to toxic soils, either to mine tailings or to their natural analogues such as serpentine-based soils, on which few species of plants can grow. Many insects have not become resistant to insecticides — nor, I suspect, will the peregrine falcon or other top carnivores that are threatened with extinction because of their susceptibility to chlorinated hydrocarbons. Although some insect pests, and the insects introduced to control them, have become established in new countries, most of the species introduced as potential agents of biological control have failed to survive. And did the American chestnut have insufficient genetic variation, over the entirety of its broad range, to resist the ravages of the chestnut blight?

These are only special instances of a general problem. Why, despite all their genetic variation, haven't natural populations become adapted to a broader range of ecological circumstances? If the geographic or ecological range of a species is limited by meteorological conditions or by competition with another species, why cannot the population adapt to those limiting factors? Why, above all, has the vast majority of species that have existed become extinct? Is it a poverty of genetic variation that renders natural selection impotent, a rigidity of genetic homeostasis that prevents adaptation, a structural and physiological organization that is intrinsically incapable of sufficient flexibility? Or is adaptability limited by the genctic and ecological features, not of the local population faced with peculiarly local vicissitudes, but of the entire species? The next chapter may shed some light on this question.

SUMMARY

Rapid evolution has been documented in quite a few natural populations, but little is known about the details of genetic changes in such instances. More information comes from studies of artificial directional selection, in which an immediate response leads to a plateau, from which there may be further progress after some delay. Populations usually resist rapid genetic change and return toward their previous genetic constitution if the new force of selection is alleviated. This resistance, or genetic homeostasis, is caused by the opposition of natural selection to the new selection pressure, for genotypes favored by the new selection regime are generally less fit in other respects. Their inferiority may be due to the linkage of deleterious alleles with those that confer the newly favored phenotype, or it may be caused by deleterious pleiotropic effects of the alleles that confer this phenotype. These alleles may affect developmental pathways disrupted by new combinations of genes that are not coadapted to interact harmoniously with each other. The intricacy of interactions among alleles in natural populations may limit the rate at which populations can adapt to changes in the environment.

FOR DISCUSSION AND THOUGHT

1 Should artificial selection for, say, increased yield in potatoes entail breeding just from the few most productive individuals, or from a larger, overall less-productive fraction of the stock? What factors might determine the best breeding program? What relevance does this question have to natural evolution?

2 What are the evolutionary implications of interpreting the cause of genetic homeostasis as the heterozygous superiority of individual pleiotropically acting loci vs. the heterozygous superiority of linked sets of polygenes?

3 Explain why a population might not adapt as quickly to each of several independent environmental factors as to a single factor. (See Pimentel and Bellotti 1976 for an example.)

4 More than one author has suggested that genetic homeostasis is a mechanism whereby populations can retain the genetic variation necessary for adapting to subsequent environmental changes. Criticize this view.

5 "Genetic assimilation" amounts to a tightening of genetic control over a character whose development was previously more responsive to the environment, a narrowing of the norm of reaction. Does this necessarily improve the population's degree of adaptation? Does it occur because a narrower reaction norm will improve the individual's survival and fecundity? Evaluate Bateson's (1963) view of the question.

6 If modifier genes can influence a major gene's expression and so cause dominance, could they also act on the various effects of a pleiotropic gene and eliminate pleiotropy? How quickly could this happen? What would be the evolutionary consequences?

7 Discuss the relation between Rendel's work and the explanations of macroevolutionary changes in developmental terms, as described in Chapter 8.

8 Does the pleiotropy model or the linkage model more plausibly explain the observation that populations which have long since reached a selection plateau are nonetheless genetically variable for the selected characteristic?

9 Distinguish between *genetic variation* and *selectable genetic variation*. Why might some genetic variation not be selectable? Does the observation that irradiation of a plateaued population may permit further response to selection mean that the population had been genetically invariant?

10 A high environmentally produced variance V_E in a character may be a maladaptive effect of poor developmental homeostasis, or it may be evidence of adaptive developmental plasticity. How might these effects be distinguished if one wanted to assess a genotype's capacity for developmental homeostasis?

11 Speculate about possible biochemical mechanisms for canalization of development.

12 Would you expect to observe genetic homeostasis, manifested as a decline in fertility and viability, in a population exposed to soft selection for a particular phenotypic characteristic? Is evolution by soft selection likely to be as rapid as by hard selection? Why or why not? What do you suppose the relative incidence of soft vs. hard selection might be in nature? What are the implications of this line of thought for evolutionary rates?

MAJOR REFERENCES

The references for Chapter 14 serve here also; a summary of major aspects of the response to selection is also given by the following reference.

Robertson, A. 1955. Selection in animals: synthesis. *Cold Spring Harbor Symp. Quant. Biol.* 20: 225–229.

The preceding chapters have explored the genetic mechanisms of evolution within populations, with the goal of understanding why the rate and direction of evolution varies among populations and among characteristics of a single population. The question of whether the limits to evolutionary rates are most commonly ecological or genetical has not been answered. But these chapters have treated primarily anagenetic change — evolution within single populations — and have ignored *cladogenesis,* the divergence of different genetic lines. The divergence that ultimately leads to the formation of higher taxa begins with the differentiation of populations and their development into different species. It is in this process that evolutionary change is often most rapid and its direction most novel.

BARRIERS TO GENE FLOW

In Chapter 9 species were defined as groups of populations that can actually or potentially exchange genes with one another and which are reproductively isolated from other such groups. The concept cannot be applied in all cases, but instances of partial reproductive isolation do not invalidate the concept in instances where it does apply. Much less do borderline cases vitiate the concept of speciation as a process; indeed they forcefully illustrate that species arise by the gradual evolution of intrinsic barriers to gene exchange.

The evolution of these barriers is the critical process in speciation, for two populations can diverge only if changes in gene frequencies in one are not immediately transmitted to the other by interbreeding. Divergence, then, depends on extrinsic or intrinsic limitations to gene flow. *Extrinsic* barriers exist if the populations reproduce in different places (are allopatric) or at different seasons (are allochronic). *Intrinsic* barriers to gene flow are either premating isolating mechanisms that prevent the formation of hybrid F_1 zygotes, or postmating barriers to gene exchange such as hy-

The Origins of Species

brid inviability or sterility, that is, low fitness (as is shown in *Table I, Chapter 9*).

The importance of such barriers is the reduction of m, the rate of gene flow from one population to the other. The factors that can cause divergence — genetic drift, which may ultimately lead to the fixation of different alleles in the populations, and selection, which can bring gene frequencies toward different values — are opposed to gene flow. The magnitude of selection s against different genotypes in the two populations must be a good deal greater than the gene flow between them to accomplish very much divergence (*Chapter 10*). If, for example, the fitnesses of genotypes A_1A_1, A_1A_2, and A_2A_2 are, respectively, 0.8, 0.9, and 1.0 in population 1, and 1.0, 0.9, and 0.8 in population 2, the populations are monomorphic for different alleles (if there is no gene exchange). But if 20 per cent of the breeding members of population 2 are derived by immigration from population 1 in each generation ($m = 0.2$), the equilibrium frequency of A_2 in population 2 is 0.66, far from the optimum, 0. Even if m is only about 0.01, the allele A_2 has a rather high frequency (0.09). Thus population 2 has a low average fitness and is unable to adapt precisely to its environment. If gene flow is bidirectional and if the populations are about equal in size, both populations have suboptimal genetic constitutions.

The populations will be better adapted, however, if a third allele A_3 should arise which confers more or less equal adaptedness on its carriers in both environments (Levins 1962, 1968). The ecologically generalized genotype A_3A_3 will replace the more specialized genotypes, as long as each genotype is frequently exposed to both environments. Highly mobile species, then, should evolve all-purpose genotypes, as do many weedy species of plants (Baker 1974), some marine invertebrates with planktonic larvae, and many of the larger, more mobile vertebrates, such as the human species.

An ecologically generalized species is not subject to strong selection for different genotypes in different environments, so such species would be expected to show little geographic variation, little genetic divergence among local populations, and little tendency to speciate. This consideration may in part explain the existence of widespread species in which little incipient speciation is evident: birds such as the brown creeper (*Certhia familiaris*) and the black-capped chickadee (*Parus atricapillus*) that are found in both Europe and North America, intertidal invertebrates such as the blue mussel (*Mytilus edulis*) that are distributed throughout the Atlantic, and large carnivores such as the mountain lion (*Felis concolor*), which is distributed throughout both North and South America.

Conversely, ecologically specialized species should respond to different environmental conditions by genetic change; they should show pronounced geographic variation and a tendency toward speciation that is accentuated by their restriction to patches of special habitat, among which they tend not to move or to move only with difficulty

(Janzen 1967). For example, many bird species in tropical rain forests are so dependent on the forest understory that they do not fly across unforested gaps of only a few hundred feet. Many of the highly specialized cichlid fishes of the Great Lakes of Africa are entirely restricted either to rocky or to sandy shores and do not cross short stretches of the inappropriate habitat (Fryer and Iles 1972).

Thus the efficacy of a topographic barrier to gene flow and the likelihood that isolated populations will genetically diverge probably depend on the mobility and the degree of ecological specialization of the species. But it is impossible to stipulate cause and effect in any simple way; a species may not disperse across seemingly trivial barriers because it is specialized, but it may be specialized in part because its limited mobility has confined it in the past to a homogeneous environment. Conversely ecological generalization permits high mobility, and this in turn may favor the evolution of generalization.

ECOLOGICAL SPECIALIZATION AND ADAPTIVE RADIATION

Ecological specialization is unquestionably the rule in species-rich groups of organisms. A most dramatic example (*Figure 1; Table I*) is that of the cichlids (Fryer and Iles 1972) which inhabit the Great Lakes of the African rift valley. Each of the three major lakes has more species of fishes than any other lake in the world, and almost all of them belong to the sunfishlike family Cichlidae. There are about 126

TABLE I. Adaptive radiation in the cichlids of Lake Malawi

Diet	Examples
Phytoplankton	*Tilapia lidole*
Organic detritus	*Tilapia shirana*
Aufwuchs (algae, etc., on rocks)	*Pseudotropheus tropheops, P. zebra, Petrotilapia tridentiger, Labeotropheus fuelleborni, Cyathochromis obliquidens, Haplochromis guentheri*
Periphyton (algae on submerged plants)	*Hemitilapia oxyrhynchus*
Vascular plants	*Haplochromis similis*
Molluscs	*Haplochromis placodon, Chilotilapia spp.*
Aquatic insects	*Aulonocera, Labidochromis vellicans, Haplochromis euchilus*
Midge larvae	*Lethrinops furcifer*
Ostracods	*Lethrinops sp.*
Zooplankton	*Cynotilapia afra, Haplochromis spp.*
Fishes	*Rhamphochromis spp., Diplotaxodon spp., Haplochromis spp.*
Fish eggs	*Haplochromis sp.*
Fish scales	*Genyochromis mento, Corematodus spp.*
Fish fins	*Docimodus johnstoni*
Fish eyes	*Haplochromis compressiceps*

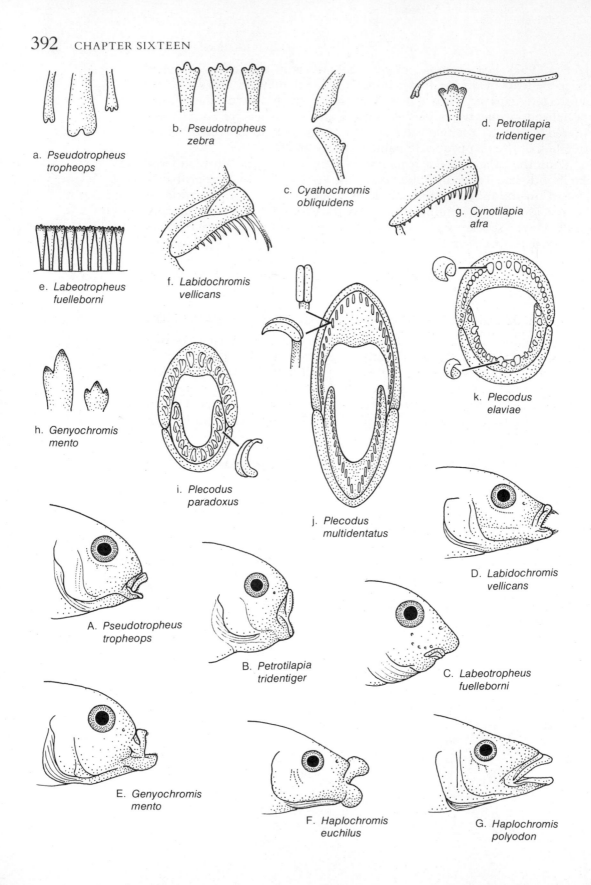

a. *Pseudotropheus tropheops*

b. *Pseudotropheus zebra*

c. *Cyathochromis obliquidens*

d. *Petrotilapia tridentiger*

e. *Labeotropheus fuelleborni*

f. *Labidochromis vellicans*

g. *Cynotilapia afra*

h. *Genyochromis mento*

i. *Plecodus paradoxus*

j. *Plecodus multidentatus*

k. *Plecodus elaviae*

A. *Pseudotropheus tropheops*

B. *Petrotilapia tridentiger*

C. *Labeotropheus fuelleborni*

D. *Labidochromis vellicans*

E. *Genyochromis mento*

F. *Haplochromis euchilus*

G. *Haplochromis polyodon*

species in Lake Tanganyika and more than 200 in Lake Malawi. The various genera are closely related to each other despite the pronounced differences among them in the morphology of their jaws and teeth, an adaptive radiation for which they were preadapted (*Chapter 8*). The cichlids of Lake Malawi alone fall into at least 11 ecological "guilds" that feed on algae, invertebrates, plankton, fishes and molluscs, and they include species with extraordinary feeding habits (such as *Genyochromis* and *Corematodus,* which feed on the scales of other cichlids, and *Haplochromis compressiceps,* which darts at a larger fish, snaps, and withdraws, leaving behind an empty eye socket). Within each of these guilds the species avoid competition by further specialization, either by foraging only in rocky or in sandy habitats or by collecting food in very specialized ways. Among the *Aufwuchs*-eaters (*Table I*), for example, *Petrotilapia* and *Petrochromis* have rows of fine movable teeth that comb large algae from the rocks, while *Labeotropheus* has closely set teeth that form a scraper (*Figure 1*). The form of the teeth in many of the cichlids is so extraordinarily modified for specialized feeding that it far transcends the range of tooth shapes found elsewhere among cichlids, or even among all other fishes (*Figure 1*). Were these species not obviously related to more orthodox forms, they would probably be assigned to different families.

But the existence of large numbers of specialized species does not in itself prove that specialized forms speciate more rapidly than generalized forms, nor that species-rich groups such as the cichlids speciate more rapidly than less diverse groups. The number of species depends not only on the speciation rate, but on the relation between the rates of speciation and extinction (*Chapter 5*). If high diversity is due to a high speciation rate rather than to a low extinction rate, it must be because intrinsic barriers to gene flow develop rapidly in species-rich groups. These barriers can evolve in several ways.

INSTANTANEOUS SPECIATION

The elementary event in speciation is the development of an intrinsic barrier to gene exchange. The first question is whether such barriers arise instantaneously or gradually. Speciation could be instantaneous only if a newly arisen mutation could immediately bar gene exchange with the rest of the population in which it arose, thereby making its bearer a reproductively isolated population in itself — a new species. But in obligately outcrossing populations, such a mutant would be effectively sterile, because it would not find a mate. The only exception

1

A sample of the diverse tooth forms and head shapes among the Cichlidae of the African Great Lakes. See Table I for feeding habits of these species. (Redrawn from Fryer 1959 and Fryer and Iles 1972, after various sources. Courtesy of the Zoological Society of London)

would be a mutation that occurred simultaneously in two individuals of opposite sex, as when two siblings inherit a mutation that arose in the germ line of their parent. Instantaneous speciation by a single mutation, if it ever occurs, is most likely in inbreeding species. Askew (1968) has suggested that this principle may help explain the high species diversity of chalcids, parasitic Hymenoptera in which mating between brothers and sisters, which emerge from a single parasitized host, is very common. This is not to suggest that these species arose by single mutations; the inbreeding process is thought merely to speed up a gradual accumulation of the several or many genes that confer reproductive isolation.

Speciation is truly instantaneous when it occurs by polyploidy in plants (*Figure 2*). A single plant formed by hybridization between two diploid species can give rise to fertile tetraploid offspring that are reproductively isolated from both diploid forms. The formation of a new allopolyploid species in one generation is especially likely in annual plants that practice self-fertilization and in long-lived perennials, in which there is great opportunity for vegetative shoots to undergo somatic doubling of chromosomes and to produce unreduced gametes in the flowers derived from these tissues (Baker 1959).

GRADUAL SYMPATRIC SPECIATION

Most genetic changes do not confer immediate, complete reproductive isolation. Thus most speciation must be gradual, as initially incomplete barriers to gene flow become progressively more effective. Whether

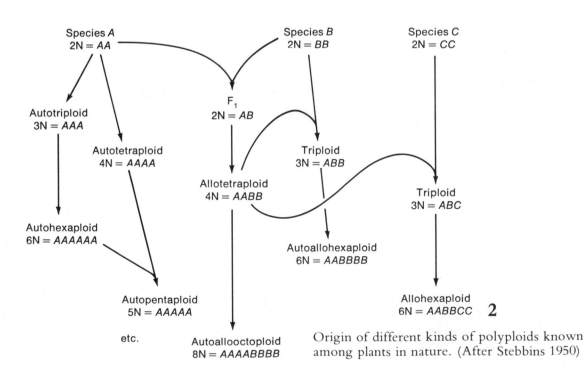

Origin of different kinds of polyploids known among plants in nature. (After Stebbins 1950)

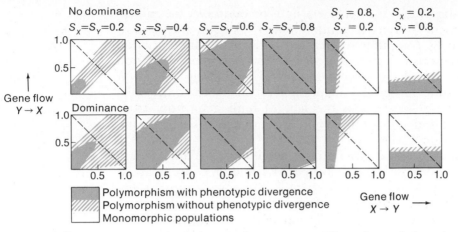

3

Genetic differentiation at a single locus when S_X is the selection coefficient against *aa* in niche X and S_Y is the selection coefficient against *AA* in niche Y, with and without dominance. In each diagram gene flow from niche Y to X, and from X to Y, forms the axes. The subpopulations in the two niches are most likely to diverge phenotypically when gene flow is restricted and/or selection is strong. (After Dickinson and Antonovics 1973, *The American Naturalist*, Copyright The University of Chicago Press)

this can happen within the confines of a single breeding population — whether speciation can be sympatric — is highly debated (Mayr 1963, Maynard Smith 1966, Bush 1975).

One model of sympatric speciation suggests that assortative mating (*Chapter 10*) may become intense enough to erect a complete barrier between two phenotypic classes. For example, some birds such as blue geese and jaegers (Cooch and Beardmore 1959, O'Donald 1959) are polymorphic in coloration and mate assortatively to some degree. If the fidelity of mate choice became perfect, the two color morphs would be different species. It is difficult to see, however, why or how the degree of assortative mating should become enhanced. Mate choice may be based on imprinting, so that the offspring of two parents with like phenotype will choose only mates with that phenotype; if such imprinting were absolute, two reproductively isolated populations would be formed (Kalmus and Maynard Smith 1966, Seiger 1967). Imprinting is known to affect mate choice in some birds, but it is far from absolute.

Most other models of sympatric speciation postulate that different genotypes are adapted to different niches. In the simplest model (*Chapter 13, Box E*) two homozygotes are adapted to different microhabitats, and the heterozygote is inferior: the fitnesses of *AA*, *AA'*, and *A'A'* are 1, $1 - s$, and 1 respectively. If there is random mating and the genotypes form a single population that is regulated as a whole, the polymorphism is unstable, and one or the other homozygote prevails. If, however, the population size is independently regulated in the two niches, the polymorphism is stable, as long as selection against genotypes in the "wrong" niche is intense enough to counteract the migration m between them (Maynard Smith 1966). Dickinson and Antonovics (1973) found by computer simulation that genetic differentiation between subpopulations in different niches persists, whether adaptation to these niches is controlled by one or by several loci (*Figure 3*). Their model describes quite nicely the adaptation of plants to toxic soils, maintained by strong selection in the face of high gene flow from neighboring populations on normal soil (Antonovics, Bradshaw, and Turner 1971; Snaydon 1970).

Such populations become different species if they become reproductively isolated. Suppose that genotypes AA and AA' are adapted to niche 1 and that $A'A'$ is adapted to niche 2. Suppose also that there is a locus B that causes assortative mating due to difference in flowering time, breeding coloration, or other reproductive characteristics. Females of genotype $B\_\_$, whatever their genotype at locus A, tend with probability α to mate with $B\_\_$ males; likewise, bb females tend to mate with bb males. The mating relationships are thus

		Male genotype	
		$B\_\_, A\_\_$ or $A'A'$	$bb, A\_\_$ or $A'A'$
Female	$B\_\_, A\_\_$ or $A'A'$	+	−
genotype	$bb, A\_\_$ or $A'A'$	−	+

If the frequency of B is initially different in the two niches, reproductive isolation can evolve rapidly (Maynard Smith 1966). Suppose that B is more common in niche 1, where allele A is favored, than in niche 2. Immigrants from niche 2, where A' is favored, will tend to have the genotype $A'A'bb$. Thus on the whole there is linkage disequilibrium: association between A and B and between A' and b. Preferential mating of B with B and of bb with bb thus also results in a nonrandom union of A with A and of A' with A', so that genotypes $A\_\_B\_\_$ and $A'A'bb$ are formed. Selection against $A'A'$ reduces the frequency of b in niche 1; similarly selection in niche 2 against allele A reduces the frequency of the associated allele B. Thus the population becomes divided into two moieties, the genotypes $AABB$ and $A'A'bb$, between which interbreeding is restricted. But complete reproductive isolation evolves only if the magnitudes of selection and of mating discrimination are great enough and if migration between the two niches is not too overwhelming. In a variation of this model Dickinson and Antonovics (1973) showed that in plants, reproductive isolation between populations in different niches could evolve another way, by the development of self-fertilization.

Evidence from both experimental and natural populations suggests that some degree of reproductive isolation can indeed evolve sympatrically, although it has not been proven that complete speciation has ever been achieved in this way. Populations of grasses adapted to toxic soils differ in flowering time from nearby genotypes on normal soil, and they are more self-fertile (Antonovics 1968, McNeilly and Antonovics 1968). Many investigators (Thoday and Gibson 1962, 1970, and references therein; Soans, Pimentel, and Soans 1974; Hurd and Eisenberg 1975) have imposed disruptive selection on laboratory populations of flies and have determined whether the two favored phenotypes mated assortatively. In the best-known such case Thoday and Gibson (1962) selected flies with the highest and lowest numbers of bristles, allowed them to mate at random, and then separated the high- and low-bristle females into separate bottles and allowed them to lay eggs. Within a few generations a tendency for assortative mating among high- and low-bristle flies developed. Perhaps the most dramatic experimental results were reported by Paterniani (1969), who for six successive generations planted mixtures of two homozygous strains of maize (corn), *white flint* and *yellow sweet*. Selection for reproductive isolation was imposed by sowing seed from the plants that bore the lowest proportion of heterozygous kernels. The level of intercrossing was initially high (35.8 per cent and 46.7 per cent in the two varieties, respectively), but after five generations it had dropped to less than 5 per cent (*Figure 4*) due to divergence in flowering time and to a reduced receptivity of the *white flint* strain to the *yellow sweet* pollen.

Entomologists have long wondered whether sympatric speciation is the mechanism by which many herbivorous species of insects, each confined to one or a few species of host plants, have arisen. Thorpe (1945) and Bush (1969, 1975) have argued that because mating in many species occurs on the host plant, a mutation that causes the animal to be attracted to a new host confers both ecological and reproductive isolation in one fell swoop. Migration from the old host to the new will be rare if the mutant allele is uncommon in the parent population, and it will be further restricted if survival on the new host requires further genetic differentiation. Once established, however, the propa-

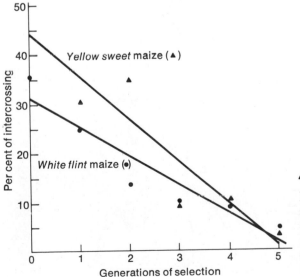

4

Reduction in interbreeding between two strains of corn due to artificial selection against intercrossing. Initially *Yellow sweet* received more pollen from *White flint* than *vice versa,* but both strains declined in the amount of intercrossing. (After Paterniani 1969)

gule could become further isolated as it adapts to its new host; for example, it may change its season of emergence and mating. Bush believes that this sequence of events is responsible for speciation in the true fruit flies of the family Tephritidae. There is some evidence, although the data are ambiguous, that choice of host plant may be controlled by only a single locus (Huettel and Bush 1972). The fruit fly *Rhagoletis pomonella* (*Figure 5*), originally associated with hawthorn, began to infest apple in the 1860s, and cherry in the 1960s. Bush (1969, 1975) believes that the apple- and cherry-associated flies are "host races" isolated from the hawthorn "race" by host plant and by time of emergence. If these are genetically differentiated forms (Futuyma and Mayer, *Systematic Zoology,* in press), they may have originated sympatrically, for the several hosts occur within distances that the flies are capable of travelling. Host races are well known in the codling moth (Phillips and Barnes 1975) and in other insects.

Mayr (1963) believes that this model entails such a concatenation of unlikely events that it is implausible. He points out, for example, that the fidelity of an insect to a single species of host is seldom as great as the model requires. He believes it more likely that a population, temporarily isolated by a geographic barrier in a locality where a new host plant is prevalent, expands its host range to include the new host. Local extinction of the old host then restricts the insect to the new host, on which it becomes dependent. Since a geographic barrier need not be very impressive to our eyes to be effective, it will probably be impossible ever to prove that sympatric speciation rather than temporary geographic isolation of a small population occurs.

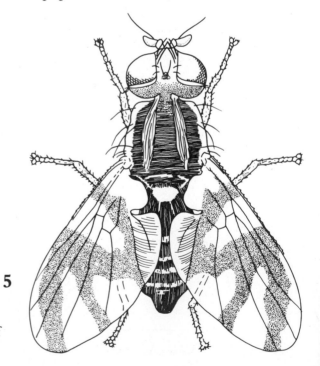

5

The apple maggot, *Rhagoletis pomonella,* a member of the family of true fruit flies, Trypetidae. (Figure by U. S. Department of Agriculture)

PARAPATRIC SPECIATION

There is little formal difference between a model of sympatric speciation by adaptation to two microhabitats or niches (such as host plants) that are intermingled with each other, and adaptations to habitats such as different soil types that are spatially separated but contiguous. The question arises whether such PARAPATRIC populations can become reproductively isolated without the action of a geographic barrier.

Endler (1973, 1977) has determined that when the average distance of migration is short compared to the distance over which the environment changes from one extreme state to another, gene frequency may change abruptly even if the environment changes gradually (*Figure 6*). This is because one homozygote is favored in all environmental conditions above a critical value, while the other homozygote is favored below the critical value. Because maladaptive alleles migrating from one direction are balanced by the input of adaptive alleles from the other, even massive gene flow has little effect on the abruptness of the genetic change.

If, however, the average distance of migration is large compared to the scale on which the environment varies, the gene frequency in any one region is determined by the average of the selection pressures in the various environments to which the genes are exposed in the course of their travels. There will be a gradual cline in gene frequencies (Slatkin 1973), and gene flow will overwhelm the genetic differences that selection would otherwise establish. This process, as Mayr (1963) has stressed, prevents the formation of local coadapted gene pools that are precisely adapted to local environments.

This gradual cline becomes steeper if there is a very effective geographic barrier to dispersal or if heterozygotes over the whole gradient are inferior in fitness to the homozygote that is locally most fit (Slatkin 1973). Migration of an AA individual to a region where $A'A'$ is predominant results in inferior AA' progeny; thus the A allele cannot increase in frequency where A' is predominant, or *vice versa*. But if the heterozygotes, despite their inferiority, reproduce at all, they introduce alleles at other loci into the population in which they reproduce, and so the populations over the whole gradient can become genetically homogeneous at loci other than A. Thus geographically distant populations may be genetically incompatible when crossed, even if they are connected by a chain of populations among which there is gene flow, so that there is no point at which distinct species can be recognized. This is the case for the "sex races" of gypsy moth described by Goldschmidt (1940), for the races of *Streptanthus* (Kruckeberg 1957) discussed in Chapter 9, and for four species of butterflies and moths studied by Oliver (1972).

This model seems to describe a case in which parapatric speciation has been suggested. White (1968) notes that in very sedentary wingless grasshoppers, parapatric populations differ in chromosome configuration, so that the fertility of hybrids is greatly reduced (*Chapter 10*).

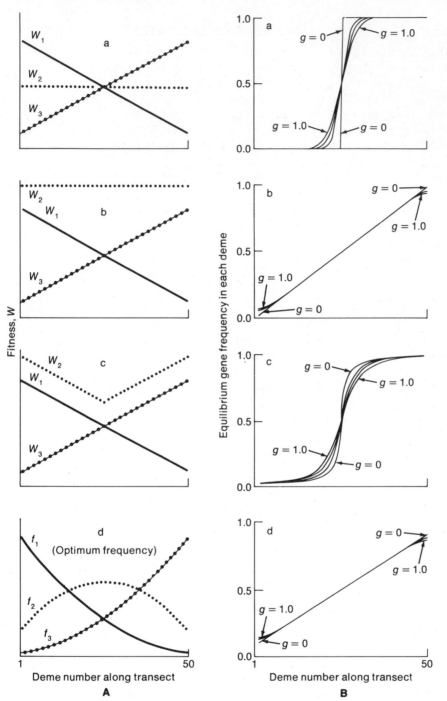

The formation of clines in gene frequency by selection and gene flow. At left (A) are four patterns of selection along a transect, in which the fitness of AA (W_1) declines, and that of $A'A'$ (W_3) increases, from deme 1 to deme 50. W_2 is the fitness of AA'. Model a is a simple gradient; in model b the heterozygote has a fixed fitness; in model c the heterozygote's fitness is a fixed increment greater than that of the more fit homozygote; in model d there is frequency-dependent selection and f_i is the expected equilibrium

Such chromosome differences among populations are also common among burrowing rodents such as gophers (Nevo et al. 1974, Bush 1975). Because there is reproductive isolation (in the form of chromosome incompatibility) among these populations, they can be considered incipient species, which White believes arose by parapatric or, as he calls it, stasipatric speciation. He supposes that a new type of chromosome arises within one population and expands its range; any one area is homozygous for one chromosome type or the other, and the border between karyotypically different populations should be (and is) very narrow.

But it is hard to see how a new chromosome type could spread in a population, for it first arises in its deleterious heterozygous form and thus should immediately be extinguished (Key 1968). It could replace the old chromosome type only by genetic drift, which can fix even very deleterious alleles in sufficiently small populations. But a population can be considered small only if it is effectively closed to gene flow, that is, if it is allopatric. There seem to be no convincing cases of parapatric speciation in nature, although it seems likely that it could occur, especially if there were both assortative mating and heterozygous inferiority. The latter ingredient is present in cases of chromosomal differences among populations, but these instances seem to illustrate allopatric speciation by genetic drift rather than parapatric speciation.

GEOGRAPHIC SPECIATION

In an extended analysis of speciation, Mayr (1963) concluded that only allopatric speciation and speciation by polyploidy are at all likely to occur. Many evolutionists are not as vehement as he is, but most accept his conclusion that geographic speciation is by far the most common mode (e.g., Dobzhansky 1970, Grant 1971, Lewontin 1974a; but see Bush 1975 for an opposing view). The evidence for this view is summarized in detail by Mayr (1963) and is treated here only briefly.

The most profuse evidence comes from the study of geographic variation. Species vary geographically in those very characteristics that can bar gene exchange among sympatric species (*Chapter 9*). Populations can be genetically incompatible; they may occupy different adaptive peaks in a genetic landscape (*Chapter 14, Figure 7*), separated by adaptive valleys — less fit genetic compositions of the kind that would arise from interbreeding.

Similarly premating isolating mechanisms vary geographically and

---

frequency of genotype i in the absence of gene flow. At right (B) are the resulting equilibrium frequencies of allele A' in each deme, when each deme exchanges immigrants equally with the neighboring demes on either side. Gene flow ranges from 0 ($g = 0$) to 100 per cent ($g = 1.0$), when the breeding population of each deme is derived entirely from neighboring demes. In these models the form of the cline is determined almost entirely by selection, very little by gene flow. (Redrawn from Endler 1973; copyright 1973 by the American Association for the Advancement of Science)

sometimes show character displacement (*Chapter 9*). Thus populations can diverge in reproductive behavior because they face the threat of hybridization with different species. Often, though, divergence in premating isolating mechanisms is likely to be a pleiotropic effect of other genetic changes. For example, artificial selection of populations of *Drosophila* and of houseflies for divergent responses to light and gravity for a few generations resulted in some sexual isolation between the selected populations (del Solar 1966; Soans, Pimentel, and Soans 1974; Hurd and Eisenberg 1975). Populations that adapt independently to a similar environmental novelty do so by different genetic changes, which carry different concomitant effects on sexual behavior. Thus, for example, Ehrman (1964) tested for sexual isolation among six populations of *Drosophila pseudoobscura,* all derived from the same stock, of which two had been kept at 16°C, two at 25°C, and two at 27°C for 4½ years. Pairs of populations were slightly but significantly isolated sexually, and such isolation occurred as often between populations adapted to the same temperature as between those adapted to different temperatures.

Thus, by the mechanisms that cause genetic change in single populations, different geographic populations of a species can diverge in characteristics that may bar interbreeding. In most cases the widely distant populations are connected by intervening populations, so that the species' characteristics change gradually from one to another of the most divergent populations. The most remote populations might well not interbreed if they came into contact. That this is likely is shown by instances of CIRCULAR OVERLAP, in which a chain of races that are believed to interbreed curves back on itself, so that the highly divergent forms at the termini come into sympatry and do not interbreed. A possible example (Fox 1951) is a group of garter snakes in western North America (*Figure 7*).

That divergence among allopatric populations is facilitated by extrinsic barriers is strongly suggested by the frequent correspondence between biological and topographical discontinuities. Regions in which a group is diverse are usually areas of great spatial heterogeneity. Freshwater animals differ at both specific and infraspecific levels from one river system to another, so they show the greatest regional diversity in mountainous regions where there are many rather well-isolated river systems. In regions of high topography, geographic variation in sedentary species can be quite staggering, every valley (or mountain top, for high-altitude species) holding a recognizably different form. In an area only 20 by 5 miles on Hawaii, Welch (1938, cited in Mayr 1963) recognized 26 subspecies and 60 "microgeographic races" in the land snail *Achatinella mustelina.*

The best-known instances of divergence among isolated populations are on islands, where a species that is rather homogeneous over the entirety of its continental range may diverge spectacularly in appearance, ecology, and behavior; the drongo (*Dicrurus*) described in Chapter 9 is an example. Of special interest are cases of "double

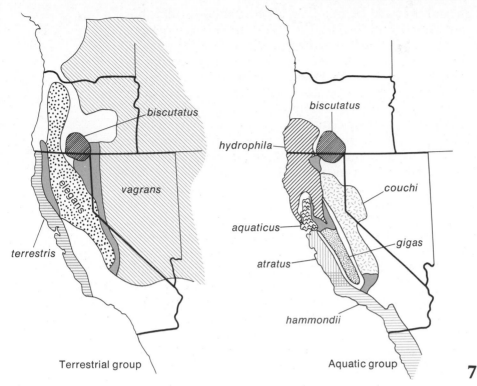

7

Complex relationships among races of garter snake (*Thamnophis*). In the aquatic group, *hammondii, gigas, couchi, hydrophila, aquaticus,* and *atratus* form a sequence of allopatric subspecies that interbreed where they meet (gray areas); but *atratus* coexists with *hammondii* without interbreeding. Moreover *hydrophila* interbreeds with *biscutatus* where they meet, but *biscutatus* also interbreeds with members of the terrestrial group, which is otherwise broadly sympatric with the aquatic group and does not interbreed with it. (Redrawn from Fox 1951)

invasion"; Mayr (1942) discusses many instances of these. For example, the bird *Acanthiza pusilla* is widespread on the Australian continent and has a slightly differentiated population on Tasmania. Tasmania also has a second species, *A. ewingi,* which is more markedly differentiated. It is plausible to suppose that during one of the Pleistocene glaciations, when the sea level was lower, *Acanthiza* invaded Tasmania. Isolated there during the interglacial period when sea level rose, the population differentiated and became what is recognized as *A. ewingi.* A second invasion by *A. pusilla,* perhaps during a later glacial period, established *A. pusilla* on the island, where *A. ewingi* had changed so much that it had become reproductively isolated.

Such multiple invasions explain why each island of an archipelago so often has several related species, while a remote island of similar size may have just one. Darwin's finches have speciated extensively within the Galápagos Archipelago, but are represented by only a single species on Cocos Island, 600 miles distant. The Cocos Island finch has been there long enough to have become morphologically distinct, but it has not split into more than one species. Within the archipelago, however, populations can diverge on different islands and recolonize the islands from which they came, having become reproductively isolated from the resident populations during their period of allopatry.

Although in principle it should be possible to trace speciation in the fossil record, it is actually almost impossible. The record is so incomplete that putative cases of speciation are not strong evidence for the allopatric model; rather the model of speciation derived from the study of living species is used to interpret the fossil record. In one such instance (Eldredge 1971), the Devonian trilobite *Phacops rana* was distributed in the warm, shallow sea that at intervals covered the central United States and in a shallow coastal sea from central New York south, where it formed marginal populations. In the lower Devonian all the populations had 18 files (rows) of eye lenses. Somewhat later the population in central New York quickly evolved to a 17-filed form, which over the next 1 to 2 million years spread west and replaced the 18-filed form in the shallow interior sea. Later still the New York population exhibited yet greater variation (15 to 17 files), and the 15-filed form, presumably a new species, appeared throughout the interior sea (*Figure 8*). Despite continuous deposits of sediment, there is no evidence of gradual change in these populations; the new form appears "instantaneously," apparently having arisen in small marginal New York populations.

Geological contributions to the theory of speciation are usually

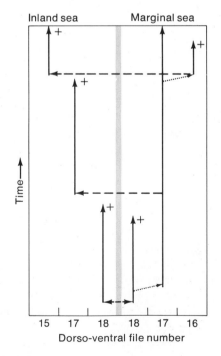

8

Allopatric speciation in Devonian trilobites of the *Phacops rana* stock. Solid lines indicate presence of a trilobite with specified number of files (rows of eye lenses) in the area indicated, broken vertical lines indicate persistence in a part of the marginal sea other than that occupied by a derived form, and crosses indicate final disappearance. Dotted lines indicate origin of a new form (with new file number) in a marginal area; horizontal broken lines show migration from one area to another. (Redrawn from Eldredge 1971)

more indirect. It is through the conjunction of paleontology and biogeography that the historical patterns of speciation have been reconstructed. The distribution of rattlesnakes cited in Chapter 6 and the hybrid zone described in Chapter 9 are examples. Such historical evidence for geographic speciation is circumstantial, but the distributions of closely related species or geographic races are so commonly consonant with those we expect from the past distribution of their environments that geographic isolation must be a very common mode of speciation.

THE EVOLUTION OF ISOLATING MECHANISMS

Assuming that geographic isolation is the most frequent prerequisite for speciation, I now treat the circumstances in which intrinsic barriers to gene flow are most likely to evolve.

If two populations that have been separated by a topographic barrier reestablish contact, they maintain their separate identities only if interbreeding is severely curtailed by premating barriers, postmating barriers, or both. These classes of barriers can evolve independently. For example, narrow hybrid zones, such as that between the *Heliconius* butterfly described in Chapter 9 (which has persisted for at least two thousand generations) indicate the existence of postmating but not premating barriers. In Europe two species of crow have formed a persistent hybrid zone since the Pleistocene (see Mayr 1963, p. 370). The narrow borders between chromosomally different populations of many species also represent genetic isolation by postmating but not premating barriers.

On the other hand, the restriction of gene flow can be due exclusively to premating barriers, without any postmating backup. Hubbs (1955, 1961) describes many cases of closely related species of freshwater fishes which throughout much of their range maintain their specific identity by differences in sexual behavior or by breeding in different habitats but which can, especially when habitats are disturbed, interbreed freely and form populations of fertile hybrids. The apparently high fertility of hybrids among species in such groups as orchids (Adams and Anderson 1958) and cichlids (Fryer and Iles 1972) suggests that speciation in such forms commonly entails the development of strong premating but only weak postmating barriers to gene exchange. Significantly these are groups in which the mechanisms and sign stimuli leading to successful pollination or insemination are very precise. In orchids the highly species-specific odor, color pattern, and form of the flower results in visitation by only a very few species of bees or other insects, upon which the pollen masses are deposited in precise positions (Dressler 1968; *Figure 9*). In cichlids patterns of movement during courtship, and especially the species-specific color pattern, are critical to successful mating (Fryer and Iles 1972). Among birds the most extravagant development of species-specific male coloration, accessory plumes, and crests occurs in groups such as the birds of paradise and hummingbirds (*Figure 10*) that do not form a pair bond; since courtship is brief and each male is competing with others for the attention of females, selection favors male phenotypes that not only communicate identity, but also provide supernormal mating stimuli.

Thus, if an isolated population included males with novel epigamic characters that constituted new supernormal stimuli, or females with different mating preferences, then it seems likely that one or more features of courtship (or of flower form in plants) could rapidly evolve to a new coevolved equilibrium of male signal and female responsiveness. It is as if the consummation of courtship in these species depended on a very delicate equilibrium between signal and response — a pattern that comes to a very different equilibrium if the fragile interplay of forces is stretched past some breaking point.

There has been some question about whether the ethological barriers to interbreeding between newly formed species arise while the populations are allopatric or whether they come into existence when they rejoin. In cases like the cichlids or the orchids, ethological barriers must develop in allopatry; for if they did not, and if postmating barriers are as undeveloped as they seem to be, there would be no barrier to interbreeding and no speciation would have occurred. But when both pre- and postmating barriers exist, the premating barrier could develop when the isolated populations rejoin. The low fitness of the hybrids would prevent genetic mixing and promote divergence of the ethological signals, because genotypes that confer distinctive characteristics or an ability to discriminate between appropriate and inappropriate partners have a strong selective advantage. Selection favors the reinforcement or character displacement of premating isolating mechanisms only to the extent that hybridization is disadvantageous, only to the degree that postmating barriers to gene flow exist.

Artificial selection against hybrids can lead to a rapid increase in mating discrimination (Koopman 1950; Knight, Robertson, and Waddington 1956; Paterniani 1959). In natural situations the distinctiveness of epigamic signals is often greater between sympatric than between

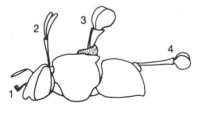

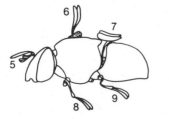

9

Typical positions of the pollinia (pollen masses) of 11 genera of neotropical orchids on the euglossine bees that pollinate them, illustrating how a mutation in flower form could alter placement and so effect reproductive isolation. (Redrawn from Dressler 1968)

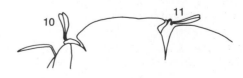

10

Secondary sexual characteristics vary greatly among the males of hummingbirds, although the females are rather similar to each other. Differences in plumage of the male facilitate reproductive isolation. Left to right, above, *Sappho sparganura, Ocreatus underwoodii, Lophornis ornata*; below, *Stephanoxis lalandi, Popelairia popelairii, Topaza pella.* (Redrawn from illustrations by A. B. Singer in Skutch 1973)

allopatric populations (see review by Levin 1970*a*). Such examples may suggest that premating isolating mechanisms evolve when formerly allopatric populations rejoin, because of their adaptive value in preventing hybridization. But such a hypothesis must face the countervailing evidence of long-persistent hybrid zones like that in the *Heliconius* butterflies, in which such selection has not given rise to sexual isolation. Moreover most pairs of related species differ in their epigamic characteristics not only where they are sympatric, but throughout their range. Hence it seems likely (Mayr 1963) that divergence in these characters arises at least incipiently while the populations are allopatric and that it is sometimes slightly accentuated by natural selection when the newly formed species become sympatric.

The likelihood that premating isolating mechanisms are reinforced is determined by the efficacy of the postmating barriers that have developed. If the hybrids can reproduce and backcross to some extent, the populations lose their distinctiveness; moreover the more genetically incompatible and hence unfit recombinant genotypes are eliminated, while compatible hybrid genotypes prosper. Thus the originally distinct gene pools become less incompatible, so selection in favor of the development of premating isolating mechanisms becomes progressively less intense, as does the strength of the postmating barrier itself.

This happens quite rapidly. For example, if the average fitness of all hybrid (including backcross) genotypes is one-fourth that of the pure parents, hybrid genotypes will make up 99.9 per cent of the whole population within eight generations (Maynard Smith 1966). Hence there is little time or opportunity for natural selection to develop highly effective sexual isolation *de novo*. Thus two populations remain distinct rather than fusing only if, when they first meet, the incidence of cross-mating is very low, or the fitness of the hybrids is very low, or both (*Figure 11*).

Postmating barriers indeed become less pronounced as hybridization proceeds. Stebbins and Daly (1961) describe a population of sunflowers in which the species *Helianthus annuus* and *H. bolanderi* began hybridizing along a road in 1941. Morphological analysis of samples taken every year from 1951 to 1958 showed that a great variety of hybrid genotypes was present and that the variation in each of five characteristics increased in successive generations. The hybrids had highly inviable pollen in 1946, but by 1955 the majority of such plants had more than 80 per cent viable pollen.

In the evolutionary literature both premating and postmating barriers have usually been included under the single rubric *isolating mechanisms*. Perhaps because a single term is so often used for two different phenomena, and perhaps because the word *mechanism* implies a function and so a *raison d'être* for a characteristic, it is commonly supposed

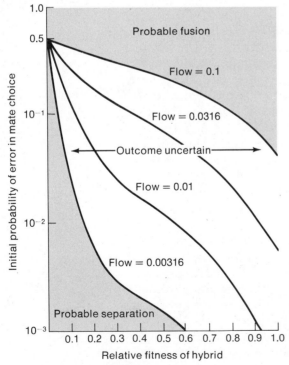

11

Likelihood of speciation vs. fusion when two previously isolated populations meet. A gene that increases discrimination in mating will increase and accomplish cessation of interbreeding (separation) if the fitness of the hybrid is low and the initial frequency of errors in mating is also low. "Flow" is the level of gene flow between the populations, at the levels of mating error and hybrid fitness indicated. The higher the initial degree of interbreeding (the weaker the initial prezygotic isolating mechanism), the lower the fitness of the hybrid (the more intense the postzygotic barrier) must be if speciation is to be consummated. (From Wilson 1965, after Bossert)

that natural selection favors the intensification of hybrid infertility as it does the accentuation of sexual isolation. But in most cases, the very opposite occurs. Genes that confer sterility on their carriers cannot increase in frequency; rather the *most* fertile of the genotypes that arise by recombination propagate their alleles. Hence postmating isolating mechanisms cannot be reinforced by natural selection. The only exception to this principle derives from parental care; if a plant or a viviparous animal bears both hybrid and nonhybrid zygotes and provides them with endosperm or yolk, it can increase its fitness by aborting the hybrid zygotes (assuming they have inferior fitness anyway) and allocating resources entirely to the nonhybrid offspring whose prospects for high fitness are better. Thus abortion of hybrid embryos in the plant genus *Gilia* is more prevalent in crosses between sympatric species than between allopatric species (Grant 1966).

MODELS OF GEOGRAPHIC SPECIATION

At first surmise it is not clear why all species are not splitting into new species at a great rate. Geographic barriers are plentiful, and populations have enough genetic variation to adapt to divergent selective pressure. Ehrlich and Raven (1969), in considering this question, argued that the traditional response, that levels of gene flow among populations are ordinarily too high to prevent divergence, was inadequate. They point to the extremely limited levels of gene flow that characterize many species (*Chapter 11*) and to cases in which indubitably isolated populations have not diverged over very long periods of time. Failure to speciate, they suggest, is caused not by excessive gene flow, but by uniform selection regimes over the entire range of the species. But according to the model by which geographic speciation probably occurs most often, gene flow and commonality of selection pressures are not alternatives but two aspects of the single factor that may retard the rate of speciation.

There are two major models of geographic speciation. Divergent genetic changes may occur in each of two widespread populations that become separated by a topographic barrier (*Figure 12a*); or genetic differentiation may occur in a small population, isolated from the main body of the species, which remains genetically rather inert (*Figure 12b*).

The first model (*Figure 12a*) probably describes the origin of few species, for large, widespread populations evolve and diverge from each other slowly (*Chapter 11*). The local demes of which they are composed vary somewhat in genetic composition, but on the whole they probably have a fairly uniform, coadapted gene pool that resists major changes (genetic homeostasis). To be sure, at some loci, such as those affecting coloration, alleles affect fitness primarily by the relation between the phenotype and the external environment; but at most loci it may well be that the fitness of an allele depends on its effect on development and so on its epistatic interactions with other loci. If epistatic interactions are important, the favored alleles in a local population subject to much gene flow are those that are compatible

A

B

12 Two models of allopatric speciation. (A) The geographic range of a species becomes fragmented when a barrier (e.g., eradication of the forest habitat) arises; the large separated populations diverge from each other and then become sympatric when the barrier is removed. (B) A propagule from a species invades an isolated habitat and diverges from its parent species; later it may reinvade the region from which it originally came and become sympatric with the parent form.

with the many genetic backgrounds they encounter in various demes; in Mayr's terms they are good mixers rather than soloists. These alleles should be much the same from one local population to another, for the variety of genetic backgrounds with which the alleles interact is much the same because of gene flow among populations.

Moreover the environment of each of two large, widespread populations may be surprisingly similar, another reason for their genetic similarity. For example, Levins (1968) found that many species of *Drosophila* occurred at both high and low altitudes, but within any single locality they occupied different microhabitats. Thus the different ecological requirements of these species were all present at a local scale; the environmental heterogeneity within a small area is almost as great as that of much broader regions. Therefore the average environment encountered by an allele carried about by gene flow from one local microhabitat to another is much the same in one large region as in another. By the law of large numbers, large regions are more similar to each other than are small areas.

If a widespread species is sundered into two large and widespread populations, they will have much the same genetic compositions; their environments, on average, will be as similar as they were previously; it is unlikely that an environmental change will transpire over the whole of one population's range but not the other's. Few environmental changes will affect the entirety of either population; except for such events as long-term climatic changes, most environmental changes happen at too local a level to evoke an evolutionary response from an entire widespread population. Thus it is not surprising that populations with broad geographic ranges seem to evolve and diverge from each other very slowly. The North American and Eurasian species of sycamore (*Platanus*) have been isolated for at least 30 million years, but they have diverged little in morphology and can be crossed to yield fertile hybrids (Stebbins 1950); the morphology of the genus has remained largely unchanged since the Cretaceous. Similarly the Mediterranean plantain *Plantago ovata* has been isolated from the American *P. insularis* for at least 20 million years, but their hybrids are quite fertile (Stebbins and Day 1967).

Species more often originate by genetic changes in small populations (Lewis 1962, Mayr 1963, Eldredge and Gould 1972, Bush 1975). New species, then, are not siblings; instead they are the offspring of parent species that remain relatively unchanged.

Mayr (1963) believes that divergence occurs primarily in PERIPHERAL POPULATIONS formed by colonization at the edge of a species' range. Lewis (1962) suggests that new species can arise by CATASTROPHIC SPECIATION from small populations within a species' range — populations that have been greatly reduced in size by environmental catastrophes that select for unusual, tolerant genotypes. In either case the populations occupy small geographic areas that are more likely than large areas to differ ecologically from one another, for they are more homogeneous internally. A small population, cut off from gene flow from other populations, faces a suite of competitors, predators, resources, and potential symbionts to which it as an evolving unit is homogeneously exposed, a set of conditions that may differ greatly from the kaleidoscopically variable biota to which its parent species has long since adapted. If the population is associated with a species-poor biota, it may encounter unprecedented ecological opportunities in the absence of a full suite of competitors and predators.

Not only are the forces of selection novel, but the genetic composition of the colony may be unusual as well. Mayr (1954, 1963) has stressed that a colony founded by just a few individuals is likely to carry less genetic variation than existed in the parent population, especially if the population remains small and genetic drift operates (*Chapter 11*). Mayr further suggests that the colonists that found populations at the edge of the range will be genetically aberrant compared with the species as a whole, because they come from genetically aberrant peripheral populations.

These conditions affect the evolution of barriers to gene flow. A new species composition of pollinators may select for changes in flower form; the new species composition of the community can select, by character displacement, for different epigamic signals. Even in the absence of such selection, changes in epigamic characters can arise by genetic drift or by correlation with other evolving characters. Once a new mode of sexual communication becomes prevalent, even if by chance, selection for responsiveness to the new signal can fix the stereotyped mating behavior in a new pattern. In short, premating isolating mechanisms can evolve rapidly.

Among postmating barriers to gene flow, hybrid sterility caused by translocations or other alterations of the chromosomes can arise rapidly by genetic drift in small populations. Postmating barriers caused by allelic differentiation may also develop rapidly in small colonies (Mayr 1963). Because of the unusual initial genetic composition of the colony, each allele finds itself in a novel genetic environment, compared to that of the species at large. At some loci, alleles that were formerly prevalent and thus interacted with each other are rare or absent, and formerly rare alleles are now common. Mayr terms this the founder effect. The genetic background in which each locus functions is more homozygous, partly because of the founder effect and subsequent inbreeding, and partly because of the cessation of gene flow. There may now be a selective advantage for soloist alleles that interact epistatically in a favorable way with this novel genetic background, alleles that formerly formed disadvantageous combinations. The population, in short, has been shifted by chance from its old adaptive peak to the slopes of a new adaptive peak in the genetic landscape, which it now climbs by natural selection to arrive at a new genetic equilibrium (*Chapter 14, Figure 7*). As gene frequencies change at each locus, the genetic environment of alleles at other loci is altered, so that there follows in rapid succession a cascade of genetic change throughout much of the genome, a genetic revolution, as Mayr has termed it. Freed from the straightjacket of genetic homeostasis, liberated from the disruption of new gene complexes by gene flow, and challenged by homogeneous, novel ecological conditions, the population evolves new developmental pathways that are incompatible with the old, and alterations of morphology, physiology, and behavior of such magnitude that the population may be classified as a new genus or family.

Such genetic revolutions may be rare events, and most colonies no doubt become extinct before they have embarked far along this path, or they are cut short in their divergence by resurgence of gene flow from the parent species. But untold numbers of such colonies are produced generation after generation, so that even improbable events are almost certain to happen eventually.

This scenario, if it is true, has sweeping implications for all of evolutionary theory. It confronts directly the problem of rates of evolution by suggesting that evolution in widespread species is typically a very slow process, restricted in speed and direction by genetic homeostasis, by gene flow among populations in different environments, and by the infrequency with which the entire body of a species is subject to directional selection in the form of a widespread change in environment. Against the slow pace of phyletic evolution, it contrasts the extreme rapidity of genetic change in small populations and the unparalleled direction it may take. Truly new evolutionary events of the kind that generate higher taxonomic categories may happen so quickly and on so restricted a geographic scale that their intermediate stages are seldom found, either among extant populations or in the fossil record. The slowness of evolution that Haldane (1957) said must follow from the "cost of selection" (*Chapter 13*) is less pronounced in small populations, in which few genetic deaths accomplish substantial change in gene frequencies. The factors that limit the rate of evolution are neither purely ecological nor purely genetical, but both; for truly novel ecological conditions that impose strong directional selection are encountered primarily on a local scale, and populations can fully adapt to them only when the genetic variation immanent in all populations can be freely expressed, unconfined by the restrictions of gene flow and genetic homeostasis. The grand story that is evolution is not a history of gradual change within species, but a history of speciation events. As Mayr (1963, p. 621) puts it (a bit teleologically, to be sure), "Speciation, the production of new gene complexes capable of ecological shifts, is the method by which evolution advances. Without speciation there would be no diversification of the organic world, no adaptive radiation, and very little evolutionary progress. The species, then, is the keystone of evolution."

THE GENETICS OF SMALL COLONIES

Only some elements of this seemingly plausible theory of speciation are supported by extensive evidence. For example, it is plausible to assume that where progressively harsher ecological conditions limit the distribution of a species, the peripheral populations are less genetically variable than those in the interior of the range (see Soulé 1973 for a review). In several species of *Drosophila,* there is less polymorphism for chromosome inversions in peripheral than in central populations (*Figure 13*). But this is almost the only evidence that peripheral populations are genetically impoverished, and this pattern could arise for a different reason. In the uncertain, stressful environments that peripheral populations occupy, inversion heterozygosity may be selected against, because the inversions prevent the formation by recombination of well-adapted genotypes. Chromosomal polymorphism is not necessarily correlated with allelic heterozygosity. Most of the evidence on genic variation comes from studies of allozymes, which in general indicate that levels of genetic variation are no lower in peripheral than in central populations (Prakash, Lewontin, and Hubby 1969; Prakash 1973).

There is good reason to believe, though, that genetic divergence most effectively leads to speciation in small, isolated populations. For example, characteristics that could serve as premating isolating mechanisms are often more divergent among small, isolated populations than they are throughout the more continuous range of a species. In the flycatcher *Petroica multicolor,* for example, the bright plumage of the male contrasts with the duller pattern of the female throughout eastern Australia; yet among various of the South Sea islands, geographic variation is so great that the males have female coloration on some islands, and the females have male coloration on others (*Figure 14*).

Speciation by the genetic divergence of small isolates has perhaps been best documented in plants of the genus *Clarkia* (Onagraceae) in California (Lewis 1962, 1966). For example, *Clarkia lingulata* exists in only two colonies, at the very southern edge of the range of its parent species, *C. biloba*. The species differ by one translocation, at least two paracentric inversions, and by a fission of one chromosome in *biloba* into two in *lingulata*. Hybrids between them are sterile, because of the differences in chromosome configuration. Morphologically the species are almost identical, differing only in the shape of the petals. Pollinating insects do not discriminate between the species; there are no premating barriers to hybridization.

The chromosomal aberrations that characterize *lingulata* are not found as polymorphisms in *biloba,* for they decrease fitness in heterozygous condition. Thus the chromosomal changes must have occurred in a very small population by genetic drift. Lewis believes that this was not a newly founded population but a preexisting population that had become nearly extinct, surviving only as a few plants with peculiar chromosomes that happened to carry adaptive gene complexes. The catastrophe that caused this was an increase of aridity, for *Clarkia*

13

Reduced chromosomal variation in peripheral populations. The diameter of each circle is proportional to the average number of heterozygous inversions in females of the fly *Drosophila willistoni*. (After da Cunha and Dobzhansky 1954)

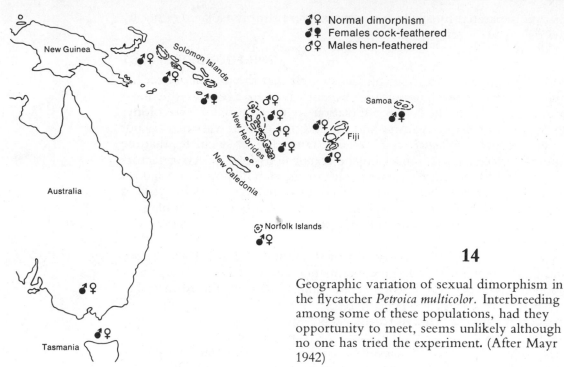

14

Geographic variation of sexual dimorphism in the flycatcher *Petroica multicolor*. Interbreeding among some of these populations, had they opportunity to meet, seems unlikely although no one has tried the experiment. (After Mayr 1942)

occupies patches of moist habitat in the otherwise arid lands of southern California, and it cannot survive more than a few years of exceptionally low rainfall; indeed extinction of some populations has been documented. *C. lingulata,* like the several other *Clarkia* species that are thought to have arisen by catastrophic selection, occupies drier habitats than its parent *C. biloba.* As the hypothesis of catastrophic selection predicts, *C. lingulata* is less polymorphic than *C. biloba* at several allozyme loci (Gottlieb 1974).

In many small populations, especially on islands, divergence in ecological role and in associated morphology is often evident. For example, the small insectivorous lizard *Uta stansburiana* exhibits only subtle geographic variation from Kansas to the Pacific coast, from Washington south to Sinaloa. But the islands of the Gulf of California have *Uta* populations that have clearly been derived from the mainland species and yet vary so greatly in body size, scalation, and coloration that they have been called not only subspecies but species (Soulé 1966). The ecological behavior of some of these populations differs not only from that of mainland *Uta,* but from that of most other species in the large family (Iguanidae) to which it belongs; for example, the *Uta* on some of the islands feed on insects and isopods in the intertidal zone. The degree of morphological divergence from mainland *Uta* is greater on islands that are separated from the mainland by deep water, and were probably colonized by a few individuals, than on those islands that are separated by shallow water and were simply isolated *in situ* by the rise in sea level.

THE FOUNDER EFFECT

Although these observations on *Uta* suggest that the founder effect has operated, its general importance is unclear. On average, the most prevalent alleles in the parent population are transported to the colony, so the level of heterozygosity and the additive genetic variance are not greatly reduced (*Chapter 11, Figure 25*). James (1971) calculates that the founder effect has an impact on the response to new selective factors if the newly favored alleles were rare in the parent population and if selection is intense. He reports that populations started with two flies responded less to artificial selection for bristle number than populations founded by 10 or 40, but only when the intensity of selection was high.

In the most widely quoted test of the founder effect (Dobzhansky and Pavlovsky 1957), a large experimental stock of *Drosophila pseudoobscura* was synthesized from 12 *PP* and 12 *AR* chromosomes; because the *AR/PP* genotype is heterotic for fitness, the chromosome frequencies are expected to reach a stable equilibrium between 0 and 1.

This mixed stock was the source of 10 populations started with 20 flies each and 10 started with 4000 each. All the populations quickly grew to large size, and all began with the same frequency (0.5) of *PP* chromosomes. They differed only in the number of founders. After about 19 generations the "small" populations differed more from one another in the frequency of *PP* than did the "large" populations (*Figure 15*). Dobzhansky and Pavlovsky interpreted this to mean that the

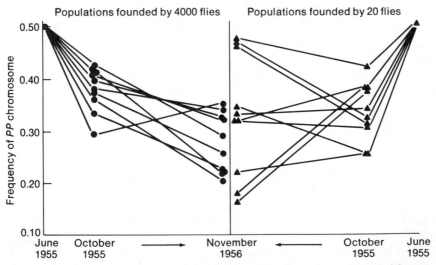

15 Experimental demonstration of the founder effect. Genetic composition varies more among populations founded by few colonists than those founded by many, in experimental cultures of *Drosophila pseudoobscura*. (After Dobzhansky and Pavlovsky 1957)

relative fitnesses of the genotypes *AR/AR, AR/PP,* and *PP/PP* depend on the rest of the genotype in which these chromosomes operate. Due to chance, this background genotype has varied more among "small" populations than among "large" populations, conferring different relative fitnesses on the inversions and hence different patterns of change in their frequencies.

There is only indirect evidence that the founder effect is important in nature. Carson and his coworkers (Carson 1970; Carson, Hardy, Spieth, and Stone 1970) believe that the founder principle has played an important role in the extraordinarily high degree of speciation in the Drosophilidae of the Hawaiian Islands, where about one-third of the world's species of drosophilids occur. They are a monophyletic group in which speciation has been rapid: at least 94 species are endemic to the island of Hawaii, which is less than a million years old. Changes in chromosome structure have played little part in the origin of these species, and closely related species (although loath to hybridize in the laboratory) form viable, fertile F_1 hybrids. The group has not greatly diversified in ecological characteristics; rather the major differentiation among these species seems to have been in mating behavior and in the secondary sexual characteristics of the males, in which peculiarities of the antennae, mouthparts, and legs play roles in courtship and are unlike any characteristics in other drosophilid flies.

The phylogeny of these species can be reconstructed from their chromosomes. For example, the inversion *Xr* appears only in *D. planitibia* on Maui and in *D. silvestris* and *D. heteroneura* on Hawaii. Thus these three species must form a monophyletic group, and there is reason to believe that *silvestris* and *heteroneura* are descendants of the Mauian population that became *planitibia* (see Carson et al. 1970 for details). In almost every case in which a population on one island can thus be shown to have been derived from a population elsewhere, the derivative population is a new species; intermediate stages of speciation are not found. Moreover a species derived from a chromosomally polymorphic ancestor is usually chromosomally monomorphic, having inherited just one of the chromosome types of its ancestor. Carson therefore believes that speciation has been exceedingly rapid and that new species are derived from very small propagules — perhaps a single fertilized female — that have brought with them only some of the chromosomal diversity in the population from which they came.

GENETIC DIFFERENCES BETWEEN RELATED SPECIES

The problem of how much genetic change occurs during speciation is one of the most interesting and important in evolutionary biology, but it is also one of the most refractory. It has remained largely unresolved, both because of the technical difficulties that attend its investigation and, I suspect, because we may not know how to frame questions properly. To begin with, the genetic differences that *distinguish* two related species may be different from those *responsible* for their species status. For example, the translocations that distinguish *Clarkia biloba* and *C. lingulata* are responsible for their reproductive isolation, but the difference in petal shape is irrelevant to their status as species; it is not a cause of their inability to exchange genes. This is not to deny, of course, that the species differ importantly in other respects, such as the characteristics that fit them to different ecological niches. But the question, How great is the genetic difference between two species? is not the same as, How much difference is required to form two species?

Moreover the genetic difference between two species includes the divergence that accompanied the speciation process and the additional genetic changes that accrued in one or both forms after they attained reproductive isolation. Thus we should best be able to quantify the genetic difference entailed in speciation by examining pairs of populations that have very recently speciated or are still in the process of speciation.

Finally, and perhaps most critically, the term *genetic difference* needs to be defined rigorously. It can mean the number or proportion of loci at which the species differ entirely in allelic constitution, or it can mean the number of loci at which alleles differ significantly in frequency. And it may be less important merely to count the number of loci that differentiate the species than to specify which loci differ and which characteristics they affect.

The two approaches to the genetics of species differences have been the classical approach of genetic crosses and the examination of genetic differences by electrophoresis. The classical approach requires that species be hybridized; but hybridization may be difficult or impossible, and the very features that prevent hybridization may be the subject of interest. Moreover analyzing the genetic architecture of a character that differs between species is subject to the same limitations as intraspecific genetic analysis. One usually arrives at the rather uninformative conclusion that it has a polygenic basis, with such-and-such a degree of heritability. How many loci affect the character, how many alleles

---

TABLE II. Measures of genetic similarity and difference between populations X and Y

Nei's index of genetic similarity (Nei 1972)

$$I_N = \frac{\sum\limits_{i=1}^{m} (p_{ix}p_{iy})}{\left[\left(\sum\limits_{i=1}^{m} p_{ix}\right)^2\left(\sum\limits_{i=1}^{m} p_{iy}\right)^2\right]^{1/2}}$$

Nei's index of genetic distance (Nei 1972)

$$D_N = -\log_e I_N$$

Rogers' index of genetic similarity (Rogers 1972)[1]

$$S_R = 1 - \left[\tfrac{1}{2}\sum\limits_{i=1}^{m} (p_{ix} - p_{iy})^2\right]^{1/2}$$

p_{ix} = frequency of allele i in population (or species) X
p_{iy} = frequency of allele i in population (or species) Y
m = number of alleles at the locus

---

[1] S_R usually gives values similar to but lower than I_N.

there are, how different the allele frequencies are in the two species, and how the alleles at different loci interact epistatically are seldom evident from such an analysis.

THE ELECTROPHORETIC APPROACH

Because of the drawbacks of the classical approach, protein electrophoresis has become a popular way of quantifying genetic divergence between species. The electrophoretic data consist, for each of two or more populations, of the frequencies of each allele (or electromorph) i at each of k loci. The GENETIC DISTANCE (or the reverse, GENETIC SIMILARITY) between the populations can be quantified for each locus by one of several formulas (*Table II*) that in essence amount to the difference in frequency of each allele, summed over alleles. The average of the separate measures for each of the k loci can serve as the genetic distance between populations.

Some results of such studies are summarized in Table III; Ayala (1975) and Avise (1976) present reviews. In some instances the electrophoretic difference between related forms is substantial. Ayala and his coworkers (1974) have found progessively greater differentiation in *Drosophila* at various levels of taxonomic divergence from local populations to morphologically distinguishable species (*Figure 16*). The electrophoretic difference between subspecies and semispecies, among which hybrids are partially sterile, is appreciable; but it is not much greater between sexually isolated semispecies. Zouros (1973), however, found that electrophoretic differences were correlated more with hybrid inviability than with hybrid sterility. It is possible, then, that the genetic changes that cause sexual isolation and hybrid sterility are rather simple; only more extensive genetic changes, those that reprogram developmental patterns, are mirrored by divergence at enzyme loci.

Species in some groups differ hardly at all at electrophoretic loci. Species of fossorial rodents such as gophers and the mole rat *Spalax* are reproductively isolated by chromosomal differences, but they differ little in allozyme gene frequencies. The same seems true of species of the plant *Clarkia*. Speciation in these instances may well have entailed almost no genic (as opposed to chromosomal) changes. In other groups, such as the African cichlids, species have diverged in ecological characters and in features that affect mating patterns, but these genic changes do not extend to enzyme loci. Species so phenotypically different that they are relegated to different genera are as electrophoretically similar as different local populations of a single species.

In these instances there has probably been little repatterning of development, so that extensive genetic reorganization is not expected. But even in cases in which development seems to have been extensively altered, gene frequencies at enzyme loci have changed little. Humans and chimpanzees are indubitably different in almost every morphological respect, yet they are electrophoretically identical at about half the enzyme loci that have been studied. Moreover the amino acid

TABLE III. Estimates of genetic similarity and difference, based on allozyme data

Comparison	Measure[1]		
	$\bar{I}_N$	$\bar{D}_N$	$\bar{S}_R$
Drosophila willistoni group			
between local conspecific populations	.970	.031	
between subspecies	.795	.230	
between semispecies	.798	.226	
between sibling species	.563	.581	
between nonsibling species	.352	1.056	
Drosophila mulleri group			
between local populations	.999		
between subspecies	.878		
between sibling species	.777		
Hawaiian *Drosophila* species			
in the *adiastola* group	.569	.573	
in the *nigra* group	.509	.715	
Lepomis (sunfish) species	.544		
Cyprinodon (pupfish) species	.894		
Lake Malawi (Africa) cichlid fishes			
between local conspecific populations	.956	.045	.909
between congeneric species	.946	.056	.893
between species in related genera	.926	.076	.877
Anolis lizard species (*roquet* group)	.667		
Dipodomys (kangaroo rat) species			.61
Thomomys (gopher) species			.845
Thomomys talpoides chromosome races	.925	.078	
Spalax ehrenbergi (mole rat) chromosome races	.978	.022	
Peromyscus (deer mouse) species			.648
Homo sapiens (human)/*Pan troglodytes* (chimpanzee)			.538
Lupinus species (Leguminosae)	.353		
Hymenopappus species (Compositae)	.896		
Clarkia biloba/*C. lingulata*			
between conspecific populations	.909		
between species	.88		

[1] These are average measures over 10 or more enzyme loci, using the formulas in Table II. Some authors use I_N, others S_R. (Data of various authors, compiled by Ayala 1975; data for Lake Malawi (Africa) cichlid fishes are those of Kornfield 1974)

sequences of several of their proteins are identical, and overall their DNA appears to differ at only about 33 out of every 3000 base pairs (King and Wilson 1975). This extraordinary genetic similarity, comparable to that between sibling species of *Drosophila,* indicates that molecular differentiation has not kept pace with the rapid anatomical evolution of the human lineage. The speciation event from which we ultimately derive did not accelerate overall genetic change very greatly. Similarly Avise and Ayala (1976) concluded that extensive speciation among minnows caused no greater rapidity of divergence in proteins than in sunfishes, which have undergone little speciation.

It appears, then, either that speciation does not entail a genetic revolution or that the revolution does not extend to enzyme loci, or perhaps to very many structural loci (those that code for proteins) at all. But even before these data were available, Mayr (1963) had pointed out that a simple quantitative approach might be misleading; species differences cannot "be expressed in terms of the genetic bits of information, the nucleotide pairs of the DNA. This would be quite as absurd as trying to express the difference between the Bible and Dante's *Divina Commedia* in terms of the difference in the frequency of the letters of the alphabet used in the two works" (p. 544). The differences between species, the major phenotypic changes that are seen in developmental patterns, in adult morphology, in sexual behavior, in ecological role, may well be due to key differences in developmental patterns — changes in the timing and organization of critical developmental sequences controlled by regulatory genes, about which we are almost entirely ignorant.

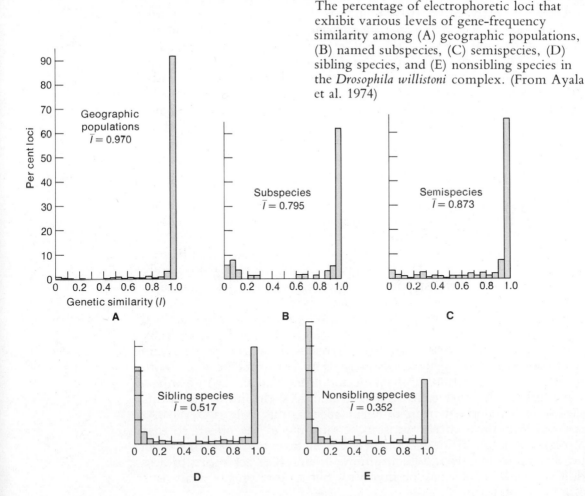

16 The percentage of electrophoretic loci that exhibit various levels of gene-frequency similarity among (A) geographic populations, (B) named subspecies, (C) semispecies, (D) sibling species, and (E) nonsibling species in the *Drosophila willistoni* complex. (From Ayala et al. 1974)

THE CLASSICAL APPROACH

If the genetic differences between newly formed species cannot be judged rightly from electrophoretic data, and if regulatory genes resist analysis, the genetics of speciation must be studied by the classical methods of genetics. Three examples show the kinds of data that may emerge.

The genes that control premating isolating mechanisms account for some critical interspecific differences. If the F_1 hybrid is fertile, the inheritance of the character can be analyzed. Danforth's (1950) work on pheasant hybrids, although based on unsatisfactorily small samples, illustrates the potentials of this approach. In the pheasants, as in hummingbirds and birds of paradise, extensive speciation appears to have occurred more by divergence in sexual characteristics and behavior than by profound genetic change; fertile hybrids can be obtained even from species relegated to different genera. Male pheasants differ in a striking array of brightly colored plumes, crests, and exaggerated wing and tail feathers, all prominently displayed during courtship.

Danforth analyzed the golden and the Lady Amherst's pheasants, *Chrysolophus pictus* and *C. amherstiae* (*Figure* 17), which differ in the color, length, and shape of the feathers of the crest and cape. The features of the crest of *C. pictus* are dominant over those of *C. amherstiae,* but in the cape feathers *C. amherstiae* characters appear to be almost dominant over *C. pictus.* From these and similar patterns of inheritance, Danforth tentatively concluded that at least three major loci affect the color of the crest and about 14 major loci affect its morphological features. The differences between the capes are due to at least two major loci. The phenotypic effect of each of these major loci is modified to some degree by an unknown number of minor loci. The general picture that emerges is much like that for many variations within species; almost every character seems to be polygenically controlled, with the major part of the variation attributable to one or a few loci whose expression is modified by an undetermined number of loci with minor effects — much like the control of the scutellar bristles of *Drosophila* that Rendel (1967) has analyzed (*Chapter 15*). In closely related species that differ strikingly in appearance, there are many such character differences, each with a moderately simple genetic basis.

The genic incompatibility that results in hybrid sterility is described for the sibling species *Drosophila pseudoobscura* and *D. persimilis* by Dobzhansky (1970). Although male hybrids are sterile, females are fertile, so backcrossing makes it possible to construct progeny with any desired combination of *persimilis* and *pseudoobscura* chromosomes. In general, to the degree that a hybrid male's X chromosome and its autosomes are derived from different species, the male is sterile. At least two loci on each chromosome contribute to the effect, so that sterility is due to an interaction between at least two sex-linked loci and six autosomal loci, which among themselves act more or less additively. More loci may be involved, but a more precise analysis is difficult to perform.

17

(A) Lady Amherst's pheasant, *Chrysolophus amherstiae*. (B) Golden pheasant, *Chrysolophus pictus*. (Redrawn from lithographs by John Gould (1804–1881), in A. Rutgers, *Birds of Asia*, Methuen and Co. Ltd., London)

It seems likely that the reduced fitness of interspecific hybrids in some instances is caused by rather profound differences in developmental processes, of which we know little. Most of the relevant analyses have been performed on hybrids between well-defined species, which may well have consolidated discrete patterns of development since their speciation; but the patterns suggest interesting directions for future research. For example, Sturtevant (1920–21) hybridized *Drosophila melanogaster* and *D. simulans*, which cannot be distinguished except by the form of the male genitalia. Only female hybrids survive in the cross ♀ *melanogaster* × ♂ *simulans*, and only males survive from the reciprocal cross, suggesting that viability depends on the nature of the egg cytoplasm. Most interesting is that the hybrids show a pronounced breakdown in developmental homeostasis; they vary greatly in numbers of bristles which in the parents are almost invariant and identical in the two species. A similar breakdown of developmental homeostasis has been described in many other hybrids, including some that occur in nature (e.g., Levin 1970*b*). Thus even the phenotypic characteristics that two species have in common can have different developmental programs, and it is not hard to imagine how these programs could clash and thus reduce a hybrid's prospects of survival and reproduction.

The investigation of these developmental programs, bringing with it, we may hope, a synthesis of developmental biology and evolutionary biology, could be one of the more exciting prospects for biology. Certainly the most persistently intractable problems in evolution require an understanding of the mechanisms of development. In this context the analysis of gene regulation in interspecific hybrids is interesting. For example, Whitt, Childers, and Cho (1973) found that in hybrids from a cross between female bass (*Micropterus salmoides*) and male sunfish (*Lepomis cyanellus*), neither paternal nor maternal alleles were expressed at several enzyme loci. In crosses between more closely related species (e.g., two species of bass), however, both parental alleles are fully expressed. Thus the genetic distance inferred from taxonomic data appears to be correlated with the differences in developmental program. Whitt et al. (1973) note that the repression of both paternal and maternal alleles requires an elaborate explanation. They hypothesize that a repressor produced by the maternal genome inhibits enzyme synthesis by maternal genes. The repressor is removed and enzyme synthesis begins if an appropriate derepressor molecule is present, but the derepressor produced by the paternal genome from the other species may be inappropriate. The same explanation in reverse might account for repression of paternal alleles (*Figure 18*).

Such schemes are all too hypothetical at this point and remain to be elucidated by developmental biologists. But it is in this realm that some of the most interesting differences between species will be found: not in the linear structure of enzymes, but in the regulatory loci that determine which cells shall divide, in what planes their rates of division will be greatest, what families of tissue-specific enzymes will be activated, and in what temporal sequence these events shall occur.

RATES OF SPECIATION

How rapidly does speciation occur? The answer, it appears, is very different for different organisms. In many cases the process is apparently extremely slow; witness the case of the Eurasian and American sycamores. Species and semispecies that developed during the Pleistocene glaciations have had at least 20,000 years, and perhaps almost a half million years, in which to diverge (*Chapter 5*). Haldane (1957) believed that species of vertebrates might differ at a minimum of 1000 loci; under his theory of the cost of selection (*Chapter 13*), complete allele replacements at this many loci would require at least 300,000 generations.

When speciation occurs by gradual allele replacement at many loci in large, widespread populations that are isolated by geographic barriers, it undoubtedly does take this long. A great many populations isolated for thousands of generations by Pleistocene glaciations did not achieve full species status, and they interbreed more or less freely where they now meet. But speciation can be a good deal faster than was once recognized, if it entails only slight genetic change in epigamic

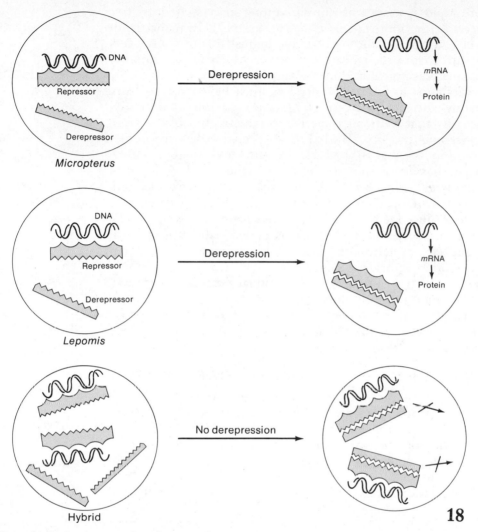

Hypothetical scheme whereby protein synthesis is inhibited in a hybrid because repressors of protein synthesis are not deactivated by derepressor molecules.

18

characters or in chromosome structure, and if it occurs by the genetic divergence of very small populations. Many of the Hawaiian species of *Drosophila* have arisen in just a few thousand years. In the Philippines, Lake Lanao contains 14 endemic species of cyprinid fishes, with modifications of the teeth and jaw utterly unlike any other members of this huge family, that have arisen since the lake was formed about 10,000 years ago (Myers 1960). The rapidity of speciation among the African cichlids is attested by the existence of five endemic species in Lake Nabugabo, a small lake that has been isolated from Lake Victoria for less than 4000 years (Fryer and Iles 1972).

Even more rapid speciation is known among insects. At least five endemic species of the moth genus *Hedylepta* in the Hawaiian Islands feed exclusively on banana, and they are distinguishable in many morphological features from their nearest relative, which feeds primarily on palms. Banana was introduced into Hawaii by the Polynesians only about 1000 years ago, so the banana-feeders must have evolved from their palm-eating ancestor since then (Zimmerman 1960). Similarly, if the *Rhagoletis* fly that feeds on apple is reproductively isolated from its hawthorn-feeding ancestor (Bush 1969), speciation has apparently taken place within the last 100 years. In one case, finally, a new biological species has arisen spontaneously in a laboratory. A strain of *Drosophila paulistorum* when first collected was fully interfertile with other strains but developed hybrid sterility after being isolated in a separate culture for just a few years (Dobzhansky and Pavlovsky 1971). The mechanism of speciation in this case is perhaps aberrant, since it is probably due to adaptation to genetic change in an intracellular symbiont that renders hybrids sterile.

The pace of speciation, then, as of evolutionary change in general, varies from the imperceptibly slow to the very rapid. At its most impressive, changes in developmental patterns and in ecological properties may occur at the speed and magnitude that paleontologists have termed QUANTUM EVOLUTION (*Chapter* 7).

FACTORS THAT AFFECT THE LIKELIHOOD OF SPECIATION

Evolutionists often assume that one group of organisms consists of more species than another because it undergoes speciation more rapidly. Ecologists, on the other hand, often assume that the number of species depends not on the speciation rate, but on competition and other ecological interactions (*Chapter* 4). In fact, both views are right in part, for the number of species, whether at equilibrium or not, is a function of the relation between the rates of speciation and extinction (*Chapter* 5).

Certainly the diversity of a group is enhanced if extinction rates are low, as they may be if newly arisen species are specialized enough to avoid competition with preexisting species. The African cichlids exemplify this principle. But they also speciate rapidly, illustrating that the diversity of a group can also follow from a high speciation rate.

In some instances, I suspect, the richness of a taxonomic group is almost entirely attributable to speciation rate. For example, it seems unlikely that all the many species of figs or orchids require unique resources for growth; species richness in these groups may be entirely a function of their specialized pollination systems, which by very slight changes can impose reproduction isolation. These species may be so ecologically similar that one species may exclude another only very slowly. The diversity of competitively equal species may then follow principles similar to the diversity of selectively equivalent alleles (Caswell 1976).

The rate of speciation, viewed by itself, depends on a complex interplay of genetic and ecological factors. If most speciation is initiated by geographic isolation, the likelihood that a topographic or ecological discontinuity will pose a barrier to gene flow depends on the ecology of the species, including how far individuals disperse, how difficult it is for them to traverse unfavorable habitats, how strictly bound they are by their "psychological" dependence on a particular habitat. Limited dispersal capacity or a social organization into local "tribes" among which there is little exchange breaks the species into inbred groups that can undergo rapid genetic divergence both by genetic drift and by selection. Aside from the effects of social behavior, ecological specialization tends to go hand in hand with fidelity to and dependence on a particular habitat, so that topographic discontinuities become effective barriers. These characteristics tend to be more pronounced in species adapted to mature, relatively stable environments than in the "weedy" colonizers of temporary, evanescent habitats.

The likelihood that populations separated by geographic barriers will become different species depends on the rapidity with which intrinsic (genetic) barriers to gene flow develop. Postmating barriers to gene exchange, in the form of drastic changes in the structure of the chromosomes, necessarily arise in very small populations, since they usually come to prevail only by genetic drift. Such chromosomal alterations are sufficient to establish new species; yet why they are commonly encountered in some groups (e.g., gophers) and not others (e.g., frogs) is not clear. It may be that they can become established only in very small populations of the kind that intense inbreeding creates.

Isolated populations can become new species simply by a change in the epigamic signals that serve as isolating mechanisms. Slight alterations of epigamic signals are most likely to cause speciation in those groups in which successful mating depends on an intricate complex of stimuli, in which any change can break off the delicately poised *rapprochement* of the sexes.

In other cases speciation depends not on slight changes in chromosomes or epigamic signals, but on genic incompatibility, changes in patterns of development that cause infertility or inviability of interspecific hybrids. It is to these genetic changes — alterations of epistatically interacting complexes of alleles that constitute postmating barriers to gene flow — that Mayr's theory of genetic revolution may most apply. It is unclear what kinds of species are the best candidates for such far-reaching changes. Wide-ranging, ecologically versatile species might be good candidates, since they might experience high levels of gene flow under their usual conditions of life; it is in such species, with gene pools adapted to the incursion of foreign genes, that an unprecedented cessation of gene flow might most drastically change the genetic environment and thus cause a genetic revolution. But, conversely, it is possible that the genotype is most subject to stabilizing selection in ecologically specialized species, adapted to an environment that is rather constant in time, subject to little gene flow, and thereby capable of precise adaptation to homogeneous local conditions. In such species, perhaps, the most intricate and delicately coadapted gene complexes develop, which when broken down by selection are reconstituted in new forms incompatible with the old. In such speculative realms, perhaps, ecology, population genetics, and developmental biology may find a synthesis.

THE SIGNIFICANCE OF SPECIES AND SPECIATION

What are the consequences of speciation? Does it enhance adaptation or affect the rate or direction of evolution in the long run?

In most instances, I suspect, the speciation event itself is not adaptive. The acquisition of reproductive isolation is often a by-product, not adaptive in itself, of other changes that may or may not be adaptive. A new species of gopher arises if chromosome translocations arise by genetic drift; a new species of tephritid fly is formed, perhaps, when a few flies become restricted by a behavioral mutation to a new host plant. In neither instance are the "mutated" individuals necessarily fitter; speciation is not a route to improved adaptation. When speciation is the consequence of the development of a new coadapted genome, the process of renewed coadaptation is one of adaptation (climbing to a new adaptive peak), but it is perhaps most often triggered by a nonadaptive event, the initial breakdown of coadaptation (being moved from an adaptive peak to a new adaptive slope).

Nevertheless the acquisition of reproductive isolation may enhance the ability of a population to attain high fitness, because it restricts recombination. Asexual forms, as Hutchinson (1968) has noted, are frequently organized into discrete units that can be defined by either morphological or ecological criteria as "asexual species" (*Figure 19*); morphology reflects adaptation to different ecological conditions. In asexual forms each of many phenotypes (genotypes) can form a persistent array, adapted to an array of resources. But in sexually reproducing forms, such an array of genotypes cannot be formed, nor can it persist, because the genotype appropriate to a particular resource is destroyed by recombination. Exceptional resources that require exceptional genotypes are therefore out of bounds, and the population is more likely to consist of generalized genotypes, each only moderately adapted to a variety of resources, than of a spectrum of specialized genotypes, each finely adapted to its special resources (Roughgarden 1972). In the largely parthenogenetic fall cankerworm, for example, different clones are adapted to different species of host plants (Mitter et al. 1979), but no such associations between genotype and host plant are evident in related sexual species (Mitter and Futuyma 1979). In sexual forms, therefore, the assembly and persistence of genotypes that can specialize in the use of particular resources or environments, and by specialization attain higher fitness, depend on the development of isolating mechanisms to hold intact those special combinations of genes. Only the reduction of recombination allows ecological and morphological diversity to arise.

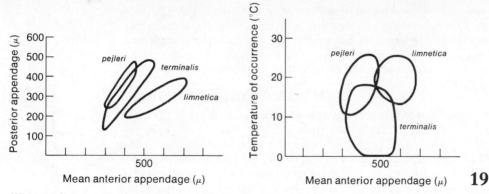

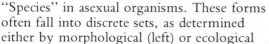

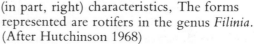

19

"Species" in asexual organisms. These forms often fall into discrete sets, as determined either by morphological (left) or ecological (in part, right) characteristics, The forms represented are rotifers in the genus *Filinia*. (After Hutchinson 1968)

As in space, so also in time. Novel and more "efficient" genotypes no doubt arise continually in large populations by recombination. And by recombination they are destroyed just as quickly, and the species has undergone little evolution of any real consequence. The wonderful gene combinations to which sex gives rise vanish in a moment, in a twinkling of an eye, unless sexual reproduction ceases or isolating mechanisms preserve them as a new species. Thus, as I noted in Chapter 7, the time-honored diagrams of evolutionary change are probably wrong. Individual species may not gradually evolve to greater heights by progressive change; the true picture of evolution may well be one of ponderous stasis, punctuated by bursts of change when new species come into being. The limits to the rate of evolution, then, are both genetic, in that genetic homeostasis and recombination militate against a response to environmental change, and ecological, in that little such environmental change is experienced except by local populations that are freed from gene exchange. It is in the process of speciation that both these limits are transcended.

SUMMARY

Speciation, the multiplication of species, entails the evolution of premating or postmating (or both) barriers to gene exchange among populations. Probably in most cases this occurs either by polyploidy or by the physical isolation of populations that then experience independent genetic events. The acquisition of barriers to interbreeding is probably most rapid in very small isolated populations, especially if they stem from a few individuals. Evolutionary change can be quite rapid in such isolates, while it remains slow in the parent species from which the isolates are derived. The genetic change required to yield a new species is likely to be very little in some cases, especially if reproductive isolation is contingent only on changes of chromosomal structure or on slight alterations of sexual behavior; but in some instances, more profound changes in developmental pathways, perhaps requiring allele substitutions at a great many loci, may be involved. In these instances the process of speciation may be slow, but in quite a few cases speciation has been accomplished in a few thousand generations, or even far less. It is during the process of speciation that shifts in ecological role are often accomplished, especially when novel ecological opportunities are available. It is during the speciation process, perhaps, that most evolutionary changes occur.

FOR DISCUSSION AND THOUGHT

1   This chapter has dealt for the most part with models of speciation favored by zoologists; one cannot appreciate from this discussion how very important hybridization and polyploidy have been in the origin of plant species. Read Chapters 15 and 20 of Grant's (1971) *Plant Speciation* for balance, and then consider the following questions. Why is it that animal species are so often genetically incompatible, while the incompatibility of related plant species is, seemingly, so often chromosomal rather than allelic? What relevance does an understanding of developmental biology have to this question? Why is polyploidy so much more common in plants than in animals? Is the higher classification of plants likely to be affected by hybridization among related species?

2   It has sometimes been proposed that diploid hybrids formed between two incipient species could themselves constitute a new, third, species. Examine possible cases of speciation by hybridization (e.g., Rao and DeBach 1969, Johnston 1969, Straw 1955, Smith and Daly 1959, Ross 1958) and evaluate the likelihood of this event.

3   Discuss the likelihood that rapid evolution can occur in hybrid populations formed when two incipient species come into contact (Anderson and Stebbins 1954, Lewontin and Birch 1966). Could this event have any long-term evolutionary consequences?

4   Hardisty (1963) has suggested that speciation in lampreys has occurred by paedomorphosis: reproduction by morphologically juvenile, almost larval, individuals. Could this happen sympatrically?

5   Hybridization between the diploid species of salamander *Ambystoma laterale* and *A. jeffersonianum* has produced populations of triploid parthenogenetic females where the ranges of the diploid species meet (Uzzell 1964). The association of parthenogenesis with hybridity is rather a common pattern in vertebrates. Why should this be?

6   Misra and Short (1967) deduced from morphological analysis of the hybrid zone between the western and eastern orioles, *Icterus bullockii* and *I. galbula,* that there has been more introgression of *bullockii* genes into *galbula* than *vice versa*. Propose a model of gene flow and/or selection that could account for such a pattern.

7   Among African cichlids, there are far fewer species of *Tilapia,* which are r-selected, ecologically generalized fishes, than of such forms as *Haplochromis,* which are K-selected specialists (Fryer and Iles 1972). How could one determine whether the disparity in species richness is attributable more to competitive limits on the numbers of generalists vs. specialists that can coexist or to differences in rates of speciation of generalized vs. specialized forms?

8   Ashburner (1969) found that the developmental timing of chromosome puffs, which are the sites of active RNA synthesis, is no more different between the species *Drosophila melanogaster* and *D. simulans* than it is among the different strains of *D. melanogaster*. What bearing has this observation on the theory of genetic revolution in speciation?

9   Axelrod (1967) has speculated that much of the diversity of the angio-

sperms is the product of quantum evolution that occurred during the Cretaceous by catastrophic speciation in increasingly arid environments. What evidence could support or refute this hypothesis?

10 Nei (1971) has proposed an equation for estimating the time since divergence of two species, based on the number of codon differences inferred from electrophoretic differences in their proteins. The formula assumes constant rates of allele substitution that have been equal in the two species. On this basis he estimates that about 500,000 years have elapsed since the formation of sibling species of *Drosophila* and about three times as many since the formation of nonsibling species. How likely are these estimates to be correct?

11 Cave populations of certain springtails (Collembola) have independently evolved such similar morphological features that Christiansen and Culver (1968) proposed that the meaningful unit of evolution is not the biological species, but the morphologically defined species. Discuss the pros and cons of this argument.

12 Discuss in detail the mechanism by which ethological isolating mechanisms might evolve, given that male display characters and female responses must both be genetically controlled. See Hoy and Paul (1973) for a genetic analysis of song specificity in crickets.

13 Alexander and Bigelow (1960) proposed that two species of crickets had arisen by allochronic speciation without geographic isolation; as the climate cooled, a species that had bred year round became split into two moieties that passed the winter in different developmental stages and thus reached reproductive age at different times of the year. *Systematic Zoology* (12:202–206, 1963) carries an argument between Alexander and Mayr, who doubts that speciation occurred in this way. Evaluate their arguments.

14 A model of sympatric speciation in parasites proposes that isolation arises by a single mutation and that the homozygotes for this mutation mate and lay eggs on a novel host. How would this model be affected if host selection were polygenically controlled?

15 Mayr argues that gene flow prevents significant genetic differentiation of any but small, strongly isolated populations, while Ehrlich and Raven argue that gene flow in most species is not great enough to cause genetic uniformity over long distances. Reconcile these arguments.

16 The existence of allopatric and parapatric populations in various stages of divergence is taken as evidence that speciation occurs by geographic isolation. We also know of instances of sympatric forms that differ from one another in varying degree, ranging from simple color dimorphisms to sympatric semispecies that hybridize to a limited extent. Why are these cases not accepted as evidence of sympatric speciation?

17 Mayr's founder effect is important to speciation insofar as epistatic interactions among loci strongly affect fitness. Yet in most statistical analyses of genetic variance, the epistatic component of genetic variance, even for fitness-related characteristics, does not appear very great; the additive and dominance components are preponderant. Does this apparent conflict imply that Mayr is wrong, that the analysis of genetic variance is invalid or irrelevant, both, or neither?

18 Provide an alternative explanation of Dobzhansky and Pavlovsky's experiment that purported to show the importance of genetic background for the operation of the founder effect.

19 Analyze the morphological and ecological diversity of a taxonomic group

to determine how frequently speciation entails major ecological shifts and how often newly formed species have only slightly altered ecological roles. Does speciation lead to adaptive breakthroughs, to quantum evolution, with any great frequency? The contrasts between speciation on islands and on continents may be informative. Is the potential for quantum evolution limited more by genetic or by ecological factors?

20 In most groups of herbivorous insects, there are a few species that are highly polyphagous, feeding on a great many kinds of plants, and a great many species that have much more specialized diets. What hypotheses can account for this pattern?

21 Would you expect speciation rates in the recent past (in the Pleistocene, for instance) in phylogenetically "old" groups such as ferns or amphibians to have been different from the rates in "young" groups such as composites or mammals? Explain, on genetic grounds, why there might or might not be a difference. What evidence bears on this question?

MAJOR REFERENCES

Mayr, E. 1963. *Animal species and evolution.* Harvard University Press, Cambridge, Mass. 797 pages. The major work on speciation in animals, arguing strongly for the role of geographic isolation.

Grant, V. 1971. *Plant speciation.* Columbia University Press, New York. 435 pages. Summarizes mechanisms of speciation in plants.

Bush, G. L. 1975. Modes of animal speciation. *Ann. Rev. Ecol. Syst.* 6: 339–364. A more pluralistic view of animal speciation than Mayr's.

Avise, J. 1976. Genetic differentiation during speciation. In *Molecular evolution,* edited by F. J. Ayala. Sinauer, Sunderland, Mass; pp. 106–122. Summarizes recent evidence on electrophoretic differences among species.

Microevolution and Macroevolution

I have now surveyed the mechanisms of microevolutionary change, those that effect genetic change within populations and species and are responsible for the origin of new species. I will now attempt, in a brief recapitulation, to determine whether the mechanisms of microevolution are sufficient explanations of macroevolutionary change. That is, I ask whether our current understanding of evolutionary principles is an adequate explanation for the diversity of the organic world and its variation in time and space. These are the major questions to be addressed: Can all features of species, adaptive and nonadaptive, be rationalized? What accounts for the origin of novel characteristics? Can we explain observed rates of evolution? Why are some traits more common among species than are others? I contend that evolutionary thought, based on both theory and empirical observation, is sufficient to provide reasonably satisfying answers to all these questions; if anything, the theory is too good. It is so versatile that it is almost impossible to imagine, let alone point to, a feature of the organic world that this theoretical structure cannot explain.

ADAPTATIONS

In any discussion of adaptation we must bear in mind a problem that I shall not attempt to resolve: how can we decide whether a feature is an adaptation? *An adaptation* presumably means a characteristic "designed" to perform a function and without which an organism would be less fit than those that possess the characteristic. But determining the function of a feature can be quite difficult, and it may well have multiple functions. Most important, as Williams (1966) points out, some features are not adaptations, but necessary consequences of the laws of physics and chemistry. The example he offers is a flying fish, which returns to the water after sailing through the air. Survival requires return to the water; but the act of returning is not an adaptation — it is a consequence of

433

gravity. Similarly, he argues, although the process of mutation seems advantageous to a species, it happens simply because DNA molecules cannot replicate themselves perfectly. We must seek an adaptive explanation, Williams urges, only when the more parsimonious hypothesis of nonadaptation is inadequate. Thus the deformation of bones that is symptomatic of rickets is to be viewed not as an adaptation to vitamin deficiency, but as a necessary pathological consequence. But the form and structure of a normal bone do not follow inevitably from the laws of chemistry and physics; they correspond to a design we envision appropriate for a structure that serves the functions of support and leverage, and they must be viewed as adaptations.

Other features, such as the color of autumn leaves in the temperate zone, are not adaptive in themselves; they are pleiotropic (*sensu lato*) manifestations of genes that have other, adaptive effects. Still other features may well be neutral, having no function whatever. The most extreme, still hypothetical, example is that of the highly repetitive short sequences of DNA that may never be transcribed into RNA. These may have a function (see below), but I would not be surprised if they did not. A segment of DNA with no effect on the phenotype replicates itself along with the rest of the genome, and its frequency changes purely by genetic drift. As more information on the rather small effective sizes of many natural populations comes to light, population biologists are becoming more and more inclined to believe that many of the variations among individuals and among species are truly neutral, having evolved by genetic drift. Most of the distinctive characteristics of higher taxa are surely adaptations that evolved by natural selection; but there may well be no adaptive significance to the differences in scale numbers among related species of snakes, in the morphology of insect genitalia, or in the precise shapes of leaves — in short, to the subtle differences by which we often distinguish related species. Evolutionary theory accounts for both adaptive and nonadaptive (neutral) differences among species, but it cannot tell us what proportion of the differences are adaptive. This question requires empirical investigation.

OPTIMALITY AND SUBOPTIMALITY

Biological thought is so permeated by the recognition that many, perhaps most, features have functions that we often forget that organisms are not perfect. In many ways they are suboptimally constructed compared with the ideal forms that an engineer might design. Such ideals, the modern versions of Plato's forms ($\epsilon\iota\delta o\iota$), underlie much of the current application of optimization theory to biology (Cody 1974b). This form of theorizing consists of asking what the "best" body size, foraging behavior, reproductive rate and so forth would be for an organism living in a particular environment. When a reasonable correspondence between predicted and observed characteristics is found, it is assumed that the theory is "right," that the selective pressures responsible for the evolution of the feature have been identified.

Although optimization theory in some instances successfully rationalizes or predicts patterns of adaptation, I think it important to recognize that organisms do not live in the best of all possible worlds, the world of Platonic ideas, but in the mundane world of materials and history. It is very hard to prove to an "idealist" that a feature is suboptimal, of course, because he or she can always argue that the feature is the best that the organism could have evolved under constraints that we do not know about. But there is nonetheless every reason to suppose that organisms are imperfect. Natural selection does not shape the best possible organisms; it merely replaces existing genotypes with whatever *better* (not best) ones arise.

Not all individuals within a population have equal fitness; some are suboptimal genotypes compared to others. Homozygotes for sickle-cell hemoglobin are a sufficient example. A population as a whole can be suboptimally adapted because of genetic properties that frustrate adaptation. Recombination breaks down ideal coadapted genotypes; gene flow from neighboring populations can prevent precise adaptation to special local environments. Sexual reproduction in general results in a disparity between the mean fitness of a population and that of the most fit of the genotypes that can be assembled.

When loci interact epistatically (form coadapted combinations) to affect fitness, the adaptive landscape envisioned by Sewall Wright can have multiple adaptive peaks. These peaks can vary in height (fitness), yet in general a population, having attained one peak, cannot move through an adaptive valley of low fitness to climb another, higher peak. Thus phylogenetic history determines subsequent evolution and may result in the development of a suboptimal constitution compared with others that are possible. Thus every species has inherited from its ancestors a form that may have been better designed for its ancestor's way of life than for that of the modern species. I have earlier cited the intersection of the respiratory and digestive tracts of tetrapods as an example. We might well profit from the ability to synthesize vitamin C, but our fruit-eating ancestors lost this ability, presumably because it was bountifully supplied in their diet.

Unnecessary features are lost either adaptively (because they are encumbrances) or by chance (by the relaxation of selection for their maintenance); but their loss is not instantaneous. Hence most species have vestigial characteristics that may serve no function. This is only one instance of the general proposition that organisms may not be up to date in all respects, especially if their way of life or the environment has recently changed. Most species are susceptible to the variety of toxins that modern industrial society has released into the environment. Less obvious examples are probably cases like the water ouzels (*Figure 1*) and certain parasitic Hymenoptera (Proctotrupidae), which structurally are like their terrestrial relatives and show little morphological evidence of their aquatic habits (see Darwin's *Origin,* chapter 6). Compensatory changes of structural components may lag behind key structural changes; for example, the jaw has become shortened in human evolution, but tooth development seems not to have evolved yet to an

1

The American dipper, or water ouzel (*Cinclus mexicanus*), which shows few special adaptations for its habit of foraging underwater in mountain streams. (Redrawn from Peterson 1961)

ideal state, as the high incidence of malocclusion, especially of wisdom teeth, suggests.

It is quite possible (although unknowable) that the "best" possible genotypes have simply never arisen by mutation and recombination. Some moths have evolved industrial melanism while others in the same habitats have not; similarly relatively few plant species have become adapted to toxic mine tailings or to their more "natural" equivalents, such as serpentine-based soils. On a longer time scale it is striking that some plants such as club mosses and ferns are attacked by far fewer insect species than other plants. It is likely that mutations conferring highly effective defense have arisen in some species but not in others.

The constraints posed by conflicting selection pressures render most species far less well adapted than they could conceivably be, although perhaps they are as well adapted as is possible. Cyanide production in *Lotus corniculatus* (*Chapter 9*) is an advantageous defense against herbivory, but in cold regions the advantage is overridden by self-poisoning when cell membranes are disrupted by frost. It may be impossible for *Lotus,* given the range of possible genotypes, to evolve a frost-resistant system of cyanogenesis. As in many other insect species, the larvae of the fall cankerworm (*Alsophila pometaria*) and of the elm spanworm (*Ennomos subsignarius*) disperse by wind on silken threads that serve as parachutes. This mechanism results in high mortality, for the larvae are unable to direct their movements to appropriate host plants; but the advantage of dispersal must outweigh the probability of mortality. Nevertheless, this does not bespeak perfect adaptation.

A genotype may have several pleiotropic effects, some advantageous and others disadvantageous; the net adaptive value of the genotype determines its fate. The *t* alleles in house mice are advantageous enough at the gametic level of selection (meiotic drive) to overcome a zygotic disadvantage (sterility or lethality) that can cause population extinction (*Chapter 12*).

Suboptimality arises, finally, when two or more species evolve at cross-purposes, as in the conflict of predators (or parasites) and their prey (or hosts). The classic example is that of plant galls (Mayr 1963, p. 196). Why should a plant make such a nice home for its enemy? A plausible answer is, Because selection in favor of the insect's ability to induce the formation of a gall on which its life depends is far stronger than selection in favor of the plant's ability to resist attack (as long as other factors keep the insect population at a level that is not devastating to the plant population.)

These, then, are reasons that organisms fall short of perfection. What problems are posed by characteristics that seem to be adaptations more or less consistent with our idea of excellence in form and function?

GROUP SELECTION AND KIN SELECTION

One problem is posed by what Williams (1966) has called biotic adaptations — characteristics that seem to benefit a whole population or species, but not the individuals that bear them. In principle such features could evolve by group selection (*Chapter 12*), which requires populations made up of individually disadvantageous genotypes to be proliferated rapidly or become extinct slowly compared with the rate at which individually advantageous genotypes replace individually disadvantageous genotypes within populations (see summary in Wilson 1975*a,* chapter 5). Group selection, in the form of spawning new populations by dispersal, may partly account for the evolution of dispersal, despite its disadvantage at the level of the individual (Van Valen 1971). It might also cause the evolution of lower predation efficiency (Gilpin 1975). Group selection in the form of species selection must certainly operate; a community contains only predator species that are not so efficient that they eliminate their prey and thereby themselves. Many authors, however, deny the importance of group selection and find kin selection (*Chapter 12*) a satisfying explanation of most group-level adaptations, including sterility in the social insects and the gamut of animal social behaviors.

FEATURES OF EXTREME PERFECTION

Other adaptations that have elicited skepticism of evolutionary theory are features of extreme perfection. Some adaptations astound us by the precision with which their form serves their function: the sensitivity of sound frequency discrimination provided by the structure of the human cochlea; the elaborate resemblance of some tropical grasshoppers to dead leaves, complete with "holes" and "fungal spots;" the elaborate design of flowers in the Orchidaceae, Aristolochiaceae, and other families; the fungi that parasitize ants and influence them to clench the top of a tall plant in their mandibles before dying, where-upon the fungi produce spores that are wafted away in the wind.

Can it really be more advantageous to a grasshopper to look like a dead leaf with fungal spots than like a dead leaf without spots? Yes; differences in fitness as slight as, say, 10^{-4} — survival of 9,999 of one genotype for every 10,000 of the other, a difference we could hardly hope to measure — can eventually fix the favorable allele on average, but only if the allele is consistently favored for a very long time. If selection is that weak, genetic drift will override selection in many species and populations, and the precision of adaptation will not evolve. But in some instances it will; and given the immense number of such imperceptibly favorable mutations that have arisen in each of the enormous number of populations and species that have existed over the course of more than 3 billion years, we will not be surprised to find quite a few instances of form most exquisitely fitted to function.

Once a structure or biochemical pathway has come into being in crude and imperfect form, it may subsequently be refined. But whence does the basic form arise?

EVOLUTIONARY NOVELTIES

Natural selection, acting in various guises at various levels, seems together with genetic drift to account for almost all features of organisms once the appropriate raw material has arisen by mutation and recombination. The real problems posed by evolution, it seems to me, lie not so much in the potency of natural selection as in the mechanisms by which the variations on which it acts arise. The entire theory of population genetics assumes variation as a given, by processes of mutation, recombination, and gene expression described in purely statistical terms. The problem of how new variations arise falls not within the province of mathematical genetical theory, but within that of molecular genetics and development biology.

Our ignorance of how genotypes produce phenotypes is, I believe, the greatest gap in our understanding of the evolutionary process — and it is a huge gap indeed. Its magnitude can be illustrated by supposing that we could alter an organism's DNA at will into any desired sequences. This ability (as opposed to selecting and recombining preexisting variation, as in artificial selection) would be genetic engineering in its purest form, whereby we could direct the course of evolution. But we could predict the phenotypic results of any such alteration only if we knew how the mutations that we induced became translated into phenotypic form. We know nothing about that; even knowing the location of genes that affect bristle development in *Drosophila melanogaster,* we have no idea what we would have to do at the molecular level to alter the number, position, or form of its bristles, much less transform *D. melanogaster* into *D. simulans,* or a fly into a flea.

That mutation and recombination give rise to variation in biochemical characters, allometric relations, growth rates, structural patterns, and behavior we know only as empirical fact. How alterations of developmental pathways might produce novel phenotypes is discussed in Chapter 8. Here I treat some largely speculative hypotheses of the molecular basis of evolutionary change. We are concerned with understanding at the genetic level the evolution of the amount of DNA and its information content, and at a phenotypic level, the translation of this information into phenotypes.

The amount of DNA per haploid complement varies, just among eukaryotes, over about three orders of magnitude — from about 0.09 picograms in fungi to more than 100 pg in some salamanders and ferns (Hinegardner 1976). Hinegardner finds a very uneven progression in DNA content from the "lower" invertebrates like sponges to "higher" invertebrates such as arthropods. But within many groups the more primitive forms often have more DNA than the more derived forms that can be considered more specialized; for example, the more bizarre fishes like flatfishes, puffers, and sea horses tend to have less DNA than the more "ordinary" fishes like herrings and minnows. Hinegardner suggests that specialized species may not need much of the genetic information that more generalized species require to cope with the greater variety of exigencies they experience. Thus organisms could contain much repressed, functionless DNA before it ultimately is lost.

Why the DNA content should increase in evolution, in some cases to astounding levels, is a greater problem. Salamanders have more DNA than any other animal (100 pg), enough for more than 16 million genes if each gene had 1200 nucleotide pairs (coded for a polypeptide of 400 amino acids). An "average" animal with about 3 pg has enough DNA for 5 million such genes. But estimates of the number of different structural genes in eukaryotes (based on numbers of known mutable loci and on the number of kinds of protein) range from about 10,000 to 400,000 at the very most. Even if much of the DNA does not code for proteins, but serves to separate structural genes and to regulate their transcription, most organisms seem to need no more than 25 per cent of their DNA.

Some of the mechanisms by which the genome can increase in size and the adaptive reasons for such an increase are clear. Polyploidy, especially prevalent in plants, multiplies genome size by a factor of two or more. Allopolyploids are often more vigorous than their diploid ancestors, possibly because of the heterotic effect that the heterozygosity of a hybrid genome confers. Such an effect also renders advantageous piecemeal increases in genome size, by gene duplication resulting from unequal crossing-over in heterotic heterozygotes. The prevalence of paralogous loci that code for isoenzymes bespeaks the high frequency of such events.

Nonetheless there is an enormous amount of redundancy in the genome, and its significance is obscure. In many animals as much as 60 per cent of the genome seems to consist of short (less than 300 nucleotide pairs) repeated sequences, some present in thousands of, even a million, copies. The African clawed frog *Xenopus* has at least 800 kinds, or "families," of repeated short sequence. The interspersion of this repetitive DNA among the structural genes leads Britten and Davidson (1971) to believe that it has some function (see also Britten and Kohne 1968, Davidson and Britten 1973, Galau et al. 1976).

From these observations Britten and Davidson (1971) have constructed a model of gene regulation (*Figure 2*); although totally speculative and not yet tested, this model has interesting implications for evolution. They note that if transcription at a structural gene locus is controlled by a regulatory locus in a manner analogous to the operon systems in bacteria, then each regulator gene must control more than one other locus or else there will be an infinite regress of one gene regulated by another, regulated by still another. Each of the loci controlled by one regulator must have a recognition site with which the regulatory gene's product interacts. They suggest that this product is

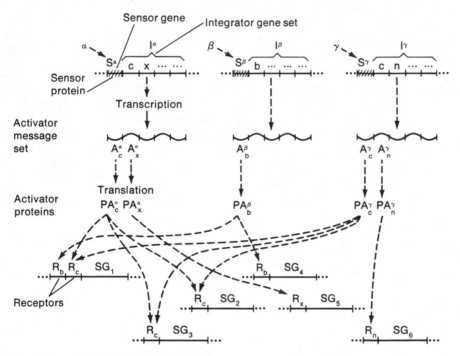

2 The Britten–Davidson model of gene regulation. External signals (e.g., molecules) α, β, γ interact with sensors S^α, S^β, S^γ, which thereby trigger transcription of associated integrator genes I^α, I^β, I^γ into messenger RNA sequences A_c^α, etc., which are then translated into activator proteins PA_c, PA_x, etc. Each sensor S is associated with several integrator genes; hence several activator proteins result from the action of each sensor. Activator proteins bind sequentially to the receptors R_b, R_c, etc.; when all receptors associated with a structural gene (SG) are bound by appropriate activator proteins, transcription of the structural gene (or bank of structural genes forming a polycistronic unit) occurs. The various structural genes associated with a given type of receptor (e.g., SG_1, SG_2, and SG_3, with receptor type R_c) are termed a battery. Integration of developmental patterns follows in part from the response of many receptors to any one activator protein, and hence to one or several integrator genes. (After Davidson and Britten 1973)

a protein or RNA sequence that derepresses the structural genes, which are otherwise repressed, perhaps by histones. The several regulated loci (a battery) must have similar or identical recognition sites, since they interact with the same derepressor. If these sites are nucleotide sequences, each such sequence must be repeated at several or many places in the genome, and most if not all unique DNA sequences (e.g., cistrons coding for enzymes) must be associated with one of a great many such repetitive sequences. Activation of a particular regulatory gene (or integrator gene, as Britten and Davidson term it) in a given cell at a given time in development would then initiate transcription at each of the several loci it regulates. The products of these loci are probably functionally related to each other. This model, then, provides one explanation for the prevalence of pleiotropy.

Britten and Davidson postulate that the integrator genes themselves are organized into contiguous sets, which are activated in concert by an interaction between inducing substances (such as hormones) and a sensor that initiates transcription of the integrator genes. If this is the case, inversions, translocations, and other chromosomal mutations may occasionally revolutionize development by inserting one or more integrator loci into another bank of integrators, controlled by a different sensor (*Figure 2*). As a hypothetical example, they imagine integrator set A controlling the development of blood cells that lack an oxygen-binding globin, and integrator set B controlling the ontogeny of the cuticle, including a heme-bearing globin. Translocation of integrators from B into set A could result in the production of the globin in the blood cells. This model, then, attributes an evolutionary significance to chromosome rearrangements greater than that envisioned by most evolutionists (see Mayr 1963, chapter 17; White 1973), to whom the primary evolutionary consequences of chromosomal change are the stabilization of coadapted combinations of alleles (e.g., by inversion polymorphism) and the imposition of meiotic barriers to gene exchange among populations and species. Wallace and Kass (1974) and Wallace (1975) have suggested ways in which the Britten-Davidson model could explain the prevalence of heterozygous superiority and thus the maintenance of variation in populations.

These models are exceedingly speculative at present; indeed, chromosome rearrangements very seldom have discernible phenotypic effects. But there is every reason to believe that whatever the mechanisms of gene regulation, alterations of these mechanisms are the most significant source of important hereditary novelties. Changes in the amino acid sequences of structural proteins and enzymes are probably the most trivial mutational events, for organisms as disparate as bacteria and mammals are surprisingly similar at the enzymatic level. Moreover organisms can diverge greatly at the level of the DNA sequence yet not at the phenotypic level. For example, two species of *Xenopus,* although differing at probably 10–15 per cent of the base pairs in the nonrepetitive portion of their DNA, are phenotypically very similar and form viable developmentally normal F_1 hybrids (Galau et al. 1976).

Conversely species such as the human and the chimpanzee, despite their pronounced phenotypic dissimilarity, are hardly distinguishable at all at the protein or DNA level (King and Wilson 1975). It seems possible that many of the mutations we know from their phenotypic effects — those that alter the proportions and sizes of structures, their fusion with or distinction from other structures, their pattern of spatial arrangement — may well entail modification in the time and site of gene action and very probably, therefore, modification of the mechanisms of gene regulation.

SPECIATION AND RATES OF EVOLUTION

What determines the rate of evolution over long periods of time? The traditional answer focuses on the relation between the amount of genetic variation and the rate of change in the environment. If the environment changes so much that the optimal genotype lies outside the range of genetic variation harbored by a population, the rate of evolution depends on how quickly new genetic variation arises by mutation. This rate is contingent (Maynard Smith 1976a) on the number of mutable sites, the mutation rate per site, and the population size; more mutations happen in larger than in smaller populations. This model could describe the rate of evolution of a character that is always suboptimal, for example the profile of defensive compounds in a plant that never achieves complete immunity to insect attack. But many characteristics seem to be so genetically variable that the rate of evolution could be limited not by variation, but by the rate of environmental change. Populations could easily "track" slow changes, such as climatic shifts.

These traditional views may be largely inapplicable to long-term evolution, however, if most evolutionary change is concentrated in speciation events (*Chapters 7, 16*). If this is so — a matter of debate — the rate of phenotypic change in a taxon over a long time depends on the amount of divergence during speciation and on the rate of turnover of species, which depends on the rates of speciation and extinction. Extinction rates appear to be stochastically constant for each major group (*Chapter 5*); speciation rates may also be stochastically constant, depending on taxon-specific patterns of mating behavior, developmental pathways, and population structure — in short, on the factors that affect reproductive isolation (*Chapter 16*).

The major question, then, is what determines the degree of phenotypic divergence during speciation. The theory of population genetics does not provide a satisfying answer to this question, since we know little of the kinds of genetic change that occur during speciation. Paleontological data, however, suggest that the rate of phenotypic change is greatest soon after a group has invaded a new adaptive zone; the average degree of divergence among species declines thereafter (*Chapter 7*). This suggests that the amount of phenotypic divergence is limited by the availability of unexploited ecological niches; as species evolve to use the major resources of an adaptive zone, the species that arise thereafter partition resources in ever more subtle ways and differ little from their progenitors in morphology. After the ecological space has become saturated with species, species continue to arise; but speciation rates are balanced by extinction rates, and the new species differ little in morphology from those they replace. The net rate of phenotypic evolution, in this view, is limited by ecological factors.

If this view is correct, the apparent differences in evolutionary rates among taxa may be due to the relative times at which the groups arose rather than to intrinsic differences in developmental patterns, chromosomal repatterning, or other properties thought to affect phenotypic divergence. For example, frogs appear to have undergone less rapid evolutionary diversification in the recent past than have mammals (A. C. Wilson 1976; *Table I*). But the two groups may have had the same time course of evolutionary diversification (*Figure 3*), from generalized ancestors to specialized descendants. The process started earlier in amphibians; so by now frogs may have diversified as fully as they can, while mammals have not yet radiated fully into the variety of niches that constitute the mammalian adaptive zone. The time courses of change in the two groups begin, conceptually, with comparably generalized ancestors, the first to invade the two adaptive zones; but the generalized frog ancestor that is comparable to the first mammal may well have been a pre-Jurassic amphibian that we would not classify as a frog, because taxonomic categories above the species level are subjectively defined. Thus the artifacts of taxonomy can confound our analyses of evolutionary rates.

DISTRIBUTIONS OF PHENOTYPES

A final question is why phenotypic traits have the frequency distributions they have. Why are there more small than large species of animals, more insect-pollinated than wind-pollinated angiosperms, more host-specific than generalized herbivorous insects, more sexual than asexual species? A closely related problem is why some traits

TABLE I. Apparent rates of evolution in frogs and placental mammals

Property	Frogs	Placental mammals
Number of known living species	3050	4600
Number of orders	1	16–20
Age of group (million years)	150	75
Rate of morphological evolution	slow	fast
Rate of albumin evolution	standard	standard
Rate of loss of hybridization potential	slow	fast
Rate of change in chromosome number	slow	fast
Rate of change in number of chromosome arms	slow	fast

(After A. C. Wilson 1976)

exhibit geographic trends; why, for example, are more of the tropical than of temperate species of swallowtail butterflies highly specialized feeders as larvae (Scriber 1973)? We also want to know why there are phenotypic gaps; for example, why are morphological intermediates between carnivores and artiodactyls no longer with us?

The usual interpretation of such observations is that they reflect patterns of adaptation, and are indeed evidence of adaptive patterns: that it is most commonly good or adaptive to be small rather than large or sexual rather than asexual, that it is better to be specialized in the tropics than in the temperate zone, that it is more advantageous to be a relatively specialized carnivore or artiodactyl than a morphologically more generalized condylarth. But these are conclusions to be demonstrated rather than assumed, for many such distributions can arise through random processes.

In Chapter 5 I described how Raup et al., simulating evolution by a computer program that allowed a lineage to speciate or become extinct at random, produced clades of hypothetical organisms that increased and decreased in species diversity in ways that mimicked the changes observed in the fossil record. Raup and Gould (1974) carried this approach further by allowing the members of such a hypothetical lineage to evolve in each of 10 characters. They assumed, following Eldredge and Gould (1972), that phenotypic change takes place only in the daughter species formed by the isolation of a small population from the parental species. Starting from state 0, each character could

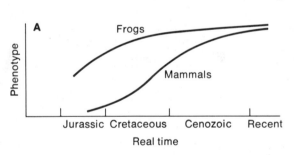

3 Diagram illustrating the problem of comparing evolutionary rates in very different groups. (A) During the real time of the Cretaceous and Cenozoic, mammals as a group seem to have diverged phenotypically from ancestral mammals more than frogs have from ancestral frogs. (B) But it is conceivable that both groups have the same curve representing divergence from an arbitrary ancestor of each group. If this process began earlier for frogs than for mammals, they will appear to have been evolving more slowly.

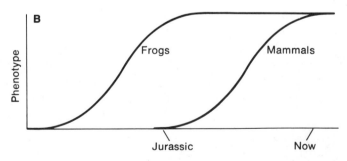

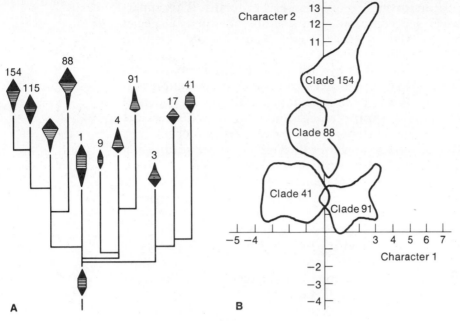

A

B

4

Some results of a computer simulation of random evolutionary changes in morphology. (A) The phylogeny and final form of each of 11 clades of imaginary animals ("triloboids"). The shape of each is based on computer-generated values of six characteristics. (B) Distribution of the species in each of the four clades that were extant at the end of the computer simulation, with respect to two characters. The four groups, although evolving randomly with respect to each character, are phenotypically distinct. Each clade consists of several species. For further explanation, see text. (From Raup and Gould 1974, in *Systematic Zoology*)

evolve during speciation by either −1 or +1 unit, equiprobably in either direction. Species could split or become extinct at random in each time unit, as in the earlier simulation. Note that each evolutionary change could be envisioned as being caused by either natural selection or genetic drift; what is critical is that the direction of evolution of each character was at all times random with respect to previous changes, with respect to the evolution of other characters in the same species, and with respect to evolution in other species.

Their results are quite unsettling (*Figure 4*). During the simulation 200 lineages (species) were formed, constituting 11 clades; only four clades were extant when the simulation was terminated. The individual species that formed each clade were, as expected, phenotypically more similar to each other than to members of other clades. The members of the four clades were phenetically distinct from each other; there were morphological gaps among them, just as there are among orders of mammals even though there was no consistent selection against intermediate ancestral forms, which simply became extinct at random. The most recently evolved species tended to have more extreme phenotypes (departing further from state 0) than earlier forms. This apparent evolution toward specialization occurs simply because a species whose ancestor had arrived at state n has a chance of only $(\frac{1}{2})^n$ of returning to state 0.

Many characteristics — most, in fact — were highly correlated with one another across species, an observation that in biological material is often taken to indicate genetic or functional interdependence. Yet this follows from the simple realization that if two ancestors randomly arrive at different extreme states for each of two characters, their descendants are also likely to differ in these characteristics. Raup and Gould found, moreover, that some characters evolved by chance much faster than others. Most important, although many characters fluctuated around 0 in time, a few evolved by chance in a unidirectional fashion, perhaps just as many as have been called trends by paleontologists. They conclude that random events can produce morphological evolution much like that described by real phylogenies, in particular that unidirectional evolutionary trends need not be caused by persistent directional selection. The only common natural phenomenon that their simulation did not yield to any significant extent was convergent evolution. These results should not be considered evidence that evolution is caused by drift rather than natural selection; selection can be responsible for each unit change, yet the forces of selection can vary at random and, together with random speciations and extinctions, still yield nonrandom patterns.

The effect of random processes rather than adaptation in yielding seemingly adaptive patterns can be even stronger if we imagine that the probability of speciation is coupled with a particular character state. Suppose, for example, that orchids with generalized flowers pollinated by a great variety of insects are as fully adapted as more florally specialized orchids that have one or a few pollinators. By equality of adaptation I mean that a florally generalized species has the same seed output, survivorship, and probability of extinction as a florally specialized species. By the mechanisms described in Chapter 16, the rate of speciation will be greater in the florally specialized forms. Thus there will be a preponderance of specialized species, attributable not to any general adaptive advantage of floral specialization, but to the rate at which stochastic events causing speciation (e.g., mutations of flower form) occur. If, moreover, the factors that influence speciation vary geographically, geographic patterns that have nothing to do with adaptation may emerge. If insect pollinators happen to be less vagile in the tropics than in the temperate zone, the disparity between the speciation rates of specialized and generalized orchids may be greater at low than at high latitudes. The consequences may be a geographic gradient in the proportion of specialized pollination systems that many would take to be evidence of a relative advantage of specialized pollination in the tropics. In a similar vein Stanley (1975b) has suggested that most species reproduce sexually not because sex is advantageous (e.g., because it lowers extinction rates), but because speciation happens faster in sexual than in asexual forms (in which speciation is indeed hard to define). I suggest that some of the conclusions about adaptive strategies in feeding habits, demographic characteristics, defense systems, and modes of reproduction that have been based on geographic patterns in the number of species employing one or another strategy need reexamination; either the functions of the characteristics must be determined exactly, or the number of convergent evolutionary origins of a characteristic must be measured.

THE STATUS OF EVOLUTIONARY BIOLOGY

Evolutionary biology, it should be apparent, is a highly dynamic field of inquiry. Within the last 15 years there have been two major challenges to the neo-Darwinian view that essentially all of evolution can be explained by the operation of natural selection on genetic variation within species. One challenge is at the microevolutionary level: the proposition that much of the newly revealed genetic variation within and among species is attributable to genetic drift rather than natural selection. The other, at the macroevolutionary level, proposes that the great events in evolution — the changes in diversity, the frequency distribution of traits, the extinction of major taxa, the seeming trends in evolution, the development of phenetic gaps between taxa — are attributable to rates of speciation and extinction rather than to consistent patterns of selection within species. The grand patterns of evolutionary history are thus viewed as consequences of group drift (or species drift) and group selection, or species selection, analogous to the drift and selection of alleles within populations. How valid these views are and what impact they will have on future evolutionary thought remains to be seen. And permeating evolutionary biology is the recognition of our continuing ignorance (although we too seldom dare admit it) of the mechanisms of development, which when fully understood may change our views of the origin of variation and the mechanisms of speciation in ways we can only dimly foresee. Although we have probably identified all the mechanisms of evolution, some of these mechanisms remain "black boxes" whose precise operation we do not understand; and the relative importance of the mechanisms we recognize will be subject to argument for a long time.

Although I wish to emphasize how much of evolution we see only as through a glass darkly, how many hallowed concepts are subjects for iconoclasm, and how fertile a field the subject is for further tilling, I do not want to leave the impression that evolutionary biology is floundering in a sea of uncertainty lest antievolutionists find fuel for their fires. Let me reiterate from Chapter 1: however much evolutionists may argue about evolutionary mechanisms, evolution is as much a fact as any principle in physics. I think there can be only two or three explanations for organic diversity: that it was created, either all at once or continually, by a sentient creator; that it stems from an immanent urge or *élan vital* toward change or perfection; that it arises from the action of selection (and to a much lesser extent chance) on randomly arisen mutations.

The *élan vital* theory has, I hope, been sufficiently discredited by this and other writings that it needs little argument. We can imagine no mechanism by which it might exist; more important, we have concrete evidence against its operation, in the randomness of mutations and the failure of populations to evolve toward adaptedness in the absence of natural selection. Besides, the *élan vital* theory has few proponents.

Creationists, on the other hand, abound and mount anti-intellectual campaigns that can have severe educational consequences. I am not sure that accepting evolution is necessary for the well-being and the humanitarian social and political performance of every member of our society; but I do think that every person's well-being and social contributions depend on his or her ability to reason clearly, to evaluate opposing arguments, and to be able to predict probable effects from known causes. One's evaluation of creationist vs. evolutionist views of the world is only one instance of the conflict between unreasoning faith and a sometimes less optimistic logic — a conflict exemplified by the difference of opinion between those who believe that God will provide for the starving millions so that political change or birth control is unnecessary, and those to whom such a nonmechanistic view of society has dire and inhumane consequences.

The conflict is between belief in mechanisms as a sufficient ground for understanding the world and belief in the inscrutable actions of a Creator who can change at will the laws of matter, motion, and mathematics that science claims to have discovered and to be discovering. If the laws of the universe can be changed at will, science is impotent and indeed nonexistent, for it can make no predictions that cannot be overturned by the whim of the Creator. No theoretical proposition can be falsified, for all possibilities, no matter how seemingly absurd, are open to the omnipotent. Thus creationism, like any theory of the universe that invokes the action of an omnipotent deity, is not a scientific theory of evolution, for it cannot be falsified in part or in whole. But it is not only an inadmissible theory; it is unnecessary, for the structure of evolutionary theory is both sufficient to explain biological phenomena and is a necessary consequence of first principles that we empirically know to be true; that variation of the most diverse kinds does indeed arise by mutation and recombination, that it is in large part inherited, and that variant genes, genotypes, and species vary in their rates of multiplication and demise.

The most serious accusation that can be leveled against evolutionary theory in fact is not that it is insufficient to explain biological phenomena, but that, like creationist theory, it can explain everything. Unlike creationism, however, the theory of evolution can in principle be falsified. The modern theory of evolution, based primarily on the natural selection of random mutations, would be false if Lamarck had been right, if inherited characteristics could be acquired. Indeed the cultural evolution of human society proceeds by this mechanism and differs strikingly from biological evolution in many ways (*Chapter 19*). And the components of the theory of biological evolution — heredity, mutation, recombination, nonrandom mating, genetic drift, natural selection — need not exist *a priori* but are known empirically to be real.

Creationists could most legitimately question not the validity of evolution as a process, but the sufficiency of the theory in explaining phenomena that do not follow necessarily from the most elementary principles. Mutation and natural selection do not in themselves predict that new species should arise by speciation or that such novelties as the turtle's shell or the human brain should evolve. But we can point to empirical evidence of speciation and to innumerable, well-documented cases of the evolution of complex organs. The resistance to evolutionary thought is grounded not in scientific argument, but in an emotional response to its implications for our view of the world and of ourselves. But I would argue, with Darwin, that "there is grandeur in this view of life," which holds that "whilst this planet has gone cycling on according to the fixed law of gravity, from so simple a beginning endless forms most beautiful and most wonderful have been, and are being evolved."

SUMMARY

I have tried in this chapter to summarize the argument, which has pervaded these many chapters, that the known mechanisms of evolution adequately account for all biological phenomena. I have stressed, however, that the most important of these mechanisms is not necessarily and universally the operation of natural selection within populations, but that some patterns emerge from the operation of the random drift of gene frequencies and from speciation and the extinction of species. The aspects of evolution of which we remain most ignorant are the relative importance of the various factors of evolution that we know exist and the details of molecular and developmental biology, which will ultimately explain how variation originates. The chapter ends with an affirmation of evolution as both a sufficient and a necessary explanation for the diversity of life.

FOR DISCUSSION AND THOUGHT

1 In what sense can structures be said to be "designed" to perform functions?

2 In what kinds of species, occupying what kinds of environments, might we expect to find the most nearly "perfect," precise adaptations?

3 Speculate on ways in which the Britten-Davidson model of gene regulation could be applied to the developmental phenomena discussed in Chapter 8.

4 Discuss the proposition that most phenotypic mutations are alterations of regulatory genes.

5 Explain the seeming paradox that although each species evolves toward greater adaptedness, extinction rates seem to have been stochastically constant throughout evolutionary time. Does evolution not increase the probability of survival?

6 Mayr (1963) and Wright (1967) have suggested that new species are to macroevolutionary change what mutations are to microevolutionary change within populations; they diverge from their ancestors in various directions, not just the direction that the group takes over the long term.

Differential extinction of these species, like selection of mutations within populations, sets whatever long-term trends might exist. See Gould and Eldredge (1977) for paleontological evidence on this hypothesis and discuss its implications.

7 Postulate ways in which mathematical population genetic models describing the genetic events during speciation might be developed. Has anyone modeled what happens when a population is cut off from a series of interconnected populations exposed to different environments?

8 No one has ever performed long-term experiments to see whether macroevolutionary changes can be produced in laboratory populations. How might such experiments best be carried out? What would they prove?

9 Debate the major arguments of creationism.

10 Discuss the practical and philosophical implications of world views that include acceptance of creation on the one hand and evolution on the other.

MAJOR REFERENCES

Mayr, E. 1963. *Animal species and evolution*. Harvard University Press, Cambridge, Mass. 797 pages. Chapter 19 treats the relation between microevolution and macroevolution.

Simpson, G. G. 1953. *The major features of evolution*. Columbia University Press, New York. 434 pages. Discusses the idea of quantum evolution.

Frazzetta, T. H. 1975. *Complex adaptations in evolving populations*. Sinauer, Sunderland, Mass. 267 pages. An attempt to relate microevolutionary processes to macroevolution.

Gould, S. J., and N. Eldredge. 1977. Punctuated equilibria: the tempo and mode of evolution reconsidered. *Paleobiology* 3:115–151. The argument for abandonment of phyletic gradualism.

POSTLUDE:
SPECIAL TOPICS
IN EVOLUTION

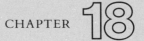

Coevolution

Among the most exciting recent developments in evolutionary biology has been the synthesis of evolutionary theory and ecology into the discipline of population biology. Population genetics and evolutionary theory have come to play an important role in ecological theory, for example in the theory of life history strategies (*Chapter 12*). Perhaps the most challenging task will be to go beyond the analysis of evolution in single species to the evolution of their interactions. Understanding species' evolutionary responses to their competitors, symbionts, enemies, and the species on which they feed is important to understanding rates of evolution and patterns of adaptive radiation. Moreover the evolutionary responses to interspecific interactions feed back on the interactions themselves; thus the species composition and structure of ecological communities can be affected by the processes of coevolution.

COEVOLUTION may be broadly defined (Roughgarden 1976) as evolution in which the fitness of each genotype depends on the population densities and genetic composition of the species itself and the species with which it interacts. In a narrower sense it is often visualized as a series of reciprocal evolutionary responses in each of two or more species, each evolutionary change having been activated by an evolutionary change in the other species. The purpose of this chapter is to point out questions that are only now being asked, to reveal our ignorance of problems that promise to be an exciting part of evolutionary biology in the near future.

SOME PROBLEMS IN COEVOLUTION

Suppose that each of two species varies genetically in traits that affect their interaction — for example, a prey species' vulnerability to a predator and a predator's ability to catch the prey. Will the gene frequencies affect their population sizes? Will evolution destabilize their interaction or promote coexistence? Will genetic change

453

continue indefinitely, or will it come to a halt? If genetic variation continues to arise by mutation, will the species continue to evolve in an evolutionary arms race, or will they come to an evolutionary equilibrium from which they do not change? Is evolution fast enough to save a species from the extinction that it might otherwise suffer from predation or competition? Is the species composition of a community actually affected by coevolution? Recognizing that each species is faced with many different species of predators and prey, does reciprocal coevolution between pairs of species actually occur, or is the evolutionary response to one species frustrated by incompatible responses required to deal with others? Can we predict the phylogenetic effects of coevolution — which groups of parasites, for example, will be associated with which groups of hosts?

There are two theoretical approaches to the ecology of interacting populations. Most of ecological theory is purely demographic and simply specifies what properties species must have to coexist. For example, two competing species cannot coexist indefinitely if they do not differ sufficiently in the resources they use. Under this theory the species composition of a community is determined by processes of immigration and extinction, a kind of species selection. A coevolutionary theory of community structure, on the other hand, allows individual selection within species to affect the species composition; if ecologically similar species evolve niche differences fast enough, the inferior competitor may escape extinction. It will be interesting to know the relative roles of individual selection and species selection in shaping the structures of communities.

SYMBIOSIS AND MUTUALISM

A symbiotic association of two species benefits at least one of them. If the other is harmed, the relation is one of predation or parasitism; if benefited, one of mutualism. Mutualistic interactions can be specialized one-on-one relationships, such as that between the moth (*Tegeticula*) whose larva feeds only on yucca fruits, and the yucca plant that depends on the moth for pollination; or the relation can be one-on-many or many-on-many, as in most pollination systems, in which neither plant nor pollinator depends exclusively on one species of mutualist. In mutualistic relationships each species uses the other as a resource; thus the degree of specialization is explained partly by the theory of the evolution of NICHE BREADTH, which I shall treat under the heading of competition. Often, however, it must be advantageous to one species to be the exclusive resource of its partners, but advantageous to each of the partners to be more generalized. A plant may avoid cross-pollination from other species if it has a specialized pollinator, but each pollinator species may find it advantageous to garner nectar and pollen from many plants. The outcome of such a conflict has not, to my knowledge, been analyzed.

Parasites and pathogens often have less harmful effects on hosts with which they have long been associated than on species that have not been exposed to them. For example, the Dutch elm disease is rather benign to European elms but is devastating the American elm population. Is this only because experienced hosts have evolved resistance, or do parasites themselves evolve to be benign? If so, can they, as many people have suggested, become so benign that they benefit their hosts?

In many interactions the exploiter cannot evolve to be avirulent; it profits a fox nothing to spare the hare. But if the fitness of an individual parasite or its offspring is lowered by the death of its host, avirulence is advantageous. The myxoma virus, introduced into Australia to control European rabbits, at first caused immense mortality. But within a few years mortality levels were lower, both because the rabbits had evolved resistance and because the virus had evolved to be less lethal (Fenner 1965). Because the virus is transmitted by mosquitoes that feed only on living rabbits, virulent virus genotypes are less likely to spread than benign genotypes. Avirulence evolves not to assure a stable future supply of hosts, but to benefit individual parasites.

A symbiont can evolve from a parasitic to a mutualistic state if it sacrifices some of the benefit it can derive from its host to further its host's and thereby its own survival (Roughgarden 1975). For example, the mutualistic mycorrhizal fungi associated with the roots of vascular plants do not extract from their hosts as much energy as do pathogenic fungi. As the symbiont comes to profit more by the association, an even greater sacrifice on behalf of its host becomes advantageous to itself, so the degree of mutualism can increase in self-reinforcing manner even without evolution on the part of the host. Of course, once the symbiont benefits the host, the host profits more by conferring greater benefit on its guest. Thus highly coevolved systems can arise, such as that between ants (*Pseudomyrmex*) and *Acacia* trees (Janzen 1966; *Figure 1*). If each species comes to depend exclusively on the other, however, rarity or extinction of one species can cause the demise of the other. For example, on Mauritius, no seeds of the endemic tree *Calvaria major* have germinated since the extinction of the dodo in 1681; germination requires passage through the gut of a large bird, and the dodo was the only such bird on the island (Temple 1977).

COMPETITION AMONG SPECIES

Two species may compete if both use at least some of the same resources. The competitive effect (α_{ij}) of one species j on the other i depends on the proportion of i's resource spectrum that j also uses (*Figure 2*) and on the interference between them when they meet at a resource; one species may be aggressive toward the other or poison it (allelopathy). Species i could lessen the impact of species j by an evolutionary shift of its resource utilization, by becoming resistant to j's interference, or by increasing its own capacity to interfere with j (by increasing α_{ji}). MacArthur and Levins (1967) and Cody (1973) have suggested that species may sometimes converge in response to competition. For example, some birds defend territories against both

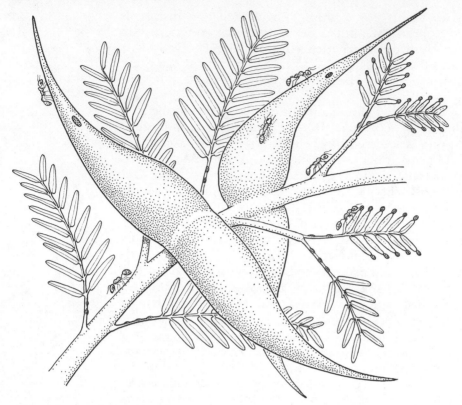

1

The mutualistic interaction between the tree *Acacia* and the ant *Pseudomyrmex*. The ants inhabit the thorns and feed at the nectaries on the petioles and on the proteinaceous Beltian bodies at the tips of the young leaflets. The tree is defended by the pugnacious ants against herbivores and competing vegetation.

conspecific individuals and those of other species and appear to have become similar to their competitors in the plumage patterns and vocalizations used in territorial display. Such convergence, as well as the evolution of increased competitive ability (interference) seems likely to destabilize the ecological interaction between the species, for the ultimately inferior competitor will become extinct.

Pimentel (1961*a*; see also Pimentel and Soans 1970), however, has suggested that continued evolution of competitive ability can stabilize such interactions. He argues that the rarer of two species is selected to improve its ability to compete with the other species, while the commoner is selected to improve its intraspecific competitive ability. As the interspecific competitive ability of the rarer species increases, it becomes numerically predominant, and the nature of selection in the two species is reversed. Pimentel suggests that the two species oscillate in abundance and in genetic composition, but that neither will become extinct even if they occupy the same ecological niche. He and his coworkers (Pimentel et al. 1965) showed that after 25 generations of competition between houseflies and blowflies, the initially inferior blowflies became genetically superior competitors and eliminated the houseflies. Thus part of the model was confirmed; but its most important prediction, stabilization of the interaction, was not. Levin (1971) considers such stabilization to be unlikely, unless the intraspecific and interspecific competitive abilities of genotypes are negatively correlated; whether this is usually the case is unknown.

THE EVOLUTION OF RESOURCE UTILIZATION

Most analyses of the coevolution of competitors have focused on changes in resource utilization. The NICHE OVERLAP between two species can be measured by d/σ (*Figure 2*), where d is the average difference in resource use and σ is the NICHE BREADTH, the variety of resources used by a species. Overlap in resource use is reduced if d increases or σ decreases. Neither event can occur unless there is genetic variation in resource use — unless different genotypes are to some extent specialized for different resources. If such variation exists, the niche breadth (measured as σ^2) of the population as a whole is a composite of the variation among genotypes (σ_B^2) and the mean niche breadth of each genotype (σ_W^2); that is, $\sigma^2 = \sigma_B^2 + \sigma_W^2$ (Roughgarden 1972). Bulmer (1974) has shown that in principle, such a polymorphism can exist if each genotype competes more intensely with itself or with phenotypically similar genotypes than with dissimilar genotypes; the gene frequencies equilibrate at values that maximize population density (Anderson 1971).

Thus the degree of polymorphism can influence a species' niche breadth; but species can differ in niche breadth simply because they are fixed for genotypes that are more or less ecologically versatile. Specialization in resource use is likely to be favored in species faced with relatively constant or abundant resources; more generalized genotypes are presumably favored when resources are inconstant or un-

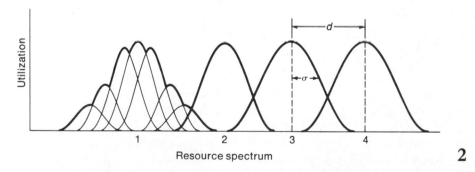

Niche width in monomorphic and polymorphic populations. Species 2, 3, and 4 are monomorphic with respect to resource utilization, but species 2 consists of a more specialized genotype than the others. Species 1 as a whole is more generalized than the others, but it consists of an assemblage of specialized genotypes. Niche overlap, measuring the potential competition between genotypes or species, is given by d/σ.

2

reliable or when they are rare (Levins 1968, MacArthur 1972). Thus the overlap in resource use between species is likely to decrease in evolution insofar as specialized genotypes are favored by avoiding competition, but it is likely to increase insofar as the unpredictability of resources favors generalized genotypes. It is therefore generally supposed that species are more specialized and overlap less in resource use in constant than in inconstant environments. This is one of several theories (*Chapter 4*) for the high diversity of species in some tropical environments, which may be (although this is a highly debatable point) more constant than most other environments. Larvae of tropical swallowtail butterflies are more specialized in host-plant utilization than temperate-zone species (Scriber 1973), but there is little other evidence that tropical species are more specialized than temperate-zone forms.

DIVERGENCE IN RESOURCE USE

Several authors (e.g., León 1974, Lawlor and Maynard Smith 1976) have affirmed by one-locus models that if there is genetic variation for resource utilization in each of two competing species, character displacement (*Chapter 9*) can usually be expected; genotypes that specialize on resources that the other species does not use are favored. By the same token, species should be more polymorphic in resource use if they are free of interspecific competition (e.g., on islands) than if they are constrained by competitors (see *Chapter 9* for an example). But remarkably few instances of character displacement in resource utilization have been described (Grant 1972; *Chapter 9*). Why is the phenomenon so rare?

Populations of sexually reproducing organisms may harbor little effective genetic variation for resource utilization. With a few exceptions (e.g., Pimentel et al. 1965, Seaton and Antonovics 1967), laboratory populations of *Drosophila* and other insects subjected to interspecific competition have shown remarkably little genetic change in competitive ability or resource partitioning (Park and Lloyd 1955, Futuyma 1970, van Delden 1970). In natural populations, moreover, very few instances of multiple-niche polymorphisms have been reported (*Chapter 13*); in a survey of several sexual species of moths, for example, Mitter and Futuyma (1979) found almost no associations between electrophoretically identified genotypes and host plants.

Theoretical considerations militate against changes in a species' niche in response to interspecific competition. If genetic variation for resource utilization exists but is polygenic, the abundance of each genotype will not be closely matched to that of the resource on which it specializes (Roughgarden 1972). Because of recombination, extreme, highly homozygous (e.g., *AABBCC*) genotypes will be rare and thus undercrowded compared to the abundance of their extreme resources (*Figure 3*). If recombination is reduced, as in asexual species, gene combinations adapted to different resources are held intact and can increase to their resource levels. Thus in the fall cankerworm (*Alsophila pometaria*) genetic variation for host-plant utilization exists; different parthenogenetic genotypes are adapted to feed on different species of trees (Mitter et al. 1979).

Because the frequencies of extreme genotypes in sexual species are set more by recombination than by the level of their special resources, their frequencies and thus the gene frequencies are not very responsive to changes in resource level, including those caused by the presence or absence of a competing species that uses those same resources (Roughgarden 1972). Thus recombination reduces the likelihood that character displacement or changes in niche breadth will occur to any appreciable degree (see also Bulmer 1974).

This conclusion and the observation that character displacement is apparently quite uncommon are rather distressing, for the history of evolution is largely one of adaptive radiations, of changes in the types of prey and other resources that species use. How and where do these shifts occur? Perhaps changes in resource use occur during speciation, in small populations with unusual genotypes faced with novel resources (*Figure 3*). Adaptation to these resources may occur independently of the effects of interspecific competition, and the niche differences among sympatric species exist largely because of species selection: when species that have evolved in isolation become sympatric, they escape extinction by competitive exclusion only if they have already become adapted to different resources.

INTERACTIONS BETWEEN PREDATORS AND PREY

Some of the most intriguing coevolutionary problems concern the relations between herbivores and plants, parasites and hosts, predators

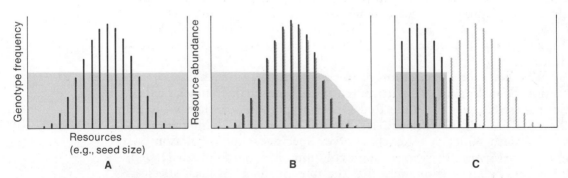

A possible view of the evolution of resource utilization. (A) The vertical lines represent equilibrium frequencies of genotypes differing in an additively inherited polygenic character that adapts each to a specific part of the spectrum of resources. The "central" genotypes are overcrowded and the "extreme" genotypes undercrowded with reference to the abundance of appropriate resources, indicated by the shaded area. (B) If a competing species that uses one end of the resource spectrum does not deplete resources to the extent that extreme genotypes experience scarcity, little evolution (character displacement) occurs. (C) But if only a portion of the resource spectrum is available to an isolated population, the genotypic composition shifts (from its former state, indicated by light lines, to a new state, indicated by heavy lines).

3

and prey — between "exploiters" and "victims." I shall focus largely on plant-herbivore interactions in this discussion; some of the principles undoubtedly apply to other exploiter-victim systems, but some probably do not.

I shall term the protective devices of plants or other victims *defenses* or *resistance mechanisms*; the adaptations of the herbivores or other exploiters that enable them to overcome these defenses are *counterdefenses* or mechanisms of *counterresistance*. Clearly there is genetic variation for both defenses and counterdefenses (Gerhold et al. 1966, van der Plank 1968, Day 1974). For example, dominant alleles at each of about five loci in wheat confer resistance to Hessian fly; corresponding to each of these resistance genes is a recessive allele in the fly that confers counterresistance (Hatchett and Gallun 1970). Genetic variation in susceptibility to herbivores is known in wild populations of *Lotus* (Jones 1973), ginger (Cates 1975), cockleburs (Hare and Futuyma 1979), and other plants. Different genotypes of the pea aphid are adapted to different strains of alfalfa (Cartier et al. 1965). Local populations of the spider mite *Tetranychus urticae* differ in their resistance to cucurbitacin, a toxic compound in cucumber leaves (F. Gould 1977).

Thus there appears to be sufficient genetic variation for rapid coevolution. It is often supposed that an evolutionary "race" between predator and prey continues indefinitely, with each species evolving new ways of foiling the other. In a well-known case (Gilbert 1971, 1975) *Passiflora* vines have evolved toxic compounds against insects. The larvae of *Heliconius* butterflies are resistant to these defenses and are specialized feeders on *Passiflora*. Some *Passiflora* species have specific defenses against *Heliconius,* such as structures that mimic *Heliconius* eggs and deter the butterflies from ovipositing on vines that appear already occupied, but these defenses are still not entirely effective. *Passiflora adenopoda,* however, has gone a step further by evolving hooked hairs that immobilize *Heliconius* larvae, which thus starve to death.

Whether such coevolution proceeds indefinitely is a matter of contention. Van Valen (1973) attributed the apparent constancy of extinction rates (*Chapter 5*) to the Red Queen hypothesis, which holds that the environment of each species deteriorates at a constant rate as the species with which it interacts evolve. Species escape extinction only if they compensate by constant evolution. Maynard Smith (1976*b*), however, believes it unlikely that the improvement of each species is precisely balanced by a decrement in the fitness of other species. If the decrease in some species' fitness is not as great as the increment achieved by others, selection for compensatory improvements is lessened, so evolution becomes slower and ultimately stops. It resumes only if environments change or if major new adaptations arise that drastically alter the framework of interspecific interactions.

A third view is that of Rosenzweig (1973), who postulates that coevolution of predator and prey continues indefinitely but seldom leads to extinction (*Figure 4*). In his view the fixation rate of mutations

Evolution of the predator isocline, as visualized by Rosenzweig. A point in the space represents the number of predator individuals P and prey individuals V. If V is low, the predator population declines; if high, it increases. Therefore there is a value $V = J$ at which the predator population does not change ($dP/dt = 0$). Similarly the prey population remains unchanged at those combinations of V and P values that fall on the curve $dV/dt = 0$. The arrows in the diagram are vectors showing changes in P and V when population densities are not on the zero isoclines ($dP/dt = 0$, $dV/dt = 0$). The system is stable if the predator zero isocline $dP/dt = 0$ falls to the right of the peak of the prey zero isocline $dV/dt = 0$; it is unstable if the predator zero isocline is to the left. J will move toward J', so that the system becomes destabilized, if the predator evolves greater

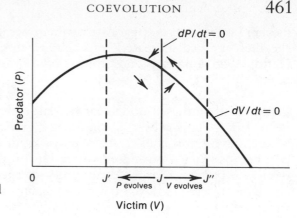

proficiency (can support itself at a lower value **4** of V). The system becomes stabilized (J moves toward J'') if the prey evolves better defenses, so that the predator needs a larger prey population to support itself.

that improve a predator's proficiency or a victim's resistance depends on how proficient or resistant the species already is. Thus proficiency evolves faster in an inefficient predator than resistance in an already resistant prey; and resistance evolves faster in a highly vulnerable victim than proficiency in an efficient predator. Thus evolutionary rates are balanced, and neither species is likely to win the race. Pimentel (1961a; see also Levin 1972, Łomnicki 1974) has similarly argued that continued evolution of predator and prey could stabilize their interaction. He reported that fluctuations in laboratory populations of house-

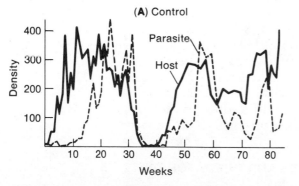
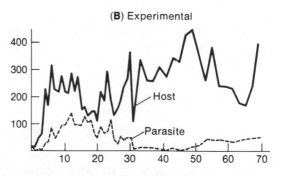

Effects of coevolution on population sizes and stability in a laboratory system of houseflies as hosts and wasps (*Nasonia vitripennis*) as parasites. (A) Neither species had a history of exposure to the other. (B) Both species had a long history of exposure to one another. **5** Fluctuations are less marked in the coevolved case, and the density of parasites is lower. (After Pimentel and Stone 1968)

flies and of a wasp that parasitized them became less pronounced after the flies evolved increased resistance to the wasp (Pimentel and Al-Hafidh 1965, Pimentel and Stone 1968; *Figure 5*).

CROSS-RESISTANCE

I suspect that these models of predator-prey coevolution are quite unrealistic — in part because they take a gradualistic view of evolution, whereas evolution is likely to be most rapid during speciation, and in part because they view a predator and its prey as an isolated pair rather than as a part of a larger community of species. Each plant, for example, is selected to develop defenses against a multitude of herbivores; each herbivore, to develop counterdefenses against a spectrum of plants. The features that provide resistance to one species may be incompatible with those required for resistance to others; if such incompatibilities are common, evolution may be very slow indeed. For example, Pimentel and Bellotti (1976) found that populations of houseflies adapted rapidly to each of six toxic substances (e.g., Cu^{++}) taken singly, but resistance did not evolve in a population subjected to all six simultaneously.

This implies that the evolution of predator-prey interactions depends on patterns of cross-resistance. For example, a plant is freed from attack by many insect species and is likely to evolve resistance rapidly if a highly generalized defense arises that is broadly effective against herbivores. Tannins, for example, appear to be broad-spectrum defenses against many kinds of insects and pathogenic microorganisms, because they bind proteins and make them undigestible (Feeny 1970). Similarly a predator can evolve a broad diet most easily if it evolves a counterresistance device that is effective against many kinds of prey defenses. In insects alleles that confer resistance to DDT often confer cross-resistance not only to other chlorinated hydrocarbons (which may be considered synthetic analogues of natural plant toxins), but to very different compounds such as organophosphate insecticides (O'Brien 1967). Spider mites that had evolved resistance to cucurbitacin, the toxic principle in cucumber, were somewhat cross-resistant to the defenses (probably alkaloids) of solanaceous plants and to several commercial insecticides as well (F. Gould 1977). Krieger, Feeny, and Wilkinson (1971) found that the activity of mixed-function oxidases was higher in polyphagous species of caterpillars than in more specialized species. These are broadly acting detoxification enzymes that may counter a wider variety of plant toxins in generalized than in specialized species.

DIVERGENCE OF PREY DEFENSES AND PREDATOR DIETS

The more different the defenses of two prey species are, the less effective the predator's counterdefense system will be against both defenses. Thus it is advantageous to the predator to specialize on one or another prey species. Sympatric species of prey therefore benefit and their coexistence is more likely if they differ in defense systems, so that each is not subject to predators harbored by the others. For example, a plant species may suffer less insect damage if grown in mixed vegetation than if planted in a monoculture (Pimentel 1961*b*, Tahvanainen and Root 1972). Ricklefs and O'Rourke (1975) found that the higher diversity of moths in tropical than temperate-zone sites is associated with a commensurately greater variety of cryptic patterns and body shapes, each presumably being a different way of reducing visibility to predators.

The advantage to any one plant of diverging in its defense properties, however, is greater if it can "hide" in the midst of chemically different vegetation than if it is more "apparent," as Feeny (1976) puts it. Perhaps for this reason plants in locally diverse vegetation communities are chemically more different from one another than those in less diverse communities. As expected, the herbivorous insects (Lepidoptera, in particular) of more diverse vegetation (early successional forbs) are more specialized than those of the less diverse vegetation in climax temperate-zone forests (Futuyma 1976). Feeny's (1976; see also Rhoades and Cates 1976) view of this pattern is that insects are more likely to adapt to the defenses of apparent (dominant, easily found) than "unapparent" plants, unless apparent plants evolve one of a few kinds of defense, such as tannins, that are not easily overcome by insects. An unapparent plant may develop any of many special defenses, such as alkaloids, to which insects may easily adapt; but rather few insect species do so if the plant is a relatively rare member of a diverse community.

Those insects that do evolve resistance to a plant's defense may actually use it as a cue for finding their host, so the same plant substance may prevent attack by some species and attract others. For example, cucurbitacin is toxic to generalized herbivores such as spider mites but is used by specialized cucumber beetles to find their host (Dacosta and Jones 1971). Such specialized insects often become so behaviorally dependent on specific host stimuli that they do not feed on plants that could be perfectly adequate hosts (e.g., Soo Hoo and Fraenkel 1966, Waldbauer 1962, Wiklund 1975). Levins and MacArthur (1969) suggested that if an insect's nervous system cannot encode the responses necessary to discriminate between all good and bad hosts, an insect may maximize fitness by responding only to a subset of the good hosts, ignoring those that it cannot distinguish from the bad, lest it make a mistake.

As plants or other kinds of hosts diverge in defense properties over evolutionary time, insects (or other parasites) may also be expected to diverge and become more specialized in diet. Related hosts should have similar defenses; thus related parasites should occupy related hosts. Ehrlich and Raven (1964) pointed to many such patterns. For example, many genera of the Pierini (cabbage butterflies) feed on the closely related families Cruciferae and Capparidaceae, which have similar mustard oil glycosides; the Danaiinae (monarch butterfly and its relatives) feed on the Asclepiadaceae and Apocynaceae, which share cardiac glycosides and certain alkaloids. Phylogenetic relationships among heliconiine butterflies more or less reflect those among the species of *Passiflora* on which their larvae feed (Benson et al. 1975).

But quite often the phylogenetic relationships among hosts are not closely mirrored by those of their parasites (Kethley and Johnston 1975), and the insect fauna of unrelated plants may be more similar than that of related plants (Futuyma and Gould 1979). Precise evolutionary resource tracking (Kethley and Johnston 1975) is more likely among specialized insects adapted to specific host defenses than among more generalized insects adapted to generalized defenses such as tannins. Unrelated hosts may converge in defense properties and thus become accessible to unrelated parasites; and speciation in host-specific insects is often associated with, and indeed may be caused by, a shift to an unrelated host plant (Bush 1969). In California, for example, the codling moth, which ordinarily attacks apple, has formed a host race on walnut (Boyce 1935, Phillips and Barnes 1975). Perhaps modern methods of phylogenetic analysis (*Chapter* 7) will help to reveal the degree to which the phylogenetic affinities of parasites follow those of their hosts, and what the patterns of host switching have been.

EVOLUTION OF FOOD-WEB COMPLEXITY

It is not clear whether the number of species of insects associated with a plant species increases or decreases over evolutionary time. Divergence among plants and the acquisition of specialized defenses should lower the diversity of associated insects; adaptation and speciation in insects should increase the diversity. An insect may enlarge its diet if its host is physically proximate to another plant with similar defense properties (Janzen 1968). Thus generalized insects may be preadapted to "apparent" plants whose defenses are similar to those of their former hosts. For example, the number of insect pest species associated with cacao is as high in the Old World tropics, to which cacao has been introduced, as in the New World, where cacao is native (Strong 1974a). But many introduced species of forbs, with more specialized defenses, are attacked only by the most generalized insects, such as certain grasshoppers (Otte 1975), or by specialized insects that feed on related plants. It is not clear how often insects become more generalized in diet, compared to the frequency with which they become more host-specific, especially during speciation.

Once an insect has adapted locally to a plant, it may then spread through the plant's range. Thus widespread plant species, which come into contact with many other hosts, are likely to have diverse insect faunas. For example, wide-ranging species of Californian oaks have a more diverse fauna of leaf-mining insects in any given locality than do geographically more restricted species (Opler 1974). This principle makes it difficult to determine whether the number of insect species associated with a plant changes over evolutionary time. Southwood (1961) presented some evidence that the insect diversity on British trees was greatest on species that had been most abundant in Britain for the longest time (*Figure 6*); but Strong (1974*b*) believes that the rich fauna of these trees is attributable entirely to their wide geographic distribution. Thus we are largely ignorant of the long-term dynamics of coevolution between prey and predator, their phylogenetic consequences, and their effects on the structure of ecological communities.

THE COURSE OF COEVOLUTION

With this background, it may be possible to develop a speculative scheme of how coevolution among plants and their insect herbivores may proceed through evolutionary time.

A species of plant, at most times in its history, is likely to serve as host to a variety of insect species. Some are highly generalized in diet; others are rather more specialized, feeding on this plant and on a few others that are related or that have similar morphological and chemical properties. A few species of insect may feed almost exclusively on this plant. The plant has several variable features that affect its herbivores to some degree. These features are likely to include "families" of related chemical compounds: a suite of related phenolic compounds, for example, which vary in relative concentration from one individual plant to another. Certain of these compounds, if increased, will deter some of the associated herbivore species, but increase susceptibility to

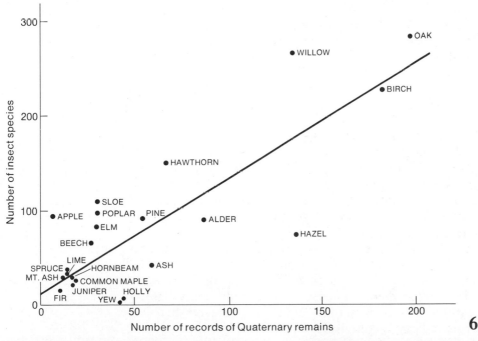

6

The relationship between the number of insect species associated with a tree species in Britain and the historical abundance of the tree, as measured by the number of records of Quaternary fossil remains. It is unclear whether the relationship depends simply on how widespread the tree is, or whether the length of time it has been available has affected the number of insect species it harbors. (After Southwood 1961)

others. Therefore the plant population is genetically variable, with frequency-dependent selection maintaining chemical diversity: as any one chemical phenotype becomes common, it loses its advantage, for those herbivore species that respond positively to it increase in population size. The gene frequencies arrive at rather stable equilibrium values, so the chemical profile of the plant population is not evolving most of the time.

Because the composition of the insect fauna may differ among localities, the plant population's defense profile may vary geographically: one or another phenol, for example, may be more or less pronounced in one population than another. The highly specialized herbivore species may be affected, either because the plant does not offer the proper complex of chemical stimuli by which they locate their host, or because the plant presents a physiological challenge. These specialized species may then be absent from some localities, having failed to adapt, or they may be genetically differentiated, attuned to the local peculiarities of their host. The generalized herbivores, however, will seldom show either of these responses; they are not likely to be affected by a slight change in the characteristics of a plant on which they do not greatly depend.

Occasionally, a species of insect that utilizes other species of plants as hosts will find itself in a locality where its usual hosts are rare and our example plant is common; if mutations in the insect's sensory system occur, it may invade and subsequently adapt to the plant, forming a specialized "host race" that may become a new species. By subsequent speciation, it may give rise to a complex of species associated with this plant and its relatives. Thus occasionally a species of insect is added to, or deleted from, the fauna associated with this plant. But at most times there is rather little evolutionary response of the plant to its insects, and genetic changes in adaptation to the host plant occur primarily in those insects that are most specialized on it.

But very rarely, a major evolutionary event will occur: a plant will evolve, through one or more truly novel mutations, a chemical (or morphological) feature that protects it against most of its fauna. The first appearance of mustard oil glycosides in the ancestors of the Cruciferae might have been such an event. It is likely that the plant will be completely freed from insect herbivores, except for a few very generalized species — perhaps those with exceptionally broadly acting detoxifying enzymes. This insect-free plant, and the species to which it gives rise, may have so great an advantage in being temporarily freed from predation that they can successfully invade many habitats, compete successfully with plants that suffer from greater herbivore attack, and become a large, ecologically "successful" group.

However, local populations of herbivorous insects occasionally become adapted to this new defense, forming specialized species by the same mechanism as described above. This will be most likely in insects adapted to plants that bear some fortuitous resemblance to the new defense — insects that are cross-resistant to the new defense, or which are attracted to chemical cues that are similar in the old and new hosts. Thus new host-specific herbivores will slowly be added to the fauna of our "remodelled" plant; these herbivores will develop specific adaptations to the new defense system and may even evolve dependence on it as a way of recognizing their host. They will give rise by speciation to a complex of related species, adapted to the complex of chemically novel plants. Because these plants are attacked by a few species of host-specific herbivores, they may subsequently develop specific counter-adaptations — alteration of the new chemical system, or slight changes in other chemical or morphological properties — that enable them to escape these very specialized enemies. At this point, each species of plant is attacked by so few kinds of insect, and each species of insect feeds on so few species of plant, that each can evolve specific adaptations to deal with the other, without becoming more susceptible to other enemies or becoming less adapted to other host plants.

Thus the overall course of coevolution may well be this: plants initially evolve characteristics that confer overall resistance to a broad spectrum of enemies, but do not "coevolve" against specific enemies to any great extent; they become so well defended as to reduce their fauna to a few very generalized and a few very specialized species of enemies; the specialized enemies evolve specific counterdefenses; and the plant may then evolve specific counter-counterdefenses against its few, most important, species of enemies. The course of coevolution may be, commonly, from generalized non-reciprocal evolutionary changes in large groups of weakly interacting species, to specific, reciprocal evolutionary responses in small groups of strongly interacting species. Probably the same course from generalized to specialized responses is typical of other predator-prey interactions and of mutualistic interactions as well.

THE EVOLUTION OF COMPLEX INTERACTIONS

Whether a feature is advantageous must depend in a very complicated way on the entire nexus of interactions among species in a community. A feature that provides protection against some predators can enhance susceptibility to others. The evolved features of a species determine the ecological impact of other species, as expressed by the coefficients a_{ij} and $b_{ij,k}$ in a generalized equation for the effects of other species on species i:

$$dN_i/dt = f(a_{i1}N_1 + a_{i2}N_2 + \cdots + a_{is}N_s \\ + b_{i1,2}N_1N_2 + b_{i1,3}N_1N_3 + \cdots + b_{i(s-1),s}N_{s-1}N_s).$$

Not only do other species individually (a_{ij}) affect the population growth of species i; the strength and indeed the form of the interaction between species i and j may depend on the abundance of a third species k, an interaction ($b_{ij,k}$) among species analogous to the epistatic interactions among genes. In Chapter 4, for example, I cited Smith's (1968) discovery that the effect of cowbirds on oropendolas depends on the abundances of wasps and botflies; cowbird nestlings are harmful to oropendola nestlings if wasps prevent botflies from entering the nest, but they can protect the oropendolas from botflies if wasps are not present. Such nonadditive interactions among species have been revealed in almost all of the few cases in which they have been sought (e.g., Hairston et al. 1969, Vandermeer 1969, Wilbur 1972, Neill 1974).

Such higher-order effects often arise when a species exploits an interaction between two or more other species. The mutualistic interaction between the cleaner wrasse and larger fishes, from which the wrasse gleans parasites, is exploited by a blenny that by mimicking the wrasse in color pattern and behavior can approach the hosts and nip off bits of flesh. Parasites with complex life cycles often exploit predator-prey interactions among their successive host species. If sequences of specialized interspecific interactions are common, the extinction of a key species must often be followed, in a kind of domino effect, by the extinction of others.

It is plausible to suppose — and the lore of natural history tends to confirm the supposition — that the strength of higher-order interactions and the number of species they encompass may be greater in constant than in inconstant environments (Futuyma 1973). If the evolution of a specialized dependence on one other species requires the resource species to be consistently available, the evolution of dependence on the joint occurrence of two other species must be even more contingent on their constancy. Moreover the more specialized species there are in a coherent interactive network, the more the persistence of the entire set may be threatened by the decline or extinction of any one species. Complex ecological networks, like complex developmental systems and epistatically interacting sets of loci, are less stable than simple systems (May 1973). Thus complex networks may have suffered less extinction in environments that have long remained constant.

Evidence bearing on these speculations is not very strong; in fact, there is virtually none. Bretsky and Lorenz (1970) showed that extinction rates, entailing whole assemblages of species, have been greater in deep-water communities of marine invertebrates than in contemporaneous shallow-water communities. Thus organisms adapted to what had presumably been the more constant environment had the greater extinction rate, but it is unclear whether they were simply more specialized in physiology or diet or were the victims of ecological domino effects. No one really knows whether species in tropical forests and coral reefs really experience more constant environments than those in less diverse communities, but ecologists have come to suspect that these communities are composed of specialized species, among which interactions are so intricate that they are "fragile ecosystems" (Farnworth and Golley 1974). Such communities may well be irreparably changed by even slight alterations of the environment, much less the human activities that now threaten to destroy them.

SUMMARY

The species composition of communities and the ecological relationships among their members are determined by immigration and extinction and by coevolutionary responses of species to one another. The rate and pattern of coevolution, the extent to which it stabilizes or destabilizes interspecific interactions, and its effects on the specificity of species associations, phylogenetic patterns of resource utilization, and the trophic structure of communities are more speculated on than demonstrated. Formulating an evolutionary theory of the development of community structure will be a challenging task for population biologists.

FOR DISCUSSION AND THOUGHT

1 Suggest ways of determining whether the characteristics of coexisting species (e.g., niche differences among competitors) are attributable more to coevolutionary responses to each other or to species selection.

2 Determine whether the literature on introduced species (e.g., biological control agents) gives evidence of rapid genetic changes in their interaction with native species. If rapid evolution is rarely found, what role can individual selection play in stabilizing otherwise unstable ecological interactions?

3 Evaluate the assumptions in the arguments of Rosenzweig (1973) and Maynard Smith (1976b).

4 The generation time of insects and pathogens is commonly far shorter than that of their hosts, so their potential rate of evolution may be greater. How is it, then, that they have not evolved such proficiency that they extinguish their hosts?

5 Some members of ancient groups of plants, such as ferns, cycads, and the ginkgo, are host to very few species of insects, whereas some of the more recently evolved groups such as composites harbor a diverse insect fauna. How common is this pattern, and what are the possible explanations for it? Does it have any implications for the dynamics of the predator-prey coevolutionary race?

6 Mutualistic interactions are said to be ecologically unstable (May 1973) because an increase or decrease in either species should engender an increase or decrease in the other. What factors can stabilize their interaction? Is coevolution likely to increase or decrease stability?

7 Discuss the similarities and differences in the likely course of coevolution between plant-insect interactions, host-parasite interactions, and other predator-prey interactions, such as those between vertebrate predators and their prey.

8 A critical feature of the theory of coevolution of all three major interactions — mutualism, competition, and predation — is the niche breadth of at least one of the species, the variety of species it uses as resources. Synthesize the factors that affect the evolution of niche breadth into a common basis for the theory of all three kinds of coevolution.

9 Discuss the factors that are likely to lead to the evolution of a broader or narrower diet in herbivorous insects. What bearing does this have on the problem of whether evolution invariably leads from generalized to specialized forms?

10 Discuss the ways in which processes of coevolution could increase or decrease the diversity of organisms over evolutionary time.

MAJOR REFERENCES

Gilbert, L. E., and P. H. Raven (eds.) 1975. *Coevolution of animals and plants.* University of Texas Press, Austin, Texas. 246 pages. Papers in this symposium treat special topics in plant-animal interactions.

Cody, M. L., and J. M. Diamond (eds.) 1975. *Ecology and evolution of communities.* Harvard University Press, Cambridge, Mass. 545 pages. Some of the papers in this symposium take an evolutionary approach to community structure, emphasizing especially competitive interactions.

I would be remiss to omit from this book a discussion of human evolution, a topic with which we are understandably fascinated. Because it seems to many to have important social and ethical implications, I will not stress the history of human evolution revealed by paleontology as much as some specific controversies about our current state: whether the human gene pool is degenerating; whether humans vary genetically in intelligence and other aspects of behavior; and whether aggression, sexual differences in behavior, and other aspects of our social behavior are products of our evolutionary past.

THE MYTH OF SCIENTIFIC OBJECTIVITY

Because of their social implications, human genetics and evolution are subjects highly charged with emotion. Scientists are as human as anyone else, and they typically discuss these topics with fervor. Consequently the literature on these topics suffers from a profusion of statements unsupported by evidence and from unspoken and largely untested assumptions. The canons of scientific rigor that are scrupulously applied to, say, analyzing selection in land snails are often not applied at all to the profoundly more important questions of human biology. For example, Jensen (1969) proceeded from a supposed demonstration of genetic variation for IQ within the white population to the conclusion that the average IQ difference between whites and blacks is largely a genetic difference — despite the well-known principle that high heritability within one population says nothing about whether differences between populations are genetic (*Chapter 14*). H. H. Goddard (1920), a pioneer in mental testing, "discovered" by administering IQ tests to immigrants that 79 per cent of the Italians, 83 per cent of the Jews, and 87 per cent of the Russians were "feebleminded" and warned against the social consequences of immigration, because of "the fixed character of mental levels," a fixity for which there was not,

CHAPTER 19

Social Issues in Human Evolution

Homo sum: humani nil a me alienum puto.
(I am human: nothing human is foreign to me.)

Terence, about 1 B.C.

and could not be, the slightest shred of evidence. Among the strongest advocates of positive eugenics (improving the gene pool) was the eminent geneticist H. J. Muller, a Nobel Prize winner. Muller (1966) proposed large-scale artificial insemination of women with sperm from men of outstanding intelligence or achievement, or at least desirable social qualities, including "joy of life, strong feelings combined with good emotional self-control and balance, the humility to be corrected and self-corrected without rancor, empathy, thrill at beholding and serving in a greater cause than one's self-interest, fortitude, patience, resilience, perceptivity, sensitivities, and gifts of musical or other artistic types, expressivity, curiosity, love of problem-solving and diverse special intellectual activities and drives."

As we shall see, Muller could not have had any evidence that variations in empathy, patience, or "thrill at beholding and serving in a greater cause than one's self-interest" have any genetic component, which would be prerequisite to his scheme; he simply assumed that they do. Such assumptions underlie much of what has been written on human evolution.

Scientific pronouncements — that IQ, sexual orientation or criminal behavior are genetically conditioned; that races, social classes, or sexes differ intrinsically in intelligence or behavioral traits; or that the human species is innately aggressive — find a ready audience, and some are ready to use these statements to justify oppressive political and social policies. The "scientific racism" of the Nazis is the most egregious example, but there are dangers closer to home. Jensen's (1969) article claiming that most variation in intelligence is inheritable was read into the *Congressional Record*. Because of the high stakes, it seems to me that scientists should demand at least as much rigorous evidence for conclusions about human behavior as they do of research on the behavior of ducks or *Drosophila*.

Scientists who argue that there are biological differences between races or sexes are not necessarily racist or sexist; in fact, they seldom are. They simply believe that they are exercising full scientific objectivity. But it is very difficult to know whether we are entirely objective, for our thinking is so conditioned that we do not often question what seems self-evident. The record of history made it obvious to the scientists of the nineteenth and early twentieth centuries that the white race (to which they themselves belonged) had achieved the most advanced civilization, so they simply assumed that whites were genetically superior. Freud and Jung merely voiced the conventional wisdom of their times when they developed psychological theories that assumed, without evidence, intrinsic differences between the sexes in aggressiveness, nurturance, and emotionality. Moreover scientists are apt to assume that their particular specialty holds the key to the great problems of the age. Given the same data on human behavior, a biologist is likely to propose a biological explanation; a social scientist or psychologist is more likely to see environmental influences. I suspect Muller assumed a genetic basis for empathy and patience simply because he was a geneticist.

Unconscious assumptions affect the interpretation of data; but sometimes a scientist's predilections seem to affect the data themselves and the way in which they are obtained. As long as homosexual behavior was believed to be pathological, it was easy to "prove" that homosexuals were neurotic and maladjusted — by studying biased samples of homosexuals who underwent psychiatric treatment (see Churchill 1967, Tripp 1975). More controlled studies, undertaken with less bias, reveal no differences in mental health between homosexuals and heterosexuals (e.g., Hooker 1957, Saghir and Robbins 1973). Whether an investigator finds evidence for genetic variation in IQ seems to depend on his or her biographical characteristics (Sherwood and Nataupsky 1968, reported by Lewontin 1975). The most comprehensive study claiming a strong genetic basis for variation in IQ may well include fraudulent data (Kamin 1974, Jensen 1974, Wade 1976). The history of scientists' pronouncements on human genetics and behavior is, to a distressing extent, a history of the conventional societal attitudes on these subjects; science has served more as a defense of the *status quo* than as a force for change.

BIOLOGICAL AND CULTURAL EVOLUTION OF THE HUMAN SPECIES

The fossil record of hominid evolution (see Simons 1972, Pilbeam 1972, Campbell 1974 for more extensive treatment), like that of many groups, is fragmentary and difficult to interpret, and the interpretations change almost yearly as new fossils are found. The hominids appear to have diverged from the line leading to the modern apes in the mid-Miocene, about 13 million years ago. In the Pliocene, 5 to 3 million years ago, a variety of forms, the australopithecines, were present in Africa; by at least 3 million years ago some of these had begun to construct simple stone tools. It is unclear which of these forms was ancestral to *Homo* (*Pithecanthropus*) *erectus,* which replaced the australopithecines about 1 million years ago, in the early Pleistocene. By the mid-Pleistocene (about half a million years ago), *H. erectus* had evolved a cranium nearly the size of modern humans' and had become widespread through the Old World. Evidence of the use of fire at about this time indicates the beginning of major cultural changes. By 75,000 to 100,000 years ago, cranial volume had evolved to its fully modern level, and the extremely rapid spread, from this time on, of increasingly complex and sophisticated cultural innovations suggests that the capacity for truly human language had evolved by then.

The evolution of the physical characteristics that are most distinctly human was thus exceedingly rapid, dating primarily from the latter half of the Pleistocene; but we have had substantial time, about 5 million years, to diverge physically, and presumably behaviorally, from our closest primate relatives. There is little question that this evolution was gradual, in the sense that our lineage has passed through a sequence of intermediate stages of physical form and mental capacity, in its divergence from the other primates. The capacity for culture, for the construction and use of tools, and for language are evident, in rudimentary form, in other modern primates. Japanese macaques, for example, have developed a variety of cultural traditions that are spread by learning, such as separating wheat from sand by floating it on water (Kawai 1965). Chimpanzees, likewise, learn from their elders how to use twigs to extract termites from their nests. Several captive chimpanzees have demonstrated a rudimentary capacity for symbolic language (see, e.g., Gardner and Gardner 1971, Premack 1971). The capacity for symbolic language is closely related to consciousness and self-awareness, and there is some evidence that the anthropoid apes have such self-awareness to some degree, for when they face their own image in a mirror, they groom themselves to adjust their appearance to what it "should" be (Gallup 1974).

Nevertheless it seems evident that humans have developed self-awareness, consciousness, and language to an extraordinarily greater degree than other species (Slobodkin 1978). Consciousness and language are the foundations of culture, which creates for us an environment that no other species experiences. It profoundly affects our behavior, our ecological relations, and our evolution.

CULTURAL EVOLUTION

It is easy to draw analogies between biological and cultural evolution, but they should not be taken too far (Harris 1975). Cultural or linguistic innovations, the analogues of mutations, are acted on by selective factors, selective in the sense that some become entrenched in the culture and others do not. At least in the evolution of languages a kind of drift seems to operate: seemingly random changes that lead, for example, to the formation of dialects. Cultures and languages diverge if geographically separated, much as populations do genetically. There exist many instances of cultural divergence, convergence, and parallelism. Many cultural traits are advantageous to the individuals who practice them; but some cultural traits change because dominant cultures impose their habits on subdominant groups, often by raw power. Perhaps more than in biological evolution, group selection of cultural traits may occur. Many cultural peculiarities, for example, seem to maintain the ecological balance between a tribe and its environment without conferring any special advantage on particular members of the group (Harris 1974). The practitioners of a cultural trait may or may not be aware of its advantages.

Cultural evolution differs from biological evolution in several important ways. Perhaps most important is that it is Lamarckian; behavior, language, or property that an individual acquires during his or her lifetime can be passed on to descendants or to other individuals, often by imitation or learning. Consequently cultural change can occur at rates that are orders of magnitude greater than biological evolution, as history bears witness. Moreover cultural evolution is far more reticulate than biological evolution and even entails "blending inheritance." Innovations are very often derived from contact between cultures, and even individual cultural traits can become blended, as is often evident in the evolution of languages (consider the pronunciation of the Spanish word *guerrilla* in English).

It is important to recognize that the selection guiding cultural change is of the traits themselves, not of the individuals practicing them; the gun replaces the spear because of its perceived advantage, not because gun carriers have more children than spear carriers. The advantage may be illusory, of course, rather than real. Many cultural traits act to the long-term (and sometimes to the short-term) disadvantage of both individuals and groups; examples range from cigarette smoking to the development of nuclear arms. It is also crucially important to recognize that cultural changes are not ordinarily caused by changes in a population's genetic composition: the recent increase in the divorce rate, for example, is a purely cultural phenomenon.

History shows that the mechanisms of cultural change are far more complicated than those of biological evolution. In genetic evolution a trait changes simply as the relative numbers of individuals with one or another genotype are altered. But cultural events do not follow so simply from numerical changes, for the social, religious, and economic institutions that culture brings into existence shape subsequent cultural events. Attitudes toward sex, for example, have changed over the last two thousand years in Western society not because people's sexual impulses and desires have changed, but because of the imposition of the Judaeo-Christian ethic on the legal system and on people's sense of propriety. Wars occur not because of changes in the proportions of aggressive and pacific individuals, but primarily because of the economic and political structures that govern their behavior. Social and economic forces interact with unique historical events and social attitudes to shape history and culture.

QUESTIONS IN THE EVOLUTION OF HUMAN BEHAVIOR

If cultural evolution has been so important in shaping our behavior, where does biological evolution fit in? To what extent is our behavior based on our genes, constrained by the inheritance of biological propensities from our prehuman ancestors? This question has been highly controversial, especially in the last few years.

The arguments concern two distinct but related problems. In some instances they deal with *differences* among individuals and groups in intelligence and the like that may or may not be based on genetic differences among them. In other cases the argument is whether a seemingly *invariant, universal* human trait (e.g. aggression) is a natural tendency encoded in the human genome. These traits include the behavioral differences between the sexes, which some authors attribute to different, more or less fixed expressions of the genome.

The two kinds of argument differ importantly as scientific hypotheses. If a trait such as intelligence varies among individuals, it is at least conceivably possible to determine whether the variation is genetically based by following its pattern of inheritance or by growing individuals in identical environments. On the other hand, one cannot say that a universal trait, such as the possession of a nose, or the capacity for language, is either genetic or environmental, for *it is the expression of genes in a series of environments* (*Chapter 3*). Genetics provides no means of investigating the inheritance of an invariant trait. Thus to postulate that it is genetic is to pose an untestable and meaningless hypothesis. The only question one can legitimately ask is, Is the trait highly canalized (*Chapter 15*), or does it vary greatly under different environmental conditions, compared to other traits? Aggression and sex role behavior could be said to be strongly canalized, to be under "tight control" by the DNA, if they were not modifiable by training or other environmental factors.

TWO VIEWS OF HUMAN BEHAVIOR

I will argue that there is insufficient evidence to conclude that normal variation in human behavioral traits has a genetic basis and that there is no evidence of strong genetic canalization of human behavior. In the absence of conclusive evidence for either a biological view or an environmental view, we can only evaluate the likelihood that one or the other will prove correct. One's inclination toward one position or the other will depend on what view of human evolution seems *a priori* more plausible (as well as on one's social biases, I suspect).

Advocates of the position that much of human behavior is controlled by the genes argue that some of the observed variation in human behavior is likely to be based on genetic variation, because they see no reason why human behavioral traits should differ from other characteristics, most of which are genetically variable in most species. Differences among populations may well be genetic, if genetic variation exists within populations, and if selection (say, by different social and ecological factors) has operated differently in different groups.

The argument that universal behavioral traits are in large degree canalized by our genes (see Tiger and Fox 1971, Ardrey 1966, Lorenz 1966) is often based on similarities between human behavior and that of other species (e.g., many mammals are territorial or show male dominance). The similar traits are presumed to be genetically homologous.

More theoretical arguments are presented by sociobiologists (e.g., Wilson 1975*a*, Barash 1977), along the following lines. (1) Suppose that genetic variation once existed for the behavior in question. (2) Imagine, from evolutionary and ecological principles, which behavior would have conferred the highest fitness on an individual and its relatives, and so postulate which genetic variant would have become fixed. (3) Compare the observed and the postulated behaviors; if they conform, the behavior is probably an adaptive genetic trait fixed by natural selection. For example, because females can be more certain than males that the children they are caring for are their own, kin selection will favor parental care more in the female than in the male (Barash 1977). Since behavior in our society more or less conforms to this expectation, it is easy to conclude that the behavior is genetically programmed. Similar arguments can rationalize incest taboos, greater aggressiveness and promiscuity of men than of women, greater co-operation among relatives than nonrelatives, indoctrinability, racial prejudice, homosexuality, love, conflict between parents and their offspring, religion — almost the whole gamut of human behavior (Wilson 1975a, Barash 1977).

Environmentalists, in contrast, hold that most variation in behavior is based on differences in education, childhood training, and other cultural and environmental factors. Human universals are genetic in the trivial sense that our behavior differs immutably from that of ants, giraffes, or monkeys; but we are so immensely capable of learning that most of our behavior is hardly canalized at all. Some few behaviors, such as suckling by infants, are highly canalized, but those that really matter to us are not. These expectations, to which I clearly subscribe, follow readily from evolutionary theory.

The evolution of a major new adaptation can drastically alter the selective pressures to which the various characteristics of a species are subjected. The evolution of flight by birds, for example, had subse-quent effects on almost every aspect of their morphology and physi-ology. Elaborate social behavior, in particular, such as the higher primates have, changes the species' environment and can alter the function of many aspects of behavior. It is generally agreed, for ex-ample, that the adaptive function of sexual intercourse in early hom-inids expanded to include not only procreation, but the maintenance of a stable pair bond during the long period of child care.

Once an elaborate social organization has evolved, it may select for still more elaborate social behavior. Whether sociality in the hominids evolved primarily because of the advantage of cooperation in tasks such as food gathering or because group cohesion provided protection against aggression from other groups, it no doubt became selectively advantageous to recognize other individuals in the coalition, to respond to each of them differently in different social contexts, and to be able to imitate behavioral innovations that provided greater immunity to the vagaries of the environment. This must have been a self-accelerating process, much like the evolution of courtship display by sexual selection (*Chapter 12*); more behaviorally rigid individuals were inferior to the behaviorally more adaptable, once the form of the social organization conferred an advantage on behavioral plasticity and learn-ing. The more elaborate the social behavior becomes, the more there is to learn, and the less fit a slow-learning genotype is. The highly developed capacity to learn provides control over the environment, and the variations in the environment that might favor the maintenance of genetic variation in a more stereotyped species are now met by the behavioral accommodations that a more versatile genotype is capable of. Thus just as we would not expect heterogeneity of environmental temperature to maintain as much genetic polymorphism in mammals or birds as in poikilotherms, we would not expect genetic variation in learning ability (or what we call intelligence) to be maintained by selection after a high degree of behavioral plasticity had evolved.

As the sophistication of the nervous system increases in a social species, the organism becomes capable of making ever more distinctions between environmental stimuli. But because it now perceives many environments, each is rare, so there is little selection in favor of a genetically programmed, specific response to each (I am indebted to Richard Levins for pointing this out). Selection favors not a limited set of genetically encoded specific responses to specific stimuli, but a flexible response system that reacts to integrated patterns of stimuli. Indeed each stimulus can occur in a great many contexts, to which different responses are appropriate (compare your response to a baseball thrown at you on a ball field and your response to a bottle thrown in a riot or demonstration). As social behavior becomes more complex, so do the patterns of stimuli, and there is stronger selection to replace automatic responses with reactions predicated on the pattern as a whole, its historical precedents, and its likely effects. With the evolution of self-consciousness as an adaptation to these selection pressures, the capacity for foresight, for imagining the likely consequences of each possible response to a contingency, became possible. But this capacity for decision making, once evolved, created a social context in which an increased capacity for decision making was ever more advantageous.

It would be naive to suppose that flexibility of response was limited to weather, fire, and the aggressive or cooperative actions of the members of the group. The brain does not develop piecemeal, but as a developmentally and functionally integrated unit. Genetic changes affect developing systems as a whole, as I have stressed throughout this book; for example, in *Drosophila* the genetic control of the development of abdominal bristles is integrated with that of scutellar bristles (*Chapter 15*). To suppose that separate neural circuits controlled sexual behavior, parental behavior, responses to siblings, responses to cousins, and so on, and that these remained unchanged while other parts of the circuitry became capable of integration and variable responses, is as naive as imagining that there is a separate gene for each of a fly's bristles — a notion that geneticists and developmental biologists abandoned long ago. It is much more likely that the remodeling of the nervous system to provide flexibility extended to the centers that had controlled sexual and other behaviors that in most species are highly "instinctive." It is precisely in social relations among people, which throughout human evolutionary history have presented each individual with an enormously varying environment, that we should expect human behavior to be most malleable, least canalized, least "genetic." It is not "human nature" to be warlike or peaceful, acquisitive or ascetic, prejudiced or egalitarian, promiscuous or monogamous. Rather, as Dobzhansky (in Dobzhansky et al. 1977) wrote, "Natural selection for educability and plasticity of behavior, rather than for genetically fixed egoism or altruism, has been the dominant directive factor in human evolution."

VARIATION IN INTELLIGENCE

I shall now examine several specific issues. I stress that the evidence is insufficient to distinguish the biological from the environmental positions, but that if anything, it tends to favor the environmentalist view.

Variation in almost every behavioral trait has been claimed at one time or another to be genetically based, in part. I shall discuss only the recent debate over the heritability of intelligence, on which there are far more data than on other behavioral traits. In what follows, "intelligence" should be read as "IQ (intelligence quotient) score," for no one really knows what intelligence is, or how to define it other than as an IQ score. IQ tests are supposed to be culture-free, but they have been strongly criticized as favoring white middle-class children (Kamin 1974, Bowles and Gintis 1973).

Few would deny that intelligence (whatever it is) has a physical and chemical basis in the brain, nor that certain genetic lesions (such as phenylketonuria) can impair intellectual development. The debate, rather, is whether groups of people within the "normal" range of IQ differ because of their genotype. People with severe genetic lesions cannot be used as evidence of a genetic basis of IQ variation any more logically than a mutant plant without chlorophyll is evidence that naturally occurring variation in chlorophyll content is genetically based. The wild plants may simply have been exposed to different environments (light intensity, trace minerals such as magnesium) that affect chlorophyll synthesis.

The strongest recent argument that variation in IQ is hereditary has been that of Jensen (1969, 1973), who concluded from other workers' data that the heritability (V_G/V_P; see *Chapter 14*) of IQ scores within Caucasian European and American populations is about 0.8; that is, about 80 per cent of the phenotypic variation is attributable to, or caused by, genotypic differences among individuals. Jensen adduced important social implications: if IQ is so highly genetically controlled, it cannot be easily improved by education, so that programs of compensatory education are doomed to failure.

If heritability within populations were indeed 0.8, this conclusion would nonetheless be unjustified. A character can show very high heritability in one environment, yet be modified greatly if the environment is changed. We might conclude from a genetic analysis of a field of wheat that variation in yield was almost entirely genetically determined and yet find thereafter that yield could be doubled by applying fertilizer. Many genetic diseases, or inborn errors of metabolism such as Wilson's disease which is caused by the retention of copper, have perfect heritability, yet they can be treated by dietary or other medical means (Lewontin 1975). Because each genotype has a norm of reaction (*Chapter 3*), a variety of phenotypes expressed under different environmental conditions, the heritability is not a fixed value; instead it depends on how greatly the environment varies. Moreover, if there is genetic variation in intelligence, it could well turn out that genotypes inferior in the present environment could be superior in different educational or cultural settings (*Figure 1*). High heritability does not imply immutability; IQ can appear highly heritable, but be raised greatly by the right social and educational programs if we are inventive and willing enough to discover and implement them.

However, there are good reasons to doubt that the heritability of IQ is as great as Jensen claims (Kamin 1974, Lewontin 1975). The relevant data are correlations between relatives. For example, if the environment had no effect at all, the correlation in IQ scores between parent and offspring should be 0.5, because they share half their genes on average; it should be 0.5 for siblings, and 1.0 for monozygotic ("identical") twins. If the correlation between monozygotic twins is 0.8, the heritability is 0.8, and 20 per cent of the variance in IQ scores is attributable to the environment (assuming that there is no interaction between genotype and environment, which makes such percentages meaningless; see *Chapter 14* and Lewontin 1974*b*).

But because members of the same family share not only genes but a common environment, a high correlation could as well be due to environment as to genetic similarity. To get around this difficulty, researchers have compared monozygotic (MZ) with dizygotic (fraternal, or DZ) twins, which might share a common environment to the same degree, so that the difference in the correlation between MZ and DZ twins should measure the genetic component of their similarity. Another method is to compare the IQ of adopted children with that of their biological mothers; similarity would be due to their genetic commonality, *if* there is no correlation between the environment of the foster home and that which their mothers would have provided. The strongest data come from MZ twins reared apart; it has always been supposed that if they are reared in uncorrelated environments, their similarity must be due to their genetic identity. However, the similarity of MZ twins in IQ and other features may be strongly influenced by whether or not they shared a single placenta (Melnick, Myrianthopoulos, and Christian 1978).

In all cases the critical assumption is that the correlation in envi-

1 Two hypothetical genotypes with different responses to different states of the environment. Neither genotype is superior in all environments. Because of the interaction between genotype and environment, analysis of phenotypic variation may misinterpret the relative contributions of genetic and environmental variation to the phenotypic variance. See also Figure 3, Chapter 14.

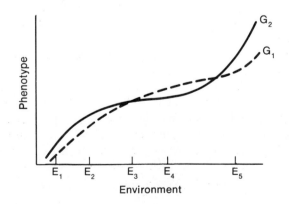

ronment has been reduced to zero; if it has not, the genetic component of similarity will be over-estimated. In all cases it is likely that this assumption has been violated (Kamin 1974, Lewontin 1975). Monozygotic twins reared together are treated more similarly (because of their physical similarity) than dizygotic twins; and when this is taken into account, the difference in IQ correlation between MZ and DZ twins yields an estimated heritability of 0.15 ± 0.16, a value not statistically different from zero and certainly far less than 0.8 (Schwartz and Schwartz 1973).

The correlations between the IQ of adopted children and their biological mothers are generally biased by the policies of adoption agencies, which place children in homes like those of their mothers; and adopted children in some studies show the same correlation with their foster parents' IQ as do those parents' own children, suggesting an important effect of environment (Kamin 1974). However, Scarr and Weinberg (1977) found that the correlation of foster parents' IQ with their adopted children was lower than with their biological children, suggesting some effect of heredity.

Studies of monozygotic twins reared apart suffer from the same criticism; the twins often remained in similar environments (usually in the same small town; one twin was often reared by relatives, and commonly the twins knew each other and sat together at school!). The sample sizes in studies of MZ twins reared apart are too small to be useful. The only large set of data is that of Cyril Burt, on whose work Jensen's (1969) conclusions depended heavily, but Burt's data are untrustworthy and are very possibly fraudulent (Kamin 1974, Jensen 1974, Wade 1976, Dorfman 1978).

Thus if there exists substantial genetic variation for intelligence, or for any other feature of human personality and behavior, it has yet to be demonstrated satisfactorily. Environmental effects, on the other hand, are clearly very important.

The most controversial of Jensen's conclusions is that the mean differences among groups of different socioeconomic standing and among "races" are largely genetic in origin. I will treat the race issue here; the considerations are much the same for socioeconomic classes. The mean IQ score of American blacks is about 15 points (one standard deviation) lower than that of whites — but there is great overlap between the groups, so a substantial fraction of blacks has higher IQ scores than the white mean. Jensen, having concluded that IQ within the white population is highly heritable, thinks it likely that the racial difference must also be largely genetically based.

This conclusion, however, suffers from two major weaknesses. Many psychologists hold that IQ tests are biased in favor of whites. The other fault, which any geneticist will recognize, is that even if a trait is 100 per cent heritable within one population, the difference between two populations may be entirely due to differences in their environment. Therefore even if IQ *is* highly heritable within the white population (for which there is no compelling evidence), the lower mean IQ of blacks could be entirely due to the differences in social conditions. Contrary to Jensen's (1973) claim, whites and blacks do not differ in IQ when they are carefully matched for such variables as family size, medical care, and other social factors (Sanday 1972). Unless these variables are carefully held constant, the groups are likely to differ in nutritional and cultural factors that even in the first few months of life can have lasting effects (Bronfenbrenner 1975). Tizard (1973) found no differences in IQ between black and white children who had spent about six months in the enriching environment of a high-quality residential nursery. The mean IQ of black children adopted by white families in Minnesota is equal to the national average, and is about 15 points higher than the black mean — strong evidence for the overwhelming effect of environment (Scarr and Weinberg 1976).

Some workers have studied people of mixed racial ancestry, whose blood group profile indicates their genetic background. Such data provide no evidence of a correlation between IQ and racial background (Loehlin, Lindzey, and Spuhler 1975; Lewontin 1976).

Why has so much effort gone into the nearly impossible task of separating genetic from environmental components of a characteristic that is hard to define, let alone measure, in a species that is so hard to study genetically? The simplest answer would be, For the potential social benefits of the research; but this cannot be so, because heritability estimates are irrelevant to any changes in social or educational programs that we might wish to institute. Several studies have shown that proper education and child care can raise the IQ of experimental groups of black or white children more than 20 points above control groups (Bronfenbrenner 1975). Because the effects of a change in environment cannot be predicted from knowing heritability values, their only use is in artificial selection — which no one seriously proposes to apply to human intelligence.

One is left speculating whether the popularity of the subject stems from an ignorance of the simplest principles of genetics or from a deep need to find an excuse for the failures of our social policies and the perpetuation of social inequalities. It is easier to lay the blame on the children than on the schools.

HUMAN RACES

Much of the search for genetic differences among races in IQ or other behavioral traits rests at least in part on the typological notion that humans can be assigned to different races and that these races should differ in behavioral as well as physical features. But as several authors (e.g., Fried and Marshall in Mead et al. 1968) have pointed out, the classification of races tends to be highly subjective. Dominant social groups have often viewed as inferior races those by whom they felt threatened; for example, New Englanders descended from the English declared the Irish immigrants a racially inferior stock in the nineteenth century; and later the same views were held of southern and eastern

Europeans. A person is often assigned to a race on sociological, not biological grounds; those of mixed ancestry are usually assigned to the socially subordinate race. Thus, although about 30 per cent of the genes in the socially defined black population of the United States are derived from white ancestry (Glass and Li 1953), most of the descendants of the original miscegenations are termed black.

The four to six commonly recognized races, based on a few physical features such as skin color, hair form, and the epicanthal fold, are categories that do not reflect real patterns of variation any more than do subspecies in other species (*Chapter 9*). These characteristics and others have rather independent patterns of distribution, so that a classification based on, say, frequencies of blood types would divide the world's population into races very differently (*Figure 2*). Thus there is no reason to expect IQ or other traits to be correlated with the features conventionally used to delimit races.

Still less is there any reason to believe that the differences among populations in bellicosity, mating patterns, economic systems, or anything else have a genetic basis. Every cultural pattern, as Harris (1975) points out, is far too heterogeneous among neighboring populations that clearly exchange genes to be accounted for by differences in gene frequency.

GENETIC CHANGES IN MODERN HUMAN POPULATIONS

The question often arises whether human populations in industrial society are presently undergoing genetic change and whether these changes might be deleterious. Two fears are often voiced: that radiation, mutagenic food additives, or other products of our technology may increase the incidence of deleterious mutations, and that less intelligent people may have a higher reproductive rate and so cause an evolutionary decline in average intelligence.

The possibility that deleterious alleles are increasing in frequency, because of increased mutation rates, is real, although such an increase has not been documented. But even a substantial increase in the rate of mutation would have almost unnoticeable effects because natural mutation rates are so low (*Chapter 10*). The burden of illness and suffering caused by inborn errors of metabolism is already enormous, because of genetic polymorphisms such as sickle-cell hemoglobin and Tay-Sachs disease, compared with any additional genetic load that an increase in mutation rates might cause. It is to the amelioration of these common genetic diseases that attention might most profitably be addressed.

How should the problem be approached? The two possible modes of action are EUGENICS, programs of selective breeding that might change the frequencies of deleterious alleles, and EUPHENICS, modifications of the phenotype by medical or dietary means that enable a person to function normally despite his or her genetic infirmity. Most eugenic programs are unlikely to lower the frequencies of deleterious genes (especially recessives) substantially, unless virtually all carriers

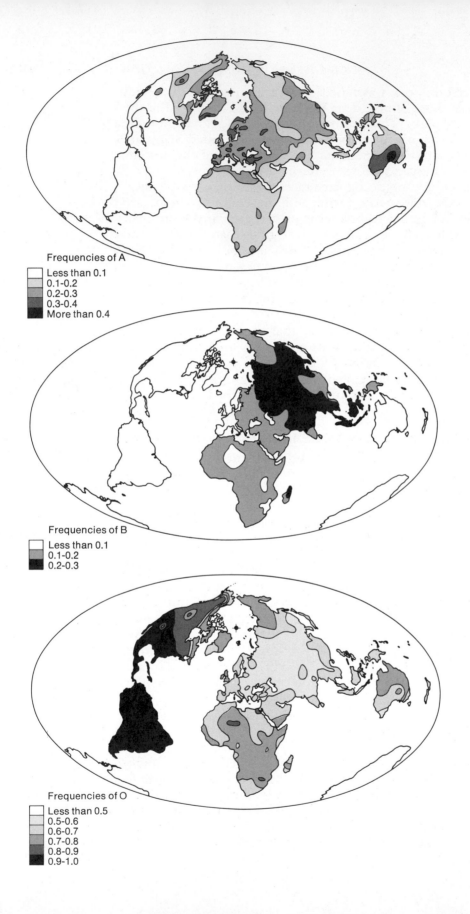

Frequencies of A

Less than 0.1
0.1-0.2
0.2-0.3
0.3-0.4
More than 0.4

Frequencies of B

Less than 0.1
0.1-0.2
0.2-0.3

Frequencies of O

Less than 0.5
0.5-0.6
0.6-0.7
0.7-0.8
0.8-0.9
0.9-1.0

are dissuaded (or prevented) from reproducing; and it would probably be impossible to implement such a program without violating human rights. GENETIC COUNSELING, however, can inform people of the chances that their children will suffer from genetic maladies that are known or suspected to run in the family, so that such individuals can decide whether to have children. Rather than attempting the impossible task of eliminating deleterious alleles from the gene pool, however, it is better to relieve the effects of the genes by euphenic methods, which should be a major object of medical research. That such programs can be successful is demonstrated by the treatment of diabetics with insulin, or of people suffering from phenylketonuria by a diet low in phenyl-alanine. We need not fear that such treatment will, by relaxing selec-tion, cause a substantial increase in the frequency of the deleterious alleles, for they will increase only as a function of the mutation rate, which by itself has hardly any effect on altering gene frequencies (*Chapter 10*). We should not advocate, of course, that levels of radiation or chemical mutagens be allowed to increase in the environment, for they are likely to cause *some* increase in genetic load, and besides, many of them are highly carcinogenic.

The question of whether other traits are evolving actively depends on whether genotypes differ in their reproductive rate, for selection requires differential mortality or fertility among genotypes. Because prereproductive mortality has been so greatly lowered by technology and medicine, about 92 per cent of the opportunity for selection (the total variance among people in the number of surviving offspring they have) in the modern U.S. population is due to difference in fecundity (number of children) (Crow 1961). But although the opportunity for selection exists, there will be no selection at all unless fecundity differs consistently among genotypes (*Chapter 12*). There is a strong sugges-tion that differential mortality and fertility have some genetic effect on certain physical traits such as stature and birth weight, but the herita-bility of these characters is so low that the effect of selection seems to be slight (Bajema 1971, Cavalli-Sforza and Bodmer 1971). But there is little evidence of substantial genetic variation in intelligence (IQ), much less any evidence that more fecund classes are genetically differ-ent in intelligence from less fecund classes (Lewontin 1975). There is no reason to suppose that average intelligence is declining because of genetic change.

2 Distribution of allele frequencies of the ABO blood group system, illustrating the arbitrary nature of "races." If races were defined by this character, they would have different distributions from those presently recognized. (From *Culture, people , nature,* 2nd ed., by Marvin Harris. Copyright © 1971, 1975 by Harper and Row Publishers, Inc. Reprinted by permission of Thomas Y. Crowell)

UNIVERSAL HUMAN TRAITS

In recent years there has been a resurgence of the opinion (Ardrey 1966, Lorenz 1966, Tiger and Fox 1971, Wilson 1975a, Barash 1977) that behavioral universals — incest taboos, aggression, sexual behavior, and the like — are "natural" qualities, inherited from our prehuman ancestors. This belief is very old; Hobbes viewed human nature as brutal and nasty, and Rousseau envisioned only good in the "noble savage," since defiled by civilization. There has existed through the centuries the conviction that there is within us a human nature trammeled by cultural bonds and struggling to get free — to reestablish the Garden of Eden or to wreak havoc, depending on one's philosophical views. But as Marston Bates (1968) has argued in a delightful book, human nature is elusive and does not exist independent of culture. Culture is not an extrinsic factor, a veneer laid over our true selves; it is an intrinsic part of our being. To develop in a state of nature like the "wild boys" that fascinated society in the eighteenth century is unnatural. If the argument that I have presented is correct, that we have been selected for flexibility in the entire range of our behavior, it is natural for us to be selfish or altruistic, aggressive or pacific, monogamous or polygamous, heterosexual or homosexual, extroverted or introverted, impulsive or reflective — depending on our interactions with the social context in which we develop.

The argument over the biological basis of human behavior has been recently revived by the controversy over sociobiology (Wilson 1975a, b, 1976, 1978; Trivers 1971; Sahlins 1976; Barash 1977; Sociobiology Study Group 1976, 1977; Lewontin 1977a, b; Caplan 1978). To illustrate the nature of the debate, I shall analyze the speculation, especially as framed by Barash (1977), that the behavioral differences between the sexes are due in large part to a genetically programmed difference in the expression of the universal human genome.

Many authors (e.g., Tiger and Fox 1971, Wilson 1975a) note that the sex differentiation of human behavior (e.g., male dominance) is similar to that in many other mammals, including other primates. They therefore postulate that the behavior is homologous and has been inherited as a conservative evolutionary trait during the evolution of the hominid lineage. But phenotypic similarity is not evidence of genetic homology (*Chapter* 7), because convergence is so common in evolution. And a trait that has been conservative throughout a group can change rapidly when one lineage in that group becomes subject to very different selective pressures; possession of two aortic arches is very conservative in the reptiles — except for those that became birds and mammals. Thus analogies between the behavior of humans and other species are just that — analogies, and not homologies — until we can analyze the material basis of behavior in the same way as for morphological characters.

The sociobiological position, however, does not depend on the argument from homology. It consists of hypothesizing what behaviors should be adaptive and finding congruence between predicted and observed behaviors. Barash's (1977, pp. 290–301) speculations rest on two principles. First, mating entails a greater investment for the female than for the male, since she may become committed to pregnancy and thus have to forego other reproductive opportunities; thus the cost (in Darwinian fitness) of a reproductive mistake is greater for the female. Second, the female of a mated pair is certain that the offspring are hers, but the male is not; therefore the coefficient of relationship is greater, on average, between offspring and female than between offspring and male, so there is greater opportunity for kin selection (*Chapter 12*) to shape the interaction between female and offspring than between male and offspring.

From these principles, Barash argues that females are likely to be more discriminating than males in choosing an appropriate mate, to be more coy and less forward in sexual encounters, and to choose males who are likely to have resources to provide for offspring. Males are expected to compete in intermale encounters for the resources that make them sexually attractive. Women will tend to mate above their socioeconomic station, often with older men, who are likely to have more resources than younger men. Because the ability to command respect and acquire resources will be selected more in males than females, men are likely to feel threatened by successful, competent women. Because a male risks wasting his energy on children who are not really his, men should have evolved to be more intolerant than women of their mates' sexual infidelity. Because of the low cost of reproduction to the male, men are expected to be more promiscuous than women and will not have as strong a sense of parental care. They will, instead, devote their energies to increasing their social status and sexual attractiveness. Thus, suggests Barash, "women have almost universally found themselves relegated to the nursery while men derive their greatest satisfaction from their jobs." The predictions that emerge from this evolutionary logic are thus remarkably similar to the sex roles with which we are too familiar, although Barash rightly points out that what may be biological is not necessarily ethical or immutable.

The criticisms of this evolutionary scenario apply to many of the sociobiological interpretations of human behavior. Western cultural patterns are not necessarily the norm for the species, for sex-differentiated behavior varies greatly within and among cultures (Mead 1935, Harris 1975). There is no evidence that sexual differentiation in behavior is caused by anything other than cultural effects. The enormous, confused psychological literature on this subject has been reviewed by Maccoby and Jacklin (1974; see also Bart 1977); they conclude that most behavioral differences between the sexes arise rather late in childhood and are probably due to differential treatment of girls and boys from an early age. But it is virtually impossible to distinguish genetic from environmental effects, because female and male infants inevitably differ both in their biological nature (potentially) and in the ways they are treated (actually). Thus even traits related to aggressiveness, which Maccoby and Jacklin (1974) believe differ biologically between the sexes because they are differentiated even in early infancy, could well be the results of very early conditioning (Birns 1976). The critical experiment, of treating both sexes exactly the same, has not been and probably cannot be done. In the closest approximation to such an experiment, individuals affected by prenatal hormones have been born with hermaphroditic sexual organs, or with those of the opposite sex. Although in such cases hormones seem to have some effect on sex-differentiated behavior in childhood, the behavior corresponds for the most part to that of the sex they are reared as, rather than to the individuals' "true," genetic, sex (Money and Ehrhardt 1972).

However, sociobiologists give little indication of how one can tell whether a trait is cultural or genetically programmed. In discussing the observation that men in polygynous human societies have only as many wives as they can support, Barash (p. 292) notes that "this phenomenon appears to be entirely cultural, but it is also adaptive, and the parallels with other animals are striking." Thus he indicates that similarity to other species is not evidence of genetic homology; but how do we know that this phenomenon is "entirely cultural"? And if this phenomenon is entirely cultural, why may not other "adaptive" behaviors, such as sex roles, also be entirely cultural?

The sociobiological argument assumes that various sex-differentiated behaviors are neurologically (and genetically) independent of each other and of other aspects of behavior, despite the lack of evidence of any genetic basis for such behavior. This assumption is prerequisite to most fundamental assumption of all — that each of the behaviors is, or at one time was, adaptive (for if the various behaviors are not genetically independent of each other to at least some degree, they cannot each be molded by natural selection into optimal states). But there is no evidence that these behaviors are adaptive. What is the evidence that men's reproductive success is enhanced by promiscuity, aggressiveness, and a negligent approach to child care, or that the fittest females are the most coy, monogamous, or submissive?

The sociobiological theory is, to a great extent, an exercise in plausibility that can be stretched to make any prediction whatever. For example, males might maximize fitness by being promiscuous and not maintaining a pair bond — or, if the survival of offspring depends on parental care, males might maximize fitness by being faithful to their mates and deriving from the pair bond the twofold advantage of helping their offspring and preventing insemination by other males. Choosing between these diametrically opposed sociobiological hypotheses would require knowledge of a whole complex of interacting factors that impinged on our ancestors (e.g., population density, risk of infant mortality, degree of kinship among competing males, and even the frequency of promiscuous vs. nonpromiscuous males), none of which we know. Moreover sociobiologists accept conflicting kinds of evidence as support for their hypotheses. For example, the universality of a trait such as female concern with child care is often taken as evidence that it is genetically canalized. Yet variation among cultures is not taken as evidence against genetic programming; Barash argues that men are biologically sexually possessive of women, despite the existence of societies in which men share wives (p. 296). This is easily explained as adaptive (it fosters cooperation).

If one assumes that everything is adaptive, it is easy to imagine a mode of selection that would favor any trait. If the behavior might benefit the individual (e.g., selfishness), ordinary natural selection will do. If it is individually deleterious, but might help relatives (e.g., old Eskimos going off to die when food is scarce), kin selection is invoked. If the behavior helps mostly nonrelatives, the principle of "reciprocal altruism" (Trivers 1971), i.e., "you scratch my back and I'll scratch yours," can be the *deus ex machina,* despite the lack of evidence that reciprocal altruism exists in nature. Trivers (1971) uses this principle to explain the existence of friendship and morality among unrelated people.

This panoply of selective agents can explain everything — so it explains nothing. One cannot imagine a trait for which a selective advantage cannot be conceived (Sociobiology Study Group 1977). The theory thereby becomes untestable, because there is no critical test that can falsify the hypothesis that the trait is adaptive. This theory of perfect adaptation has the trappings of modern population biology, yet it is not scientifically rigorous and it ignores much of developmental biology, population genetics, and evolutionary theory, which have demonstrated that organisms are rife with characteristics that are not and never were adaptive (*Chapter 17*).

SEXUAL BEHAVIOR

It is reasonable to argue that if any human behavior should be highly canalized and instinctive, sexual behavior should be, since it is intimately related to Darwinian fitness. If sexual behavior is not strongly canalized, we should expect most of the rest of our behavior to be even less so. But sexual behavior seems to have few, if any, instinctual controls.

Consider sexual orientation: whether an individual is erotically inclined toward members of the same or the opposite sex, or both. Modern Western culture has traditionally viewed homosexuality as a "crime against nature," or at best a pathology, and has visited the most appalling oppression on homosexuals. But this view is highly culture-bound and is more typical of the sexually repressive Judaeo-Christian tradition than of most other cultures, in which homosexual behavior is often as much a part of the cultural norm as heterosexual behavior, as a reading of Plato's *Symposium* or familiarity with the sexual mores of some Moslem cultures will illustrate.

The incidence of homosexual behavior nonetheless appears to be remarkably constant across cultures, social strata, and historical periods (Churchill 1967), a fact that suggests to some the existence of a genetic polymorphism in sexual orientation. Wilson (1975a, 1978) poses a sociobiological explanation for the high incidence of putative genes for homosexuality: even though "of course homosexual men marry much less frequently and have far fewer children" (an assertion for which there is no evidence, given the high incidence of homosexual behavior among "heterosexual" men), kin selection could favor such genes if homosexuals helped with child care. Generations of psychologists, seeking the causes of this "aberration" so that it could be "cured," have looked for hormonal differences between homosexual and heterosexual people and for evidence of transmission within families, but they have been as unsuccessful in finding clear evidence of a biological basis of sexual orientation as in finding simple psychological bases such as a dominant mother (Marmor 1965, Hoffman 1976, Tripp 1975).

The argument evaporates if we abandon the notion that heterosexuality is normal or natural and that homosexuality is a deviation from normality. Rather it is likely (Ford and Beach 1951, Tripp 1975) that we are born with a generalized, undifferentiated sexual urge, at first directed toward anyone; as we experience enormous numbers of subtle positive and negative reinforcing stimuli, our erotic interest becomes more and more focused — toward one sex or the other, toward older or younger men or women, toward those of higher or heavier stature, or of dark or light complexion. It is only because homosexuality has been considered "abnormal" that anyone worries what causes it; no one does comparable research on the question of why some men are attracted to blondes and others to brunettes or postulates that these different sexual orientations are caused by different genes. Our sexual orientation is based on learning, in the widest sense; and if this is true of behavior so intimately related to fitness, it is even more likely to be true of the rest of our behavior.

AGGRESSION

Largely because of the perceived similarities between some human behavior and that of other species, some popular writers (e.g., Ardrey 1966) have claimed that we are descended from territorial ancestors and so are naturally inclined to be possessive, warlike, and generally nasty. Ethologists (e.g., Lorenz 1966) have proposed that like other species, we evolved ritualized submissive behaviors that stayed the hand of the aggressor, but that modern depersonalized warfare prevents the expression of these behaviors. Sociobiologists (e.g., Wilson 1975a) argue that we are not altogether red in tooth and claw, but that we evolved, by kin selection, to be altruistic to members of our kin group and by extension to the whole of whatever group we identify with, although we remain hostile to aliens.

But these are, again, nonrigorous or even meaningless propositions. Similarity does not necessarily imply genetic homology; and there is no reason to believe that territoriality in animals is at all the same phenomenon as individual human aggressiveness, much less warfare. Indeed, as Harris (1975) and Sahlins (1976) point out, the feelings that impel a person to kill in war are less likely to be anger or territoriality than fear of authority or desire for status. We have manifest evidence that appeasement behavior often does not deter humans from killing one another. Some cultures are highly aggressive, while others foster pacifism, yet there is no evidence that they differ genetically in this respect (Montagu 1973). All in all, since each of us clearly has the capacity to be kind or mean, aggressive or pacific, as the situation warrants and as our training has inclined us, it is as meaningless to say that we are naturally aggressive as to say that we are naturally peaceful. I for one do not believe that the pacifism of the Quakers or the savagery of the Huns lay in their genes. We are each born a potential Buddha and a potential Hitler; and it is up to us in our molding of our society to determine which shall prevail.

SUMMARY

Whether evolutionary biology is important for understanding human behavior is a highly controversial question. Some authors attribute much of the behavioral variation within and between groups to genetic variation; others point out that there is no unequivocal evidence of substantial genetic variation in human behavior and argue that variation arises from social and cultural conditions. Likewise some view supposedly universal human traits — "human nature" — as genetically canalized, evolved adaptations; others see these traits as common responses to common environmental conditions and point to rapid historical changes and cultural variation among peoples as evidence of the immense behavioral flexibility of which all people are capable. On balance, the evidence for the modifiability of human behavior is so great that genetic constraints on our behavior hardly seem to exist. The dominant factor in recent human evolution has been the evolution of behavioral flexibility, the ability to learn and transmit culture.

CODA

I have taken a strong environmentalist position in this chapter and argued that the biological determinist position is not strongly supported either by evolutionary theory or by evidence. I have stressed these points as an antidote to the strongly genetic, selectionist point of view that has gained wide currency both in biology books and in the popular press. For it is, I believe, a *hubris,* an overweening pride, on the part of biologists to assume that biology necessarily can offer more insights into human behavior than can the many other disciplines that take humanity as their special concern — as if anthropology, psychology, sociology, history, and yes, philosophy, ethics, and the arts were hopelessly lost without the pure light of evolution. Evolution, and natural selection in particular, are immensely important and useful concepts. They explain much of the natural world, have numerous practical applications, and have philosophical value in emphasizing that humanity is part of nature. But like all other concepts and knowledge, evolutionary theory should be used for our good. It bears within it no moral force or obligation and should not be used to rationalize violations of the ethical codes that we, as sentient, empathic beings who can act as if we possess free will, decide on. Evolutionary biology should not constrain our ethics, demean the visions of our poets, philosophers, and spiritual leaders, or restrain us from the ideals to which we aspire.

> *O holde Ruhe, steig hernieder,*
> *kehr in der Menschen Herzen wieder;*
> *dann ist die Erd' ein Himmelreich,*
> *und Sterbliche sind Göttern gleich.*

Mozart, *Die Zauberflöte*

FOR DISCUSSION AND THOUGHT

1 Many excellent treatises on human genetics (e.g., Dobzhansky 1962, Cavalli-Sforza and Bodmer 1971, Lerner 1968) accept the view that there is a fairly high genetic component to individual variation in IQ and in various behavioral traits. After reading one or more of these treatments, discuss the differences in the conclusions reached by those authors and this book.

2 Suppose that the heritability of IQ within populations is indeed greater than zero; suppose further that within the current social and educational environment there are genetic differences among races or other groups in mean IQ levels. What practical differences should this make to social and educational policy?

3 Suppose that I am wrong, and that sexual orientation is in part genetically based, or that men are indeed biologically more prone to aggressive, dominant behavior than women. Should such conclusions have practical consequences for social policy? Exactly what would they be, and why?

4 It is often claimed that if we could arrive at a deep understanding of the evolution of human behavior, we would be in a better position to shape our social institutions to our benefit. I am dubious of this. Exactly which social institutions could profit from such understanding, and how?

5 Wilson (1975a), having argued that much of our social behavior has evolved through kin selection, expresses the fear that the high mobility of people in the modern world will lower coefficients of consanguinity, so that there could be an "eventual lessening of altruistic behavior through the maladaptation and loss of group-selected genes" (p. 575). Analyze the assumptions of this argument.

6 To deny that human behavior is genetically controlled seems to many people to imply the primacy of social conditioning — that our behavior is determined entirely by conditioned reflexes, as the behavioral psychologist B. F. Skinner believes. Is there a middle course between the Scylla of biological determinism and the Charybdis of environmental determinism?

7 It is hard to deny that some human behaviors, such as nursing behavior in infants, are innate, or highly canalized. Certain others seem very probably homologous with those in other primates, for example the facial expressions associated with certain emotions such as anger or fear. It is even possible that all languages share a common deep structure, as Noam Chomsky (1972) has argued, determined by the genetic architec-

ture of the brain. Can it be argued from such examples that the behaviors entailed in social intercourse are genetically constrained? Can we distinguish socially important and socially unimportant traits, or those that are more vs. less genetically canalized? What is the relation between the two classifications?

8 For each of the following observations about Western culture, provide an evolutionary (sociobiological) explanation and a nonevolutionary (cultural) explanation, and judge their relative merits. (a) Young people are often more adventurous and less conservative than older people. (b) Most people would rather rear their own children than adopt those of others. (c) Incest is considered immoral. (d) Men are more sexually promiscuous than women. (e) Many people accept religious and political doctrines with little question. (f) Many people are suspicious of or hostile to those unlike themselves in religious or cultural background, skin color, or sexual orientation. (g) Monogamy in heterosexual relationships is held as an ideal. (h) Men are more forward in courtship, women are considered more coy.

9 It can be argued that the distinction between political conservatives and liberals is a distinction between those who hold to the doctrine of original sin, which says that we are by nature nasty creatures requiring the rule of law to govern our antisocial impulses, and those who hold to a doctrine of human perfectability. Trace the history of the ways in which evolutionary theory has been invoked to support each of these positions, and interpret recent writings on human evolution in light of this history.

MAJOR REFERENCES

Dobzhansky, Th. 1962. *Mankind evolving.* Yale University Press, New Haven, Conn. 381 pages. A somewhat dated but excellent account of human genetics and evolution; the discussion of genetic variation in human behavior needs to be considered in light of recent developments.

Cavalli-Sforza, L. L., and W. F. Bodmer. 1971. *The genetics of human populations.* Freeman, San Francisco. 965 pages. An excellent integration of population genetic theory with data on human genetics.

Lewontin, R. C. 1975. Genetic aspects of intelligence. *Ann. Rev. Genet.* 9: 387–405. An analysis of the inadequacies of the methods and data on which claims of genetic variation in intelligence are based.

Wilson, E. O. 1975. *Sociobiology: the new synthesis.* Harvard University Press, Cambridge, Mass. 697 pages. The outstanding treatise on the evolution of social behavior in animals. Chapter 27, on humans, is highly controversial.

Harris, M. 1975. *Culture, people, nature. An introduction to general anthropology.* 2nd ed. Thomas Y. Crowell, New York. A comprehensive introduction to physical and cultural anthropology, with an enlightened understanding of evolution.

Lewontin, R. C. 1977. Biological determinism as a social weapon. In *Biology as a Social Weapon,* edited by The Ann Arbor Science for the People Editorial Collective. Burgess, Minneapolis, Minn.; pp. 6–18. A history of the political uses of concepts of human genetics and evolution. See also "Sociobiology — A new biological determinism," by the Sociobiology Study Group, in the same volume, for a vigorous critique of human sociobiology.

Means, Variances, and Correlations

This appendix introduces the statistical concepts and notation used in the text for readers to whom such material is unfamiliar. More extensive treatments are found in introductory books on statistics; *Quantitative Zoology* by Simpson, Roe, and Lewontin (1960) is an excellent introductory treatment that stresses biological applications.

THE ARITHMETIC MEAN

Let X be the value of some measured variable, for example the length of a snake's tail. X_i is the value for the ith observation (snake 3 might have the value $X_3 = 10$ cm). If there are n observations, the sum of the n tail lengths is denoted $\sum_{i=1}^{n} X_i$ (or simply ΣX_i). The ARITHMETIC MEAN (commonly known as the AVERAGE) is then

$$\bar{x} = \frac{\Sigma X_i}{n} \, .$$

(Other kinds of means can be calculated and are used in evolutionary theory, but I have used them very little in this book.) If X is a discrete variable such as the number of scales, there may be n_1 individuals with X_1 scales, n_2 with X_2, and so on, where $n_1 + n_2 + \ldots + n_k = n$. Then the arithmetic mean is

$$\bar{x} = \frac{n_1 X_1 + n_2 X_2 + \ldots + n_k X_k}{n_1 + n_2 + \ldots + n_k} = \frac{n_1 X_1 + n_2 X_2 + \ldots + n_k X_k}{n}$$

$$= \frac{n_1}{n} X_1 + \frac{n_2}{n} X_2 + \ldots + \frac{n_k}{n} X_k.$$

If we set $n_i/n = f_i$, the *frequency* of individuals with value X_i, this becomes

$$\bar{x} = \sum_{i=1}^{k} (f_i X_i).$$

A special case that is most important in genetics is the binomial distribution, in which the probability that an event of the ith type will occur is p_i. Let there be two possible events, 0 (*e.g.*, heads if a coin is tossed) and 1 (tails) with probabilities q and p respectively. Since there are only two possible events, $q = 1 - p$. The weighted sum of the values in a series of n trials (coin tosses) is then $n(1 - p)(0) + np(1) = np$. Dividing by n, we find the mean of the probability distribution of the two events, $\bar{x} = p$.

VARIATION

We are also interested in the variation represented by a series of measurements. The most useful measure of variation is the VARIANCE, defined as the mean value of the square of an observation's deviation from the arithmetic mean of the population or sample:

$$V = \frac{(X_1 - \bar{x})^2 + (X_2 - \bar{x})^2 + \ldots + (X_n - \bar{x})^2}{n} = \frac{1}{n} \sum_{i=1}^{n} (X_i - \bar{x})^2.$$

Each X_i value might occur several times, with frequency f_i; if there are k different values (classes) of X_i, the variance can be written

$$V = \frac{n_1(X_1 - \bar{x})^2 + n_2(X_2 - \bar{x})^2 + \ldots + n_k(X_k - \bar{x})^2}{n} = \sum_{i=1}^{k} f_i(X_i - \bar{x})^2,$$

or

$$V = \left(\sum_{i=1}^{k} f_i X_i^2 \right) - \bar{x}^2.$$

Note that a variance is always positive; that the farther an observation is from the mean, the more it contributes to the variance; and that the more observations there are that deviate greatly from the mean, the greater the variance is. V is thus more sensitive to variation than the simple range between the smallest and largest observations. If $X_i - \bar{x}$ is written as the deviation d_i, the variance is the same if the mean is set to 0 by subtracting $\bar{x}$ from each X_i, in which case the value of each observation is $X_i - \bar{x} = d_i$.

For the binomial distribution, the probability, or expected frequency f_i, of 0 is $1 - p$, and that of 1 is p. Then because $\bar{x} = p$,

$$V = \Sigma f_i(X_i - \bar{x}) = (1 - p)(0 - p)^2 + p(1 - p)^2 = p(1 - p).$$

This is the variance of the probability distribution of heads and tails, for example. If we toss the coin n times, we expect the proportion of tails to be p (the mean of the probability distribution), but in practice it may not be exactly p. In repeated sets of n tosses, p will vary from one set to another; and the smaller n is, the larger the variation in p will be. The variance of p, in repeated sets of n tosses, is

$$V = \frac{p(1 - p)}{n}.$$

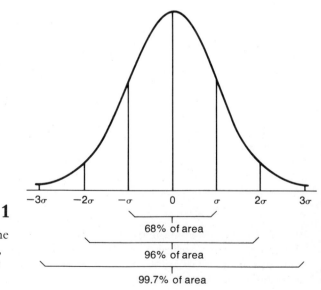

1

The normal distribution curve, showing the fraction of the area embraced by one, two, and three standard deviations (σ) on either side of the mean.

TABLE I. Additivity of variances

	Group 1	Group 2
X values	4, 4, 5, 6, 6	1, 2, 3, 4, 5
Group means	5	3
Sum of squared deviations from group means	4	10
Within-group variances	0.8	2.0

The mean within-group variance is $(0.8 + 2.0)/2 = 1.4$. The mean of the group means is $(5 + 3)/2 = 4$, so the variance between group means is $[(5 - 4)^2 + (3 - 4)^2]/2 = 1.0$. The sum of within-group and between-group variances is thus $1.4 + 1.0 = 2.4$.

This is identical to the variance calculated for the data as a whole, for which $\bar{x} = 4$. $V = [(4 - 4)^2 + (4 - 4)^2 + \ldots + (5 - 4)^2]/10 = 2.4$.

This is important in genetics. If the proportion of A alleles in a population is p, repeated samples of n individuals ($2n$ genes) will vary in allele frequency, with variance $p(1 - p)/2n$.

Because it is expressed in squared units (cm², for example), the variance is not as easily visualized as a related measure of variation, the standard deviation S (often denoted s or σ, just as the variance is often denoted s^2 or σ^2). This is the square root of the variance: $S = \sqrt{V}$. It is most easily visualized if the frequency distribution of X values forms a bell-shaped, or normal, curve (*Figure 1*). Because of the mathematical form of the distribution, S constitutes a fixed fraction of the area under the curve; for example, 68 per cent of the observations fall within one standard deviation on either side of the mean. If S is large, this fraction of observations must spread out farther from the mean than if S is small; thus a large S implies a broad, variable distribution.

An important property of variances is their additivity. If a set of observations falls into several groups, the variance of the entire set is the sum of the within-group and among-group variances. A simple example is provided in Table I.

There may be many components to the total variance. Table II offers some

TABLE II. Hypothetical effects of fertilizers on plant weights

	(a) No interaction PHOSPHORUS			(b) Interaction PHOSPHORUS	
NITROGEN	low	high	NITROGEN	low	high
low	9,11	14,16	low	9,11	14,16
high	19,21	24,26	high	19,21	29,31

Increased phosphorus adds 5, on average

Increased nitrogen adds 10, on average

(High N, high P) − (low N, low P) = 15, on average

High phosphorus adds 5 if nitrogen is low, 10 if nitrogen is high

High nitrogen adds 10 if phosphorus is low, 15 if phosphorus is high

hypothetical weights of eight individual plants grown under four combinations of low and high levels of nitrogen and phosphorus. In part *a*, the total variance is the sum of (1) the average variance within each treatment, (2) the variance among rows (nitrogen levels), and (3) the variance among columns (phosphorus treatments). In part *b*, however, there is a fourth component of the variance, that due to the interaction between the two fertilizers ("the whole is greater than the sum of its parts"). In statistical terms the fertilizers do not interact to influence weight in case *a*; they are said to have additive effects.

CORRELATIONS AMONG VARIABLES

Table III presents hypothetical data on the body length X and tail length Y of each of five individuals. The mean total length $\bar{z}$ is the sum of the separate means $\bar{x}$ and $\bar{y}$. The variance of the total length V_Z would be the sum of the separate variances V_X and V_Y, except that there is a clear correlation between X and Y. Rather, V_Z is found to be

$$V_Z = \frac{1}{n} \sum_{i=1}^{n} [(X_i + Y_i) - (\bar{x} + \bar{y})]^2$$

$$= \frac{1}{n} \sum (X_i - \bar{x} + Y_i - \bar{y})^2$$

$$= \frac{1}{n} [\sum(X_i - \bar{x})^2 + \sum (Y_i - \bar{y})^2 + 2 \sum (X_i - \bar{x})(Y_i - \bar{y})]$$

$$= V_X + V_Y + 2 \operatorname{Cov}(X, Y).$$

That is, the variance is inflated by twice the term $(1/n) \sum (X_i - \bar{x})(Y_i - \bar{y})$, the covariance, which measures the dispersion of values around the joint mean $\bar{x}, \bar{y}$.

The covariance is closely related to the CORRELATION COEFFICIENT between the variables X and Y, which may be written

$$r_{XY} = \frac{1}{n} \frac{\sum_i (X_i - \bar{x})(Y_i - \bar{y})}{S_X S_Y} \; ,$$

where S_X and S_Y are the standard deviations of X and Y respectively. The

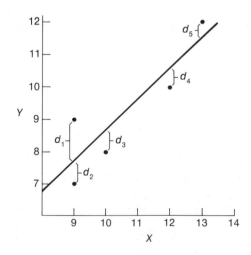

2

Example of linear regression of a dependent variable Y on an independent variable X. The regression line has a slope b and Y-intercept a such that the sum of squared deviations of the points from the line ($\Sigma_i d_i^2$) is minimized.

TABLE III. Correlation and regression

Specimen	Tail length (Y)	Body length (X)	Total length (Z)
1	7	9	16
2	9	9	18
3	8	10	18
4	10	12	22
5	12	13	25
Σ	46	53	99
Mean	$\bar{y} = 9.2$	$\bar{x} = 10.6$	$\bar{z} = 19.8$
Variance	$s_Y^2 = 2.96$	$s_X^2 = 2.64$	$s_Z^2 = 10.56$

Covariance $\qquad\qquad\qquad\quad$ $cov(X,Y) = 4.96$
Correlation coefficient $\qquad\quad$ $r_{X \cdot Y} = 0.89$
Regression coefficient, Y on X $\quad$ $b_{Y \cdot X} = 0.838$
Y-intercept $\qquad\qquad\qquad\quad$ $a = 1.106$
Regression equation $\qquad\qquad$ $Y = 1.106 + 0.838\,X$

correlation coefficient r_{XY} ranges from +1 for variables that are perfectly positively correlated to −1 for those that are perfectly negatively correlated. It measures the degree of association of the two variables.

Figure 2 plots the data from Table III and introduces another measure of association between Y and X. If Y depends on or is caused by X, it is a dependent variable, and X is the independent variable; for example, Y might be the phenotype of offspring, X that of parents. In this case it is possible not only to specify that they are correlated, but to predict Y from X. If the points approximate a straight line, the predictive equation is $Y = a + bX$, where b is the COEFFICIENT OF REGRESSION of Y on X, the slope of the line illustrated in the figure. This is the line from which the sum of squared deviations of the Y values is minimal. The regression coefficient $b_{Y \cdot X}$ is related to the correlation coefficient:

$$b_{Y \cdot X} = r_{XY}\,\frac{S_Y}{S_X};$$

but they are not identical unless X and Y have equal standard deviations. The regression coefficient describing the relation between the phenotypes of parents (X) and their offspring (Y) is used to calculate the heritability of a trait (*Chapter 15*). Because the phenotypes of parents and offspring are usually not perfectly correlated $(r_{XY} < 1)$, $b_{Y \cdot X}$ and therefore the heritability is usually less than 1.

The regression coefficient can be calculated as

$$b_{Y \cdot X} = \frac{\sum\limits_{i=1}^{n} [(X_i - \bar{x})(Y_i - \bar{y})]^2}{\sum\limits_{i=1}^{n} (X_i - \bar{x})^2}.$$

List of Symbols

A An allele (also A', a) or locus; area (*Chapter 5*); used as a subscript in V_A, additive genetic variance

a An allele; allometric coefficient (*Chapter 8*); the phenotypic effect of an allele in homozygous form (*Chapter 11*)

α_{ij} Competitive effect of species i on species j (*Chapter 18*)

b Per capita birth rate (*Chapter 4*); constant in the allometric equation (*Chapter 8*); an allele (likewise B, B')

$b_{Y \cdot X}$ Regression coefficient

Cov   Covariance

D Coefficient of linkage disequilibrium; D_N, genetic distance between populations (*Chapter 16*); frequency of dominant homozygote (*Chapter 10*)

d Derivative (as in dq/dt); phenotypic value of a heterozygote (*Chapter 11*); distance between niches (*Chapter 18*)

δ Per capita death rate (*Chapter 4*)

Δ Difference, or change in (as in $\Delta q / \Delta t$)

E Extinction rate

e Base of natural logarithms, 2.718

ϵ Coefficient of epistasis (*Chapter 14*)

F Inbreeding coefficient of a population; F_t, F at time t

f Inbreeding coefficient of an individual; $f(\ \ )$, function of; f_i, frequency of event or item i

G, g Gamete frequencies

H Frequency of heterozygotes; H_E, H_F, frequencies of heterozygotes in randomly mating and inbred populations, respectively; H_t, frequency of heterozygotes at time t

h^2 Heritability; h, degree of dominance, with respect to fitness, of an allele with fitness $1 - s$ in homozygous condition (*Chapter 13*)

I Immigration rate; I_N, genetic similarity of two populations (*Chapter 16*)

i, j Counters; for example, A_i and A_j, the ith and jth alleles at locus A

K Equilibrium density of a population, or "carrying capacity"

L Genetic load

l_x Probability of survival from birth to age x

500

m	Fraction of reproducing individuals in a population that are immigrants, in a given generation; *i.e.*, rate of gene flow
m_x	Average fecundity of a female of age x
N	Usually population size (N_0 and N_t, population sizes at times zero and t, respectively); also number of chromosome complements in the genome (e.g., $2N$ = diploid)
N_e	Effective population size
n	Any number (of generations, individuals, and so on)
O	Origination rate
P	Density of predators (*Chapter 18*)
p, q	Allele frequencies; $\hat{p}$, $\hat{q}$, allele frequencies at equilibrium
q_m	Allele frequency among immigrants into a population
R	Frequency of crossing-over; the replacement rate, or ratio of population sizes in successive generations (hence also the fitness of a genotype); the phenotypic response to selection in one generation; rate of increase of species diversity; frequency of recessive homozygote
r	The instantaneous rate of increase of a population; occasionally an allele frequency; fitness of an allele or genotype; correlation coefficient
r_m	The intrinsic rate of increase of a population, i.e., the instantaneous rate of increase (r) when a population has a stable age distribution and is at low density
S	Number of species; selection differential (*Chapter 14*); rate of speciation (*Chapter 7*); S_R, genetic similarity of populations (*Chapter 16*)
s	Selection coefficient; rarely a standard deviation or, as s^2, a variance (*Appendix I*)
Σ	Addition symbol; for example, $\Sigma_{i=0}^{k} X_i$, the sum of the values X_i, from X_0 to X_k
σ	Standard deviation: the square root of σ^2, the variance; also, breadth of niche (*Chapter 18*)
t	Time; a selection coefficient (*Chapter 13*); the name of an allele (*Chapter 12*)
u	Mutation rate
V	Variance ($= \sigma^2$); V_G, V_E, components of total variance (*Chapter 14*); also, density of prey (*Chapter 18*)
v	Rate of back mutation
W	Fitness of a genotype
$\bar{w}$	Average fitness of an individual in a population
$\bar{x}$	The arithmetic mean of some variable
z	Slope of the relation between species number and area on a log-log plot

Glossary

allele One of several forms of the same gene, presumably differing by mutation of the DNA sequence, and capable of segregating as a unit Mendelian factor.

allopatric Of a population or species; occupying a geographic region different from that of another population or species. See *parapatric, sympatric.*

allopolyploid A *polyploid* in which the several chromosome sets are derived from more than one species.

allozyme One of several forms of an enzyme coded for by different alleles at a locus. See *isozyme.*

apomixis Parthenogenetic reproduction in which an individual develops from an unfertilized egg or somatic cell.

aposematic Coloration or other features that advertise noxious properties; warning coloration.

Artenkreis A group of similar, closely related species; sometimes used to denote a *superspecies.* See also *Rassenkreis.*

autopolyploid A *polyploid* in which the several chromosome sets are derived from the same species.

autosome A chromosome other than a sex chromosome.

autotroph An organism that synthesizes the compounds it requires for energy.

cistron A gene that differs in location on the chromosome, and usually in function, from other genes. Usually a length of DNA that codes for a polypeptide.

clade In the narrow sense, the set of species descended from a particular ancestral species. Sometimes used more loosely, as a set of related species, from which some descendants are excluded.

cladistic Pertaining to branching patterns; a cladistic classification classifies organisms on the basis of the historical sequences by which they have diverged from common ancestors.

cleistogamy Self-pollination within a flower that does not open.

clone A lineage of individuals produced asexually by parthenogenesis or vegetative reproduction, by mitotic division.

coefficient of variation (C.V.) The standard deviation divided by the mean, multiplied by 100. C.V. $= 100 \times (s/\bar{x})$.

deme A local population.

502

diploid   A cell or organism possessing two chromosome complements; *ploidy* thus refers to the number of chromosome complements (see *haploid, polyploid*).

dominance   Of an allele, the extent to which it produces when heterozygous the same phenotype as when homozygous. Of a species, the extent to which it is numerically (or otherwise) predominant in a community.

efficient cause   Aristotle's term for the mechanical reason for an event.

endemic   Of a species, restricted to a specified region or locality.

epigenesis   The developmental processes whereby the genotype is expressed as a phenotype.

epistasis   A synergistic effect, on the phenotype or fitness, of two or more gene loci, whereby their joint effect differs from the sum of the loci taken separately.

equilibrium   A condition of stasis, as of population size or genetic composition. Also the value (of population size, gene frequency) at which stasis occurs. See also *stability, unstable equilibrium*.

evolution   In a broad sense, the origin of entities possessing different states of one or more characteristics, and changes in their proportions over time. *Organic evolution,* or *biological evolution,* is a change over time of the proportions of individual organisms differing genetically in one or more traits; such changes transpire by the origin and subsequent alteration of the frequencies of alleles or genotypes from generation to generation within populations, by the alteration of the proportions of genetically differentiated populations of a species, or by changes in the numbers of species with different characteristics, thereby altering the frequency of one or more traits within a higher taxon. See Chapter 1.

final cause   Aristotle's term for a goal, attainment of which is the reason for being of an event or process.

fitness   The average contribution of one allele or genotype to the next generation or to succeeding generations, compared with that of other alleles or genotypes. See Chapter 12.

fixation   Attainment of a frequency of 1 (i.e., 100 per cent) by an allele in a population, which thereby becomes monomorphic for the allele.

forb   An herbaceous plant other than a grass.

gene   The functional unit of heredity; a cistron.

gene frequency   The proportion of gene copies in a population that an allele accounts for; i.e., the probability of finding this allele when a gene is taken randomly from the population. Equals *allele frequency*.

genetic drift   Random changes in the frequencies of two or more alleles within a population.

genic selection   The differential propagation of different alleles within a population; i.e., a form of natural selection in which the frequency of an allele is determined by its rate of propagation relative to that of other alleles, averaged over the variety of genotypes in which it occurs. See *individual selection, kin selection, natural selection*.

genotype   The set of genes possessed by an individual organism; often, its genetic composition at a specific locus or set of loci singled out for discussion.

geometric mean   The nth root of the product of n values. $G = (x_1 \cdot x_2 \cdot \ldots \cdot x_n)^{1/n}$.

grade   A level of phenotypic organization attained by one or more species during evolution.

group selection   The differential rate of origination or extinction of whole populations (or species, if the term is used broadly) on the basis of differences among them in one or more characteristics. See *interdemic selection, species selection.*

haploid   A cell or organism possessing a single chromosome complement, hence a single gene copy at each locus.

heterosis   Equivalent to *hybrid vigor:* the superiority in one or more characteristics (e.g., size, yield) of crossbred organisms compared with correspondingly inbred organisms, as a result of differences in the genetic constitutions of the uniting parental gametes. Sometimes used to describe the higher fitness of heterozygous than homozygous genotypes, which is better termed *euheterosis,* and which is distinguished from *luxuriance,* a superiority in size, etc., that does not increase fitness. See *heterozygous advantage, overdominance.*

heterotroph   An organism that does not synthesize the compounds it uses for energy.

heterozygote   An individual organism that possesses different alleles at a locus.

heterozygous advantage   The manifestation of higher fitness by heterozygotes at a specific locus than by homozygotes.

histone   One of a class of proteins that are constituents of the chromosomes in eukaryotes.

homology   Possession by two or more species of a trait derived, with or without modification, from their common ancestor. See Chapter 7. Also, *homologous chromosomes*: the members of a chromosome complement that bear the same genes.

homoplasy   Possession by two or more species of a similar or identical trait that has not been derived by both species from their common ancestor; embraces convergence, parallel evolution, and evolutionary reversal.

homozygote   An individual organism that has the same allele at each of its copies of a gene locus.

host race   An ill-defined term in entomology, denoting a differentiated form, which may or may not interbreed with other host races, that feeds on a specific host plant.

individual selection   A form of natural selection consisting of non-random differences among different genotypes within a population in their contribution to subsequent generations. See *genic selection, natural selection.*

interdemic selection   Group selection of populations within a species.

intrinsic rate of natural increase   The potential rate of increase of a population with a stable age distribution, whose growth is not depressed by the negative effects of density.

inversion   A 180°-reversal of the orientation of a part of a chromosome, relative to some standard chromosome.

isozyme (isoenzyme)   One of several forms of an enzyme, produced by different, nonallelic, parts of an individual organism's genome. Often misused in place of *allozyme*.

kin selection   A form of genic selection whereby alleles differ in their rate of propagation by influencing the survival or reproduction of individuals (kin) who carry the same alleles by common descent.

linkage   Occurrence of two loci on the same chromosome; they are functionally linked only if they are so close together that they do not segregate independently in meiosis.

linkage equilibrium and **linkage disequilibrium**   If two alleles at two or more loci are associated more or less frequently than predicted by their individual frequencies, they are in linkage disequilibrium; if not, they are in linkage equilibrium.

locus   A site on a chromosome occupied by a specific gene; more loosely, the gene itself, in all its allelic states.

macroevolution   A vague term for the evolution of great phenotypic changes, usually great enough to allocate the changed lineage and its descendants to a distinct genus or higher taxon.

mean   Usually the arithmetic mean or average; the sum of n values, divided by n. The mean $\bar{x} = (x_1 + x_2 + \ldots + x_n)/n$.

meiotic drive   Used broadly to denote a preponderance (>50 per cent) of one allele among the gametes produced by a heterozygote; results in genic selection.

microevolution   A vague term for slight evolutionary changes within species.

migration   Used in theoretical population genetics as a synonym for gene flow among populations; in other contexts, directed large-scale movement of organisms that does not necessarily result in gene flow.

monomorphic   A population in which virtually all individuals have the same genotype at a locus. Cf. *polymorphism*.

mutualism   A symbiotic relation in which each of two species benefits by their interaction.

mycorrhiza   The structure formed by a mutualistic association between certain fungi and the roots of vascular plants.

natural selection   The differential survival and/or reproduction of classes of entities that differ in one or more hereditary characteristics; the difference in survival and/or reproduction is not due to chance, and it must have the potential consequence of altering the proportions of the different entities, to constitute natural selection. Thus natural selection is also definable as a partly or wholly deterministic difference in the contribution of hereditarily different classes of entities to subsequent generations. The entities may be alleles, genotypes or subsets of genotypes, populations, or in the broadest sense, species. A complex concept; see Chapter 12. Cf. also *genic selection, individual selection, kin selection, group selection*.

negative feedback   A dynamic relation whereby the product of a process inhibits the process that produces it, usually enhancing stability.

norm of reaction   The set of phenotypic expressions of a genotype under different environmental conditions.

ocelli　The simple eyes of insects and other arthropods.

ontogeny　The development of an individual organism, from fertilized zygote until death.

overdominance　The expression by two alleles in heterozygous condition of a phenotypic value for some characteristic that lies outside the range of the two corresponding homozygotes; a possible basis for *heterosis,* but not the only one. Higher fitness of a heterozygote than of homozygotes at that locus (*heterozygous advantage*) is often termed *overdominance for fitness.*

paralogous　Two or more gene loci, or their polypeptide products, derived by duplication of an ancestral locus, and occurring together in a haploid chromosome complement.

parapatric　Populations that have contiguous but non-overlapping geographic distributions.

parthenogenesis　Virgin birth; development from an egg to which there has been no paternal contribution of genes.

phenetic　Pertaining to phenotypic similarity, as in a phenetic classification (Chapter 7).

phenocopy　A phenotype, developed in response to an environmental stimulus, that resembles one known to be produced by a gene mutation.

phenotype　The morphological, physiological, biochemical, behavioral, and other properties of an organism, manifested throughout its life, that develop through the action of genes and environment; or any subset of such properties, especially those affected by a particular allele or other portion of the *genotype.*

phylogeny　The evolutionary history, or genealogy, of a group of organisms.

pleiotropy　The phenotypic effect of a gene on more than one characteristic.

polymorphism　The existence within a population of two or more genotypes for a given trait, the rarest of which exceeds some arbitrarily low frequency (say 1 per cent); more rarely, the existence of phenotypic variation within a population, whether or not genetically based.

polyphagous　Feeding on many kinds of food; usually used to describe insects that feed on many plants.

polyploid　Possessing more than two entire chromosome complements.

polytopy　Geographic variation in which each of one or more distinctive forms is found in each of several separate localities, between which other forms are distributed.

polytypy　The existence of named geographic races or subspecies within a species.

population　A group of conspecific organisms that occupy a more or less well defined geographic region and exhibit reproductive continuity from generation to generation; it is generally presumed that ecological and reproductive interactions are more frequent among these individuals than between them and the members of other populations of the same species.

race　A poorly defined term for a set of populations occupying a particular region that differ in one or more characteristics from populations elsewhere; equivalent to *subspecies*. In some writings, a distinctive phenotype, whether or not allopatric from others (cf. *host race*).

Rassenkreis   A species composed of geographic races or subspecies.

ratite   Any of several large flightless (or almost so) birds, for example the ostrich.

relict   A species that has been "left behind," for example, the last survivor of an otherwise extinct group. Sometimes, a species or population left in a locality after earlier climatic events.

relictual   The geographic distribution of a species or group that persists in localities that it occupied at an earlier time, but which has been extinguished over much of its former range.

reproductive value   Of an individual of a specific age, its likely contribution to the growth of the population.

semispecies   One of several groups of populations that are partially but not entirely reproductively isolated from each other by biological factors (isolating mechanisms).

sex-linked   A gene carried by one of the sex chromosomes; it can be expressed phenotypically in both sexes.

sexual reproduction   Production of offspring whose genetic constitution is a mixture of that of two potentially genetically different gametes.

species   In the sense of biological species, the members in aggregate of a group of populations that interbreed or potentially interbreed with each other under natural conditions; a complex concept (see Chapter 9). Also, a basic taxonomic category to which individual specimens are assigned, which often but not always corresponds to the biological species.

species selection   A form of *group selection* in which sets of species with different characteristics increase (by speciation) or decrease (by extinction) at different rates, because of a difference in their characteristics.

stability   Often used to mean constancy; more often in this book, the propensity to return to a condition (a stable equilibrium) after displacement from that condition.

standard deviation   The square root of the *variance*.

subspecies   A named geographic race; a set of populations of a species that share one or more distinctive features and occupy a different geographic area from other subspecies.

substitution   The complete replacement of one allele by another within a population or species; in the term *nucleotide substitution,* the complete replacement of one nucleotide pair by another within a lineage over evolutionary time. Cf. *fixation*.

supergene   A group of two or more loci between which recombination is so reduced that they are usually inherited together as a single entity.

superspecies   A group of semispecies; sometimes, an *Artenkreis*.

symbiosis   An interaction between two or more species that benefits at least one of them.

sympatric   Of two species or populations, occupying the same geographic locality so that the opportunity to interbreed is presented.

systematics   In a restricted sense, the study of the historical evolutionary and genetic relationships among organisms, and of their phenotypic similarities and differences.

taxon (pl. taxa)   The named taxonomic unit (e.g., *Homo sapiens,* Hominidae, or Mammalia) to which individuals, or sets of species, are assigned. *Higher taxa* are those above the species level.

taxonomy   The naming and assignment of organisms to taxa.

unstable equilibrium   An unchanging state, to which a system (e.g., a population density or gene frequency) does not return if disturbed.

variance (σ^2 or s^2)   The average squared deviation of an observation from the arithmetic mean; hence, a measure of variation. $s^2 = [\Sigma(x_i - \bar{x})^2]/(n - 1)$, where $\bar{x}$ is the mean and n the number of observations. See Appendix I.

wild type   The allele, genotype, or phenotype that is most prevalent (if there is one) in wild populations; with reference to the wild type allele, other alleles are often termed mutations. See Chapter 9.

Literature Cited

Adams, H., and E. Anderson. 1958. A conspectus of hybridization in the Orchidaceae. *Evolution* 12:512–518.

Alexander, R. D. 1974. The evolution of social behavior. *Ann. Rev. Ecol. Syst.* 5:325–383.

Alexander, R. D., and R. S. Bigelow. 1960. Allochronic speciation in field crickets, and a new species, *Acheta veletis*. *Evolution* 14:334–346.

Allard, R. W. 1960. *Principles of plant breeding*. Wiley, New York.

Allen, M. K., and C. Yanofsky. 1963. A biochemical and genetic study of reversion with the A-gene A-protein system of *Escherichia coli* tryptophan synthetase. *Genetics* 48:1065–1083.

Allison, A. C. 1961. Genetic factors in resistance to malaria. *Ann. N.Y. Acad. Sci.* 91:710–729.

Anderson, E. 1949. *Introgressive hybridization*. Wiley, New York.

Anderson, E., and G. L. Stebbins. 1954. Hybridization as an evolutionary mechanism. *Evolution* 8:378–388.

Anderson, W. W. 1966. Genetic divergence in M. Vetukhiv's experimental populations of *Drosophila pseudoobscura*. 3. Divergence in body size. *Genet. Research* 7:255–266.

Anderson, W. W. 1971. Genetic equilibrium and population growth under density-regulated selection. *Amer. Natur.* 105:489–498.

Anderson, W. W., Th. Dobzhansky, O. Pavlovsky, J. Powell, and D. Yardley. 1975. Genetics of natural populations. XLII. Three decades of genetic change in *Drosophila pseudoobscura*. *Evolution* 29:24–36.

Andrewartha, H. B., and L. C. Birch. 1954. *The distribution and abundance of animals*. Univ. of Chicago Press, Chicago.

Antonovics, J. 1968. Evolution in closely adjacent plant populations. V. Evolution of self-fertility. *Heredity* 23:219–238.

Antonovics, J., A. D. Bradshaw, and R. G. Turner. 1971. Heavy metal tolerance in plants. *Adv. Ecol. Research* 7:1–85.

Ardrey, R. 1966. *The territorial imperative*. Dell, New York.

Arnold, C. A. 1947. *An introduction to paleobotany*. McGraw-Hill, New York.

Ashburner, M. 1969. On the problem of genetic similarity between sibling species — puffing patterns in *Drosophila melanogaster* and *Drosophila simulans*. *Amer. Natur.* 103:189–191.

Askew, R. R. 1968. Considerations on speciation in Chalcidoidea (Hymenoptera). *Evolution* 22:642–645.

Atwood, K. C., L. K. Schneider, and F. J. Ryan. 1951. Periodic selection in *Escherichia coli. Proc. Natl. Acad. Sci. U.S.A.* 37:145–155.

Avise, J. 1976. Genetic differentiation during speciation. *In* F. J. Ayala (ed.), *Molecular evolution,* pp. 106–122. Sinauer, Sunderland, Mass.

Avise, J. C., and F. J. Ayala. 1976. Genetic differentiation in speciose versus depauperate phylads: evidence from the California minnows. *Evolution* 30:46–58.

Avise, J. C., and M. H. Smith. 1974. Biochemical genetics of sunfish. I. Geographic variation and subspecific intergradation in the bluegill, *Lepomis macrochirus. Evolution* 28:42–56.

Axelrod, D. I. 1966. Origin of deciduous and evergreen habits in temperate forests. *Evolution* 20:1–15.

Axelrod, D. I. 1967. Drought, diastrophism, and quantum evolution. *Evolution* 21:201–209.

Ayala, F. J. 1966. Evolution of fitness. I. Improvement in the productivity and size of irradiated populations of *Drosophila serrata* and *Drosophila birchii. Genetics* 53:883–895.

Ayala, F. J. 1968. Genotype, environment, and population numbers. *Science* 162:1453–1459.

Ayala, F. J. 1975. Genetic differentiation during the speciation process. *Evol. Biol.* 8:1–78.

Ayala, F. J. (ed.). 1976. *Molecular evolution.* Sinauer, Sunderland, Mass.

Ayala, F. J., M. L. Tracy, D. Hedgecock, and R. C. Richmond. 1974. Genetic differentiation during the speciation process in *Drosophila. Evolution* 28:576–592.

Bajema, C. (ed.). 1971. *Natural selection in human populations. The measurement of ongoing genetic evolution in contemporary societies.* Wiley, New York.

Baker, H. G. 1959. Reproductive methods as factors in speciation in flowering plants. *Cold Spring Harbor Symp. Quant. Biol.* 24:177–191.

Baker, H. G. 1974. The evolution of weeds. *Ann. Rev. Ecol. Syst.* 5:1–24.

Band, H. T., and P. T. Ives. 1963. Genetic structure of populations. I. On the nature of the genetic load in the South Amherst population of *Drosophila melanogaster. Evolution* 17:198–215.

Banks, H. P. 1970a. Major evolutionary events and the geological record of plants. *Biol. Rev.* 45:451–454.

Banks, H. P. 1970b. *Evolution and plants of the past.* Wadsworth, Belmont, Cal.

Barash, D. P. 1977. *Sociobiology and behavior.* Elsevier, New York.

Bart, P. B. 1977. Biological determinism and sexism: is it all in the ovaries? *In* The Ann Arbor Science for the People Editorial Collective (eds.), *Biology as a social weapon,* pp. 69–83. Burgess, Minneapolis.

Barzun, J. 1958. *Darwin, Marx, Wagner: a critique of a heritage.* Doubleday, New York.

Bateman, A. J. 1947a. Contamination of seed crops. I. Insect pollination. *J. Genet.* 48:257–275.

Bateman, A. J. 1947b. Contamination of seed crops. II. Wind pollination. *Heredity* 1:235–246.

Bates, M. 1960. Ecology and evolution. *In* S. Tax (ed.), *The evolution of life,* pp. 547–568. Univ. of Chicago Press, Chicago.

Bates, M. 1968. *Gluttons and libertines: human problems of being natural.* Random House, New York.

Bateson, G. 1963. The role of somatic change in evolution. *Evolution* 17:529–539.

Battaglia, B. 1958. Balanced polymorphism in *Tisbe reticulata,* a marine copepod. *Evolution* 12:358–364.

Bellairs, A. 1970. *The life of reptiles.* Universe Books, New York.

Bennett, J. 1960. A comparison of selective methods and a test of the preadaptation hypothesis. *Heredity* 15:65–77.

Benson, W. W., K. S. Brown, Jr., and L. E. Gilbert. 1975. Coevolution of plants and herbivores: passion flower butterflies. *Evolution* 29:659–680.

Bentley, B. L. 1976. Plants bearing extrafloral nectaries and the associated ant community: interhabitat differences in the reduction of herbivore damage. *Ecology* 54:815–820.

Bentley, B. L. 1977. The protective function of ants visiting the extrafloral nectaries of *Bixa orellana* L. (Bixaceae). *J. Ecol.* 65:27–38.

Berger, E. 1976. Heterosis and the maintenance of enzyme polymorphisms. *Amer. Natur.* 110:823–839.

Birch, L. C., Th. Dobzhansky, P. O. Elliott, and R. C. Lewontin. 1963. Relative fitness of geographic races of *Drosophila serrata*. *Evolution* 17:72–83.

Birns, B. 1976. The emergence and socialization of sex differences in the earliest years. *Merrill-Palmer Quarterly* 22:229–254.

Björkman, O., and P. Holmgren. 1963. Adaptability of the photosynthetic apparatus to light intensity in ecotypes from exposed and shaded habitats. *Physiol. Plant.* 16:889–914.

Bock, W. J. 1959. Preadaptation and multiple evolutionary pathways. *Evolution* 13:194–211.

Bock, W. J. 1970. Microevolutionary sequences as a fundamental concept in macroevolutionary models. *Evolution* 24:704–722.

Bock, W. J. 1973. Philosophical foundations of classical evolutionary classification. *Syst. Zool.* 22:375–392.

Bodmer, W. F., and P. A. Parsons. 1962. Linkage and recombination in evolution. *Adv. Genet.* 11:1–100.

Bonnell, M. L., and R. K. Selander. 1974. Elephant seals: genetic variation and near extinction. *Science* 184:908–909.

Borror, D. J., and D. M. DeLong. 1960. *An introduction to the study of insects.* Holt, Rinehart and Winston, New York.

Bos, M., and W. Scharloo. 1973. The effects of disruptive and stabilizing selection on body size in *Drosophila melanogaster*. I. Mean values and variances. *Genetics* 75:679–693.

Bowles, S., and H. Gintis. 1972–1973. I.Q. in the U.S. class structure. *Social Policy* 3:65–96.

Boyce, A. M. 1935. The codling moth in Persian walnuts. *J. Econ. Entomol.* 28:864–873.

Bretsky, P. W. 1969. Evolution of Paleozoic benthic marine invertebrate communities. *Palaeogeography, Palaeoclimatology, Palaeoecology* 6:45–59.

Bretsky, P. W., and D. M. Lorenz. 1970. An essay on genetic-adaptive strategies and mass extinctions. *Geol. Soc. Amer. Bull.* 81:2449–2456.

Briggs, B. G. 1962. Hybridization in *Ranunculus. Evolution* 16:372–390.

Britten, R. J., and E. H. Davidson. 1971. Repetitive and non-repetitive DNA and a speculation on the origin of evolutionary novelty. *Quart. Rev. Biol.* 46:111–133.

Britten, R. J., and D. E. Kohne. 1968. Repeated sequences in DNA. *Science* 161:529–540.

Bronfenbrenner, U. 1975. Is early intervention effective? Some studies of early education in familial and extra-familial settings. *In* A. Montagu (ed.), *Race and IQ,* pp. 287–322. Oxford University Press, London.

Brown, A. W. A. 1964. Animals in toxic environments: resistance of insects to insecticides. *In Handbook of physiology,* Section 4: Adaptation to the environment, pp. 773–793. American Physiological Society, Washington, D.C.

Brown, J. H. 1971. Mechanisms of competitive exclusion between two species of chipmunks. *Ecology* 52:305–319.

Brown, W. L., Jr. 1959. General adaptation and evolution. *Syst. Zool.* 7:157–168.

Brown, W. L., and E. O. Wilson. 1956. Character displacement. *Syst. Zool.* 5:49–64.

Brundin, L. 1965. On the real nature of transantarctic relationships. *Evolution* 19:496–505.

Brussard, P. F., P. R. Ehrlich, and M. C. Singer. 1974. Adult movements and population structure in *Euphydryas editha. Evolution* 28:408–415.

Bryant, E. H. 1974. On the adaptive significance of enzyme polymorphisms in relation to environmental variability. *Amer. Natur.* 108:1–19.

Bulmer, M. G. 1971. The effect of selection on genetic variability. *Amer. Natur.* 105:201–212.

Bulmer, M. G. 1973. Inbreeding in the great tit. *Heredity* 30:313–325.

Bulmer, M. G. 1974. Density-dependent selection and character displacement. *Amer. Natur.* 108:45–58.

Bumpus, H. C. 1899. The elimination of the unfit as illustrated by the introduced sparrow. *Biol. Lec. Mar. Biol. Woods Hole* 11:209–226.

Buri, P. 1956. Gene frequency drift in small populations of mutant *Drosophila. Evolution* 10:367–402.

Bush, G. L. 1969. Sympatric host race formation and speciation in frugivorous flies of the genus *Rhagoletis* (Diptera, Tephritidae). *Evolution* 23:237–251.

Bush, G. L. 1975. Modes of animal speciation. *Ann. Rev. Ecol. Syst.* 6:339–364.

Cain, A. J., and P. M. Sheppard. 1954a. Natural selection in *Cepaea. Genetics* 39:89–116.

Cain, A. J., and P. M. Sheppard. 1954*b*. The theory of adaptive polymorphism. *Amer. Natur.* 88:321–326.

Cain, S. A. 1944. *Foundations of plant geography.* Harper, New York.

Camin, J. H., and P. R. Ehrlich. 1958. Natural selection in water snakes (*Natrix sipedon* L.) on islands in Lake Erie. *Evolution* 12:504–511.

Camin, J. H., and R. R. Sokal. 1965. A method for deducing branching sequences in phylogeny. *Evolution* 19:311–326.

Campbell, B. (ed.). 1972. *Sexual selection and the descent of man 1871–1971.* Aldine, Chicago.

Campbell, B. G. 1974. *Human evolution.* Second edition. Aldine, Chicago.

Caplan, A. L. (ed.). 1978. *The sociobiology debate: readings on ethical and scientific issues.* Harper & Row, New York.

Capranica, R. R., L. S. Frishkoff, and E. Nevo. 1973. Encoding of geographic dialects in the auditory system of the cricket frog. *Science* 182:1272–1275.

Carson, H. L. 1959. Genetic conditions which promote or retard the formation of species. *Cold Spring Harbor Symp. Quant. Biol.* 24:87–105.

Carson, H. L. 1970. Chromosome tracers of the origin of species. *Science* 168:1414–1418.

Carson, H. L., D. E. Hardy, H. T. Spieth, and W. S. Stone. 1970. The evolutionary biology of the Hawaiian Drosophilidae. *In* M. K. Hecht and W. C. Steere (eds.), *Essays in evolution and genetics in honor of Theodosius Dobzhansky,* pp. 437–543. Appleton-Century-Crofts, New York.

Cartier, J. J., A. Isaak, R. H. Painter, and E. L. Sorensen. 1965. Biotypes of pea aphid *Acyrthosiphon pisum* (Harris) in relation to alfalfa clones. *Canad. Ent.* 97:754–760.

Caswell, H. 1976. Community structure: a neutral model analysis. *Ecol. Monogr.* 46:327–357.

Cates, R. G. 1975. The interface between slugs and wild ginger: some evolutionary aspects. *Ecology* 56:391–400.

Cavalli-Sforza, L. L., and W. F. Bodmer. 1971. *The genetics of human populations.* Freeman, San Francisco.

Cavalli-Sforza, L. L., and A. W. F. Edwards. 1967. Phylogenetic analysis: models and estimation procedures. *Evolution* 21:550–570.

Charlesworth, B., and D. Charlesworth. 1973. A study of linkage disequilibrium in populations of *Drosophila melanogaster. Genetics* 73:351–359.

Chinnici, J. P. 1971. Modification of recombination frequency in *Drosophila.* I. Selection for increased and decreased crossing over. II. The polygenic control of crossing over. *Genetics* 69:71–83; 85–96.

Chomsky, N. 1972. *Language and mind.* Harcourt, Brace, Jovanovich, New York.

Christiansen, K., and D. Culver. 1968. Geographical variation and evolution in *Pseudosinella hirsuta. Evolution* 22:237–255.

Churchill, W. 1967. *Homosexual behavior among males. A cross-cultural and cross-species investigation.* Prentice-Hall, Englewood Cliffs, N.J.

Cisne, J. L. 1974. Evolution of the world fauna of aquatic free-living arthropods. *Evolution* 28:337–366.

Clarke, B. 1962. Balanced polymorphism and the diversity of sympatric species. *Syst. Assoc. Publ.* 4:47–70.

Clarke, B., and P. O'Donald. 1964. Frequency-dependent selection. *Heredity* 19:201–206.

Clarke, C. A., C. G. C. Dickson, and P. M. Sheppard. 1963. Larval color pattern in *Papilio demodocus. Evolution* 17:130–137.

Clarke, C. A., and P. M. Sheppard. 1960. Supergenes and mimicry. *Heredity* 14:175–185.

Clarke, P. H. 1974. The evolution of enzymes for the utilisation of novel substrates. *In* M. J. Carlile and J. J. Skehel (eds.), *Evolution in the microbial world,* pp. 183–217. Cambridge Univ. Press, Cambridge, England.

Clausen, J. 1951. *Stages in the evolution of plant species.* Cornell Univ. Press, Ithaca, N.Y.

Clausen, J., D. D. Keck, and W. M. Hiesey. 1940. Experimental studies on the nature of species. I. Effect of varied environments on western North American plants. *Carnegie Inst. Wash. Publ.* 520.

Clausen, J., D. D. Keck, and W. M. Hiesey. 1947. Heredity of geographically and ecologically isolated races. *Amer. Natur.* 81:114–133.

Clayton, G. A., G. R. Knight, J. A. Morris, and A. Robertson. 1957. An experimental check on quantitative genetical theory. III. Correlated responses. *J. Genet.* 55:171–180.

Clayton, G. A., J. A. Morris, and A. Robertson. 1957. An experimental check on quantitative genetical theory. I. Short-term responses to selection. *J. Genet.* 55:131–151.

Clayton, G. A., and A. Robertson. 1955. Mutation and quantitative variation. *Amer. Natur.* 89:151–158.

Clayton, G. A., and A. Robertson. 1957. An experimental check on quantitative genetical theory. II. The long-term effects of selection. *J. Genet.* 55:152–170.

Cleland, R. E. 1972. *Oenothera: cytogenetics and evolution.* Academic Press, New York.

Cloud, P. E. 1968. Pre-metazoan evolution and the origin of the metazoa. *In* E. T. Drake (ed.), *Evolution and environment,* pp. 1–72. Yale Univ. Press, New Haven, Conn.

Cloud, P. 1976. Beginnings of biospheric evolution and their biochemical consequences. *Paleobiology* 2:351–387.

Cock, A. G. 1966. Genetical aspects of metrical growth and form in animals. *Quart. Rev. Biol.* 41:131–190.

Cody, M. L. 1973. Character convergence. *Ann. Rev. Ecol. Syst.* 4:189–211.

Cody, M. L. 1974a. *Competition and the structure of bird communities.* Princeton Univ. Press, Princeton, N.J.

Cody, M. L. 1974b. Optimization in ecology. *Science* 183:1156–1164.

Cody, M. L., and J. M. Diamond (eds.). 1975. *Ecology and evolution of communities.* Harvard Univ. Press, Cambridge, Mass.

Colbert, E. H. 1949. Progressive adaptations as seen in the fossil record. *In* G. L. Jepsen, G. G. Simpson, and E. Mayr (eds.), *Genetics, paleontology, and evolution,* pp. 390–402. Princeton Univ. Press, Princeton, N.J.

Cold Spring Harbor Symposia on Quantitative Biology. 1957. *Population studies: animal ecology and demography.* The Biological Laboratory, Cold Spring Harbor, New York.

Cole, L. C. 1954. The population consequences of life history phenomena. *Quart. Rev. Biol.* 29:103–137.

Collins, J. 1959. Darwin's impact on philosophy. *Thought* 34:185–248.

Colwell, R. K., and D. J. Futuyma. 1971. On the measurement of niche breadth and overlap. *Ecology* 52:567–576.

Conant, R. 1958. *A field guide to reptiles and amphibians.* Houghton Mifflin Co., Boston.

Connell, J. H. 1970. On the role of natural enemies in preventing competitive exclusion in some marine animals and in rain forest trees. *In* P. J. den Boer and G. R. Gradwell (eds.), *Dynamics of populations.* Proc. Adv. Study Inst. Dynamics Numbers Popul. (Oosterbeek), 298–312.

Connell, J. H., and R. O. Slatyer. 1977. Mechanisms of succession in natural communities and their role in community stability and organization. *Amer. Natur.* 111:1119–1144.

Cooch, F. G., and J. A. Beardmore. 1959. Assortative mating and reciprocal differences in the Blue-Snow Goose complex. *Nature* 183:1833–1834.

Cook, L. M. 1971. *Coefficients of natural selection.* Hutchinson & Co., London.

Cook, S. A., and M. P. Johnson. 1968. Adaptation to heterogeneous environments. I. Variation in heterophylly in *Ranunculus flammula* L. *Evolution* 22:496–516.

Cooke, F., and F. Cooch. 1968. The genetics of polymorphism in the goose *Anser caerulescens. Evolution* 22:289–300.

Coyne, J. 1976. Lack of genic similarity between two sibling species of Drosophila as revealed by varied techniques. *Genetics* 84:593–607.

Cracraft, J. 1974a. Continental drift and vertebrate distribution. *Ann. Rev. Ecol. Syst.* 5:215–261.

Cracraft, J. 1974b. Phylogenetic models and classification. *Syst. Zool.* 23:71–90.

Croizat, L., G. Nelson, and D. E. Rosen. 1974. Centers of origin and related concepts. *Syst. Zool.* 23:265–287.

Crompton, W. A. 1963. On the lower jaw of *Diarthrognathus* and the origin of the mammalian lower jaw. *Proc. Zool. Soc. London* 140:697–753.

Crow, J. F. 1957. Genetics of insect resistance to chemicals. *Ann. Rev. Entomol.* 2:227–246.

Crow, J. F. 1958. Some possibilities for measuring selection intensities in man. *Hum. Biol.* 30:1–13.

Crow, J. F. 1961. Population genetics. *Amer. J. Hum. Genet.* 13:137–150.

Crow, J. F., and M. Kimura. 1965. Evolution in sexual and asexual populations. *Amer. Natur.* 99:439–450.

Crow, J. F., and M. Kimura. 1970. *An introduction to population genetics theory.* Harper & Row, New York.

Crow, J. F., and R. G. Temin. 1964. Evidence for the partial dominance of recessive lethal genes in natural populations of Drosophila. *Amer. Natur.* 98:21–33.

Cruden, R. W. 1966. Birds as agents of long-distance dispersal for disjunct plant groups of the temperate western hemisphere. *Evolution* 20:517–532.

Crumpacker, D. W. 1967. Genetic loads in maize (*Zea mays* L.) and other cross-fertilized plants and animals. *Evol. Biol.* 1:306–324.

Dacosta, C. P., and C. M. Jones. 1971. Cucumber beetle resistance and mite susceptibility controlled by the bitter gene in *Cucumis sativus*. *Science* 172:1145–1146.

da Cunha, A. B. 1949. Genetic analysis of the polymorphism of color pattern in *Drosophila polymorpha*. *Evolution* 3:239–251.

da Cunha, A. B., and Th. Dobzhansky. 1954. A further study of chromosomal polymorphism in *Drosophila willistoni* in its relation to the environment. *Evolution* 8:119–134.

Daday, H. 1954. Gene frequencies in wild populations of *Trifolium repens* L. I. Distribution by latitude. *Heredity* 8:61–78.

Danforth, C. H. 1950. Evolution and plumage traits in pheasant hybrids, *Phasianus* X *Chrysolophus*. *Evolution* 4:301–315.

Darlington, P. J., Jr. 1957. *Zoogeography: the geographical distribution of animals*. John Wiley, New York.

Darlington, P. J. 1965. *Biogeography of the southern end of the world*. Harvard University Press, Cambridge, Mass.

Davidson, E. H., and R. J. Britten. 1973. Organization, transcription and regulation in the animal genome. *Quart. Rev. Biol.* 48:565–613.

Davies, R. W. 1971. The genetic relationship of two quantitative characters in *Drosophila melanogaster*. II. Location of the effects. *Genetics* 69:363–375.

Davies, R. W., and P. L. Workman. 1971. The genetic relationship of two quantitative characters in *Drosophila melanogaster*. I. Responses to selection and whole chromosome analysis. *Genetics* 69:353–361.

Dawson, P. S. 1970. Linkage and the elimination of deleterious mutant genes from experimental populations. *Genetica* 41:147–169.

Day, P. R. 1974. *Genetics of host-parasite interaction*. Freeman, San Francisco.

DeBach, P. 1966. The competitive displacement and coexistence principles. *Ann. Rev. Entomol.* 11:183–212.

Deely, J. N., and R. J. Nogar. 1973. *The problem of evolution. A study of the philosophical repercussions of evolutionary science*. Appleton-Century-Crofts, New York.

Deevey, E. S. 1949. Biogeography of the Pleistocene. *Bull. Geol. Soc. Amer.* 60:1315–1416.

Delevoryas, T. 1962. *Morphology and evolution of fossil plants*. Holt, Rinehart, Winston, New York.

del Solar, E. 1966. Sexual isolation by selection for positive and negative phototaxis and geotaxis in *Drosophila pseudoobscura*. *Proc. Natl. Acad. Sci. U.S.A.* 56:484–487.

den Boer, P. J., and G. R. Gradwell (eds.). 1970. *Dynamics of populations*. Proc. Adv. Study Inst. Dynamics Numb. Pop. (Oosterbeek). Wageningen, Netherlands.

Dewey, J. 1910. *The influence of Darwin on philosophy, and other essays in contemporary thought*. P. Smith, New York.

Diamond, J. M. 1969. Avifaunal equilibria and species turnover rates on the Channel Islands of California. *Proc. Natl. Acad. Sci. U.S.A.* 64:57–63.

Dickinson, H., and J. Antonovics. 1973. Theoretical considerations of sympatric divergence. *Amer. Natur.* 107:256–274.

Diver, C. 1929. Fossil records of Mendelian mutants. *Nature* 124:183.

Dobzhansky, Th. 1936. Studies on hybrid sterility. II. Localization of sterility factors in *Drosophila pseudoobscura* hybrids. *Genetics* 21:113–135.

Dobzhansky, Th. 1944. Chromosomal races in *Drosophila pseudoobscura* and *D. persimilis*. *Carneg. Inst. Wash. Publ. No.* 554:47–144. Washington, D.C.

Dobzhansky, Th. 1948. Genetics of natural populations. XVIII. Experiments on chromosomes of *Drosophila pseudoobscura* from different geographic regions. *Genetics* 33:588–602.

Dobzhansky, Th. 1950. Evolution in the tropics. *Amer. Sci.* 38:208–221.

Dobzhansky, Th. 1955. A review of some fundamental concepts and problems of population genetics. *Cold Spring Harbor Symp. Quant. Biol.* 20:1–15.

Dobzhansky, Th. 1956. What is an adaptive trait? *Amer. Natur.* 90:337–347.

Dobzhansky, Th. 1958. Genetics of natural populations. XXVII. The genetic changes in populations of *Drosophila pseudoobscura* in the American southwest. *Evolution* 12:385–401.

Dobzhansky, Th. 1962. *Mankind evolving*. Yale Univ. Press, New Haven.

Dobzhansky, Th. 1970. *Genetics of the evolutionary process*. Columbia Univ. Press, New York.

Dobzhansky, Th., F. J. Ayala, G. L. Stebbins, and J. W. Valentine. 1977. *Evolution*. Freeman, San Francisco.

Dobzhansky, Th., C. Krimbas, and M. G. Krimbas. 1960. Genetics of natural populations. XXX. Is the genetic load in *Drosophila pseudoobscura* a mutational or a balanced one? *Genetics* 45:741–753.

Dobzhansky, Th., and H. Levene. 1955. Genetics of natural populations. XXIV. Developmental homeostasis in natural populations of *Drosophila pseudoobscura*. *Genetics* 40:797–808.

Dobzhansky, Th., and O. Pavlovsky. 1957. An experimental study of interaction between genetic drift and natural selection. *Evolution* 11:311–319.

Dobzhansky, Th., and O. Pavlovsky. 1971. An experimentally created incipient species of *Drosophila*. *Nature* 23:289–292.

Dobzhansky, Th., and B. Spassky. 1944. Genetics of natural populations. XI. Manifestation of genetic variants in *Drosophila pseudoobscura* in different environments. *Genetics* 29:270–290.

Dobzhansky, Th., and S. Wright. 1943. Genetics of natural populations. X. Dispersal rates in *Drosophila pseudoobscura*. *Genetics* 28:304–340.

Dorfman, D. D. 1978. The Cyril Burt question: new findings. *Science* 201:1177–1186.

Drake, J. W. 1970. *The molecular basis of mutation*. Holden-Day, San Francisco.

Drake, J. W. 1974. The role of mutation in microbial evolution. *In* M. J. Carlile and J. J. Skehel (eds.), *Evolution in the microbial world,* pp. 41–58. Cambridge Univ. Press, Cambridge, England.

Dressler, R. E. 1968. Pollination by euglossine bees. *Evolution* 22:202–210.

Durham, J. W. 1971. The fossil record and the origin of the Deuterostomata. N. Amer. Paleontol. Conv., Chicago, 1969, Proc., Pt. H, pp. 1104–1131.

Durrant, A. 1962. The environmental induction of heritable change in *Linum*. *Heredity* 17:27–61.

Eakin, R. M. 1968. Evolution of photoreceptors. *Evol. Biol.* 2:194–242.

Edwards, A. W. F., and L. L. Cavalli-Sforza. 1964. Reconstruction of evolutionary trees. *In* V. H. Heywood and J. McNeill (eds.), *Phenetic and phylogenetic classification,* pp. 67–76. Systematics Assoc. Publ. No. 6, London.

Ehrlich, P. R. 1965. The population biology of the butterfly, *Euphydryas editha*. II. The structure of the Jasper Ridge colony. *Evolution* 19:327–336.

Ehrlich, P. R., and P. H. Raven. 1964. Butterflies and plants: a study in coevolution. *Evolution* 18:586–608.

Ehrlich, P. R., and P. H. Raven. 1969. Differentiation of populations. *Science* 165:1228–1232.

Ehrlich, P. R., R. R. White, M. C. Singer, S. W. McKechnie, and L. E. Gilbert. 1975. Checkerspot butterflies: a historical perspective. *Science* 188:221–228.

Ehrman, L. 1962. Hybrid sterility as an isolating mechanism in the genus *Drosophila*. *Quart. Rev. Biol.* 37:279–302.

Ehrman, L. 1964. Genetic divergence in M. Vetukhiv's experimental populations of *Drosophila pseudoobscura*. 1. Rudiments of sexual isolation. *Genet. Res.* 5:150–157.

Ehrman, L. 1965. Direct observation of sexual isolation between allopatric and between sympatric strains of the different *Drosophila paulistorum* races. *Evolution* 19:459–464.

Ehrman, L. 1967. Further studies on genotype frequency and mating success in *Drosophila*. *Amer. Natur.* 101:415–424.

Eiseley, L. 1958. *Darwin's century*. Doubleday, New York.

Eisen, E. J. 1975. Population size and selection intensity effects on long-term selection response in mice. *Genetics* 79:305–323.

Eisen, E. J., J. P. Hanrahan, and J. E. Legates. 1973. Effects of population size and selection intensity on correlated responses to selection for post-weaning gain in mice. *Genetics* 74:157–170.

Eldredge, N. 1971. The allopatric model and phylogeny in Paleozoic invertebrates. *Evolution* 25:156–167.

Eldredge, N., and S. J. Gould. 1972. Punctuated equilibria: an alternative to phyletic gradualism. *In* T. J. M. Schopf (ed.), *Models in paleobiology,* pp. 82–115. Freeman, Cooper & Co., San Francisco.

Elton, C. S. 1942. *Voles, mice and lemmings: problems in population dynamics*. Oxford University Press, Oxford, England.

Emerson, A. E. 1961. Vestigial characters of termites and processes of regressive evolution. *Evolution* 15:115–131.

Emerson, S. 1939. A preliminary survey of the *Oenothera organensis* population. *Genetics* 24:524–537.

Emlen, J. M. 1966. The role of time and energy in food preference. *Amer. Natur.* 100:611–617.

Emlen, J. M. 1973. *Ecology: an evolutionary approach.* Addison-Wesley, Reading, Mass.

Endler, J. A. 1973. Gene flow and population differentiation. *Science* 179:243–250.

Endler, J. A. 1977. *Geographic variation, speciation and clines.* Princeton Univ. Press, Princeton, N.J.

Engler, A. 1930. *Die natürlichen Pflanzenfamilien.* Wilhelm Engelmann, Leipzig.

Eshel, J., and M. W. Feldman. 1970. On the evolutionary effect of recombination. *Theoret. Pop. Biol.* 1:88–100.

Faegri, K., and L. van der Pijl. 1971. *The principles of pollination ecology.* Second edition. Pergamon Press, New York.

Falconer, D. S. 1953. Selection for large and small size in mice. *J. Genet.* 51:470–501.

Falconer, D. S. 1954. Validity of the theory of genetic correlation. An experimental test with mice. *J. Hered.* 45:42–44.

Falconer, D. S. 1960. *Introduction to quantitative genetics.* Ronald Press, New York.

Falconer, D. S., and J. W. B. King. 1953. A study of selection limits in the mouse. *J. Genet.* 51:561–581.

Farnworth, E. G., and F. B. Golley (eds.). 1974. *Fragile ecosystems: evaluation of research and applications in the neotropics.* Springer-Verlag, New York.

Farris, J. S. 1967. The meaning of relationship and taxonomic procedure. *Syst. Zool.* 16:44–51.

Farris, J. S. 1969. A successive approximations approach to character weighting. *Syst. Zool.* 18:374–385.

Farris, J. S. 1970. Methods for computing Wagner trees. *Syst. Zool.* 19:83–92.

Farris, J. S. 1973. A probability model for inferring evolutionary trees. *Syst. Zool.* 22:250–256.

Farris, J. S., A. G. Kluge, and M. J. Eckardt. 1970. A numerical approach to phylogenetic systematics. *Syst. Zool.* 19:172–189.

Feeny, P. P. 1970. Seasonal changes in oak leaf tannins as a cause of spring feeding by winter moth caterpillars. *Ecology* 51:656–681.

Feeny, P. P. 1976. Plant apparency and chemical defense. *In* J. Wallace and R. Mansell (eds.), *Biochemical interactions between plants and insects. Rec. Adv. Phytochem.* 10:1–40.

Feldman, M. W., and R. C. Lewontin. 1975. The heritability hang-up. *Science* 190:1163–1168.

Felsenstein, J. 1965. The effect of linkage on directional selection. *Genetics* 52:349–363.

Felsenstein, J. 1971. Inbreeding and variance effective number in populations with overlapping generations. *Genetics* 68:581–597.

Felsenstein, J. 1974. The evolutionary advantage of recombination. *Genetics* 78:737–756.

Felsenstein, J. 1976. The theoretical population genetics of variable selection and migration. *Ann. Rev. Genet.* 10:253–280.

Felsenstein, J., and S. Yokoyama. 1976. The evolutionary advantage of recombination. II. Individual selection for recombination. *Genetics* 83:845–859.

Fenner, F. 1965. Myxoma virus and *Oryctolagus cuniculus*: two colonizing species. *In* H. G. Baker and G. L. Stebbins (eds.), *The genetics of colonizing species,* pp. 485–501. Academic Press, New York.

Ficken, R. W., M. S. Ficken, and D. H. Morse. 1968. Competition and character displacement in two sympatric pine-dwelling warblers (*Dendroica,* Parulidae). *Evolution* 22:307–314.

Fisher, R. A. 1930. *The genetical theory of natural selection.* Clarendon Press, Oxford, England.

Fitch, W. M. 1970. Distinguishing homologous from analogous proteins. *Syst. Zool.* 19:99–113.

Fitch, W. M., and E. Margoliash. 1970. The usefulness of amino acid and nucleotide sequences in evolutionary studies. *Evol. Biol.* 4:67–109.

Flessa, K. W., K. V. Powers, and J. L. Cisne. 1975. Specialization and evolutionary longevity in the Arthropoda. *Paleobiology* 1:71–81.

Ford, C. S., and F. A. Beach. 1951. *Patterns of sexual behavior.* Harper & Row, New York.

Ford, E. B. 1940. Genetic research in the Lepidoptera. *Ann. Eugen.* 10:227–252.

Ford, E. B. 1971. *Ecological genetics.* Third edition. Chapman and Hall, London.

Fox, W. 1951. Relationships among the garter snakes of the *Thamnophis elegans Rassenkreis.* Univ. California Publ. Zool. 50:485–530.

Franklin, I., and R. C. Lewontin. 1970. Is the gene the unit of selection? *Genetics* 65:707–734.

Frazzetta, T. H. 1975. *Complex adaptations in evolving populations.* Sinauer, Sunderland, Mass.

Freedman, L. 1962. Growth of muzzle length relative to calvaria length in *Papio. Growth* 26:117–128.

Fryer, G. 1959. Some aspects of evolution in Lake Nyasa. *Evolution* 13:440–451.

Fryer, G., and T. D. Iles. 1969. Alternative routes to evolutionary success as exhibited by African cichlid fishes of the genus *Tilapia* and the species flocks of the Great Lakes. *Evolution* 23:359–369.

Fryer, G., and T. D. Iles. 1972. *The cichlid fishes of the great lakes of Africa.* T. F. H. Publications, Neptune City, N.J.

Futuyma, D. J. 1970. Variation in genetic response to interspecific competition in laboratory populations of *Drosophila. Amer. Natur.* 104:239–252.

Futuyma, D. J. 1973. Community structure and stability in constant environments. *Amer. Natur.* 107:443–446.

Futuyma, D. J. 1976. Food plant specialization and environmental predictability in Lepidoptera. *Amer. Natur.* 110:285–292.

Futuyma, D. J., and F. Gould. 1979. Associations of plants and insects in a deciduous forest. *Ecology,* in press.

Gadgil, M., and W. H. Bossert. 1970. Life historical consequences of natural selection. *Amer. Natur.* 104:1–24.

Galau, G. A., M. E. Chamberlin, B. R. Hough, R. J. Britten, and E. H. Davidson. 1976. Evolution of repetitive and non-repetitive DNA. *In* F. J. Ayala (ed.), *Molecular evolution,* pp. 200–224. Sinauer, Sunderland, Mass.

Gallup, G. G., Jr. 1974. Toward an operational definition of self-awareness. *In* R. H. Tuttle (ed.), *IX Internat. Congr. Anthropol. Ethnol. Sci., Primatology Session.* Mouton Press, The Hague.

Gans, C. 1974. *Biomechanics: an approach to vertebrate biology.* Lippincott, Philadelphia.

Gardner, B. T., and R. A. Gardner. 1971. Two-way communication with a chimpanzee. *In* A. Schrier and F. Stollnitz (eds.), *Behavior of non-human primates,* pp. 117–184. Academic Press, New York.

Gerhold, H. D., R. E. McDermott, E. J. Shreiher, and J. A. Winieski (eds.). 1966. *Breeding pest-resistant trees.* Pergamon Press, New York.

Ghiselin, M. T. 1969. *The triumph of the Darwinian method.* Univ. of California Press, Berkeley, Calif.

Gibson, T. C., M. L. Scheppe, and E. C. Cox. 1970. Fitness of an *Escherichia coli* mutator gene. *Science* 169:686–688.

Giesel, J. T. 1970. On the maintenance of a shell pattern and behavior polymorphism in *Acmaea digitalis,* a limpet. *Evolution* 24:98–119.

Giesel, J. T. 1971. The relations between population structure and rate of inbreeding. *Evolution* 25:491–496.

Gilbert, J. J., and J. K. Waage. 1967. *Asplanchna, Asplanchna*-substance, and posterolateral spine length variation of the rotifer *Brachionus calyciflorus* in a natural environment. *Ecology* 48:1027–1031.

Gilbert, L. E. 1971. Butterfly-plant coevolution: has *Passiflora adenopoda* won the selectional race with heliconiine butterflies? *Science* 172:585–586.

Gilbert, L. E. 1975. Ecological consequences of a coevolved mutualism between butterflies and plants. *In* L. E. Gilbert and P. H. Raven (eds.), *Coevolution of animals and plants.* Univ. Texas Press, Austin.

Gilbert, L. E., and M. C. Singer. 1973. Dispersal and gene flow in a butterfly species. *Amer. Natur.* 107:58–72.

Gillespie, J. H., and C. H. Langley. 1974. A general model to account for enzyme variation in natural populations. *Genetics* 76:837–884.

Gilpin, M. E. 1975. *Group selection in predator-prey communities.* Princeton Univ. Press, Princeton, N.J.

Glass, B., and C. C. Li. 1953. The dynamics of racial intermixture: an analysis based on the American Negro. *Amer. J. Hum. Genet.* 5:1–20.

Goddard, H. H. 1920. *Human efficiency and levels of intelligence.* Princeton Univ. Press, Princeton, N.J.

Goldschmidt, R. 1938. *Physiological genetics.* McGraw-Hill, New York.

Goldschmidt, R. 1940. *The material basis of evolution.* Yale Univ. Press, New Haven, Conn.

Good, R. 1947. *The geography of the flowering plants.* Longmans, Green & Co., London.

Goodman, M., G. W. Moore, J. Barnabas, and G. Matsuda. 1974. The phylogeny of the human globin genes investigated by the maximum parsimony method. *J. Molec. Evol.* 3:1–48.

Gottlieb, L. D. 1974. Genetic confirmation of the origin of *Clarkia lingulata*. *Evolution* 28:244–250.

Gould, F. 1977. The evolution of adaptation to host plants and pesticides in a polyphagous herbivore, *Tetranychus urticae* (Koch). Ph.D. dissertation, State University of New York at Stony Brook.

Gould, S. J. 1966. Allometry and size in ontogeny and phylogeny. *Biol. Rev.* 41:587–640.

Gould, S. J. 1972. Allometric fallacies and the evolution of *Gryphaea*: a new interpretation based on White's criterion of geometric similarity. *Evol. Biol.* 6:91–119.

Gould, S. J. 1974. The origin and function of "bizarre" structures: antler size and skull size in the "Irish elk," *Megaloceros giganteus. Evolution* 28:191–220.

Gould, S. J. 1976. Palaeontology plus ecology as palaeobiology. *In* R. M. May (ed.), *Theoretical ecology: principles and applications,* pp. 218–236. Saunders, Philadelphia.

Gould, S. J. 1977. *Ontogeny and phylogeny.* Harvard Univ. Press, Cambridge, Mass.

Gould, S. J. 1978. An early start. *Natural History* 87(2):10–24.

Gould S. J., and N. Eldredge. 1977. Punctuated equilibria: the tempo and mode of evolution reconsidered. *Paleobiology* 3:115–151.

Gould, S. J., and R. F. Johnston. 1971. Geographic variation. *Ann. Rev. Ecol. Syst.* 3:457–498.

Gould, S. J., D. M. Raup, J. J. Sepkoski, Jr., T. J. M. Schopf, and D. S. Simberloff. 1977. The shape of evolution: a comparison of real and random clades. *Paleobiology* 3:23–40.

Gowen, J. W., and L. E. Johnson. 1946. On the mechanism of heterosis. I. Metabolic capacity of different races of *Drosophila melanogaster* for egg production. *Amer. Natur.* 80:149–179.

Graham, A. 1964. Origin and evolution of the biota of southeastern North America: evidence from the fossil plant record. *Evolution* 18:571–585.

Grant, K. A., and V. Grant. 1964. Mechanical isolation of *Salvia apiana* and *Salvia mellifera* (Labiatae). *Evolution* 18:196–212.

Grant, P. R. 1972. Convergent and divergent character displacement. *Biol. J. Linn. Soc.* 4:39–68.

Grant, V. 1966. The selective origin of incompatibility barriers in the plant genus *Gilia. Amer. Natur.* 100:99–118.

Grant, V. 1971. *Plant speciation.* Columbia Univ. Press, New York.

Greene, J. C. 1959. *The death of Adam: evolution and its impact on Western thought.* Iowa State Univ. Press, Ames, Iowa.

Greene, J. C. 1961. *Darwin and the modern world view.* Louisiana State Univ. Press, Baton Rouge, La.

Greenslade, P. J. M. 1968. Island patterns in the Solomon Islands bird fauna. *Evolution* 22:751–761.

Greenwood J. J. D. 1974. Effective population numbers in the snail *Cepaea nemoralis. Evolution* 28:513–526.

Griffiths, I. 1963. The phylogeny of the Salientia. *Biol. Rev.* 38:241–292.

Grun, P. 1976. *Cytoplasmic genetics and evolution.* Columbia Univ. Press, New York.

Guthrie, R. D. 1965. Variability in characters undergoing rapid evolution, an analysis of *Microtus* molars. *Evolution* 19:214–233.

Hairston, N. G. 1951. Interspecies competition and its probable influence upon the vertical distribution of Appalachian salamanders of the genus *Plethodon. Ecology* 32:266–274.

Hairston, N. G., J. D. Allan, R. K. Colwell, D. J. Futuyma, J. Howell, M. D. Lubin, J. Mathias, and J. H. Vandermeer. 1969. The relationship between species diversity and stability: an experimental approach with Protozoa and bacteria. *Ecology* 49:1091–1101.

Hairston, N. G., F. E. Smith, and L. B. Slobodkin. 1960. Community structure, population control, and competition. *Amer. Natur.* 94:421–425.

Haldane, J. B. S. 1949. Suggestions as to the quantitative measurement of rates of evolution. *Evolution* 3:51–56.

Haldane, J. B. S. 1956. The relation between density regulation and natural selection. *Proc. Roy. Soc. Lond. (B)* 145:306–308.

Haldane, J. B. S. 1957. The cost of natural selection. *J. Genet.* 55:511–524.

Haldane, J. B. S., and S. D. Jayakar. 1963. Polymorphism due to selection of varying direction. *J. Genet.* 58:318–323.

Hamilton, W. D. 1964. The genetical evolution of social behavior, I and II. *J. Theoret. Biol.* 7:1–52.

Hamilton, W. D. 1967. Extraordinary sex ratios. *Science* 156:477–488.

Hampé, A. 1960. La compétition entre les éléments osseux du zeugupode de poulet. *J. Embryol. Exper. Morph.* 8:241–245. (Cited in Waddington 1962.)

Hamrick, J. L., and R. W. Allard. 1972. Microgeographical variation in allozyme frequencies in *Avena barbata. Proc. Natl. Acad. Sci. U.S.A.* 69:2100–2104.

Hansche, P. E. 1975. Gene duplication as a mechanism of genetic adaptation in *Saccharomyces cerevisiae. Genetics* 79:661–674.

Hardisty, M. W. 1963. Fecundity and speciation in lampreys. *Evolution* 17:17–22.

Hare, J. D., and D. J. Futuyma. 1978. Different effects of variation in *Xanthium strumarium* L. (Compositae) on two insect seed predators. *Oecologia* 39:109–120.

Harlan, J. R., and J. M. J. de Wet. 1963. The compilospecies concept. *Evolution* 17:497–501.

Harland, W. B., C. H. Holland, M. R. House, N. F. Hughes, A. B. Reynolds, M. J. S. Rudwick, G. E. Satterthwaite, L. B. H. Tarlo and E. C. Wiley. 1967. *The fossil record.* Geological Society of London.

Harris, H. 1966. Enzyme polymorphisms in man. *Proc. Roy. Soc. Lond. (B)* 164:298–310.

Harris, M. 1974. *Cows, pigs, wars and witches. The riddles of culture.* Random House, New York.

Harris, M. 1975. *Culture, people, nature. An introduction to general anthropology.* Thomas Y. Crowell, New York.

Hatchett, J. H., and R. Gallun. 1970. Genetics of the ability of the Hessian fly, *Mayetiola destructor,* to survive on wheats having different genes for resistance. *Ann. Ent. Soc. Amer.* 63:1400–1407.

Haverschmidt, F. 1968. *Birds of Surinam.* Oliver & Boyd, London.

Hebert, P. 1974. Enzyme variability in natural populations of *Daphnia magna.* II. Genotypic frequencies in permanent populations. *Genetics* 77:323–334.

Hecht, M. K. 1952. Natural selection in the lizard genus *Aristelliger. Evolution* 6:112–124.

Hecht, M. K. 1965. The role of natural selection and evolutionary rates in the origin of higher levels of organization. *Syst. Zool.* 14:301–317.

Hedgpeth, J. W. 1957. Marine biogeography. *Geol. Soc. Amer. Mem.* 67, 1:359–382.

Hedrick, P. W., M. E. Ginevan, and E. P. Ewing. 1976. Genetic polymorphism in heterogeneous environments. *Ann. Rev. Ecol. Syst.* 7:1–32.

Henbest, L. G. *et al.* 1952. Distribution of evolutionary explosions in geologic time. *J. Paleontol.* 26:298–394.

Hennig, W. 1965. Phylogenetic systematics. *Ann. Rev. Ent.* 10:97–116.

Hennig, W. 1966. *Phylogenetic systematics.* Translated by D. D. Davis and R. Zangerl. Univ. of Illinois Press, Chicago.

Henry, G. M. 1971. *Birds of Ceylon.* Oxford University Press, Oxford.

Hersh, A. H. 1930. The facet-temperature relation in the Bar series of *Drosophila. J. Exp. Zool.* 57:283–306.

Herskowitz, I. H. 1977. *Principles of genetics.* Second edition. Macmillan, New York.

Heslop-Harrison, J. 1964. Forty years of genecology. *Adv. Ecol. Res.* 2:159–247.

Hibbard, C. W. 1963. The origin of the P_3 pattern of *Sylvilagus, Caprolagus,* and *Lepus. J. Mamm.* 44:1–15.

Hill, J. 1967. The environmental induction of heritable changes in *Nicotiana rustica* parental and selection lines. *Genetics* 55:735–754.

Himmelfarb, G. 1962. *Darwin and the Darwinian revolution.* Doubleday, Garden City, New York.

Hinegardner, R. 1976. Evolution of genome size. *In* F. J. Ayala (ed.), *Molecular evolution,* pp. 179–199. Sinauer, Sunderland, Mass.

Hiraizumi, Y., and J. F. Crow. 1960. Heterozygous effects on viability, fertility, rate of development, and longevity of Drosophila chromosomes that are lethal when homozygous. *Genetics* 45:1071–1083.

Hochachka, P. W., and G. N. Somero. 1973. *Strategies of biochemical adaptation.* W. B. Saunders, Philadelphia.

Hoffman, M. 1976. Homosexuality. *In* F. A. Beach (ed.), *Human sexuality in four perspectives,* pp. 164–189. Johns Hopkins Univ. Press, Baltimore, Md.

Hoffstetter, R. 1972. Relationships, origins, and history of the ceboid monkeys and caviomorph rodents: a modern reinterpretation. *Evol. Biol.* 6:323–348.

Hofstadter, R. 1955. *Social Darwinism in American thought.* Beacon Press, Boston.

Holliday, R. 1970. The organization of DNA in eukaryotic chromosomes. *Symp. Soc. Gen. Microbiol.* XX:359–380.

Hooker, E. 1957. The adjustment of the male overt homosexual. *J. of Projective Techniques* 21:18–31.

Hooper, E. T. 1957. Dental patterns in mice of the genus *Peromyscus. Univ. Mich. Mus. Zool. Misc. Publ.* No. 99.

Horn, H. S. 1975. Markovian properties of forest succession. *In* M. L. Cody and J. M. Diamond (eds.), *Ecology and evolution of communities,* pp. 196–211. Harvard Univ. Press, Cambridge, Mass.

Horn, H. S., and R. H. MacArthur. 1972. Competition among fugitive species in a harlequin environment. *Ecology* 53:749–752.

Howard, W. E. 1949. Dispersal, amount of inbreeding, and longevity in a local population of prairie deermice on the George Reserve, southern Michigan. *Contr. Lab. Vert. Biol. Univ. Mich.* 43:1–50.

Hoy, R. R., and R. C. Paul. 1973. Genetic control of song specificity in crickets. *Science* 180:82–83.

Hubbs, C. L. 1955. Hybridization between fish species in nature. *Syst. Zool.* 4:1–20.

Hubbs, C. (ed.). 1958. Zoogeography. *Amer. Assoc. Adv. Sci. Publ.* 51.

Hubbs, C. 1961. Isolating mechanisms in the speciation of fishes. *In* W. F. Blair (ed.), *Vertebrate speciation,* pp. 5–23. Univ. of Texas Press, Austin.

Huettel, M. D., and G. L. Bush. 1972. The genetics of host selection and its bearing on sympatric speciation in *Procecidochares* (Diptera: Tephritidae). *Ent. Exp. Appl.* 15:465–480.

Hurd, L. E., and R. M. Eisenberg. 1975. Divergent selection for geotactic response and evolution of reproductive isolation in sympatric and allopatric populations of houseflies. *Amer. Natur.* 109:353–358.

Hutchinson, G. E. 1951. Copepodology for the ornithologist. *Ecology* 32:571–577.

Hutchinson, G. E. 1957. Concluding remarks. *Cold Spring Harbor Symp. Quant. Biol.* 22:415–427.

Hutchinson, G. E. 1968. When are species necessary? *In* R. C. Lewontin (ed.), *Population biology and evolution,* pp. 177–186. Syracuse Univ. Press, Syracuse, N.Y.

Hutchinson, J. 1969. *Evolution and phylogeny of flowering plants.* Academic Press, New York.

Huxley, J. S. 1932. *Problems of relative growth.* MacVeagh, London. (Second edition 1972, Dover, New York.)

Huxley, T. H. 1893. *Evolution and ethics,* The Romanes Lecture, Oxford University. Reprinted in T. H. Huxley and J. Huxley (1947), *Touchstone for ethics,* Harper and Bros., New York.

Imms, A. D. 1957. *A general textbook of entomology.* Ninth edition. Revised by O. W. Richards and R. G. Davies. Methuen, London.

Inger, R. F. 1967. The development of a phylogeny of frogs. *Evolution* 21:369–384.

Inglis, W. G. 1966. The observational basis of homology. *Syst. Zool.* 15:219–228.

Ingram, V. M. 1963. *The hemoglobins in genetics and evolution*. Columbia Univ. Press, New York.

Istock, C. A. 1967. The evolution of complex life history phenomena: an ecological perspective. *Evolution* 21:592–605.

Ives, P. T. 1950. The importance of mutation rate genes in evolution. *Evolution* 4:236–252.

Jain, S. K., and D. R. Marshall. 1967. Population studies on predominantly self-pollinating species. X. Variation in natural populations of *Avena fatua* and *A. barbata*. *Amer. Natur.* 101:19–33.

James, J. W. 1971. The founder effect and response to artificial selection. *Genet. Research* 16:241–250.

Janzen, D. H. 1966. Coevolution of mutualism between ants and acacias in Central America. *Evolution* 20:249–275.

Janzen, D. H. 1967. Why mountain passes are higher in the tropics. *Amer. Natur.* 101:233–249.

Janzen, D. H. 1968. Host plants as islands in evolutionary and contemporary time. *Amer. Natur.* 102:592–595.

Janzen, D. H. 1970. Herbivores and the number of tree species in tropical forests. *Amer. Natur.* 104:501–528.

Jensen, A. R. 1969. How much can we boost IQ and scholastic achievement? *Harvard Educational Review* 39:1–123.

Jensen, A. R. 1973. *Educability and group differences*. Harper & Row, New York.

Jensen, A. R. 1974. Kinship correlations reported by Sir Cyril Burt. *Behav. Genet.* 4:1–28.

Jerison, H. J. 1973. *Evolution of the brain and intelligence*. Academic Press, New York.

Jinks, J. L. 1964. *Extrachromosomal inheritance*. Prentice-Hall, Englewood Cliffs, N.J.

Jinks, J. L., and K. Mather. 1955. Stability in development of heterozygotes and homozygotes. *Proc. Roy. Soc. Lond.* (B) 143:561–578.

Jinks, J. L., J. M. Perkins, and H. S. Pooni. 1973. The incidence of epistasis in normal and extreme environments. *Heredity* 31:263–269.

Johnson, G. B. 1976. Genetic polymorphism and enzyme function. *In* F. J. Ayala (ed.), *Molecular evolution,* pp. 46–59. Sinauer, Sunderland, Mass.

Johnston, R. F. 1969. Taxonomy of house sparrows and their allies in the Mediterranean basin. *Condor* 71:129–139.

Johnston, R. F., D. M. Niles, and S. A. Rohwer. 1972. Hermon Bumpus and natural selection in the house sparrow *Passer domesticus*. *Evolution* 26:20–31.

Johnston, R. F., and R. K. Selander. 1964. House sparrows: rapid evolution of races in North America. *Science* 144:548–550.

Jones, D. A. 1966. On the polymorphism of cyanogenesis in *Lotus corniculatus* L. I. Selection by animals. *Can. J. Genet. Cytol.* 8:556–567.

Jones, D. A. 1973. Co-evolution and cyanogenesis. *In* V. H. Heywood (ed.), *Taxonomy and ecology,* pp. 213–242. Academic Press, London and New York.

Jones, D. F. 1924. The attainment of homozygosity in inbred strains of maize. *Genetics* 9:405–418.

Jones, J. S., and T. Yamazaki. 1974. Genetic background and the fitness of allozymes. *Genetics* 78:1185–1189.

Kalmus, H., and S. Maynard Smith. 1966. Some evolutionary consequences of pegmatypic mating systems (imprinting). *Amer. Natur.* 100:619–635.

Kamin, L. J. 1974. *The science and politics of IQ.* Wiley, New York.

Karn, M. N., and L. S. Penrose. 1951. Birth weight and gestation time in relation to maternal age, parity, and infant survival. *Ann. Eugen.* 16: 147–164.

Kawai, M. 1965. Newly acquired pre-cultural behavior of the natural troop of Japanese monkeys on Koshima islet. *Primates* 6:1–30.

Kearsey, M. J., and K. Kojima. 1967. The genetic architecture of body weight and egg hatchability in *Drosophila melanogaster*. *Genetics* 56:23–37.

Keast, A., F. C. Erk, and B. Glass (cds.). 1972. *Evolution, mammals, and southern continents.* State Univ. of New York Press, Albany.

Kellogg, D. A. 1973. The role of phyletic change in the evolution of *Pseudocubus vema* (Radiolaria). *Paleobiology* 1:359–370.

Kerfoot, W. C., and A. G. Kluge. 1971. Impact of the lognormal distribution on studies of phenotypic variation and evolutionary rates. *Syst. Zool.* 20:459–464.

Kerr, W. E., and S. Wright. 1954. Experimental studies of the distribution of gene frequencies in very small populations of *Drosophila melanogaster*. I. Forked. *Evolution* 8:172–177.

Kerster, H. W. 1964. Neighborhood size in the rusty lizard, *Sceloporus olivaceus*. *Evolution* 18:445–457.

Kethley, J. B., and D. E. Johnston. 1975. Resource tracking patterns in bird and mammal ectoparasites. *Misc. Publ. Entomol. Soc. Amer.* 9:231–236.

Kettlewell, H. B. D. 1973. *The evolution of melanism.* Clarendon Press, Oxford, England.

Kettlewell, H. B. D., and D. L. T. Conn. 1977. Further background-choice cxpcriments on cryptic Lepidoptera. *J. Zool., Lond.* 181:371–376.

Key, K. H. L. 1968. The concept of stasipatric speciation. *Syst. Zool.* 17:14–22.

Kidd, K. K., and L. L. Cavalli-Sforza. 1974. The role of genetic drift in the differentiation of Icelandic and Norwegian cattle. *Evolution* 28:381–395.

Kimura, M. 1955. Solution of a process of random genetic drift with a continuous model. Proc. Natl. Acad. Sci. U.S.A. 41:144–150.

Kimura, M. 1968. Evolutionary rates at the molecular level. *Nature* 217:624–626.

Kimura, M., and J. F. Crow. 1964. The number of alleles that can be maintained in a finite population. *Genetics* 49:725–738.

Kimura, M., and T. Ohta. 1971. *Theoretical aspects of population genetics.* Princeton Univ. Press, Princeton, N.J.

Kindred, B. 1967. Selection for an invariant character, vibrissa number in the house mouse. V. Selection on non-Tabby segregants from Tabby selection lines. *Genetics* 55:365–373.

King, J. L. 1967. Continuously distributed factors affecting fitness. *Genetics* 55:483–492.

King, J. L., and T. H. Jukes. 1969. Non-Darwinian evolution. *Science* 164: 788–798.

King, M.-C., and A. C. Wilson. 1975. Evolution at two levels: molecular similarities and biological differences between humans and chimpanzees. *Science* 188:107–116.

Klauber, L. M. 1972. *Rattlesnakes: their habits, life histories, and influence on mankind.* Second edition. Univ. of California Press, Berkeley.

Kluge, A. G., and J. S. Farris. 1969. Quantitative phyletics and the evolution of anurans. *Syst. Zool.* 18:1–32.

Kluge, A. G., and W. C. Kerfoot. 1973. The predictability and regularity of character divergence. *Amer. Natur.* 107:426–442.

Knight, G. R., A. Robertson, and C. H. Waddington. 1956. Selection for sexual isolation within a species. *Evolution* 10:14–22.

Koehn, R. K., J. E. Perez, and R. B. Merritt. 1971. Esterase enzyme function and genetical structure of populations of the freshwater fish *Notropis stramineus. Amer. Natur.* 105:51–69.

Koehn, R. K., and D. I. Rasmussen. 1967. Polymorphic and monomorphic serum esterase heterogeneity in catostomid fish populations. *Biochem. Genet.* 1:131–144.

Koehn, R. K., F. J. Turano, and J. B. Mitton. 1973. Population genetics of marine pelecypods. II. Genetic differentiation in microhabitats of *Modiolus demissus. Evolution* 27:100–105.

Koopman, K. F. 1950. Natural selection for reproductive isolation between *Drosophila pseudoobscura* and *Drosophila persimilis. Evolution* 4:135–145.

Kornfield, I. R. 1974. Evolutionary genetics of endemic cichlid fishes (Pisces: Cichlidae) in Lake Malawi, Africa. Ph.D. Dissertation, State University of New York at Stony Brook.

Krebs, C. J. 1972. *Ecology: the experimental analysis of distribution and abundance.* Harper & Row, New York.

Krieger, R. I., P. P. Feeny, and C. F. Wilkinson. 1971. Detoxification enzymes in the guts of caterpillars: an evolutionary answer to plant defenses? *Science* 172:579–581.

Kropotkin, P. 1902. *Mutual aid. A factor of evolution.* Reprinted 1955 by Extending Horizons, Boston.

Kruckeberg, A. R. 1957. Variation in fertility of hybrids between isolated populations of the serpentine species, *Streptanthus glandulosus* Cook. *Evolution* 11:185–211.

Kuhn, T. S. 1962. *The structure of scientific revolutions.* Univ. of Chicago Press, Chicago.

Kummel, B. 1970. *History of the earth: an introduction to historical geology.* Second edition. W. H. Freeman, San Francisco.

Kurtén, B. 1959. Rates of evolution in fossil mammals. *Cold Spring Harbor Symp. Quant. Biol.* 24:205–215.

Kurtén, B. 1963. Return of a lost structure in the evolution of the felid dentition. *Soc. Scient. Fenn., Comment. Biol.* 26:3–11.

Kurtén, B. 1971. *The age of mammals*. Weidenfeld & Nicolson, London.

Lack, D. 1947. *Darwin's finches*. Cambridge Univ. Press, Cambridge, England.

Lack, D. 1954. *The natural regulation of animal numbers*. Oxford Univ. Press, Oxford.

Lack, D. 1969. Tit niches in two worlds; or homage to Evelyn Hutchinson. *Amer. Natur.* 103:43–50.

Lamotte, M. 1959. Polymorphism of natural populations of *Cepaea nemoralis*. *Cold Spring Harbor Symp. Quant. Biol.* 24:65–86.

Langley, C. H., and W. M. Fitch. 1974. An examination of the constancy of the rate of molecular evolution. *J. Mol. Evol.* 3:161–177.

Lawlor, R., and J. Maynard Smith. 1976. The coevolution and stability of competing species. *Amer. Natur.* 110:79–99.

Lebrun, D. 1961. Evolution de l'appareil genital dans les diverses castes de *Calotermes flavicollis*. *Bull. Soc. Zool. France* 86:235–242. (Cited in Gould 1977.)

Leigh, E. 1970. Natural selection and mutability. *Amer. Natur.* 104:301–305.

Leigh, E. G., Jr. 1973. The evolution of mutation rates. *Genetics Suppl.* 73:1–18.

Lemon, R. E. 1975. How birds develop song dialects. *Condor* 77:385–406.

León, J. A. 1974. Selection in contexts of interspecific competition. *Amer. Natur.* 108:739–757.

Lerner, I. M. 1954. *Genetic homeostasis*. Oliver & Boyd, Edinburgh.

Lerner, I. M. 1958. *The genetic basis of selection*. Wiley, New York.

Lerner, I. M. 1968. *Heredity, evolution and society*. Freeman, San Francisco.

Lerner, I. M., and C. A. Gunns. 1952. Egg size and reproductive fitness. *Poultry Sci.* 31:537–544.

Levene, H. 1953. Genetic equilibrium when more than one ecological niche is available. *Amer. Natur.* 87:331–333.

Levin, B. 1971. The operation of selection in situations of interspecific competition. *Evolution* 25:249–264.

Levin, B. R., and W. L. Kilmer. 1974. Interdemic selection and the evolution of altruism: a computer simulation study. *Evolution* 28:527–545.

Levin, D. A. 1970a. Reinforcement of reproductive isolation: plants versus animals. *Amer. Natur.* 104:571–581.

Levin, D. A. 1970b. Developmental instability in species and hybrids of *Liatris*. *Evolution* 24:613–624.

Levin, D. A., and H. W. Kerster. 1967. Natural selection for reproductive isolation in *Phlox*. *Evolution* 21:679–687.

Levin, D. A., and H. W. Kerster. 1968. Local gene dispersal in *Phlox*. *Evolution* 22:130–139.

Levin, D. A., and H. W. Kerster. 1974. Gene flow in seed plants. *Evol. Biol.* 7:139–220.

Levin, S. A. 1972. A mathematical analysis of the genetic feedback mechanism. *Amer. Natur.* 106:145–164.

Levin, S. A. 1974. Dispersion and population interactions. *Amer. Natur.* 108:207–228.

Levins, R. 1962. Theory of fitness in a heterogeneous environment. I. The fitness set and adaptive function. *Amer. Natur.* 96:361–373.

Levins, R. 1968. *Evolution in changing environments.* Princeton Univ. Press, Princeton, N.J.

Levins, R., and D. Culver. 1971. Regional coexistence of species, and competition between rare species. *Proc. Nat. Acad. Sci. U.S.A.* 68:1246–1248.

Levins, R., and R. MacArthur. 1969. An hypothesis to explain the incidence of monophagy. *Ecology* 50:910–911.

Levinton, J. S., and R. K. Bambach. 1975. A comparative study of Silurian and recent deposit-feeding bivalve communities. *Paleobiology* 1:97–124.

Levi-Setti, R. 1975. *Trilobites: a photographic atlas.* Univ. of Chicago Press, Chicago.

Levitan, M. 1959. Non-random associations of inversions. *Cold Spring Harbor Symp. Quant. Biol.* 23:251–268.

Lewis, H. 1962. Catastrophic selection as a factor in speciation. *Evolution* 16:257–271.

Lewis, H. 1966. Speciation in flowering plants. *Science* 152:167–172.

Lewontin, R. C. 1962. Interdeme selection controlling a polymorphism in the house mouse. *Amer. Natur.* 96:65–78.

Lewontin, R. C. 1965. Selection for colonizing ability. *In* H. G. Baker and G. L. Stebbins (eds.), *The genetics of colonizing species,* pp. 77–94. Academic Press, New York.

Lewontin, R. C. 1966. Is nature probable or capricious? *Bioscience* 16:25–27.

Lewontin, R. C. 1970. The units of selection. *Ann. Rev. Ecol. Syst.* 1:1–18.

Lewontin, R. C. 1972. The apportionment of human diversity. *Evol. Biol.* 6:381–398.

Lewontin, R. C. 1974a. *The genetic basis of evolutionary change.* Columbia Univ. Press, New York.

Lewontin, R. C. 1974b. The analysis of variance and the analysis of causes. *Amer. J. Hum. Genet.* 26:400–411.

Lewontin, R. C. 1975. Genetic aspects of intelligence. *Ann. Rev. Genet.* 9:387–405.

Lewontin, R. C. 1976. Review of *Race differences in intelligence,* by J. C. Loehlin, G. Lindzey, and J. N. Spuhler. *Amer. J. Hum. Genet.* 28:92–97.

Lewontin, R. C. 1977a. Biological determinism as a social weapon. *In* Ann Arbor Science for the People Editorial Collective (eds.), *Biology as a social weapon,* pp. 6–18. Burgess, Minneapolis.

Lewontin, R. C. 1977b. Sociobiology — a caricature of Darwinism. PSA 2, Philosophy of Science Association.

Lewontin, R. C., and L. C. Birch. 1966. Hybridization as a source of variation for adaptation to new environments. *Evolution* 20:315–336.

Lewontin, R. C., L. R. Ginzburg, and S. D. Tuljapurkar. 1978. Heterosis as an explanation for large amounts of genetic polymorphism. *Genetics* 88:149–169.

Lewontin, R. C., and J. L. Hubby. 1966. A molecular approach to the study of genic heterozygosity in natural populations. II. Amount of variation and degree of heterozygosity in natural populations of *Drosophila pseudoobscura*. *Genetics* 54:595–609.

Lewontin, R. C., and K. Kojima. 1960. The evolutionary dynamics of complex polymorphisms. *Evolution* 14:458–472.

Lewontin, R. C., and M. J. D. White. 1960. Interaction between inversion polymorphisms of two chromosome pairs in the grasshopper, *Moraba scurra*. *Evolution* 14:116–129.

Li, C. C. 1955. *Population genetics*. Univ. of Chicago Press, Chicago.

Licht, P., and A. F. Bennett. 1972. A scaleless snake: tests of the role of reptilian scales in water loss and heat transfer. *Copeia* 1972:702–707.

Liem, K. F. 1973. Evolutionary strategies and morphological innovations: cichlid pharyngeal jaws. *Syst. Zool.* 22:425–441.

Linhart, Y. B. 1973. Ecological and behavioral determinants of pollen dispersal in hummingbird-pollinated *Heliconia*. *Amer. Natur.* 107:511–523.

Lipps, J. H. 1970. Plankton evolution. *Evolution* 24:1–21.

Littlejohn, M. J. 1965. Premating isolation in the *Hyla ewingi* complex (Anura.Hylidae). *Evolution* 19.234–243.

Livingstone, F. B. 1964. The distributions of the abnormal hemoglobin genes and their significance for human evolution. *Evolution* 18:685–699.

Lloyd, J. E. 1966. Studies on the flash communication system in *Photinus* fireflies. *Misc. Publ. Mus. Zool. Univ. Michigan* 130:1–95.

Lloyd, M., and H. S. Dybas. 1966. The periodical cicada problem. II. Evolution. *Evolution* 20:466–505.

Loehlin, J. C., G. Lindzey, and J. N. Spuhler. 1975. *Race differences in intelligence*. Freeman, San Francisco.

Łomnicki, A. 1974. Evolution of the herbivore-plant, predator-prey, and parasite-host systems: a theoretical model. *Amer. Natur.* 108:167–180.

Lorenz, K. Z. 1966. *On aggression*. Harcourt, Brace, and World, New York.

Lovejoy, A. O. 1936. *The great chain of being: a study of the history of an idea*. Harvard University Press, Cambridge, Mass.

Lovejoy, A. O. 1959. The argument for organic evolution before the *Origin of Species*, 1830–1858. *In* B. Glass, O. Temkin, and W. Straus, Jr. (eds.), *Forerunners of Darwin, 1745–1859*. Johns Hopkins Univ. Press, Baltimore.

Løvtrup, S. 1974. *Epigenetics, a treatise on theoretical biology*. Wiley, New York.

Lowther, J. K. 1961. Polymorphism in the white-throated sparrow, *Zonotrichia albicollis* (Gmelin). *Canad. J. Zool.* 39:281–292.

Ludwin, I. 1951. Natural selection in *Drosophila melanogaster* under laboratory conditions. *Evolution* 5:231–242.

Lundberg, J. G. 1972. Wagner networks and ancestors. *Syst. Zool.* 21:398–413.

Luria, S. E., and M. Delbrück. 1943. Mutations of bacteria from virus sensitivity to virus resistance. *Genetics* 28:491–511.

Lynch, J. F., and D. B. Wake. 1975. Systematics of the *Chiropterotriton bromeliacia* group (Amphibia: Caudata), with descriptions of two new species from Guatemala. *Contrib. Sci. Nat. Hist. Mus. Los Angeles County No.* 265:1–45.

MacArthur, R. H. 1972. *Geographical ecology. Patterns in the distribution of species.* Harper & Row, New York.

MacArthur, R. H., and R. Levins. 1964. Competition, habitat selection and character displacement in a patchy environment. *Proc. Nat. Acad. Sci. U.S.A.* 51:1207–1210.

MacArthur, R. H., and R. Levins. 1967. The limiting similarity, convergence, and divergence of coexisting species. *Amer. Natur.* 101:377–385.

MacArthur, R. H., H. Recher, and M. Cody. 1966. On the relation between habitat selection and species diversity. *Amer. Natur.* 100:319–332.

MacArthur, R. H., and E. O. Wilson. 1967. *The theory of island biogeography.* Princeton Univ. Press, Princeton, N.J.

Maccoby, E. E., and C. N. Jacklin. 1974. *The psychology of sex differences.* Stanford Univ. Press, Stanford, Cal.

Maglio, V. J. 1972. Evolution of mastication in the Elephantidae. *Evolution* 26:638–658.

Mark, G. A., and K. W. Flessa. 1977. A test for evolutionary equilibria: Phanerozoic brachiopods and Cenozoic mammals. *Paleobiology* 3:17–22.

Marmor, J. 1965. Introduction. *In* J. Marmor (ed.), *Sexual inversion: the multiple roots of homosexuality,* pp. 1–24. Basic Books, New York.

Martin, P. S., and H. E. Wright, Jr. 1967. *Pleistocene extinctions: the search for a cause.* Yale Univ. Press, New Haven.

Mason, L. G. 1964. Stabilizing selection for mating fitness in natural populations of *Tetraopes. Evolution* 18:492–497.

Mather, K. 1949. *Biometrical genetics: the study of continuous variation.* Methuen, London.

Mather, K., and B. J. Harrison. 1949. The manifold effect of selection. *Heredity* 3:1–52; 131–162.

Matthew, W. D. 1915. Climate and evolution. *Ann. N.Y. Acad. Sci.* 24:171–318.

May, R. M. 1973. *Stability and complexity in model ecosystems.* Princeton Univ. Press, Princeton, N.J.

Maynard Smith, J. 1966. Sympatric speciation. *Amer. Natur.* 100:637–650.

Maynard Smith, J. 1970. Genetic polymorphism in a varied environment. *Amer. Natur.* 104:487–490.

Maynard Smith, J. 1971. What use is sex? *J. Theoret. Biol.* 30:319–335.

Maynard Smith, J. 1972. Game theory and the evolution of fighting. *In* J. Maynard Smith, *On evolution,* pp. 8–28. Edinburgh Univ. Press, Edinburgh.

Maynard Smith, J. 1975. *The theory of evolution.* Third edition. Penguin Books, Baltimore, Md.

Maynard Smith, J. 1976a. What determines the rate of evolution? *Amer. Natur.* 110:331–338.

Maynard Smith, J. 1976b. A comment on the Red Queen. *Amer. Natur.* 110:325–330.

Maynard Smith, J., and J. Haigh. 1974. The hitchhiking effect of a favourable gene. *Genet. Research* 23:23–35.

Maynard Smith, J., and K. Sondhi. 1960. The genetics of a pattern. *Genetics* 45:1039–1050.

Mayr, E. 1942. *Systematics and the origin of species.* Columbia Univ. Press, New York.

Mayr, E. 1954. Change of genetic environment and evolution. *In* J. Huxley, A. C. Hardy, and E. B. Ford (eds.), *Evolution as a process,* pp. 157–180. Macmillan, New York.

Mayr, E. 1960. The emergence of evolutionary novelties. *In* S. Tax (ed.), *The evolution of life,* pp. 349–380. Univ. of Chicago Press, Chicago.

Mayr, E. 1963. *Animal species and evolution.* Belknap Press of Harvard Univ. Press, Cambridge, Mass.

Mayr, E. 1965. Numerical phenetics and taxonomic theory. *Syst. Zool.* 14:73–97.

Mayr, E. 1969a. *Principles of systematic zoology.* McGraw-Hill, New York.

Mayr, E. 1969b. *Populations, species, and evolution.* Harvard Univ. Press, Cambridge, Mass.

Mayr, E., and C. Vaurie. 1948. Evolution in the family Dicruridae (birds). *Evolution* 3:238–265.

McNab, B. K. 1971. On the ecological significance of Bergmann's rule. *Ecology* 52:845–854.

McNeill, W. H. 1976. *Plagues and peoples.* Doubleday, New York.

McNeilly, T., and J. Antonovics. 1968. Evolution in closely adjacent plant populations. IV. Barriers to gene flow. *Heredity* 23:99–108.

Mead, M. 1935. *Sex and temperament in three primitive societies.* William Morrow, New York.

Mead, M., Th. Dobzhansky, E. Tobach, and R. E. Light (eds.). 1968. *Science and the concept of race.* Columbia Univ. Press, New York.

Meeuse, B. J. D. 1961. *The story of pollination.* Ronald Press, New York.

Melnick, M., N. C. Myrianthopoulos, and J. C. Christian. 1978. The effects of chorion type on variation in IQ in the NCPP twin population. *Amer. J. Hum. Genet.* 30:425–433.

Merritt, R. B. 1972. Geographic distribution and enzymatic properties of lactate dehydrogenase allozymes in the fathead minnow, *Pimephales promelas. Amer. Natur.* 196:173–184.

Mettler, L. E., and T. G. Gregg. 1969. *Population genetics and evolution.* Prentice-Hall, Englewood Cliffs, N.J.

Milkman, R. D. 1967. Heterosis as a major cause of heterozygosity in nature. *Genetics* 55:493–495.

Milkman, R. 1973. Electrophoretic variation in *Escherichia coli* from natural sources. *Science* 182:1024–1026.

Misra, R. K., and L. L. Short. 1974. A biometric analysis of oriole hybridization. *Condor* 76:137–146.

Mitter, C., and D. Futuyma. 1977. Parthenogenesis in the fall cankerworm, *Alsophila pometaria* (Lepidoptera, Geometridae). *Ent. Exp. Appl.* 21:192–198.

Mitter, C., and D. J. Futuyma. 1979. Population genetic consequences of feeding habits in some forest Lepidoptera. *Genetics,* in press.

Mitter, C., D. J. Futuyma, J. C. Schneider, and J. D. Hare. 1979. Genetic variation and host plant relations in a parthenogenetic moth. *Evolution,* in press.

Money, J., and A. A. Ehrhardt. 1972. *Man and woman, boy and girl: the differentiation and dimorphism of gender identity from conception to maturity.* Johns Hopkins Univ. Press, Baltimore, Md.

Montagu, A. (ed.). 1973. *Man and aggression.* Second edition. Oxford Univ. Press, London.

Moore, J. A. 1961. A cellular basis for genetic isolation. *In* W. F. Blair (ed.), *Vertebrate speciation,* pp. 62–68. Univ. Texas Press, Austin.

Moran, P. A. P. 1964. On the nonexistence of adaptive topographies. *Ann. Human Genet.* 27:383–393.

Morris, D. 1967. *The naked ape: a zoologist's study of the human animal.* McGraw-Hill, New York.

Morton, N. E., J. F. Crow, and H. J. Muller. 1956. An estimate of the mutational damage in man from data on consanguineous marriages. *Proc. Nat. Acad. Sci. U.S.A.* 42:855–863.

Mosquin, T. 1967. Evidence for autopolyploidy in *Epilobium angustifolium* (Onagraceae). *Evolution* 21:713–719.

Mukai, T. 1964. The genetic structure of natural populations of *Drosophila melanogaster.* I. Spontaneous mutation rate of polygenes controlling viability. *Genetics* 50:1–19.

Mukai, T., S. I. Chigusa, L. E. Mettler, and J. F. Crow. 1972. Mutation rate and dominance of genes affecting viability in *Drosophila melanogaster. Genetics* 72:335–355.

Mukai, T., I. Yoshikawa, and K. Sano. 1966. The genetic structure of natural populations of *Drosophila melanogaster.* IV. Heterozygous effects of radiation-induced mutations on viability in various genetic backgrounds. *Genetics* 53:513–527.

Muller, H. J. 1950. Our load of mutations. *Amer. J. Hum. Genet.* 2:111–176.

Muller, H. J. 1966. What genetic course will man steer? *Proc. III Internat. Congr. Genet.* 521–543. Johns Hopkins Univ. Press, Baltimore.

Murdoch, W. W., and A. Oaten. 1975. Predation and population stability. *Adv. Ecol. Res.* 9:2–131.

Murray, J. J. 1962. Factors affecting gene-frequencies in some populations of Cepaea. Ph.D. Thesis, Oxford; cited in Ford 1971.

Murray, J. 1972. *Genetic diversity and natural selection.* Hafner, New York.

Myers, G. S. 1960. The endemic fish fauna of Lake Lanao, and the evolution of higher taxonomic categories. *Evolution* 15:323–333.

Nei, M. 1971. Interspecific gene differences and evolutionary time estimated from electrophoretic data on protein identity. *Amer. Natur.* 105:385–398.

Nei, M. 1972. Genetic distance between populations. *Amer. Natur.* 106:283–292.

Nei, M. 1975. *Molecular population genetics and evolution.* American Elsevier Publishing Co., New York.

Nei, M., T. Maruyama, and R. Chakraborty. 1975. The bottleneck effect and genetic variability in populations. *Evolution* 29:1–10.

Nei, M., and A. K. Roychoudhury. 1972. Gene differences between Caucasian, Negro, and Japanese populations. *Science* 177:434–436.

Neill, W. E. 1974. The community matrix and interdependence of the competition coefficients. *Amer. Natur.* 108:399–408.

Nelson, G. J. 1972. Comments on Hennig's "phylogenetic systematics" and its influence on ichthyology. *Syst. Zool.* 21:364–374.

Nelson, G. 1973. The higher-level phylogeny of verbebrates. *Syst. Zool.* 22:87–91.

Nestmann, E. R., and R. F. Hill. 1973. Population changes in continuously growing mutator cultures of *Escherichia coli*. *Genetics* 73 (Suppl.):41–44.

Nevo, E., Y. J. Kim, C. R. Shaw, and C. S. Thaler, Jr. 1974. Genetic variation, selection, and speciation in *Thomomys talpoides* pocket gophers. *Evolution* 28:1–23.

Newell, N. D. 1949. Phyletic size increase, an important trend illustrated by fossil invertebrates. *Evolution* 3:103–124.

Newell, N. D. 1967. Revolutions in the history of life. *Geol. Soc. Amer. Special Papers* 89:63–91.

Nicholson, A. J. 1958. The self-adjustment of populations to change. *Cold Spring Harbor Symp. Quant. Biol.* 22:153–173.

Noble, G. K. 1931. *The biology of the Amphibia.* McGraw-Hill, New York. Reprinted 1954, Dover Publications, New York.

O'Brien, R. D. 1967. *Insecticides: action and metabolism.* Academic Press, New York.

O'Donald, P. 1959. Possibility of assortative mating in the Arctic skua. *Nature* 183:1210–1211.

Ohno, S. 1970. *Evolution by gene duplication.* Springer-Verlag, New York.

Ohno, S., C. Stenius, L. Christian, and G. Schipmann. 1969. *De novo* mutation-like events observed at the 6PGD locus of the Japanese quail, and the principle of polymorphism breeding more polymorphism. *Biochem. Genet.* 3:417–428.

Ohta, T., and M. Kimura. 1975. Theoretical analysis of electrophoretically detectable polymorphisms: models of very slightly deleterious mutations. *Amer. Natur.* 109:137–145.

Oliver, C. G. 1972. Genetic and phenotypic differentiation and geographic distance in four species of Lepidoptera. *Evolution* 26:221–241.

Olson, E. C. 1959. The evolution of mammalian characters. *Evolution* 13:344–353.

Olson, E. C., and R. L. Miller. 1958. *Morphological integration.* Univ. of Chicago Press, Chicago.

Opler, P. A. 1974. Oaks as islands for leaf-mining insects. *Amer. Sci.* 62:67–73.

Orians, G. H., *et al.* 1974. Tropical population ecology. *In* E. G. Farnworth and F. B. Golley (eds.), *Fragile ecosystems: evaluation of research and applications in the neotropics.* Springer-Verlag, New York.

Osborn, H. F. 1929. The titanotheres of ancient Wyoming, Dakota, and Nebraska. U.S. Geol. Surv. Monograph 55.

Otte, D. 1975. Plant preference and plant succession: a consideration of evolution of plant preference in *Schistocerca*. *Oecologia* 18:129–144.

Paine, R. T. 1966. Food web complexity and species diversity. *Amer. Natur.* 100:65–75.

Palmer, E. J. 1948. Hybrid oaks of North America. *Journ. Arnold Arboretum* 29:1–48.

Park, T., and M. Lloyd. 1955. Natural selection and the outcome of competition. *Amer. Natur.* 89:235–240.

Paterniani, E. 1969. Selection for reproductive isolation between two populations of maize, *Zea mays* L. *Evolution* 23:534–547.

Patterson, B., and R. Pascual. 1972. The fossil mammal fauna of South America. *In* A. Keast, F. C. Erk, and B. Glass (eds.), *Evolution, mammals, and southern continents,* pp. 247–309. State Univ. of New York Press, Albany.

Persons, S. (ed.). 1950. *Evolutionary thought in America.* Yale Univ. Press, New Haven.

Peterson, R. T. 1961. *A field guide to western birds.* Houghton-Mifflin Co., Boston.

Phillips, P. A., and M. M. Barnes. 1975. Host race formation among sympatric apple, walnut, and plum populations of the codling moth, *Laspeyresia pomonella*. *Ann. Ent. Soc. Amer.* 68:1053–1060.

Pianka, E. R. 1978. *Evolutionary ecology.* Second edition. Harper & Row, New York.

Pilbeam, D. R. 1972. *The ascent of man: an introduction to human evolution.* Macmillan, New York.

Pimentel, D. 1961a. Animal population regulation by the genetic feedback mechanism. *Amer. Natur.* 95:65–79.

Pimentel, D. 1961b. Species diversity and insect population outbreaks. *Ann. Ent. Soc. Amer.* 54:76–86.

Pimentel, D., and R. Al-Hafidh. 1965. Ecological control of a parasite population by genetic evolution in the parasite-host system. *Ann. Ent. Soc. Amer.* 58:1–6.

Pimentel, D., and A. C. Bellotti. 1976. Parasite-host population systems and genetic stability. *Amer. Natur.* 110:877–888.

Pimentel, D., E. H. Feinberg, D. W. Wood, and J. T. Hayes. 1965. Selection, spatial distribution, and the co-existence of competing fly species. *Amer. Natur.* 99:97–108.

Pimentel, D., and A. B. Soans. 1970. Animal population regulated to carrying capacity of plant host by genetic feedback. *In* P. J. den Boer and G. R. Gradwell (eds.), *Dynamics of populations,* pp. 313–326. Proc. Adv. Study Inst. Dynamics Numbers Population, Oosterbeek. Wageningen, Netherlands.

Pimentel, D., and F. A. Stone. 1968. Evolution and population ecology of parasite-host systems. *Canad. Entomol.* 100:655–662.

Plapp, F. W. 1976. Biochemical genetics of insecticide resistance. *Ann. Rev. Entomol.* 21:179–197.

Popham, E. J. 1942. Further experimental studies on the selective action of predators. *Proc. Zool. Soc. Lond.* (A) 112:105–117.

Postlethwait, J. H., and H. A. Schneiderman. 1973. Developmental genetics of *Drosophila* imaginal discs. *Ann. Rev. Genet.* 7:381–433.

Pough, R. H. 1951. *Audubon water bird guide.* Doubleday, New York.

Powell, J. R. 1971. Genetic polymorphism in varied environments. *Science* 174:1035–1036.

Prakash, S. 1973. Patterns of gene variation in central and marginal populations of *Drosophila robusta. Genetics* 75:347–369.

Prakash, S., and R. C. Lewontin. 1968. A molecular approach to the study of genic heterozygosity in natural populations. III. Direct evidence of co-adaptation in gene arrangements of *Drosophila. Proc. Natl. Acad. Sci. U.S.A.* 59:398–405.

Prakash, S., R. C. Lewontin, and J. L. Hubby. 1969. A molecular approach to the study of genic heterozygosity in natural populations. IV. Patterns of genic variation in central, marginal, and isolated populations of *Drosophila pseudoobscura. Genetics* 61:841–858.

Premack, D. 1971. On the assessment of language competence in the chimpanzee. *In* A. Schrier and F. Stollnitz (eds.), *The behavior of non-human primates,* pp. 185–228. Academic Press, New York.

Price, G. R. 1972. Fisher's "fundamental theorem" made clear. *Ann. Hum. Genet.* 36:129–140.

Proctor, M., and P. Yeo. 1972. *The pollination of flowers.* Collins, London.

Prosser, C. L., and F. A. Brown, Jr. 1961. *Comparative animal physiology.* Saunders, Philadelphia.

Radinsky, L. B. 1966. The adaptive radiation of the phenacodontid condylarths and the origin of the Perissodactyla. *Evolution* 20:408–417.

Rao, S. V., and P. DeBach. 1969. Experimental studies on hybridization and sexual isolation between some *Aphytis* species (Hymenoptera: Aphelinidae). III. The significance of reproductive isolation between interspecific hybrids and parental species. *Evolution* 23:525–533.

Raup, D. M. 1962. Computer as aid in describing form in gastropod shells. *Science* 138:150–152.

Raup, D. M. 1966. Geometric analysis of shell coiling: general problems. *J. Paleont.* 40:1178–1190.

Raup, D. M. 1972. Taxonomic diversity during the Phanerozoic. *Science* 177:1065–1071.

Raup, D. M. 1975. Taxonomic survivorship curves and Van Valen's Law. *Paleobiology* 1:82–96.

Raup, D. M., and S. J. Gould. 1974. Stochastic simulation and evolution of morphology — towards a nomothetic paleontology. *Syst. Zool.* 23:305–322.

Raup, D. M., S. J. Gould, T. J. M. Schopf, and D. S. Simberloff. 1973.

Stochastic models of phylogeny and the evolution of diversity. *J. Geology* 81:525–542.

Raup, D. M., and S. M. Stanley. 1971. *Principles of paleontology*. Freeman, San Francisco.

Raven, P. H. 1963. Amphitropical relationships in the floras of North and South America. *Quart. Rev. Biol.* 38:151–177.

Reanney, D. 1976. Extrachromosomal elements as possible agents of adaptation and development. *Bact. Rev.* 40:552–590.

Reeve, E. C. R., and J. S. Huxley. 1945. Some problems in the study of allometric growth. *In* W. E. le Gros Clark and P. B. Medawar (eds.), *Essays on growth and form presented to D'Arcy Wentworth Thompson,* pp. 121–156. Clarendon Press, Oxford.

Reichle, D. E. 1966. Some pselaphid beetles with boreal affinities and their distribution along the post-glacial fringe. *Syst. Zool.* 15:330–334.

Remington, C. L. 1968. Suture-zones of hybrid interaction between recently joined biotas. *Evol. Biol.* 2:321–428.

Rendel, J. M. 1951. Mating of ebony vestigial and wild type *Drosophila melanogaster* in light and dark. *Evolution* 5:226–230.

Rendel, J. M. 1953. Variations in the weights of hatched and unhatched ducks' eggs. *Biometrika* 33:48–58.

Rendel, J. M. 1967. *Canalisation and gene control*. Logos Press, London.

Rensch, B. 1959. *Evolution above the species level*. Columbia Univ. Press, New York.

Rensch, B. 1960. The laws of evolution. *In* S. Tax (ed.), *The evolution of life,* pp. 95–116. Univ. of Chicago Press, Chicago.

Reynoldson, T. B. 1966. The distribution and abundance of lake-dwelling triclads — towards a hypothesis. *Adv. Ecol. Res.* 3:1–71.

Rhoades, D. F., and R. G. Cates. 1976. Toward a general theory of plant antiherbivore chemistry. *In* J. W. Wallace and R. L. Mansell (eds.), *Biochemical interaction between plants and insects,* pp. 168–213. *Rec. Adv. Phytochem.* 10. Plenum Press, New York.

Ricklefs, R. E. 1976. *Ecology*. Chiron Press, Portland, Ore.

Ricklefs, R. E., and G. W. Cox. 1972. Taxon cycles in the West Indian avifauna. *Amer. Natur.* 106:195–219.

Ricklefs, R. E., and K. O'Rourke. 1975. Aspect diversity in moths: a temperate-tropical comparison. *Evolution* 29:313–324.

Riley, H. P. 1938. A character analysis of colonies of *Iris fulva* and *I. hexagona* var. *giganticaerulea. Amer. J. Bot.* 29:323–331.

Robertson, A. 1955. Selection in animals: synthesis. *Cold Spring Harbor Symp. Quant. Biol.* 20:225–229.

Robertson, A. 1962. Selection for heterozygotes in small populations. *Genetics* 47:1291–1300.

Robertson, F. W. 1954. Studies in quantitative inheritance. V. Chromosome analysis of crosses between selected and unselected lines of different body size in *Drosophila melanogaster. J. Genet.* 52:494–520.

Robertson, F. W., and E. C. R. Reeve. 1952. Heterozygosity, environmental variation and heterosis. *Nature* 170:296.

Rogers, J. S. 1972. Measures of genetic similarity and genetic distance. Univ. Tex. Publ. 7213:145–153.

Rohlf, F. J., and G. D. Schnell. 1971. An investigation of the isolation-by-distance model. *Amer. Natur.* 105:295–324.

Romer, A. S. 1949. Time series and trends in animal evolution. *In* G. L. Jensen, E. Mayr, and G. G. Simpson (eds.), *Genetics, paleontology, and evolution,* pp. 103–120. Princeton Univ. Press, Princeton, N.J.

Romer, A. S. 1956. *Osteology of the reptiles.* Univ. of Chicago Press, Chicago.

Romer, A. S. 1960. *Vertebrate paleontology.* Univ. of Chicago Press, Chicago.

Rosenzweig, M. L. 1973. Evolution of the predator isocline. *Evolution* 27:84–94.

Ross, H. H. 1958. Evidence suggesting a hybrid origin for certain leafhopper species. *Evolution* 12:337–446.

Roughgarden, J. 1972. Evolution of niche width. *Amer. Natur.* 106:683–718.

Roughgarden, J. 1975. Evolution of marine symbiosis — a simple cost-benefit model. *Ecology* 56:1201–1208.

Roughgarden, J. 1976. Resource partitioning among competing species — a coevolutionary approach. *Theor. Pop. Biol.* 9:388–424.

Rutgers, A. 1969. *Birds of Asia.* Methuen, London.

Saghir, M. T., and E. Robbins. 1973. *Male and female homosexuality.* Williams & Wilkins, Baltimore.

Sahlins, M. 1976. *The use and abuse of biology: an anthropological critique of sociobiology.* Univ. of Michigan Press, Ann Arbor, Mich.

Sanday, P. R. 1972. On the causes of IQ differences between groups and implications for social policy. *Human Organization* 31:411–424. Reprinted in A. Montagu (ed.), *Race and IQ,* Oxford Univ. Press, London (1975).

Sanders, H. L. 1969. Benthic marine diversity and the stability-time hypothesis. *In* G. M. Woodwell and H. H. Smith (eds.), *Diversity and stability in ecological systems,* pp. 71–81. Brookhaven National Laboratory, Upton, N.Y.

Sargent, T. D. 1969. Background selections of the pale and melanic forms of the cryptic moth *Phigalia titea* (Cramer). *Nature* 222:585–586.

Savage, J. M. 1973. The geographic distribution of frogs: patterns and predictions. *In* J. L. Vial (ed.), *Evolutionary biology of the anurans,* pp. 351–445. Univ. of Missouri Press, Columbia, Mo.

Scarr, S., and R. A. Weinberg. 1976. IQ test performance of black children adopted by white families. *Amer. Psychol.* 31:726–739.

Scarr, S., and R. A. Weinberg. 1977. Intellectual similarities within families of both adopted and biological children. *Intelligence* 1:170–191.

Schaeffer, B. 1952. Rates of evolution in coelacanth and dipnoan fishes. *Evolution* 6:101–111.

Schaeffer, B. 1956. Evolution in the subholostean fishes. *Evolution* 10:201–212.

Schaeffer, B. 1965. The role of experimentation in the origin of higher levels of organization. *Syst. Zool.* 14:318–336.

Schaeffer, B., M. K. Hecht, and N. Eldredge. 1972. Phylogeny and paleontology. *Evol. Biol.* 6:31–46.

Schaffer, W. M. 1974. Optimal reproductive effort in fluctuating environments. *Amer. Natur.* 108:783–790.

Scharloo, W., A. Zweep, K. A. Schuitema, and J. G. Wijnstra. 1972. Stabilizing and disruptive selection on a mutant character in *Drosophila*. IV. Selection on sensitivity to temperature. *Genetics* 71:551–566.

Schmalhausen, I. I. 1949. *Factors of evolution*. Blakiston, Philadelphia.

Schoener, T. W. 1968. The *Anolis* lizards of Bimini: resource partitioning in a complex fauna. *Ecology* 49:704–726.

Schoener, T. W. 1974. Resource partitioning in ecological communities. *Science* 185:27–39.

Schopf, J. W. 1974. Paleobiology of the Precambrian: the age of blue-green algae. *Evol. Biol.* 7:1–43.

Schopf, T. J. M. 1974. Permo-Triassic extinctions: relation to sea-floor spreading. *J. Geol.* 82:129–143.

Schopf, T. J. M., D. M. Raup, S. J. Gould, and D. S. Simberloff. 1975. Genomic versus morphologic rates of evolution: influence of morphologic complexity. *Paleobiology* 1:63–70.

Schull, W. J., and J. V. Neel. 1965. *The effects of inbreeding on Japanese children*. Harper & Row, New York.

Schwartz, D., and W. J. Laughner. 1969. A molecular basis for heterosis. *Science* 166:626–627.

Schwartz, M., and J. Schwartz. 1973. Evidence against a genetical component to performance in IQ tests. *Nature* 248:84–85.

Scossiroli, R. E. 1954. Artificial selection of a quantitative trait in *Drosophila melanogaster* under increased mutation rate. *Atti IX Contr. Intern. Genet. Caryologia* 4 (Suppl.): 861–864.

Scriber, J. M. 1973. Latitudinal gradients in larval feeding specialization of the world Papilionoidea (Lepidoptera). *Psyche* 80:355–373.

Seaton, A. P. C., and J. Antonovics. 1967. Population inter-relationships. I. Evolution in mixtures of *Drosophila* mutants. *Heredity* 22:19–33.

Seiger, M. B. 1967. A computer simulation study of the influence of imprinting on population structure. *Amer. Natur.* 101:47–57.

Selander, R. K. 1965. On mating systems and sexual selection. *Amer. Natur.* 99:129–141.

Selander, R. K. 1966. Sexual dimorphism and differential niche utilization in birds. *Condor* 68:113–151.

Selander, R. K. 1970. Behavior and genetic variation in natural populations. *Amer. Zool.* 10:53–66.

Selander, R. K. 1976. Genic variation in natural populations. *In* F. J. Ayala (ed.), *Molecular evolution*, pp. 21–45. Sinauer, Sunderland, Mass.

Selander, R. K., and D. W. Kaufman. 1973. Genic variability and strategies of adaptation in animals. *Proc. Nat. Acad. Sci. U.S.A.* 70:1875–1877.

Sepkoski, J. J., Jr. 1975. Stratigraphic biases in the analysis of taxonomic survivorship. *Paleobiology* 1:343–355.

Sheldon, B. L., and M. K. Milton. 1972. Studies on the scutellar bristles of *Drosophila melanogaster*. II. Long-term selection for high bristle number in the Oregon RC strain and correlated responses in abdominal chaetae. *Genetics* 71:567–595.

Sheppard, P. M. 1959. *Natural selection and heredity*. Harper and Brothers, New York.

Sherman, P. W. 1977. Nepotism and the evolution of alarm calls. *Science* 197:1246–1253.

Sherwood, J. J., and M. Nataupsky. 1968. Predicting the conclusions of negro-white intelligence research from biographical characteristics of the investigators. *J. Pers. Soc. Psychol.* 8:53–58.

Sibley, C. G. 1954. Hybridization in the red-eyed towhees of Mexico. *Evolution* 8:252–290.

Sibley, C. G. 1957. The evolutionary and taxonomic significance of sexual selection and hybridization in birds. *Condor* 59:166–191.

Simberloff, D. S. 1974. Permo-Triassic extinctions: effects of area on biotic equilibrium. *J. Geol.* 82:267–274.

Simberloff, D. 1978. Using island biogeographic distributions to determine if colonization is stochastic. *Amer. Natur.* 112:713–726.

Simons, E. L. 1972. *Primate evolution*. McGraw-Hill, New York.

Simpson, G. G. 1950. History of the fauna of Latin America. *Amer. Sci.* 38:361–389.

Simpson, G. G. 1952. How many species? *Evolution* 6:342.

Simpson, G. G. 1953. *The major features of evolution*. Columbia Univ. Press, New York.

Simpson, G. G. 1959. The nature and origin of supraspecific taxa. *Cold Spring Harbor Symp. Quant. Biol.* 24:255–271.

Simpson, G. G. 1961a. Historical zoogeography of Australian mammals. *Evolution* 15:431–446.

Simpson, G. G. 1961b. *Principles of animal taxonomy*. Columbia Univ. Press, New York.

Simpson, G. G., A. Roe, and R. C. Lewontin. 1960. *Quantitative zoology*. Harcourt, Brace and Co., New York.

Sinclair, D. 1969. *Human growth after birth*. Oxford Univ. Press, New York.

Singh, R., R. C. Lewontin, and A. Felton. 1976. Genetic heterogeneity within electrophoretic "alleles" of xanthine dehydrogenase in *Drosophila pseudoobscura*. *Genetics* 84:609–629.

Sinnott, E. W., L. C. Dunn, and Th. Dobzhansky. 1958. *Principles of genetics*. Fifth edition. McGraw-Hill, New York.

Skutch, A. F. 1973. *The life of the hummingbird*. Crown Publishers, New York.

Slatkin, M. 1973. Gene flow and selection in a cline. *Genetics* 75:733–756.

Slobodkin, L. B. 1961. *Growth and regulation of animal populations*. Holt, Rinehart and Winston, New York.

Slobodkin, L. B. 1968. Toward a predictive theory of evolution. *In* R. C. Lewontin (ed.), *Population biology and evolution*, pp. 187–205. Syracuse Univ. Press, Syracuse, N.Y.

Slobodkin, L. B. 1974. Prudent predation does not require group selection. *Amer. Natur.* 108:665–678.

Slobodkin, L. B. 1978. Is history a consequence of evolution? *Perspectives in Ethology* 3:223–255. Plenum, New York.

Slobodkin, L. B., and A. Rapoport. 1974. An optimal strategy of evolution. *Quart. Rev. Biol.* 49:181–200.

Slobodkin, L. B., and H. L. Sanders. 1969. On the contribution of environmental predictability to species diversity. *In* G. M. Woodwell and H. H. Smith (eds.), *Diversity and stability in ecological systems,* pp. 82–95. Brookhaven National Laboratory, Upton, New York.

Slobodkin, L. B., F. E. Smith, and N. G. Hairston. 1967. Regulation in terrestrial ecosystems, and the implied balance of nature. *Amer. Natur.* 101:104–124.

Smith, H. H., and K. Daly. 1959. Discrete populations derived by interspecific hybridization and selection in *Nicotiana. Evolution* 13:476–487.

Smith, N. G. 1966. Evolution of some arctic gulls (*Larus*): an experimental study of isolating mechanisms. Amer. Ornithologists' Union Ornithol. Monogr. No. 4.

Smith, N. G. 1968. The advantage of being parasitized. *Nature* 219:690–694.

Snaydon, R. W. 1970. Rapid population differentiation in a mosaic environment. I. The response of *Anthoxanthum odoratum* populations to soils. *Evolution* 24:257–269.

Sneath, P. H. A., and R. R. Sokal. 1973. *Numerical taxonomy: the principles and practice of numerical classification.* Freeman, San Francisco.

Soans, A. B., D. Pimentel, and J. S. Soans. 1974. Evolution of reproductive isolation in allopatric and sympatric populations. *Amer. Natur.* 108: 116–124.

Sociobiology Study Group. 1976. Sociobiology — another biological determinism. *Bioscience* 26:182–186.

Sociobiology Study Group. 1977. Sociobiology — a new biological determinism. *In* Ann Arbor Science for the People Editorial Collective (eds.), *Biology as a social weapon,* pp. 133–149. Burgess, Minneapolis.

Sokal, R. R. 1973. The species problem reconsidered. *Syst. Zool.* 22:360–374.

Sokal, R. R. 1976. The Kluge-Kerfoot phenomenon reexamined. *Amer. Natur.* 110:1077–1091.

Sokal, R. R., and T. J. Crovello. 1970. The biological species concept: a critical evaluation. *Amer. Natur.* 104:127–154.

Sokal, R. R., and P. H. A. Sneath. 1963. *Principles of numerical taxonomy.* Freeman, San Francisco.

Solbrig, O. T. 1971. The population biology of dandelions. *Amer. Sci.* 59:686–694.

Sondhi, K. C. 1962. The evolution of a pattern. *Evolution* 16:186–191.

Sondhi, K. C. 1963. The biological foundations of animal patterns. *Quart. Rev. Biol.* 38:289–327.

Soo Hoo, C. F., and G. Fraenkel. 1966. The consumption, digestion, and utilization of food plants by a polyphagous insect, *Prodenia eridania* (Craner). *J. Insect Physiol.* 12:711–730.

Sorenson, F. 1969. Embryonic genetic load in coastal Douglas fir, *Pseudotsuga menziesii* var. *menziesii*. *Amer. Natur.* 103:389–398.

Soulé, M. 1966. Trends in the insular radiation of a lizard. *Amer. Natur.* 100:47–64.

Soulé, M. 1967. Phenetics of natural populations. II. Asymmetry and evolution in a lizard. *Amer. Natur.* 101:141–160.

Soulé, M. 1973. The epistasis cycle: a theory of marginal populations. *Ann. Rev. Ecol. Syst.* 4:165–188.

Soulé, M. 1976. Allozyme variation: its determinants in space and time. *In* F. J. Ayala (ed.), *Molecular evolution,* pp. 60–77. Sinauer, Sunderland, Mass.

Soulé, M., S. Y. Yang, M. G. W. Weiler, and G. C. Gorman. 1973. Island lizards: the genetic-phenetic variation correlation. *Nature* 242: 191–193.

Southwood, T. R. E. 1961. The number of species of insects associated with various trees. *J. Animal Ecol.* 30:1–8.

Spencer, W. P. 1957. Genetic studies on *Drosophila mulleri*. I. Genetic analysis of a population. Tex. Univ. Publ. 5721:186–205.

Spiess, E. B. 1977. *Genes in populations*. Wiley, New York.

Spuhler, J. N. 1968. Assortative mating with respect to physical characteristics. *Eugen. Quart.* 15:128–140.

Srb, A. M., R. D. Owen, and R. S. Edgar. 1965. *General genetics*. Freeman, San Francisco.

Stahl, F. W. 1969. One way to think about gene conversion. *Genetics* 61 (Suppl.):1–13.

Stanley, S. M. 1973. An explanation for Cope's rule. *Evolution* 27:1–26.

Stanley, S. M. 1974. Relative growth of the titanothere horn: a new approach to an old problem. *Evolution* 28:447–457.

Stanley, S. M. 1975a. A theory of evolution above the species level. *Proc. Natl. Acad. Sci. U.S.A.* 72:646–650.

Stanley, S. M. 1975b. Clades versus clones in evolution: why we have sex. *Science* 190:382–383.

Stearns, S. C. 1976. Life-history tactics: a review of the data. *Quart. Rev. Biol.* 51:3–47.

Stearns, S. C. 1977. The evolution of life history traits: a critique of the theory and a review of the data. *Ann. Rev. Ecol. Syst.* 8:145–171.

Stebbins, G. L. 1950. *Variation and evolution in plants*. Columbia Univ. Press, New York.

Stebbins, G. L. 1957. Self fertilization and population variability in the higher plants. *Amer. Natur.* 91:337–354.

Stebbins, G. L. 1970. Variation and evolution in plants: progress during the past twenty years. *In* M. K. Hecht and W. C. Steere (eds.), *Essays in evolution and genetics in honor of Theodosius Dobzhansky,* pp. 173–208. Appleton-Century-Crofts, New York.

Stebbins, G. L. 1971a. *Processes of organic evolution*. Second edition. Prentice-Hall, Englewood Cliffs, N.J.

Stebbins, G. L. 1971b. *Chromosomal evolution in higher plants*. Arnold, London.

Stebbins, G. L. 1974. *Flowering plants: evolution above the species level.* Belknap Press of Harvard Univ. Press, Cambridge, Mass.

Stebbins, G. L., and K. Daly. 1961. Changes in the variation pattern of a hybrid population of *Helianthus* over an eight-year period. *Evolution* 15:60–71.

Stebbins, G. L., and A. Day. 1967. Cytogenetic evidence for long continued stability in the genus *Plantago. Evolution* 21:409–428.

Stebbins, R. C. 1954. *Amphibians and reptiles of western North America.* McGraw-Hill, New York.

Stehli, F. G., R. G. Douglas, and I. A. Kafescioglu. 1972. Models for the evolution of planktonic Foraminifera. *In* T. J. M. Schopf (ed.), *Models in paleobiology,* pp. 116–128. Freeman, Cooper and Co., San Francisco.

Stehli, F. G., R. G. Douglas, and N. D. Newell. 1969. Generation and maintenance of gradients in taxonomic diversity. *Science* 164:947–949.

Steiner, W. W. 1977. Niche width and genetic variation in Hawaiian *Drosophila.* Amer. Natur. 111:1037–1045.

Sterba, G. 1962. *Freshwater fishes of the world.* Longacre Press, London.

Stern, C. 1968. *Genetic mosaics and other essays.* Harvard Univ. Press, Cambridge, Mass.

Stern, C. 1973. *Principles of human genetics.* Third edition. W. H. Freeman, San Francisco.

Straw, R. M. 1955. Hybridization, homogamy, and sympatric speciation. *Evolution* 9:441–444.

Strickberger, M. W. 1968. *Genetics.* Macmillan, New York.

Strong, D. R., Jr. 1974a. Rapid asymptotic species accumulation in phytophagous insect communities: the pests of cacao. *Science* 185:1064–1066.

Strong, D. R., Jr. 1974b. Nonasymptotic species richness models and the insects of British trees. *Proc. Natl. Acad. Sci. U.S.A.* 71:2766–2769.

Sturtevant, A. H. 1920–21. Genetic studies on *Drosophila simulans. Genetics* 5:488–500; 6:179–207.

Sved, J. A. 1968. Possible rates of gene substitution in evolution. *Amer. Natur.* 102:283–293.

Sved, J. A., and O. Mayo. 1970. The evolution of dominance. *In* K. Kojima (ed.), *Mathematical topics in population genetics,* pp. 289–316. Springer-Verlag, New York.

Sved, J. A., T. E. Reed, and W. F. Bodmer. 1967. The number of balanced polymorphisms that can be maintained in a natural population. *Genetics* 55:469–481.

Swanson, C. P. 1957. *Cytology and cytogenetics.* Prentice-Hall, Englewood Cliffs, N.J.

Tahvanainen, J. O., and R. B. Root. 1972. The influence of vegetational diversity on the population ecology of a specialized herbivore, *Phyllotreta cruciferae. Oecologia* 10:321–346.

Taylor, C. E., and J. B. Mitton. 1974. Multivariate analysis of genetic variation. *Genetics* 76:575–585.

Tebb, G., and J. M. Thoday. 1954. Stability in development and relational balance of X-chromosomes in *Drosophila melanogaster*. *Nature* 174:1109–1110.

Temple, S. A. 1977. Plant-animal mutualism: coevolution with dodo leads to near extinction of plant. *Science* 197:885–886.

Thoday, J. M. 1953. Components of fitness. *Symp. Soc. Exp. Biol.* 7:96–113.

Thoday, J. M. 1955. Balance, heterozygosity and developmental stability. *Cold Spring Harbor Symp. Quant. Biol.* 20:318–326.

Thoday, J. M. 1959. Effects of disruptive selection. I. Genetic flexibility. *Heredity* 13:187–203.

Thoday, J. M., and T. B. Boam. 1959. Effects of disruptive selection. II. Polymorphism and divergence without isolation. *Heredity* 13:205–218.

Thoday, J. M., and J. B. Gibson. 1962. Isolation by disruptive selection. *Nature* 193:1164–1166.

Thoday, J. M., and J. B. Gibson. 1970. The probability of isolation by disruptive selection. *Amer. Natur.* 104:219–230.

Thomas, P. A. 1968a. Geographic variation of the rabbit tick, *Haemaphysalis leporispalustris,* in North America. *Univ. Kansas Sci. Bull.* 47:787–828.

Thomas, P. A. 1968b. Variation and covariation in characters of the rabbit tick, *Haemaphysalis leporispalustris. Univ. Kansas Sci. Bull.* 47:829–862.

Thompson, D. W. 1917. *On growth and form.* Cambridge Univ. Press, Cambridge, England.

Thorpe, W. H. 1945. The evolutionary significance of habitat selection. *J. Animal Ecol.* 14:67–70.

Thorpe, W. H. 1956. *Learning and instinct in animals.* Methuen, London.

Throckmorton, L. H. 1965. Similarity versus relationship in *Drosophila. Syst. Zool.* 14:221–236.

Tiger, L., and R. Fox. 1971. *The imperial animal.* Holt, Rinehart and Winston, New York.

Timofeeff-Ressovsky, N. 1940. Mutations and goegraphical variation. *In* J. Huxley (ed.), *The new systematics,* pp. 73–136. Oxford Univ. Press, London.

Tinbergen, L. 1960. The natural control of insects in pinewoods. I. Factors affecting the intensity of predation by songbirds. *Arch. Neerl. Zool.* 13:265–336.

Tinkle, D. W. 1969. The concept of reproductive effort and its relation to the evolution of life histories of lizards. *Amer. Natur.* 103:501–516.

Tizard, B. 1973. I.Q. and race. *Nature* 247:316.

Toulmin, S. 1957. Contemporary scientific mythology. *In* S. Toulmin, R. W. Hepburn, and A. MacIntyre, *Metaphysical beliefs,* pp. 50–65. SCM Press, London.

Tripp, C. A. 1975. *The homosexual matrix.* McGraw-Hill, New York.

Trivers, R. L. 1971. The evolution of reciprocal altruism. *Quart. Rev. Biol.* 46:35–57.

Turesson, G. 1922. The genotypical response of the plant species to the habitat. *Hereditas* 3:211–350.

Turner, J. R. G. 1967. Why does the genome not congeal? *Evolution* 21: 645–656.

Turner, J. R. G. 1970. Changes in mean fitness under natural selection. *In* K. Kojima (ed.), *Mathematical topics in population genetics,* pp. 32–78. Springer-Verlag, New York.

Turner, J. R. G. 1971. Two thousand generations of hybridisation in a *Heliconius* butterfly. *Evolution* 25:471–482.

Turner, J. R. G. 1972. Selection and stability in the complex polymorphism of *Moraba scurra*. *Evolution* 26:334–343.

Udvardy, M. D. F. 1969. *Dynamic zoogeography*. Van Nostrand Reinhold Co., New York.

Uzzell, T. M., Jr. 1964. Relations of the diploid and triploid species of the *Ambystoma jeffersonianum* complex (Amphibia, Caudata). *Copeia* 1964: 257–300.

Uzzell, T., and D. Pilbeam. 1971. Phyletic divergence dates of hominoid primates: a comparison of fossil and molecular data. *Evolution* 25:615–635.

Valentine, J. W. 1970. How many marine invertebrate fossil species? A new approximation. *J. Paleontol.* 44:410–415.

Valentine, J. W. 1973. *Evolutionary paleoecology of the marine biosphere*. Prentice-Hall, Englewood Cliffs, N.J.

van Delden, W. 1970. Selection for competitive ability. *Dros. Inf. Serv.* 45:169.

Vandermeer, J. H. 1969. The competitive structure of communities: an experimental approach with Protozoa. *Ecology* 50:362–371.

van der Plank, J. E. 1968. *Disease resistance in plants*. Academic Press, New York.

Van Valen, L. 1962. Growth fields in the dentition of *Peromyscus*. *Evolution* 16:272–277.

Van Valen, L. 1965. Morphological variation and width of ecological niche. *Amer. Natur.* 99:377–390.

Van Valen, L. 1971. Group selection and the evolution of dispersal. *Evolution* 25:591–598.

Van Valen, L. 1973. A new evolutionary law. *Evolutionary Theory* 1:1–30.

Vasek, F. C. 1964. The evolution of *Clarkia unguiculata* derivatives adapted to relatively xeric environments. *Evolution* 18:26–42.

von Borstel, R. C., S.-K. Quah, C. M. Steinberg, F. Flury, and D. J. C. Gottlieb. 1973. Mutants of yeast with enhanced spontaneous mutation rates. *Genetics* 73(Suppl.):141–151.

Waddington, C. H. 1953. Genetic assimilation of an acquired character. *Evolution* 7:118–126.

Waddington, C. H. 1956a. *Principles of embryology*. Allen and Unwin, London.

Waddington, C. H. 1956b. Genetic assimilation of the *bithorax* phenotype. *Evolution* 10:1–13.

Waddington, C. H. 1962. *New patterns in genetics and development*. Columbia Univ. Press, New York.

Wade, M. J. 1977. An experimental study of group selection. *Evolution* 31:134–153.

Wade, N. 1976. I.Q. and heredity: suspicion of fraud beclouds classic experiment. *Science* 194:916–919.

Wagner, R. P., and H. K. Mitchell. 1964. *Genetics and metabolism.* Wiley, New York.

Wagner, W. H. 1961. Problems in the classification of ferns. *Recent Advances in Botany* 1:841–844. Univ. of Toronto Press, Toronto.

Waldbauer, G. P. 1962. The growth and reproduction of maxillectomized tobacco hornworms feeding on normally rejected non-solanaceous plants. *Ent. Exp. Appl.* 5:147–158.

Walker, E. P. 1975. *Mammals of the world.* Third edition. Johns Hopkins Univ. Press, Baltimore.

Walker, K. R., and L. F. Laporte. 1970. Congruent fossil communities from Ordovician and Devonian carbonates of New York. *J. Paleontol.* 44: 928–944.

Wallace, B. 1958. The average effect of radiation-induced mutations on viability in *Drosophila melanogaster. Evolution* 12:532–552.

Wallace, B. 1963. Further data on the over-dominance of induced mutations. *Genetics* 48:633–651.

Wallace, B. 1966a. On the dispersal of Drosophila. *Amer. Natur.* 100:551–564.

Wallace, B. 1966b. Distance and the allelism of lethals in a tropical population of *Drosophila melanogaster.* Amer. Natur. 100:565–578.

Wallace, B. 1968a. *Topics in population genetics.* W. W. Norton, New York.

Wallace, B. 1968b. Polymorphism, population size, and genetic load. *In* R. C. Lewontin (ed.), *Population biology and evolution,* pp. 87–108. Syracuse Univ. Press, Syracuse, N.Y.

Wallace, B. 1975. Gene control mechanisms and their possible bearing on the neutralist-selectionist controversy. *Evolution* 29:193–202.

Wallace, B., and Th. Dobzhansky. 1962. Experimental proof of balanced genetic loads in Drosophila. *Genetics* 47:1027–1042.

Wallace, B., and T. L. Kass. 1974. On the structure of gene control regions. *Genetics* 77:541–558.

Wasserman, A. O. 1957. Factors affecting interbreeding in sympatric species of spadefoots (genus *Scaphiopus*). *Evolution* 11:320–338.

Watson, J. D. 1976. *Molecular biology of the gene.* Third edition. W. A. Benjamin, Menlo Park, Cal.

Watt, W. B. 1972. Intragenic recombination as a source of population genetic variability. *Amer. Natur.* 106:737–753.

Webb, S. D. 1969. Extinction-origination equilibria in late Cenozoic land mammals of North America. *Evolution* 23:688–702.

Weinrich, J. D. 1977. Human sociobiology: pair-bonding and resource predictability (effects of social class and race). *Behav. Ecol. Sociobiol.* 2:91–118.

Welch, D. A. 1938. Distribution and variation of *Achatinella mustelina* Mighels in the Waianae Mountains, Oahu. *Bishop Mus. Bull., Honolulu* 152:1–164.

Westoll, T. S. 1949. On the evolution of the Dipnoi. *In* G. L. Jepsen, G. G.

Simpson, and E. Mayr (eds.), *Genetics, paleontology, and evolution,* pp. 121–184. Princeton Univ. Press, Princeton, N.J.

Westoll, T. S. 1954. Mountain revolutions and organic evolution. *In* J. Huxley, A. C. Hardy, and E. B. Ford (eds.), *Evolution as a process,* pp. 291–305. Macmillan, New York.

White, M. J. D. 1968. Models of speciation. *Science* 159:1065–1070.

White, M. J. D. 1973. *Animal cytology and evolution.* Third edition. Cambridge Univ. Press, London.

Whitt, G. S., W. F. Childers, and P. L. Cho. 1973. Allelic expression at enzyme loci in an intertribal hybrid sunfish. *J. Hered.* 64:55–61.

Whittaker, R. H. 1975. *Communities and ecosystems.* Macmillan, New York.

Wickler, W. 1968. *Mimicry in plants and animals.* McGraw-Hill, New York.

Wicklund, C. 1975. The evolutionary relationship between adult oviposition preferences and larval host plant range in *Papilio machaon. Oecologia* 18:185–197.

Wilbur, H. M. 1972. Competition, predation, and the structure of the *Ambystoma-Rana sylvatica* community. *Ecology* 53:3–21.

Williams, E. E. 1969. The ecology of colonization as seen in the zoogeography of anoline lizards on small islands. *Quart. Rev. Biol.* 44:345–389.

Williams, G. C. 1957. Pleiotropy, natural selection, and the evolution of senescence. *Evolution* 11:398–411.

Williams, G. C. 1966. *Adaptation and natural selection: a critique of some current evolutionary thought.* Princeton Univ. Press, Princeton, N.J.

Williams, G. C. 1975. *Sex and evolution.* Princeton Univ. Press, Princeton, N.J.

Williams, G. C., and J. B. Mitton. 1973. Why reproduce sexually? *J. Theoret. Biol.* 39:545–554.

Willis, E. O. 1974. Populations and local extinctions of birds on Barro Colorado Island, Panamá. *Ecol. Monogr.* 44:153–169.

Willis, J. C. 1922. *Age and area. A study in geographical distribution and origin of species.* Cambridge Univ. Press, Cambridge, England.

Wills, C. 1968. Three kinds of genetic variability in yeast populations. *Proc. Natl. Acad. Sci. U.S.A.* 61:937–944.

Wilson, A. C. 1976. Gene regulation in evolution. *In* F. J. Ayala (ed.), *Molecular evolution,* pp. 225–236. Sinauer, Sunderland, Mass.

Wilson, D. S. 1977. Structured demes and the evolution of group-advantageous traits. *Amer. Natur.* 111:157–185.

Wilson, E. O. 1958. Patchy distributions of ant species in New Guinea rain forests. *Psyche* 65:26–38.

Wilson, E. O. 1959. Adaptive shift and dispersal in a tropical ant fauna. *Evolution* 13:122–144.

Wilson, E. O. 1961. The nature of the taxon cycle in the Melanesian ant fauna. *Amer. Natur.* 95:169–193.

Wilson, E. O. 1965. The challenge from related species. *In* H. G. Baker and G. L. Stebbins (eds.), *The genetics of colonizing species,* pp. 7–27. Academic Press, New York.

Wilson, E. O. 1969. The species equilibrium. *In* G. M. Woodwell and H. H. Smith (eds.), *Diversity and stability in ecological systems,* pp. 38–47. Brookhaven National Laboratory, Upton, N.Y.

Wilson, E. O. 1975*a*. *Sociobiology: the new synthesis*. Harvard Univ. Press, Cambridge, Mass.

Wilson, E. O. 1975*b*. Human decency is animal. *The New York Times Magazine,* October 12, pp. 38–50.

Wilson, E. O. 1976. Academic vigilantism and the political significance of sociobiology. *BioScience* 26:183–190.

Wilson, E. O. 1978. *On human nature*. Harvard Univ. Press, Cambridge, Mass.

Wilson, E. O., and W. H. Bossert. 1971. *A primer of population biology*. Sinauer, Sunderland, Mass.

Wilson, E. O., and W. L. Brown. 1953. The subspecies concept and its taxonomic applications. *Syst. Zool.* 2:97–111.

Wood, A. E. 1959. Eocene radiation and phylogeny of the rodents. *Evolution* 13:354–360.

Wood, A. E., and B. Patterson. 1971. Relationships among hystricognathous and hystricomorphous rodents. *Mammalia* 34: 628–639.

Woodson, R. E., Jr. 1962. Butterflyweed revisited. *Evolution* 16:168–185.

Wright, S. 1921. Systems of mating. *Genetics* 6:111–178.

Wright, S. 1931. Evolution in Mendelian populations. *Genetics* 16:97–159.

Wright, S. 1932. The roles of mutation, inbreeding, crossbreeding, and selection in evolution. *Proc. XI Internat. Congr. Genet.* 1:356–366.

Wright, S. 1937. The distribution of gene frequencies in populations. *Proc. Natl. Acad. Sci. U.S.A.* 23:307–320.

Wright, S. 1965. The interpretation of population structure by *F*-statistics with special regard to systems of mating. *Evolution* 19:395–420.

Wright, S. 1967. Comments on the preliminary working papers of Eden and Waddington. *In* P. S. Moorehead and M. M. Kaplan (eds.), *Mathematical challenges to the neo-Darwinian theory of evolution*. Wistar Inst. Symp. 5:117–120.

Wright, S. 1968–1978. *Evolution and the genetics of populations*. Vol. 1, *Genetic and biometric foundations;* Vol. 2, *The theory of gene frequencies;* Vol. 3, *Experimental results and evolutionary deductions*; Vol. 4, *Variability within and among natural populations*. Univ. of Chicago Press, Chicago.

Wright, S. 1969. [See Wright 1968–1978.]

Wright, S., Th. Dobzhansky, and W. Hovanitz. 1942. Genetics of natural populations. VII. The allelism of lethals in the third chromosome of *Drosophila pseudoobscura*. *Genetics* 27:363–394.

Wright, T. H. F. 1970. The genetics of embryogenesis in *Drosophila*. *Adv. Genet.* 15:261–395.

Wynne-Edwards, V. C. 1962. *Animal dispersion in relation to social behaviour*. Oliver & Boyd, Edinburgh.

Yamazaki, T. 1971. Measurement of fitness at the esterase-5 locus in *Drosophila pseudoobscura*. *Genetics* 67:579–603.

Yang, S. Y., and R. K. Selander. 1968. Hybridization in the grackle *Quiscalus quiscala* in Louisiana. *Syst. Zool.* 17:107–143.

Zamenhof, S., and H. H. Eichhorn. 1967. Study of microbial evolution through loss of biochemical function: establishment of "defective" mutants. *Nature* 216:456–458.

Zaret, T. M., and R. T. Paine. 1973. Species introduction in a tropical lake. *Science* 182:449–455.

Zimmerman, E. C. 1960. Possible evidence of rapid evolution in Hawaiian moths. *Evolution* 14:137–138.

Zouros, E. 1973. Genic differentiation associated with the early stages of speciation in the *Mulleri* subgroup of *Drosophila*. *Evolution* 27:601–621.

Index

ABOUT THE BOOK
This book was set in VIP Bembo by DEKR Corporation. The original font was cut by Francesco Griffo for the office of Aldus, and was first used in 1495 for the tract *De Aetna,* by Cardinal Peter Bembo.

Joseph Vesely designed the book and supervised its production. The artwork was created by Fredric J. Schoenborn and Sarah Landry. The book was manufactured by The Murray Printing Company.